Oxide Free Nanomaterials for Energy Storage and Conversion Applications

Oxide Free Nanomaterials for Energy Storage and Conversion Applications

Edited By

Prabhakarn Arunachalam

Jayaraman Theerthagiri

Abdullah M. Al-Mayouf

Myong Yong Choi

Madhavan Jagannathan

ELSEVIER

Elsevier
Radarweg 29, PO Box 211, 1000 AE Amsterdam, Netherlands
The Boulevard, Langford Lane, Kidlington, Oxford OX5 1GB, United Kingdom
50 Hampshire Street, 5th Floor, Cambridge, MA 02139, United States

Copyright © 2022 Elsevier Inc. All rights reserved.

No part of this publication may be reproduced or transmitted in any form or by any means, electronic or mechanical, including photocopying, recording, or any information storage and retrieval system, without permission in writing from the publisher. Details on how to seek permission, further information about the Publisher's permissions policies and our arrangements with organizations such as the Copyright Clearance Center and the Copyright Licensing Agency, can be found at our website: www.elsevier.com/permissions.

This book and the individual contributions contained in it are protected under copyright by the Publisher (other than as may be noted herein).

Notices
Knowledge and best practice in this field are constantly changing. As new research and experience broaden our understanding, changes in research methods, professional practices, or medical treatment may become necessary.

Practitioners and researchers must always rely on their own experience and knowledge in evaluating and using any information, methods, compounds, or experiments described herein. In using such information or methods they should be mindful of their own safety and the safety of others, including parties for whom they have a professional responsibility.

To the fullest extent of the law, neither the Publisher nor the authors, contributors, or editors, assume any liability for any injury and/or damage to persons or property as a matter of products liability, negligence or otherwise, or from any use or operation of any methods, products, instructions, or ideas contained in the material herein.

Library of Congress Cataloging-in-Publication Data
A catalog record for this book is available from the Library of Congress

British Library Cataloguing-in-Publication Data
A catalogue record for this book is available from the British Library

ISBN: 978-0-12-823936-0

For information on all Elsevier publications
visit our website at https://www.elsevier.com/books-and-journals

Publisher/Acquisitions Editor: Matthew Deans
Editorial Project Manager: Joshua Mearns
Production Project Manager: Anitha Sivaraj
Cover Designer: Christian J. Bilbow

Typeset by STRAIVE, India

Contents

Contributors   xiii

1. Nanostructured nonoxide nanomaterials an introduction

Prabhakarn Arunachalam, Chenrayan Senthil, and Ganesan Elumalai

1. Introduction	1
2. Transition-metal phosphides	3
3. Transition-metal carbides	4
4. Transition-metal nitrides	5
5. Transition-metal borides	6
6. Transition-metal sulfides	6
6.1 Transition-metal sulfides synthesis	8
6.2 Applications of metal sulfides	11
6.3 High entropy alloys	13
7. Conclusions	15
References	15

2. Novel synthesis methods of nanostructured oxide-free materials for energy storage and conversion applications

Arun Prasad Murthy

1. Introduction	26
1.1 Template synthesis method	26
1.2 A common strategy to synthesize CoP and CoX_2 (X = S, Se, and Te) nanoframes	27
1.3 Charge transfer engineering through multiheteroatom doping	29
1.4 Novel single-step synthesis of 1D nanoarchitecture	30
1.5 Synthesis of different electrocatalysts through annealing gas regulation	31
1.6 Multicomponent hybrid materials as multifunctional electrodes for energy applications	32
1.7 Strategy of direct growth of metal-organic frameworks on conducting substrates	34

vi Contents

	1.8	Synthetic strategy of excluding toxic NH_3 gas in the nitridation of transition metals	34
	1.9	Facile synthesis of multimetallic aerogels using a strong salting-out agent	35
	1.10	General synthesis of 1D multicomponent hybrid heterostructures through modulated electronic state	36
	1.11	Comparison between hydrothermal and electrochemical methods of preparing catalysts	37
	1.12	Role of ultra-sonication in the synthesis of noble metal electrocatalysts	38
	1.13	Strategy to achieve both high dispersibility and conductivity of polyaniline in aqueous media	39
	1.14	Aerogel preparation through anion exchange method	40
	1.15	Synthesis of carbon-supported Pt nanoparticle through a modified borohydride method in a mixed solvent	41
	1.16	Synthetic strategy to prepare a new class of advanced nanoalloy aerogel	41
	1.17	Synthesis of multimetallic anode electrocatalyst for direct methanol fuel cell in high quantity	43
	1.18	Exfoliant-assisted liquid phase exfoliation	44
	1.19	Significance of carbon supports in the syntheses of catalysts for energy applications	45
2.	**Conclusions**		46
	References		47

3. Fundamentals, basic components and performance evaluation of energy storage and conversion devices

Durai Govindarajan and Karthik Kumar Chinnakutti

1.	**Introduction**		52
2.	**Batteries**		52
	2.1	Principles of the Li-S battery	53
	2.2	Fundamental challenges of Li-S batteries	54
3.	**Supercapacitors**		54
	3.1	Principle of supercapacitors	55
	3.2	Characteristics of a supercapacitors	56
	3.3	Classification of supercapacitors	57
	3.4	Material selection for supercapacitors	60
	3.5	Electrolytes	60
4.	**Solar energy and solar cells**		61
	4.1	Principle of photovoltaic	61
	4.2	Charge carrier's generation	62
	4.3	Subsequent separation of the photo-generated charge carriers in the junction	63
	4.4	Collection of the photo-generated charge carriers at the terminals of the junction	64
	4.5	Solar cells types	64

Contents **vii**

4.6	First generation solar cells (silicone cells)	65
4.7	Second-generation solar cells (thin-film solar cells)	66
4.8	Third-generation solar cells	67
4.9	Overview of thin-film solar cell	68
4.10	Thin-film solar cell challenges	71
5.	**Conclusions**	71
	References	72

4. Oxides free materials for symmetric capacitors

Govindhasamy Murugadoss, Sunitha Salla, and
Prabhakarn Arunachalam

1.	**Introduction**	75
	1.1 Electrochemical energy storage systems	76
	1.2 SCs and capacitor	76
	1.3 Structure of SCs	77
	1.4 Supercapacitor mechanism and its type	77
2.	**TMO-based SCs-PCs**	81
3.	**TMOs merits and drawbacks**	81
4.	**Metal-based chalcogenides on electrochemical SCs**	82
5.	**Sulfide based SCs**	82
	5.1 Types of TMS based SCs	83
	5.2 Other transition TMS	86
6.	**Graphene, CNT, and carbon-based materials for SCs**	87
	6.1 Graphene with metal oxides and metal sulfides nanocomposites	88
	6.2 CNTs for high performance of SCs	89
	6.3 AC	89
7.	**Summary and conclusion**	90
	References	90

5. Oxides free materials for asymmetric capacitor

Prabhakarn Arunachalam, Mabrook S. Amer,
Govindhasamy Murugadoss, and Pugalenthi Ramesh

1.	**Introduction**	95
2.	**Transition metal chalcogenides for asymmetric supercapacitors**	96
	2.1 Transition metal sulfides for asymmetric supercapacitors	97
	2.2 Transition metal selenides for asymmetric supercapacitors	98
3.	**Transition metal nitrides for electrochemical SCs**	102
	3.1 Transition metal carbides for asymmetric supercapacitors	103
	3.2 Transition metal phosphides for asymmetric supercapacitors	105
4.	**Conclusions**	107
	References	107

viii Contents

6. Oxides free materials for flexible and paper-based supercapacitors
C. Justin Raj, Hyun Jung, and Byung Chul Kim

1.	**Introduction**	115
2.	**Supercapacitor designs**	117
	2.1 Flexible substrates	117
	2.2 Flexible solid-state electrolytes	119
	2.3 Device architectures	120
3.	**Electrode materials for flexible supercapacitors**	122
	3.1 Metal carbide	122
	3.2 Metal nitride	127
	3.3 Metal chalcogenides	128
	3.4 Metal phosphides	137
4.	**Conclusion and future prospects**	137
	Acknowledgments	140
	References	140

7. Metal oxides-free anodes for lithium-ion batteries
A. Nichelson, Bradha Madhavan,
Ganesh Kumar Veerasubramani, Waqas Hassan Tanveer,
Jayaraman Theerthagiri, A.G. Ramu,
Dhanasekaran Vikraman, K. Karuppasamy, and
Sung-Chul Yi

1.	**Introduction**	149
2.	**Components and operating mechanism of LIBs**	152
3.	**Anode materials for LIB**	152
	3.1 Graphene and its composites as anode materials	153
	3.2 Selenide-based anodes	154
	3.3 Sulfide-based anodes	159
	3.4 Phosphide-based anodes	162
4.	**Summary, outlook, and future scopes**	167
	References	167

8. Oxides free materials as anodes for sodium-ion batteries
Chelladurai Karuppiah, Dhayanantha Prabu Jaihindh,
Balamurugan Thirumalraj, Ahmed S. Haidyrah, and
Chun-Chen Yang

1.	**Introduction**	177
2.	**Carbon nanostructure based anode materials for SIBs**	180
3.	**Carbon nanotube (CNTs)**	182
	3.1 Carbon nanofibers (CNFs)	182
	3.2 Graphene-based materials	185

Contents **ix**

4. Carbon materials with transition metal chalcogenides for SIBs anode 186
 4.1 Carbon based transition metal sulfide composites 186
5. Carbon based transition metal selenide composites 187
6. Carbon based transition metal telluride composites 189
7. Transition metal phosphides and their carbon composites 191
8. Summary 193
 References 193

9. Oxides free materials as anodes for zinc-bromine batteries

Prabhakarn Arunachalam, Mabrook S. Amer,
Govindhasamy Murugadoss, and Abdullah M. Al-Mayouf

1. Introduction 201
2. Principle and structure of $Zn-Br_2$ RFBs 203
3. Electrodes in ZBBs 205
 3.1 Carbon electrode materials in ZBBs 205
4. Future prospects for $ZnBr_2$ RFBs 211
5. Conclusions 212
 References 212

10. Metal nitrides and carbides as advanced counter electrodes for dye-sensitized solar cells

Meenakshamma Ambapuram, Gurulakshmi Maddala,
and Raghavender Mitty

1. Introduction 219
 1.1 Characterization of counter electrodes 220
 1.2 Photovoltaic measurements 227
2. Advancement of metal nitrides and carbides as a counter electrode for dye-sensitized solar cells 228
 2.1 Metal nitrides 229
 2.2 Metal carbides 235
3. Summary 249
 References 255

11. Metal chalcogenide-based counter electrodes for dye-sensitized solar cells

Subalakshmi Kumar, Senthilkumar Muthu, Sankar Sekar,
Chinna Bathula, Ashok Kumar Kaliamurthy, and
Sejoon Lee

1. Introduction 259
2. Sulfide based counter electrodes 261
3. Selenides and tellurides based counter electrodes 267

x Contents

4.	Composite based counter electrodes	272
5.	Conclusion and future prospects	279
	References	280

12. Oxide free materials for perovskite solar cells

Ramya Krishna Battula, Easwaramoorthi Ramasamy,
P. Bhyrappa, C. Sudakar, and Ganapathy Veerappan

1.	Introduction	287
2.	Oxide free-electron transport materials	289
	2.1 Fullerene and its derivatives	289
	2.2 Others	290
3.	Oxide free hole transporting materials (HTMs)	291
	3.1 Spiro-OMeTAD	291
	3.2 Poly [bis(4-phenyl)(2,5,6-trimethylphenyl)amine] (PTAA)	292
	3.3 Copper thiocyanate (CuSCN)	292
	3.4 Copper iodide (CuI)	293
	3.5 Copper chalcogenides	294
	3.6 Carbazole based HTMs	295
	3.7 PEDOT:PSS (PEDOT =poly(3,4-ethylene dioxythiophene) and PSS =polystyrene sulfonate)	296
	3.8 Poly (3-hexylthiophene) (P3HT) based HTMs	297
	3.9 Lead sulfide (PbS)	298
4.	HTM free carbon-based PSCs	299
5.	Oxide free TCO for PSCs	301
6.	Conclusion	302
	Acknowledgments	302
	References	302

13. Multijunction solar cells based on III–V and II–VI semiconductors

Raja Arumugam Senthil, Jayaraman Theerthagiri,
S.K. Khadheer Pasha, Madhavan Jagannathan,
Andrews Nirmala Grace, and Sivakumar Manickam

1.	Introduction	307
2.	Multijunction solar cells	310
3.	Multijunction solar cells based on III–V semiconductors	312
4.	Multijunction solar cells based on II–VI semiconductors	321
5.	Conclusion and future prospects	321
	References	322

Contents **xi**

14. **Recent advances in nanostructured nonoxide materials — Borides, borates, chalcogenides, phosphides, phosphates, nitrides, carbides, alloys, and metal-organic frameworks**
Leticia S. Bezerra, Bibiana K. Martini, Eduardo S.F. Cardoso, Guilherme V. Fortunato, and Gilberto Maia

1.	Introduction	329
2.	Existing technology	330
	2.1 Alkaline electrolysis cell	331
	2.2 Proton exchange membrane electrolysis cell	331
	2.3 Solid oxide electrolysis cell	331
3.	Methodology or research design to overcome the drawbacks of existing protocols	331
	3.1 Borides	332
	3.2 Borates	332
	3.3 Chalcogenides (sulfides and selenides)	333
	3.4 Phosphides	334
	3.5 Phosphates	335
	3.6 Nitrides	336
	3.7 Carbides	337
	3.8 Alloys	338
	3.9 MOFs	339
4.	Findings or results with implications for managers and decision-makers	341
5.	Discussion, limitations, future research, and conclusion	350
	References	353

15. **Oxides free nanomaterials for (photo) electrochemical water splitting**
Lakshmana Reddy Nagappagari, Santosh S. Patil, Kiyoung Lee, and Shankar Muthukonda Venkatakrishnan

1.	General introduction	369
2.	Metal nitrides	370
	2.1 General synthesis methods of metal nitrides	372
	2.2 Mono metal nitrides for PEC water splitting	373
	2.3 Binary metal nitrides for PEC water splitting	374
3.	Metal phosphides	378
	3.1 Synthesis methods of metal phosphides	379
	3.2 Mono metal phosphides for PEC water splitting	383
	3.3 Binary metal phosphides for PEC water splitting	383
4.	Metal sulfides	386
	4.1 Synthesis methods of metal sulfides	386
	4.2 Mono metal sulfides for PEC water splitting	391
	4.3 Binary metal sulfides for PEC water splitting	391

xii Contents

5.	Summary and future prospects	393
	Acknowledgment	396
	References	396

16. Oxides free materials for photocatalytic water splitting
M.L. Aruna Kumari

1.	Introduction	409
2.	Fundamental aspects of photocatalytic water splitting	412
3.	Experimental needs for photocatalytic water splitting	413
4.	Oxide-free materials for photocatalytic water splitting	414
	4.1 Metal chalcogenides	414
	4.2 Metal pnictides	419
	4.3 Metal carbides	424
5.	Conclusion and future aspects	427
	References	427

17. Oxide-free materials for thermoelectric and piezoelectric applications
Jayaraman Theerthagiri, Seung Jun Lee, and Myong Yong Choi

1.	Introduction	435
2.	Oxide-free materials for thermoelectric applications	436
	2.1 Basic working function of a thermoelectric device	436
	2.2 Metal chalcogenides (MX; X = S, Se, Te)	437
	2.3 Thermoelectric metal carbides and nitrides	442
3.	Oxide-free materials for piezoelectric applications	443
	3.1 Basic working function of a piezoelectric device	443
	3.2 Piezoelectric metal chalcogenides and nitrides	443
4.	Conclusions	446
	Acknowledgments	446
	References	446

18. Future prospects of oxide-free materials for energy-related applications
Dhandapani Balaji, Kumar Premnath, and Madhavan Jagannathan

1.	Introduction	451
2.	Transition metal-based oxide-free materials for hydrogen evolution reaction	452
	2.1 Transition metal carbides	453
	2.2 Transition metal nitrides	456
	2.3 Transition metal phosphides	457
	2.4 Transition metal sulfides	459
3.	Conclusions and future prospects	462
	References	463

Index	467

Contributors

Numbers in parentheses indicate the pages on which the authors' contributions begin.

Abdullah M. Al-Mayouf (201), Electrochemical Science Research Chair, Chemistry Department, College of Science, King Saud University, Riyadh, Saudi Arabia

Meenakshamma Ambapuram (219), Department of Physics, Yogi Vemana University, Kadapa, Andhra Pradesh, India

Mabrook S. Amer (95, 201), Electrochemical Science Research Chair, Chemistry Department, College of Science, King Saud University, Riyadh, Saudi Arabia

M.L. Aruna Kumari (409), Department of Chemistry, Ramaiah College of Arts, Science and Commerce, Bengaluru, India

Prabhakarn Arunachalam (1, 75, 95, 201), Electrochemical Science Research Chair, Chemistry Department, College of Science, King Saud University, Riyadh, Saudi Arabia

Dhandapani Balaji (451), Solar Energy Lab, Department of Chemistry, Thiruvalluvar University, Vellore, India

Chinna Bathula (259), Division of Electronics and Electrical Engineering, Dongguk University-Seoul, Seoul, Republic of Korea

Ramya Krishna Battula (287), Centre for Solar Energy Materials, International Advanced Research Centre for Powder Metallurgy and New Materials (ARCI), Hyderabad; Department of Chemistry; Multifunctional Materials Lab, Department of Physics, Indian Institute of Technology Madras, Chennai, India

Leticia S. Bezerra (329), Institute of Chemistry, Federal University of Mato Grosso do Sul, Campo Grande, MS, Brazil

P. Bhyrappa (287), Department of Chemistry, Indian Institute of Technology Madras, Chennai, India

Eduardo S.F. Cardoso (329), Institute of Chemistry, Federal University of Mato Grosso do Sul, Campo Grande, MS, Brazil

Karthik Kumar Chinnakutti (51), Department of Chemistry, Vinayaka Mission's Kirupananda Variyar Arts and Science College, Vinayaka Mission's Research Foundation (Deemed to be University), Salem, India

Myong Yong Choi (435), Department of Chemistry and Research Institute of Natural Sciences, Gyeongsang National University, Jinju, South Korea

Byung Chul Kim (115), Department of Printed Electronics Engineering, Sunchon National University, Suncheon-si, Jellanamdo, Republic of Korea

xiii

xiv Contributors

Ganesan Elumalai (1), International Center for Materials Nanoarchitectonics (MANA), National Institute for Materials Science (NIMS), Tsukuba, Ibaraki, Japan

Guilherme V. Fortunato (329), São Carlos Institute of Chemistry, University of São Paulo, São Carlos, SP, Brazil

Durai Govindarajan (51), Centre for Nanoscience and Nanotechnology, Sathyabama Institute of Science and Technology, Chennai, India

Andrews Nirmala Grace (307), Centre for Nanotechnology Research, Vellore Institute of Technology, Vellore, Tamil Nadu, India

Ahmed S. Haidyrah (177), Nuclear and Radiological Control Unit, King Abdulaziz City for Science and Technology (KACST), Riyadh, Saudi Arabia

Madhavan Jagannathan (307, 451), Solar Energy Lab, Department of Chemistry, Thiruvalluvar University, Vellore, India

Dhayanantha Prabu Jaihindh (177), Department of Chemical Engineering, National Taiwan University, Taipei City, Taiwan, ROC

Hyun Jung (115), Department of Chemistry, Dongguk University, Seoul, Republic of Korea

C. Justin Raj (115), Department of Chemistry, Dongguk University, Seoul, Republic of Korea

Ashok Kumar Kaliamurthy (259), Department of Nuclear Physics, University of Madras, Chennai, Tamil Nadu, India; Department of Energy and Materials Engineering, Dongguk University-Seoul, Seoul, Republic of Korea

K. Karuppasamy (149), Division of Electronics and Electrical Engineering, Dongguk University-Seoul, Seoul, Republic of Korea

Chelladurai Karuppiah (177), Battery Research Center of Green Energy, Ming Chi University of Technology, New Taipei City, Taiwan, ROC

Subalakshmi Kumar (259), Department of Nuclear Physics, University of Madras, Chennai, Tamil Nadu, India

Kiyoung Lee (369), School of Nano & Materials Science and Engineering, Kyungpook National University, Sangju, Gyeongbuk; Research Institute of Environmental Science & Technology, Kyungpook National University, Daegu, South Korea

Sejoon Lee (259), Division of Physics & Semiconductor Science; Quantum-functional Semiconductor Research Center, Dongguk University-Seoul, Seoul, Republic of Korea

Seung Jun Lee (435), Department of Chemistry and Research Institute of Natural Sciences, Gyeongsang National University, Jinju, South Korea

Gurulakshmi Maddala (219), Department of Physics, Yogi Vemana University, Kadapa, Andhra Pradesh, India

Bradha Madhavan (149), Department of Science and Humanities, Rathinam Technical Campus, Coimbatore, Tamil Nadu, India

Gilberto Maia (329), Institute of Chemistry, Federal University of Mato Grosso do Sul, Campo Grande, MS, Brazil

Contributors **xv**

Sivakumar Manickam (307), Petroleum and Chemical Engineering, Faculty of Engineering, Universiti Teknologi Brunei, Bandar Seri Begawan, Brunei Darussalam

Bibiana K. Martini (329), Institute of Chemistry, Federal University of Mato Grosso do Sul, Campo Grande, MS, Brazil

Raghavender Mitty (219), Department of Physics, Yogi Vemana University, Kadapa, Andhra Pradesh, India

Arun Prasad Murthy (25), Department of Chemistry, School of Advanced Sciences, Vellore Institute of Technology, Vellore, Tamil Nadu, India

Govindhasamy Murugadoss (75, 95, 201), Centre for Nanoscience and Nanotechnology, Sathyabama Institute of Science and Technology (Deemed to be University), Chennai, Tamil Nadu, India

Senthilkumar Muthu (259), Crystal Growth Centre, Anna University, Chennai, India

Shankar Muthukonda Venkatakrishnan (369), Nanocatalysis and Solar Fuels Research Laboratory, Department of Materials Science & Nanotechnology, Yogi Vemana University, Kadapa, Andhra Pradesh, India

Lakshmana Reddy Nagappagari (369), School of Nano & Materials Science and Engineering, Kyungpook National University, Sangju, Gyeongbuk; Research Institute of Environmental Science & Technology, Kyungpook National University, Daegu, South Korea

A. Nichelson (149), Department of Physics, National Engineering College, Kovilpatti, Thoothukudi, Tamil Nadu, India

S.K. Khadheer Pasha (307), Department of Physics, Vellore Institute of Technology (Amaravati Campus), Amaravati, Guntur, Andhra Pradesh, India

Santosh S. Patil (369), School of Nano & Materials Science and Engineering, Kyungpook National University, Sangju, Gyeongbuk; Research Institute of Environmental Science & Technology, Kyungpook National University, Daegu, South Korea

Kumar Premnath (451), Solar Energy Lab, Department of Chemistry, Thiruvalluvar University, Vellore, India

Easwaramoorthi Ramasamy (287), Centre for Solar Energy Materials, International Advanced Research Centre for Powder Metallurgy and New Materials (ARCI), Hyderabad, India

Pugalenthi Ramesh (95), Department of Chemistry, Vivekanandha College of Arts & Sciences for Women, Tirchengode, Tamil Nadu, India

A.G. Ramu (149), Department of Materials Science and Engineering, Hongik University, Sejong-City, Republic of Korea

Sunitha Salla (75), Department of Chemistry, Sathyabama Institute of Science and Technology (Deemed to be University), Chennai, Tamil Nadu, India

xvi Contributors

Sankar Sekar (259), Division of Physics & Semiconductor Science; Quantum-functional Semiconductor Research Center, Dongguk University-Seoul, Seoul, Republic of Korea

Chenrayan Senthil (1), Department of Chemical Engineering, College of Engineering, Kyung Hee University, Yongin-si, South Korea

Raja Arumugam Senthil (307), Faculty of Materials and Manufacturing, Beijing University of Technology, Beijing, PR China

C. Sudakar (287), Multifunctional Materials Lab, Department of Physics, Indian Institute of Technology Madras, Chennai, India

Waqas Hassan Tanveer (149), Energy Safety Research Institute, College of Engineering, Swansea University, Bay Campus, Fabian Way, Swansea, United Kingdom; Department of Mechanical and Manufacturing Engineering, National University of Sciences and Technology (NUST), Islamabad, Pakistan

Jayaraman Theerthagiri (149, 307, 435), Centre of Excellence for Energy Research, Centre for Nanoscience and Nanotechnology, Sathyabama Institute of Science and Technology (Deemed to be University), Chennai, Tamil Nadu, India; Department of Chemistry and Research Institute of Natural Sciences, Gyeongsang National University, Jinju, South Korea

Balamurugan Thirumalraj (177), Department of Energy & Mineral Resources Engineering, Sejong University, Seoul, South Korea

Ganapathy Veerappan (287), Centre for Solar Energy Materials, International Advanced Research Centre for Powder Metallurgy and New Materials (ARCI), Hyderabad, India

Ganesh Kumar Veerasubramani (149), Department of Chemical Engineering, Hanyang University, Seoul, Republic of Korea

Dhanasekaran Vikraman (149), Division of Electronics and Electrical Engineering, Dongguk University-Seoul, Seoul, Republic of Korea

Chun-Chen Yang (177), Battery Research Center of Green Energy; Department of Chemical Engineering, Ming Chi University of Technology, New Taipei City, Taiwan, ROC

Sung-Chul Yi (149), Department of Chemical Engineering; Department of Hydrogen and Fuel Cell Technology, Hanyang University, Seoul, Republic of Korea

Chapter 1

Nanostructured nonoxide nanomaterials an introduction

Prabhakarn Arunachalam[a], Chenrayan Senthil[b], and Ganesan Elumalai[c]

[a]*Electrochemical Science Research Chair, Chemistry Department, College of Science, King Saud University, Riyadh, Saudi Arabia,* [b]*Department of Chemical Engineering, College of Engineering, Kyung Hee University, Yongin-si, South Korea,* [c]*International Center for Materials Nanoarchitectonics (MANA), National Institute for Materials Science (NIMS), Tsukuba, Ibaraki, Japan*

Chapter outline

1. **Introduction**	1	6.1 Transition-metal sulfides synthesis	8
2. **Transition-metal phosphides**	3	6.2 Applications of metal sulfides	11
3. **Transition-metal carbides**	4	6.3 High entropy alloys	13
4. **Transition-metal nitrides**	5	7. **Conclusions**	**15**
5. **Transition-metal borides**	6	**References**	**15**
6. **Transition-metal sulfides**	6		

1. Introduction

The accessibility of a clean, economical, and copious energy source is one of humankind's primary challenges in the 21st century. Out of concern for natural source depletion and ecological disputes, the developments of commercially sustainable eco-friendly energy technologies are necessary as a matter of urgency [1]. The future of energy sources is determined by advanced breakthroughs in designing a cost-effective, sustainable, and efficient system for converting and storing renewable energy sources [2]. Solar power offers a clean, renewable energy source with minimal environmental impact because it is a decentralized and virtually everlasting resource [3–7]. The solar power reaching the earth's surface relates to providing by 130 million power plants of 500 MW. The sunlight delivers us with a wide range of uses such as solar heating, photovoltaics, photoelectrochemical (PEC) water-splitting, photosynthesis, and photocatalysis [8–15]. Further, the future use of solar energy as a primary source to meet global demand depends on achieving important fundamental and technological milestones [16, 17]. Notably, the more effective harnessing of the sunlight potential still remains a hurdle, requiring new innovative

---

Oxide Free Nanomaterials for Energy Storage and Conversion Applications
https://doi.org/10.1016/B978-0-12-823936-0.00014-0
Copyright © 2022 Elsevier Inc. All rights reserved.

materials or device design. Although these sources can encounter the global energy necessities, these energy sources' discontinuous behavior is an inevitable concern that considerably encourages the research on energy storage devices.

During the last decades, nanomaterials are becoming industrialized. Current theoretical and experimental investigations focused on further increasing the effect that nanomaterials can play in industrial advances. During the past few decades, new kinds of nanomaterials and devices derived from transition metal oxide and oxyhydride materials and their composites with organic materials have been demonstrated to have a current interest in energy-related systems [13, 18–21]. Also, the greater families of inorganic-based nonoxide materials containing transition metal carbides, nitrides, sulfides, etc., are of tremendous benefits owing to their wide-range functional natures. Particularly, transition metal carbides and nitrides, namely SiC, TiC, WC, and Si_3N_4, are considered high-temperature ceramics. ZnS, CdSe, GaN, InP, GaAs, Bi_2Te_3, and $CuInGaSe_2$ are the widely investigated semiconductors generally employed in luminance devices, bio-labeling, photovoltaic and thermoelectric devices, and photocatalytic water splitting.

For photocatalytic applications, various classes of n-type semiconductors, namely TiO_2, Fe_2O_3, WO_3, and $BiVO_4$ electrode materials, are utilized to harvest the visible-light photons [22–29]. In this regard, the energy level positions are not suitable for harvesting the photons and permitting these materials inefficient in the visible-light spectrum region. In recent years, certain (oxy) nitrides electrodes have been demonstrated as a potential candidate to replace metal-oxide-based electrodes to grasp visible-light photons.

For energy conversion skills, converting renewable energy to effortlessly stowed chemical fuels (i.e., H_2) by electrochemical water splitting is of unlimited potential; a catalyst for H_2 evolution reaction (HER) happens at the cathode, whereas O_2 evolution reaction (OER) happens at the anode [30–33]. Consequently, to exploit H_2 as a usable fuel in fuel cells, O_2 reduction reaction (ORR) also desires to improve the performance [34–38]. Though, the present state-of-the-art electrocatalytic materials for these applications are still built on expensive noble metals (i.e., Pt for HER, [39] ORR, and RuO_2/IrO_2 for OER [31]), which prevented its commercial usage. In this regard, cost-efficient and highly abundant substitutes with good electrocatalytic features and durability over time are greatly anticipated. The grouping of transition-metals and elemental phosphorus (P), carbon (C), nitrogen (N), and boron (B) provides a wide range of compounds, such as phosphides, nitrides, carbides, and borides. These oxide-free materials are considered to be alternatives for oxide-based semiconductors. These materials notably reveal significant rewards of being cost-effective, producing commercially and exciting physicochemical features, which obtain an enormous potential for electrocatalytic and energy-related applications.

Herein, we examine and discuss the improvement in energy storages (Lithium-ion batteries (LIBs)/sodium-ion batteries (SIBs) and supercapacitors (SCs)) and conversion applications (HER, OER, ORR) that involves the nanostructured oxide-free nanomaterials of transition metal nitrides, carbides and phosphides are elucidated in detail. Further, we relatively discuss all recent advances of transition metal sulfides in the synthetic methods, seeing issues such as their accessibility and their usage to energy-related applications employing oxide-free nanomaterials. Finally, we advise our personal outlook on these oxide-free nanomaterials to pave the approach for further studies.

2. Transition-metal phosphides

Transition-metal phosphides (TMPs) are recognized as attractive electrode materials concerning their inherent material features such as great melting points, high hardness, great electrical conductivity, and chemical inertness [40–42]. In this regard, electrocatalytic materials have been demonstrated as robust and effective electrocatalyst materials for energy-related applications [43]. By considering their crystal structure, enormous differences are reported between TMPs and Transition-metal carbides (TMCs)/nitrides (TMNs). This TMPs group has been employed to develop different types of electrode materials involving phosphides of Ni, Cu, Co, Fe, W, Mo, and so on. TMPs electrode materials were employed as energy applications, and their electrocatalytic features were superior to other electrodes built on TMCs/TMPs/TMNs owed to their electronic richness over the metal surface in the phosphides [44]. For instance, TMPs with metalloid behavior and remarkable electronic conductivity have gathered extensive consideration as electrode candidates for electrochemical supercapacitors [43]. Further, these TMPs are also vastly abundant and environmentally friendly [45, 46]. In comparison with transition metal oxides, they are kinetically favorable for the rapid electron transport obligatory for great power density [47, 48].

In terms of applications, TMPs have been usually applied as catalytic materials for hydrodenitrogenation, hydrodesulfurization, HER, and ORR [1]. Pralong et al. reported the CoP_3 electrodes for LIBs, and it offered favorable Li-storage behavior with its conversion mechanism and enjoyed extensive consideration owing to its efficient nature and great theoretical capacitance [49]. Subsequently, numerous classes of TMPs were involved as electrode materials for LIBs. Also, TMPs preserve the metalloid features and greater specific capacitances; thereby it is generally applied in SCs. Till to date, numerous remarkable reports and reviews on the fabrication of nanostructured TMPs [43, 50–52], and their use in HERs and magnetic and catalytic fields [53].

Liu and co-workers [54] revealed a supercapacitors electrode involving nanostructured biphasic Ni_5P_4-Ni_2P (Ni_xP_y) nanosheets. Further, they revealed the ultrahigh-specific capacitance of $1270 \, F \, g^{-1}$ at $2 \, A \, g^{-1}$ with enhanced rate capability and durability (Fig. 1). Asymmetric supercapacitors comprised of Ni_xP_y and activated carbon electrodes exposed significant power and energy densities ($67.2 \, W \, h \, kg^{-1}$ at $0.75 \, kW \, kg^{-1}$ and $20.4 \, W \, h \, kg^{-1}$ at $15 \, kW \, kg^{-1}$, correspondingly).

4 Oxide free nanomaterials for energy storage and conversion applications

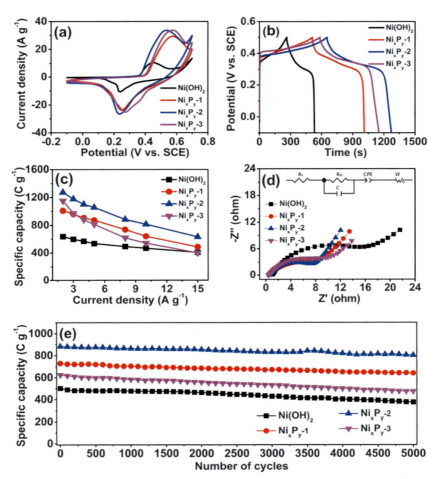

FIG. 1 Electrochemical features of TMPs-based electrode materials: (A) comparative cyclic voltammogram plots obtained at 5 mV s^{-1}, (B) comparative galvanostatic charge-discharge curve at 2 A g^{-1}, (C) specific capacity with respect to discharge current density, (D) electrochemical impedance spectroscopy plots along with an equivalent circuit, and (E) cyclic stability for 5000 cycles at 8 A g^{-1}. *(Reproduced from S. Liu, K.V. Sankar, A. Kundu, M. Ma, J. Y. Kwon, S.C. Jun, Honeycomblike interconnected network of nickel phosphide heteronanoparticles with superior electrochemical performance for supercapacitors, ACS Appl. Mater. Interfaces 9(26) (2017), 21829–21838. Copyright, American Chemical Society.)*

3. Transition-metal carbides

Due to the benefits of superior electrical conductivity, exceptional chemical durability, and inherent catalytic features, transition-metal carbides (TMCs) have been extensively explored as active candidates for numerous uses, such as LIBs/SIBs, electrocatalysis, and fuel cell. During the past decades, the importance of innovative 2-D candidates MXenes (M-signifies a family of early transition-metals, X signifies carbon), viz. early TMCs and/or carbonitrides, has been hugely explored in the energy-related methods, which profit from greater electrical conductivities and hydrophilic surfaces [55–58]. Generally, most of the transition metals from the families of IVB to VIB can create TMCs, excluding the Pt-group metals. These TMCs have identical crystalline features concerning parent metals since the smaller carbon atoms tend to localize at the parent metals' interstitial sites. Owing to metals' dissimilar inherent physicochemical features, the TMCs typically have three types of methods, such as MC, M_2C, and M_3C. Accordingly, the TMCs are greatly reliant on their crystal and electronic structures. Overall, the TMCs display the integrated features of covalent solids, transition metals, and ionic crystals [59]. Generally, TMCs also reveal similar electronic conductivity as parent metals and have identical electrocatalytic features as Pt-group metals. As such, this group of these electrode materials has been receiving considerable recognition in fundamental studies and commercial usages in numerous fields. Mainly, these TMCs are promising materials for energy-related systems with cost-effective features and eco-friendliness.

Till to date, the MXenes groups contain Ti_2CT_x, $(Ti_{0.5}, Nb_{0.5})_2CT_x$, Nb_2CT_x, Ti_3CNT_x, V_2CT_x, $Ti_3C_2T_x$, $(V_{0.5}, Cr_{0.5})_3C_2T_x$, $Ta_4C_3T_x$, and $Nb_4C_3T_x$, amongst which $Ti_3C_2T_x$ has been the best explored [60]. In terms of theoretic point of view as well as an experimental examination has previously revealed that these electrodes are probable electrode candidates for LIBs/SIB and supercapacitors [61, 62].

4. Transition-metal nitrides

Nitrogen is the maximum abundant constituent existing in the earth's atmosphere. Identical to TMCs, nitrogen can create transition metal nitrides (TMNs) with practically all transition metals. With the insertion of nitrogen, the distance between metal atoms is expanded. But the TMNs with standard formulations of MN, M_2N, and M_4N can still uphold their cubic/hexagonally close-packed structure [63]. The TMNs are recognized by their greater conductivity, refractoriness, and hardness. These features create the TMNs materials exclusively appropriate for developing crucibles, thermocouple sheaths, and cutter coating. During the last century, the usage of TMNs in heterogeneous catalytic reactions are revealed. For example, iron nitrides as Fischer-Tropsch catalysts [39]. The TMNs are of specific interest due to their benefits of exceptional electrical conductivities and eco-friendly longevity, and more excellent selectivity. Numerous research works have been dedicated to multiple applications during the past decades, namely, energy storage devices, optical storage systems, reforming catalysts, and electrical uses [64–67]. For instance, the TMNs are favorable electrode candidates in electrochemical supercapacitors owed to their physicochemical and mechanical features [1, 68]. Amongst the transition-metal

6 Oxide free nanomaterials for energy storage and conversion applications

compounds, TMNs such as VN, [69] TiN, [70] CrN, [71] Mo_xN, [72] RuN, [73] and Ni_3N [74] are exciting electrode materials for SCs.

In recent years, there have been numerous distinguished ways employed to fabricate the TMNs, comprising direct nitriding of the metal with N_2 or NH_3 or ammonolysis of oxides and chlorides. Liu and co-workers [75] demonstrated a new kind of approach to fabricating greatly porous and single-crystal-like Mo_2N nanobelts. They also reported the uppermost specific capacitance $160\,Fg^{-1}$ at $5\,mVs^{-1}$; CV curves through respect to numerous sweep rates and their estimated specific capacitances are presented in Fig. 2A. Meanwhile, these developed highly porous Mo_2N delivered exceptionally higher electrochemical stability even after 1000 cycles with greater capability, presented in Fig. 2B.

5. Transition-metal borides

Compared to TMPs, TMCs, and TMNs, transition-metal borides (TMBs) are less well examined for energy-related systems. Similar to TMPs/TMNs, boron can create borides with all the transition metals [76]. Most of TMBs share (M—B) bonds with a stronger covalent component. Generally, the existing M—B bonds are stronger than those seen in TMCs and TMNs. TMBs families show several combinations (from M_3B to MB_{66}), with the characteristics varying from metallic to semiconducting state with increasing boron content. Amongst, TMBs reveal distinctive physicochemical features, similar to TMNs discussed earlier, which have fascinated much consideration [77]. Further, TMBs reveal exciting electronic ground states, with superconductivity and greater chemical and thermal stability [78–81].

Till now, numerous research efforts on TMBs are involved as electrocatalytic materials for HER are on Mo, Co, and nickel borides [82–84]. In particular, Hu and co-workers demonstrated initially by employing polycrystalline α-MoB for HER [85]. Whereas, in terms of ORR, TMBs are less investigated. There are only a few research efforts have been carried out [86].

6. Transition-metal sulfides

Sulfur (S), a nonmetallic element with an electronic configuration [Ne] $3s^1$ belongs to the oxygen group (Group 16) in the periodic table of elements. The reactive and multivalent nature of sulfur enables it to react with almost all the metals, except gold (Au) and platinum (Pt), and nonmetals, to form corresponding sulfides and compounds. Metal sulfides are composed of either sulfide (S^{2-}) or persulfide (S_2^{2-}) anion bonded with a metal or semimetal cation to form mono-metal or bi-metallic sulfides represented as M_xS_y or $A_{1-x}B_xS_y$, respectively. The atomic arrangements of metal and sulfur in metal sulfides mostly confine to a close-packed system governed by their ionic size and charge. Further, the occupancy of anions surrounding the metal cations classifies sulfides

Nanostructured nonoxide nanomaterials **Chapter | 1** 7

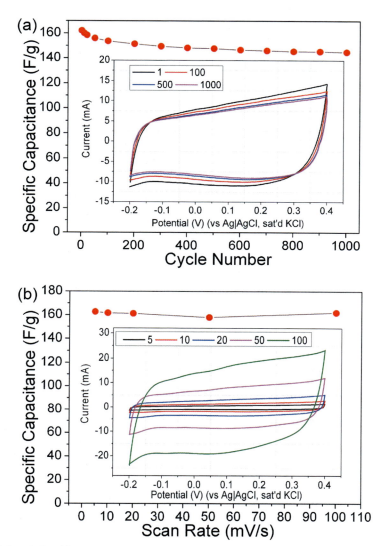

FIG. 2 (A) Specific capacitance of highly porous Mo$_2$N nanobelts under altered sweep rates and their equivalent cyclic voltammogram curves (inset) and (B) cycling behavior as SC electrodes and its CV plots for selected cycles (inset). *(Reproduced from J. Liu, K. Huang, H.L. Tang, M. Lei, Porous and single-crystalline-like molybdenum nitride nanobelts as a non-noble electrocatalyst for alkaline fuel cells and electrode materials for supercapacitors, Int. J. Hydrogen Energy 41(2) (2016), 996–1001. Copyright Elsevier.)*

8 Oxide free nanomaterials for energy storage and conversion applications

into six basic structures, namely sodium chloride (NaCl), sphalerite (ZnS), fluorite (CaF_2), antifluorite, layered, and spinel structure. The first four structures are covalently bonded sulfides, where the latter two structures exhibit metallic properties. Considering the covalently bonded sulfides, if an array of anions surrounds the cation and vice-versa with each ion in a 6-coordinate octahedral geometry, then the structure corresponds to NaCl type structure, whereas the coordination to tetrahedral geometry leads to sphalerite (ZnS) structure. In the case of fluorite and antifluorite, the cation is surrounded by eight anions, where the cation lattice is face-centered cubic (FCC) with the anions filling the tetrahedral interstices to give rise to fluorite structure; upon the reverse arrangement leads to the antifluorite structure possessing cubic close-packed (CCP) array of the anion with cations located in tetrahedral sites [1, 87].

Naturally, sulfides occur in all rock types formed either as sulfide melt during the early crystallization of magma or deposits beneath the seafloor at a relatively high temperature and pressure. Generally, pyrites, pyrrhotites, pentlandites, and chalcopyrites are formed through the former process and most other sulfides in the latter process. Some of the commonly reported metal sulfides are depicted in Table 1.

6.1 Transition-metal sulfides synthesis

Numerous physical and chemical synthesis methods have been advanced to realize high-quality, defined size, shape, and metal sulfides. Based on the common synthesis methods, the synthesis of metal sulfides is broadly categorized as top-down and bottom-up methods, as depicted in Fig. 3.

6.1.1 Top-down approaches

The top-down approach refers to downsizing the materials to nanostructures through the successive etching or removal of the layers or planes. Some of the methods widely employed in the top-down approach include electrospinning,

TABLE 1 Commonly reported metal sulfides.

CrS	Cr_2S_3	FeS	Fe_3S_4	FeS_2	MnS
CoS	CoS_2	Co_2S_3	Co_3S_4	Co_4S_3	Co_9S_8
NiS	NiS_2	Ni_3S_2	Ni_3S_4	Ni_6S_5	Ni_7S_6
CuS	CuS_2	Cu_2S	SnS	SnS_2	Sn_2S_3
Sn_3S_4	Sn_4S_5	In_2S_3	NbS	NbS_2	PdS
PdS_2	IrS_2	Ir_2S_3	MoS_2	MoS_3	ZnS
CaS	CdS	VS_2	ZrS	PtS	Bi_2S_3

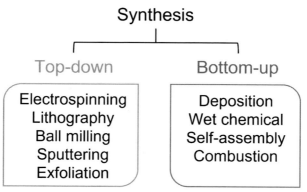

FIG. 3 Broad classification of synthesis of metal sulfides.

lithography, sputtering, ball milling, and exfoliation. Obviously, the type of synthesis method adopted for the fabrication determines the morphological features desired for specific applications. For instance, electrospinning is an active method for the fabrication of nanowires achieved through a voltage-driven process governed by electrohydrodynamic phenomena. Lithography is yet another technique to achieve metal sulfide nanostructures where the process overcomes the etching process. Electron beam nanolithography was employed to prepare arbitrarily size and shaped ZnS nanostructures through single-source molecular resist [88]. Further, to improve the catalytic activity of MoS_2, defective layered MoS_2 crystals were realized through lithography and found to be an efficient catalyst [89].

A mechano-chemical route like the ball milling process produces highly-deformed particles achieved through collisions with the stainless vessel and balls to yield nanoscale materials. Quantum dots of CuS possessing high surface area and defects were prepared through a ball milling process, which improved catalytic performances [90]. Exfoliation is yet another method that involves layer by layer removal of the surfaces to produce nanostructures. Mechanical and liquid exfoliation are widely followed techniques; however, the choice of size, shape, number of layers, and yield depend on the exfoliation method. Generally, the quality and yield of metal sulfides are important, whereas the liquid exfoliation process produces a higher yield with low quality; on the contrary, the mechanical exfoliation method produces high quality with a low yield of metal sulfide materials. A variety of metal sulfides such as $ZnIn_2S_4$, $CdIn_2S_4$, and In_2S_3 were prepared through mechanical exfoliation for hydrogen generation applications [91]. In terms of fabricating large-area metal sulfides, deposition through sputtering is applied to develop thin films consisting of nanoparticles. Magnetron sputtering is allowed to fabricate monolayers of MoS_2 thin films on various substrates [92].

6.1.2 Bottom-up approaches

The bottom-up approach denotes the fabrication of nanostructures build-up from the basic units such as atom by atom or molecule by molecule. Primarily, bottom-up fabrication involves either gas-phase synthesis or liquid-phase formation, and the methods include deposition techniques like chemical vapor deposition (CVD), atomic layer deposition (ALD), pulsed layer deposition (PLD), and thermal deposition, combustion, wet synthesis, and self-assembly processes [93].

CVD is a widely used synthesis technique for the preparation of metal sulfides, where the substrate is exposed to one or many volatile precursors produced at high temperatures to form thin films. The CVD technique enables the creation of good quality and possesses control over the metal sulfides. Monolayers of $Mo_{1-x}W_xS_2$, a ternary metal sulfide, were prepared through the CVD to control the atomic ratio of Mo/W to achieve good quality films [94]. Moreover, the use of various substrates from plastic to glass can be used in recent CVD owing to the lower temperature of 150–300°C against the usual temperature range of 700–1000°C [95]. Another subclass of the CVD technique is the atomic layer deposition, which enables fabricating thin films of metal sulfides from the vapor phase, termed ALD. The key advantage of ALD being the layer-by-layer control and precision of the technique has been exploited to prepare several metal sulfides that include CoS, CuS, CaS, In_2S_3, and PbS [96, 97].

PLD is a physical vapor deposition that possesses good control of the metal sulfides' morphology and thickness. Highly crystalline thin films of WS_2 with hybrid 1T and 2H phases were grown on a silicon substrate coated with Ag [98]. Also, a simple thermal evaporation technique was used to fabricate ZnS nanostructures with various morphologies like nanosheets, nanorods, and nanobelts [99]. In terms of rapid fabrication, the combustion method is used to prepare metal sulfides utilizing the exothermic reaction at high temperatures through an appropriate oxidant/fuel mixture ratio. Nanoparticles of CdS and SnS_2 were produced through a self-propagating combustion reaction using a preheated furnace [100].

Metal sulfides are attractive inorganic materials for numerous usages owing to the variety of crystal structures, tunable morphology, and attractive electrochemical properties. The occurrence of naturally available sulfides like pyrites (FeS_2), heazelwoodite (Ni_3S_2), chalcocite (Cu_2S), tungstenite (WS_2), etc., makes them an abundant and less-expensive material for employing in prototype scales. In addition, the synthesized metal sulfides like zinc sulfide (ZnS), cobalt sulfide (Cu_2S), nickel sulfide (Ni_3S_2), molybdenum sulfide (MoS_2), etc., serve as an important class of materials in several technological applications. Zinc sulfide, a semiconducting material possessing a wide bandgap of $3.7\,eV$, finds its application in display panels, light-emitting diodes (LEDs), photocatalyst, and so on. Further, transition metal sulfides such as Co_9S_8 and Ni_3S_2 are widely employed as catalysts for hydrodesulfurization in the petroleum-based industry. The large direct bandgap and high carrier mobility of MoS_2 enables it to use in areas such as photodetectors, chemical sensors, valleytronic devices, and energy storage applications. Following the brief outline, the following sections explore the applications of metal sulfides.

6.2 Applications of metal sulfides

Metal sulfides of various combinations such as pristine, composites, hybrids, and heterostructures have gained significant attention in renewable energy conversion and storage technologies such as electrocatalyst for ORR/HER, supercapacitors, batteries, and solar cells.

6.2.1 Electrocatalyst for ORR/HER

The development of bi-functional catalysts towards an efficient ORR and HER was prompted due to the sluggish kinetics of the electrochemical system. Despite the notable ORR electroactivity of Pt-based electrocatalysts, the formation of oxides at higher over potential restricts the catalytic ability of Pt. Also, the highly efficient OER electrocatalytic materials are generally less efficient for the HER and are vice-versa. Metal sulfides such as MoS_2 and WS_2 with high surface area and surface-exposed atoms have increased electrocatalytic activity [101, 102]. However, it is well evidenced that the pristine metal sulfides and their morphologically varied nanostructures expose poor electrocatalytic activity, which requires further improvement. For instance, CoS_2 nanoparticles of 30.7 nm have been prepared via a one-pot hydrothermal route that improved ORR catalytic activity with an open circuit potential of 0.94 V [103]. Morphologically varied NiS like rod-like and flower-like structures have been fabricated through the single-step hydrothermal method and evaluated as air catalysts for $Li-O_2$ batteries [104]. Hybrid structures composed of metal sulfides of MS@G/NSC (M: Fe, Co, and Ni) with nitrogen-sulfur doped carbon matrix were prepared and found to be highly electroactive [105]. Furthermore, $NiCo_2S_4$ solvothermal prepared exhibited good OER activity with a low onset potential of 1.45 V [106].

HER based-on metal sulfides has been under intense investigation, which involves MoS_2, WS_2, FeS_2, CoS_2, and NiS_2. Owing to the pronounced effect of nanoscale materials, the structural engineering of metal sulfides to nano-level is interesting. Thin films of MoS_2 have been prepared through electrodeposition and sulfurization, which showed enhanced catalytic activity attributed to the presence of exposed edges of MoS_2 serving as active sites [107]. Further, WS_2 nanosheets have been fabricated through a prelithiation method, which showed enhanced activity with a current exchange density of $2 \times 10^{-5} A cm^{-1}$ [108]. Several sulfides of vanadium, namely vanadium disulfide (VS_2) have been explored as monolayer and nanostructures like nanosheets, nanoplates, etc., to show efficient HER activity. Mixed metal sulfides are also an effective

12 Oxide free nanomaterials for energy storage and conversion applications

HER catalyst, where $Zn_{0.30}Co_{2.70}S_4$, Cu_2MoS_4, $Fe_{0.9}Co_{0.1}S_2$/CNT composites, etc., exhibited higher HER catalytic activity [109–111].

6.2.2 Photocatalyst

The development of photocatalyst materials is based on the bandgap of the employed materials, where the few metal sulfides possess a narrow bandgap that is feasible to oxidize and reduce dyes to serve as photocatalyst. Nanostructures of CuS were employed to degrade Methylene Blue at 663 nm, and similarly, mesoporous FeS_2 nanoparticles were used to degrade Rose Bengal dye [112, 113]. Metal-semiconductor heterojunctions consisting of mixed-phase MoS_2, i.e., 1T metallic and 2H semiconducting phases, were investigated for selective oxidation of benzylamines [114]. Apart, room-temperature synthesis of mixed-phase CdS has been shown to be an effective photocatalyst for aerobic oxidation of benzyl alcohol. Further, morphologically varied In_2S_3, such as microspheres and hierarchical architectures, were also shown to be photocatalysts [115, 116]. Mixed metal sulfide $CdIn_2S_4$ and $ZnIn_2S_4$ were used as a photocatalyst to selective oxidation of benzyl alcohol to benzaldehyde and aerobic oxidation of alcohols, respectively [117].

6.2.3 Supercapacitors

Carbon-based materials have been the prime electrode materials for supercapacitors, although several metal sulfides have been investigated as electrodes considering their redox nature. Since carbon-based two-dimensional (2D) materials like graphene and activated carbon are favorable electrodes, several layered 2D metal sulfides such as MoS_2 and VS_2 are attractive electrode materials. Nanosheets of MoS_2 obtained through exfoliation of bulk MoS_2 were used as an electrode, which showed capacitance of $400–710\,F\,cm^{-3}$ in an aqueous electrolyte [118]. Further, modified MoS_2 electrodes like MoS_2/graphene, MoS_2/polypyrrole nanocomposites were developed to improve the long-term cycling performance [119, 120]. Besides, layered VS_2 nanosheets are an effective electrode material for supercapacitors [121]. Further, morphologically varied Ni_3S_2 like nanosheet arrays and mushroom-like structures exhibited good pseudocapacitive property [122, 123]. Sulfides of Mn and Cu have also been employed as supercapacitor electrodes to show high specific capacitances with the enhanced cycling life [124, 125].

6.2.4 Batteries

Metal sulfides are widely explored as alternative anode materials for lithium (Li)- and sodium (Na)-ion batteries due to their capability to involve in multiple electron transfer reactions to deliver a high theoretical capacity, however, the volume changes propagate poor electronic and ionic conductivity which hinders the commercial exploitation of metal sulfides. Amongst several candidates explored as anodes, metal sulfides such as MoS_2, SnS_2, SnS, SnS_2/SnS hybrid, and WS_2 have shown promising results. The two-dimensional and layered sulfides structures are advantageous for better battery operations in terms of the interlayers and their distance, enabling for favorable insertion/des-insertion of the guest alkali ions, such as Li^+ and Na^+ into/from the structure. Pristine MoS_2 was investigated as a Li-storage electrode in late 1980 [126]; since then, several variations with morphology and the nanostructure were studied for batteries. Interestingly, the electrochemical lithiation in MoS_2 undergoes a well-defined intercalation and conversion reaction mechanism, as notable from the cyclic voltammogram curves. The Li^+ ions intercalate into the MoS_2 structure until 1.1 V (vs Li^+) and thereafter, the Li^+ reacts with the Li_xMoS_2 leading to the formation of Mo metal and Li_2S through a conversion reaction [127]. Obviously, the involvement of the conversion process stimulates concerns related to volume change, thereby, capacity fading makes pristine MoS_2 vulnerable to direct battery operation. Heterostructures of MoS_2/nitrogen-doped carbon were studied as anode materials for Li-ion batteries, which showed improved specific capacity of above $860 \, mAh \, g^{-1}$ than the pristine MoS_2 and commercially used graphite anodes [128]. Similarly, varied heterostructures of MoS_2/nitrogen-doped carbon showed improved specific capacity as an anode for sodium-ion batteries [129].

Apart, SnS_2, SnS, and hybrids of SnS_2/SnS are interesting electrode materials for Li-ion and Na-ion batteries that could deliver a theoretical capacity of 782 and $1136 \, mAh \, g^{-1}$ and 790 and $1022 \, mAh \, g^{-1}$, respectively. In addition to the volume changes of SnS_2 and SnS electrodes during the electrochemical reaction, hybrids with several carbon-based materials such as graphene, graphitic oxide, and carbon nanotube have been reported [130]. The reversible capacities of tin sulfides could reach close to theoretical capacities through the hybrids and morphological variation. Similarly, nonlayered metal sulfides like FeS_2, NiS, Ni_2S_3, Ni_3S_4, and CoS_2 have been investigated based on their theoretical capacity of 894, 590, 445, 704, and $872 \, mAh \, g^{-1}$ (vs Li^+). In most cases, the pristine metal sulfide electrodes experienced poor reversible capacity, which is alleviated through the employment of carbon-based materials that mitigates the volume change by acting as a buffer and as a conductive agent.

6.2.5 Solar cells

Semiconducting metal and mixed metal sulfides like Cu_2S, $CuInS_2$, and $CuGaS_2$ have profound applications in solar cells. Thin-film solar cells based on CdS/Cu_2S were developed in the late 1980s, which showed an efficiency of up to $\sim 10\%$ [131]. Later, $CuInS_2$ nanocrystals were developed to overcome the instability caused by Cu^{2+} ions [132]. Nanostructures based on FeS, FeS_2, MoS_2, and WS_2 have been investigated as photocathodes in solar cells.

6.3 High entropy alloys

Over the decades, the alloying strategy was limited to adding small fractions of other elements into a base element to achieve the new kinds of composite material with enhanced features. For instance, the limited variety of possible alloys can be prepared by employing numerous base elements. High Entropy Alloys (HEAs) are alloying systems composed of at least 5 elements in near-equiatomic quantities. This method viewpoints in sharp contrast to the traditional practice and has, consequently, fascinated much consideration. Rely on the maximization of Configurational Entropy, which can stabilize solid solution phases against intermetallic compounds. In 1996, HEAs were initially demonstrated by Huang [133]; this field's interest didn't develop until 2004. Afterward, significant works have been reported. More importantly, the definition of HEAs expanded to alloys comprised of five or more elements (Fig. 4) with compositions between 5% and 35% (composition-based definition) and multielemental alloys (entropy-based definition) [134].

In particular, Yeh et al. [135] investigating these: they theorized that the incidence of multiple elements in equiatomic quantities would upsurge the configurational entropy of mixing by an amount sufficient to overcome the enthalpies of compound formation, thus preventing the development of possibly damaging intermetallics. Recent reports have also demonstrated that numerous HEAs can tolerate a range of extremely corrosive atmospheres even at higher

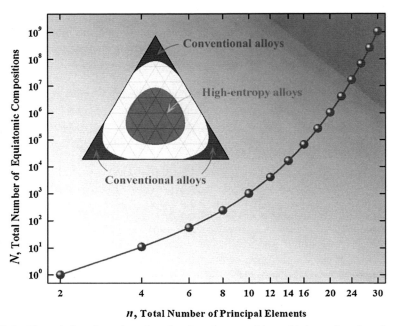

FIG. 4 The variation of a total number of equiatomic compositions with the total number of principal elements. *(Reproduced from Y.F. Ye, Q. Wang, J. Lu, C.T. Liu, Y. Yang, High-entropy alloy: challenges and prospects, Mater. Today 19(6) (2016) 349–362. Copyright © 2016 Materials Today.)*

temperatures. It should be evidenced that through their incredible features, HEAs have not yet established general application, excluding the protective coatings and high-temperature components. The combined mixture of elements resulting in HEA with exceptional features as functional materials in energy-related systems. Particularly, HEAs have been explored as potential electrocatalysts for ORR, HER, OER, methanol, ethanol, CO_2 reduction reactions in different electrolyte media and for anodic oxidation of fuels in direct methanol fuel cell (DMFC), direct formic acid fuel cell (DFAFC) and direct ethanol fuel cell (DEFC) [136–145].

These HEAs have distinctive features and excellent properties when associated with classical alloys. HEAs have the potential to substitute numerous costly alloys used currently in the manufacturing industry. There is an enormous possibility of further research in HEAs that can give birth to new alloys

7. Conclusions

The chapters are titled to give an overview of recent nonoxide nanostructured materials beyond the realm of science fiction and their role in the development of energy-related devices. Therefore, in-depth knowledge of nonoxide materials and their energy-level mechanisms and the latest nanotechnology-based applications in energy-related applications will help the scientific community understand the selection of nonoxide materials and design new kinds of innovative materials.

The small insertion of P, C, N, or B over the transition metals resulting in different kinds of transition metal-based nanostructures with flexibility and can apply in energy storage and conversion applications. Lastly, we have broadly summarized and discussed transition metal-based materials for energy-related systems applications in terms of synthetic methodologies with their own merits and demerits of each method and approaches to advance the features or overcome their limits. We have also presented some recently innovative research reports for enhancing the efficiency of the materials in corresponding applications.

References

[1] P. Simon, Y. Gogotsi, Materials for electrochemical capacitors, in: Nanoscience and Technology: A Collection of Reviews from Nature Journals, 2010, pp. 320–329.

[2] A.K. Das, S. Sahoo, P. Arunachalam, S. Zhang, J.J. Shim, Facile synthesis of Fe_3O_4 nanorod decorated reduced graphene oxide (RGO) for supercapacitor application, RSC Adv. 6 (108) (2016) 107057–107064.

[3] X. Zhao, B.M. Sánchez, P.J. Dobson, P.S. Grant, The role of nanomaterials in redox-based supercapacitors for next generation energy storage devices, Nanoscale 3 (3) (2011) 839–855.

[4] S. Dong, X. Chen, X. Zhang, G. Cui, Nanostructured transition metal nitrides for energy storage and fuel cells, Coord. Chem. Rev. 257 (13–14) (2013) 1946–1956.

16 Oxide free nanomaterials for energy storage and conversion applications

[5] J. Theerthagiri, R. Senthil, B. Senthilkumar, A.R. Polu, J. Madhavan, M. Ashokkumar, Recent advances in MoS_2 nanostructured materials for energy and environmental applications—a review, J. Solid State Chem. 252 (2017) 43–71.

[6] K. Thiagarajan, J. Theerthagiri, R. Senthil, P. Arunachalam, J. Madhavan, M.A. Ghanem, Synthesis of $Ni_3V_2O_8$@graphene oxide nanocomposite as an efficient electrode material for supercapacitor applications, J. Solid State Electrochem. 22 (2017) 527–536.

[7] J. Theerthagiri, R. Sudha, K. Premnath, P. Arunachalam, J. Madhavan, A.M. Al-Mayouf, Growth of iron diselenide nanorods on graphene oxide nanosheets as advanced electrocatalyst for hydrogen evolution reaction, Int. J. Hydrogen Energy 42 (2017) 13020–13030.

[8] A. Malathi, J. Madhavan, M. Ashokkumar, P. Arunachalam, A review on BiVO4 photocatalyst: activity enhancement methods for solar photocatalytic applications, Appl. Catal. A. Gen. 555 (2018) 47–74.

[9] C. Karthikeyan, P. Arunachalam, K. Ramachandran, A.M. Al-Mayouf, S. Karuppuchamy, Recent advances in semiconductor metal oxides with enhanced methods for solar photocatalytic applications, J. Alloys Compd. 828 (2020) 154281.

[10] J. Theerthagiri, S. Salla, R.A. Senthil, P. Nithyadharseni, A. Madankumar, P. Arunachalam, H.S. Kim, A review on ZnO nanostructured materials: energy, environmental and biological applications, Nanotechnology 30 (39) (2019) 392001.

[11] P. Ramesh, S. Amalraj, P. Arunachalam, M. Gopiraman, A.M. Al-Mayouf, S. Vasanthkumar, Covalent intercalation of hydrazine derived graphene oxide as an efficient 2D material for supercapacitor application, Synth. Met. 272 (2021) 116656.

[12] P. Arunachalam, K. Nagai, M.S. Amer, M.A. Ghanem, R.J. Ramalingam, A.M. Al-Mayouf, Recent developments in the use of heterogeneous semiconductor photocatalyst based materials for a visible-light-induced water-splitting system—a brief review, Catalysts 11 (2) (2021) 160.

[13] J.R. Rajabathar, H.A. Al-lohedan, P. Arunachalam, Z.A. Issa, M.K. Gnanamani, J.N. Appaturi, W.M. Dahan, Synthesis and characterization of metal chalcogenide modified graphene oxide sandwiched manganese oxide nanofibers on nickel foam electrodes for high performance supercapacitor applications, J. Alloys Compd. 850 (2021) 156346.

[14] P. Arunachalam, M.S. Amer, M.A. Ghanem, A.M. Al-Mayouf, D. Zhao, Activation effect of silver nanoparticles on the photoelectrochemical performance of mesoporous TiO2 nanospheres photoanodes for water oxidation reaction, Int. J. Hydrogen Energy 42 (16) (2017) 11346–11355.

[15] A. Priya, P. Arunachalam, A. Selvi, J. Madhavan, A.M. Al-Mayouf, M.A. Ghanem, A low-cost visible light active $BiFeWO_6/TiO_2$ nanocomposite with an efficient photocatalytic and photoelectrochemical performance, Opt. Mater. 81 (2018) 84–92.

[16] S. Zhang, P. Arunachalam, T. Abe, T. Iyoda, K. Nagai, Photocatalytic decomposition of N-methyl-2-pyrrolidone, aldehydes, and thiol by biphase and p/n junction-like organic semiconductor composite nanoparticles responsive to nearly full spectrum of visible light, J. Photochem. Photobiol. A Chem. 244 (2012) 18–23.

[17] P. Arunachalam, S. Zhang, T. Abe, M. Komura, T. Iyoda, K. Nagai, Weak visible light ($\sim mW/cm^2$) organophotocatalysis for mineralization of amine, thiol and aldehyde by biphasic cobalt phthalocyanine/fullerene nanocomposites prepared by wet process, Appl. Catal. Environ. 193 (2016) 240–247.

[18] P. Arunachalam, M.N. Shaddad, M.A. Ghanem, A.M. Al-Mayouf, M.T. Weller, Zinc tantalum Oxynitride ($ZnTaO_2N$) photoanode modified with cobalt phosphate layers for the photoelectrochemical oxidation of alkali water, Nanomaterials 8 (1) (2018) 48.

Nanostructured nonoxide nanomaterials **Chapter | 1** **17**

[19] M.N. Shaddad, P. Arunachalam, A.M. Al-Mayouf, M.A. Ghanem, A.I. Alharthi, Enhanced photoelectrochemical oxidation of alkali water over cobalt phosphate (Co-Pi) catalyst-modified $ZnLaTaON_2$ photoanodes, Ionics 25 (2) (2019) 737–745.

[20] J. Theerthagiri, K. Thiagarajan, B. Senthilkumar, Z. Khan, R.A. Senthil, P. Arunachalam, M. Ashokkumar, Synthesis of hierarchical cobalt phosphate nanoflakes and their enhanced electrochemical performances for supercapacitor applications, ChemistrySelect 2 (1) (2017) 201–210.

[21] M. Naushad, T. Ahamad, M. Ubaidullah, J. Ahmed, A.A. Ghafar, K.M. Al-Sheetan, P. Arunachalam, Nitrogen-doped carbon quantum dots (N-CQDs)/Co_3O_4 nanocomposite for high performance supercapacitor, J. King Saud Univ. Sci. 33 (1) (2021) 101252.

[22] M.N. Shaddad, P. Arunachalam, J. Labis, M. Hezam, A.M. Al-Mayouf, Fabrication of robust nanostructured (Zr) $BiVO_4$/nickel hexacyanoferrate core/shell photoanodes for solar water splitting, Appl. Catal. Environ. 244 (2019) 863–870.

[23] M.N. Shaddad, P. Arunachalam, M. Hezam, A.M. Al-Mayouf, Cooperative catalytic behavior of SnO_2 and $NiWO_4$ over $BiVO_4$ photoanodes for enhanced photoelectrochemical water splitting performance, Catalysts 9 (11) (2019) 879.

[24] A. Malathi, V. Vasanthakumar, P. Arunachalam, J. Madhavan, M.A. Ghanem, A low cost additive-free facile synthesis of $BiFeWO_6$/$BiVO_4$ nanocomposite with enhanced visible-light induced photocatalytic activity, J. Colloid Interface Sci. 506 (2017) 553–563.

[25] M.N. Shaddad, D. Cardenas-Morcoso, P. Arunachalam, M. García-Tecedor, M.A. Ghanem, J. Bisquert, S. Gimenez, Enhancing the optical absorption and interfacial properties of $BiVO_4$ with Ag_3PO_4 nanoparticles for efficient water splitting, J. Phys. Chem. C 122 (22) (2018) 11608–11615.

[26] M.N. Shaddad, P. Arunachalam, A.A. Alothman, A.M. Beagan, M.N. Alshalwi, A.M. Al-Mayouf, Synergetic catalytic behavior of AgNi-OH-Pi nanostructures on Zr: $BiVO_4$ photoanode for improved stability and photoelectrochemical water splitting performance, J. Catal. 371 (2019) 10–19.

[27] A. Priya, R.A. Senthil, A. Selvi, P. Arunachalam, C.S. Kumar, J. Madhavan, A.M. Al-Mayouf, A study of photocatalytic and photoelectrochemical activity of as-synthesized WO_3/g-C_3N_4 composite photocatalysts for AO7 degradation, Mate. Sci. Energy Technol. 3 (2020) 43–50.

[28] A. Priya, P. Arunachalam, A. Selvi, J. Madhavan, A.M. Al-Mayouf, Synthesis of $BiFeWO_6$/WO_3 nanocomposite and its enhanced photocatalytic activity towards degradation of dye under irradiation of light, Colloids Surf. A Physicochem. Eng. Asp. 559 (2018) 83–91.

[29] R.J. Ramalingam, P. Arunachalam, J.N. Appaturai, P. Thiruchelvi, Z.A. Al-othman, A.G. Alanazi, H.A. Al-lohedan, Facile sonochemical synthesis of nanoparticle modified Bi-MnO_x and Fe_3O_4 deposited Bi-MnO_x nanocomposites for sensor and pollutant degradation application, J. Alloys Compd. 859 (2021) 158263.

[30] I. Roger, M.A. Shipman, M.D. Symes, Earth-abundant catalysts for electrochemical and photoelectrochemical water splitting, Nat. Rev. Chem. 1 (1) (2017) 1–13.

[31] K.N. Dinh, P. Zheng, Z. Dai, Y. Zhang, R. Dangol, Y. Zheng, Q. Yan, Ultrathin porous NiFeV ternary layer hydroxide nanosheets as a highly efficient bifunctional electrocatalyst for overall water splitting, Small 14 (8) (2018) 1703257.

[32] C.F. Du, K.N. Dinh, Q. Liang, Y. Zheng, Y. Luo, J. Zhang, Q. Yan, Self-assemble and in situ formation of $Ni1 - xFexPS_3$ nanomosaic-decorated MXene hybrids for overall water splitting, Adv. Energy Mater. 8 (26) (2018) 1801127.

18 Oxide free nanomaterials for energy storage and conversion applications

[33] Z. Wang, X. Li, H. Ling, C.K. Tan, L.P. Yeo, A.C. Grimsdale, A.I.Y. Tok, 3D FTO/FTO-nanocrystal/TiO_2 composite inverse opal photoanode for efficient photoelectrochemical water splitting, Small 14 (20) (2018) 1800395.

[34] M.A. Ghanem, A.M. Al-Mayouf, P. Arunachalam, T. Abiti, Mesoporous cobalt hydroxide prepared using liquid crystal template for efficient oxygen evolution in alkaline media, Electrochim. Acta 207 (2016) 177–186.

[35] M.A. Ghanem, P. Arunachalam, A. Almayouf, M.T. Weller, Efficient bi-functional electrocatalysts of strontium iron oxy-halides for oxygen evolution and reduction reactions in alkaline media, J. Electrochem. Soc. 163 (6) (2016) H450.

[36] M.S. Amer, M.A. Ghanem, A.M. Al-Mayouf, P. Arunachalam, Low-symmetry mesoporous titanium dioxide (lsm-TiO_2) electrocatalyst for efficient and durable oxygen evolution in aqueous alkali, J. Electrochem. Soc. 165 (7) (2018) H300.

[37] M.S. Amer, P. Arunachalam, M.A. Ghanem, M. Al-Shalwi, A. Ahmad, A.I. Alharthi, A.M. Al-Mayouf, Synthesis of iron and vanadium co-doped mesoporous cobalt oxide: an efficient and robust catalysts for electrochemical water oxidation, Int. J. Energy Res. 45 (11) (2021) 16963–16972, https://doi.org/10.1002/er.6934.

[38] M.S. Amer, P. Arunachalam, M.A. Ghanem, A.M. Al-Mayouf, M.A. Shar, Enriched active surface structure in nanosized tungsten-cobalt oxides electrocatalysts for efficient oxygen redox reactions, Appl. Surf. Sci. 513 (2020) 145831.

[39] K.N. Dinh, Q. Liang, C.F. Du, J. Zhao, A.I.Y. Tok, H. Mao, Q. Yan, Nanostructured metallic transition metal carbides, nitrides, phosphides, and borides for energy storage and conversion, Nano Today 25 (2019) 99–121.

[40] H.O. Pierson, Handbook of Refractory Carbides & Nitrides: Properties, Characteristics, Processing and Applications, William Andrew, 1996.

[41] S.T. Oyama, Introduction to the chemistry of transition metal carbides and nitrides, in: The Chemistry of Transition Metal Carbides and Nitrides, Springer, Dordrecht, 1996, pp. 1–27.

[42] H. Yu, K.N. Dinh, Y. Sun, H. Fan, Y. Wang, Y. Jing, Q. Yan, Performance-improved Li-O_2 batteries by tailoring the phases of Mo_xC porous nanorods as an efficient cathode, Nanoscale 10 (31) (2018) 14877–14884.

[43] S. Carenco, D. Portehault, C. Boissiere, N. Mezailles, C. Sanchez, Nanoscaled metal borides and phosphides: recent developments and perspectives, Chem. Rev. 113 (10) (2013) 7981–8065.

[44] Y. Shu, S.T. Oyama, Synthesis, characterization, and hydrotreating activity of carbon-supported transition metal phosphides, Carbon 43 (7) (2005) 1517–1532.

[45] X. Wang, H.M. Kim, Y. Xiao, Y.K. Sun, Nanostructured metal phosphide-based materials for electrochemical energy storage, J. Mater. Chem. A 4 (39) (2016) 14915–14931.

[46] Y. Lu, J.K. Liu, X.Y. Liu, S. Huang, T.Q. Wang, X.L. Wang, S.X. Mao, Facile synthesis of Ni-coated Ni_2P for supercapacitor applications, CrstEngComm 15 (35) (2013) 7071–7079.

[47] S. Faraji, F.N. Ani, Microwave-assisted synthesis of metal oxide/hydroxide composite electrodes for high power supercapacitors—a review, J. Power Sources 263 (2014) 338–360.

[48] H. Ma, J. He, D.B. Xiong, J. Wu, Q. Li, V. Dravid, Y. Zhao, Nickel cobalt hydroxide@ reduced graphene oxide hybrid nanolayers for high performance asymmetric supercapacitors with remarkable cycling stability, ACS Appl. Mater. Interfaces 8 (3) (2016) 1992–2000.

[49] V. Pralong, D.C.S. Souza, K.T. Leung, L.F. Nazar, Reversible lithium uptake by CoP_3 at low potential: role of the anion, Electrochem. Commun. 4 (6) (2002) 516–520.

[50] X. Zhang, L. Zhang, G. Xu, A. Zhao, S. Zhang, T. Zhao, Template synthesis of structure-controlled 3D hollow nickel-cobalt phosphides microcubes for high-performance supercapacitors, J. Colloid Interface Sci. 561 (2020) 23–31.

Nanostructured nonoxide nanomaterials **Chapter | 1 19**

[51] M. Cui, X. Meng, Overview of transition metal-based composite materials for supercapacitor electrodes, Nanoscale Adv. 2 (2020) 5516–5528, https://doi.org/10.1039/D0NA00573H.

[52] S. He, Z. Li, H. Mi, C. Ji, F. Guo, X. Zhang, J. Qiu, 3D nickel-cobalt phosphide heterostructure for high-performance solid-state hybrid supercapacitors, J. Power Sources 467 (2020) 228324.

[53] J.F. Callejas, C.G. Read, C.W. Roske, N.S. Lewis, R.E. Schaak, Synthesis, characterization, and properties of metal phosphide catalysts for the hydrogen-evolution reaction, Chem. Mater. 28 (17) (2016) 6017–6044.

[54] S. Liu, K.V. Sankar, A. Kundu, M. Ma, J.Y. Kwon, S.C. Jun, Honeycomb-like interconnected network of nickel phosphide heteronanoparticles with superior electrochemical performance for supercapacitors, ACS Appl. Mater. Interfaces 9 (26) (2017) 21829–21838.

[55] S.L. Brock, K. Senevirathne, Recent developments in synthetic approaches to transition metal phosphide nanoparticles for magnetic and catalytic applications, J. Solid State Chem. 181 (7) (2008) 1552–1559, https://doi.org/10.1016/j.jssc.2008.03.012.

[56] M. Naguib, O. Mashtalir, J. Carle, V. Presser, J. Lu, L. Hultman, M.W. Barsoum, Two-dimensional transition metal carbides, ACS Nano 6 (2) (2012) 1322–133162.

[57] M. Naguib, V.N. Mochalin, M.W. Barsoum, Y. Gogotsi, 25th anniversary article: MXenes: a new family of two-dimensional materials, Adv. Mater. 26 (7) (2014) 992–1005.

[58] M. Naguib, R.R. Unocic, B.L. Armstrong, J. Nanda, Large-scale delamination of multi-layers transition metal carbides and carbonitrides "MXenes", Dalton Trans. 44 (20) (2015) 9353–9358.

[59] H.H. Hwu, J.G. Chen, Surface chemistry of transition metal carbides, Chem. Rev. 105 (1) (2005) 185–212.

[60] M. Naguib, M. Kurtoglu, V. Presser, J. Lu, J. Niu, M. Heon, M.W. Barsoum, Two-dimensional nanocrystals produced by exfoliation of Ti_3AlC_2, Adv. Mater. 23 (37) (2011) 4248–4253.

[61] Y. Dall'Agnese, P.L. Taberna, Y. Gogotsi, P. Simon, Two-dimensional vanadium carbide (MXene) as positive electrode for sodium-ion capacitors, J. Phys. Chem. Lett. 6 (12) (2015) 2305–2309.

[62] M. Naguib, V.N. Mochalin, M.W. Barsoum, Y. Gogotsi, Two-dimensional materials: 25th anniversary article: MXenes: a new family of two-dimensional materials (Adv. Mater. 7/2014), Adv. Mater. 26 (7) (2014) 982.

[63] N.N. Greenwood, A. Earnshaw, Chemistry of the Elements, Elsevier, 2012.

[64] C. Chen, D. Zhao, X. Wang, Influence of addition of tantalum oxide on electrochemical capacitor performance of molybdenum nitride, Mater. Chem. Phys. 97 (1) (2006) 156–161.

[65] A. Guerrero-Ruiz, Q. Xin, Y.J. Zhang, A. Maroto-Valiente, I. Rodriguez-Ramos, Microcalorimetric study of H adsorption on molybdenum nitride catalysts, Langmuir 15 (14) (1999) 4927–4929.

[66] K.H. Lee, Y.W. Lee, A.R. Ko, G. Cao, K.W. Park, Single-crystalline mesoporous molybdenum nitride nanowires with improved electrochemical properties, J. Am. Ceram. Soc. 96 (1) (2013) 37–39.

[67] T. Palaniselvam, R. Kannan, S. Kurungot, Facile construction of non-precious iron nitride-doped carbon nanofibers as cathode electrocatalysts for proton exchange membrane fuel cells, Chem. Commun. 47 (10) (2011) 2910–2912.

[68] D. Yang, J. Zhu, X. Rui, H. Tan, R. Cai, H.E. Hoster, Q. Yan, Synthesis of cobalt phosphides and their application as anodes for lithium ion batteries, ACS Appl. Mater. Interfaces 5 (3) (2013) 1093–1099.

20 Oxide free nanomaterials for energy storage and conversion applications

[69] R. Lucio-Porto, S. Bouhtiyya, J.F. Pierson, A. Morel, F. Capon, P. Boulet, T. Brousse, VN thin films as electrode materials for electrochemical capacitors, Electrochim. Acta 141 (2014) 203–211.

[70] A. Achour, R.L. Porto, M.A. Soussou, M. Islam, M. Boujtita, K.A. Aissa, T. Brousse, Titanium nitride films for micro-supercapacitors: effect of surface chemistry and film morphology on the capacitance, J. Power Sources 300 (2015) 525–532.

[71] O. Banakh, P.E. Schmid, R. Sanjines, F. Levy, High-temperature oxidation resistance of Cr1 − xAlxN thin films deposited by reactive magnetron sputtering, Surf. Coat. Technol. 163 (2003) 57–61.

[72] S.I.U. Shah, A.L. Hector, J.R. Owen, Redox supercapacitor performance of nanocrystalline molybdenum nitrides obtained by ammonolysis of chloride-and amide-derived precursors, J. Power Sources 266 (2014) 456–463.

[73] S. Bouhtiyya, R.L. Porto, B. Laïk, P. Boulet, F. Capon, J.P. Pereira-Ramos, J.F. Pierson, Application of sputtered ruthenium nitride thin films as electrode material for energy-storage devices, Scr. Mater. 68 (9) (2013) 659–662.

[74] M.S. Balogun, Y. Zeng, W. Qiu, Y. Luo, A. Onasanya, T.K. Olaniyi, Y. Tong, Three-dimensional nickel nitride (Ni_3N) nanosheets: free standing and flexible electrodes for lithium ion batteries and supercapacitors, J. Mater. Chem. A 4 (25) (2016) 9844–9849.

[75] J. Liu, K. Huang, H.L. Tang, M. Lei, Porous and single-crystalline-like molybdenum nitride nanobelts as a non-noble electrocatalyst for alkaline fuel cells and electrode materials for supercapacitors, Int. J. Hydrogen Energy 41 (2) (2016) 996–1001.

[76] U.B. Demirci, P. Miele, Cobalt in $NaBH_4$ hydrolysis, Phys. Chem. Chem. Phys. 12 (44) (2010) 14651–14665.

[77] J.M.V. Nsanzimana, Y. Peng, Y.Y. Xu, L. Thia, C. Wang, B.Y. Xia, X. Wang, An efficient and earth-abundant oxygen-evolving electrocatalyst based on amorphous metal borides, Adv. Energy Mater. 8 (1) (2018) 1701475.

[78] L. Xu, S. Li, Y. Zhang, Y. Zhai, Synthesis, properties and applications of nanoscale nitrides, borides and carbides, Nanoscale 4 (16) (2012) 4900–4915.

[79] X. Wang, G. Tai, Z. Wu, T. Hu, R. Wang, Ultrathin molybdenum boride films for highly efficient catalysis of the hydrogen evolution reaction, J. Mater. Chem. A 5 (45) (2017) 23471–23475.

[80] L. Chen, Y. Gu, L. Shi, Z. Yang, J. Ma, Y. Qian, Synthesis and oxidation of nanocrystalline HfB_2, J. Alloys Compd. 368 (1–2) (2004) 353–356.

[81] H. Park, A. Encinas, J.P. Scheifers, Y. Zhang, B.P. Fokwa, Boron-dependency of molybdenum boride electrocatalysts for the hydrogen evolution reaction, Angew. Chem. 129 (20) (2017) 5667–5670.

[82] Z. Chen, X. Duan, W. Wei, S. Wang, Z. Zhang, B.J. Ni, Boride-based electrocatalysts: emerging candidates for water splitting, Nano Res. 13 (2) (2020) 293–314.

[83] M. Zeng, H. Wang, C. Zhao, J. Wei, K. Qi, W. Wang, X. Bai, Nanostructured amorphous nickel boride for high-efficiency electrocatalytic hydrogen evolution over a broad pH range, ChemCatChem 8 (4) (2016) 708–712.

[84] M. Sheng, Q. Wu, Y. Wang, F. Liao, Q. Zhou, J. Hou, W. Weng, Network-like porous Co-Ni-B grown on carbon cloth as efficient and stable catalytic electrodes for hydrogen evolution, Electrochem. Commun. 93 (2018) 104–108.

[85] H. Vrubel, X. Hu, Molybdenum boride and carbide catalyze hydrogen evolution in both acidic and basic solutions, Angew. Chem. Int. Ed. 51 (2012) 12703–12706.

[86] J. Masa, I. Sinev, H. Mistry, E. Ventosa, M. De La Mata, J. Arbiol, W. Schuhmann, Ultrathin high surface area nickel boride (Ni_xB) nanosheets as highly efficient electrocatalyst for oxygen evolution, Adv. Energy Mater. 7 (17) (2017) 1700381.

[87] A.R. West, Solid State Chemistry and its Applications, John Wiley & Sons, 2014.

[88] M.S. Saifullah, M. Asbahi, M. Binti-Kamran Kiyani, S. Tripathy, E.A. Ong, A. Ibn Saifullah, K.S. Chong, Direct patterning of zinc sulfide on a sub-10 nanometer scale via electron beam lithography, ACS Nano 11 (10) (2017) 9920–9929.

[89] C.B. Roxlo, H.W. Deckman, J. Gland, S.D. Cameron, R.R. Chianelli, Edge surfaces in lithographically textured molybdenum disulfide, Science 235 (4796) (1987) 1629–1631.

[90] S. Li, Z.H. Ge, B.P. Zhang, Y. Yao, H.C. Wang, J. Yang, Y.H. Lin, Mechanochemically synthesized sub-5 nm sized CuS quantum dots with high visible-light-driven photocatalytic activity, Appl. Surf. Sci. 384 (2016) 272–278.

[91] M.Q. Yang, Y.J. Xu, W. Lu, K. Zeng, H. Zhu, Q.H. Xu, G.W. Ho, Self-surface charge exfoliation and electrostatically coordinated 2D hetero-layered hybrids, Nat. Commun. 8 (1) (2017) 1–9.

[92] J. Tao, J. Chai, X. Lu, L.M. Wong, T.I. Wong, J. Pan, S. Wang, Growth of wafer-scale MoS_2 monolayer by magnetron sputtering, Nanoscale 7 (6) (2015) 2497–2503.

[93] J.N. Tiwari, R.N. Tiwari, K.S. Kim, Zero-dimensional, one-dimensional, two-dimensional and three-dimensional nanostructured materials for advanced electrochemical energy devices, Prog. Mater. Sci. 57 (4) (2012) 724–803.

[94] S. Zheng, L. Sun, T. Yin, A.M. Dubrovkin, F. Liu, Z. Liu, H.J. Fan, Monolayers of WxMo1 − xS2 alloy heterostructure with in-plane composition variations, Appl. Phys. Lett. 106 (6) (2015), 063113.

[95] Y.R. Lim, W. Song, J.K. Han, Y.B. Lee, S.J. Kim, S. Myung, J. Lim, Wafer-scale, homogeneous MoS_2 layers on plastic substrates for flexible visible-light photodetectors, Adv. Mater. 28 (25) (2016) 5025–5030.

[96] H. Gleiter, Nanostructured materials: state of the art and perspectives, Nanostruct. Mater. 6 (1–4) (1995) 3–14.

[97] S. Chandrasekaran, L. Yao, L. Deng, C. Bowen, Y. Zhang, S. Chen, P. Zhang, Recent advances in metal sulfides: from controlled fabrication to electrocatalytic, photocatalytic and photoelectrochemical water splitting and beyond, Chem. Soc. Rev. 48 (15) (2019) 4178–4280.

[98] T.A. Loh, D.H. Chua, Origin of hybrid 1T-and 2H-WS2 ultrathin layers by pulsed laser deposition, J. Phys. Chem. C 119 (49) (2015) 27496–27504.

[99] C. Liang, Y. Shimizu, T. Sasaki, H. Umehara, N. Koshizaki, Au-mediated growth of wurtzite ZnS nanobelts, nanosheets, and nanorods via thermal evaporation, J. Phys. Chem. B 108 (28) (2004) 9728–9733.

[100] N.M. Hosny, A. Dahshan, Synthesis, structure and optical properties of SnS_2, CdS and HgS nanoparticles from thioacetate precursor, J. Mol. Struct. 1085 (2015) 78–83.

[101] Q. Liu, Y. Wang, L. Dai, J. Yao, Scalable fabrication of nanoporous carbon fiber films as bifunctional catalytic electrodes for flexible Zn-air batteries, Adv. Mater. 28 (15) (2016) 3000–3006.

[102] X.J. Chua, J. Luxa, A.Y.S. Eng, S.M. Tan, Z. Sofer, M. Pumera, Negative electrocatalytic effects of p-doping niobium and tantalum on MoS2 and WS2 for the hydrogen evolution reaction and oxygen reduction reaction, ACS Catal. 6 (9) (2016) 5724–5734.

[103] M. Kim, M. Kobayashi, H. Kato, M. Kakihana, Enhancement of luminescence properties of a $KSrPO_4$: Eu^{2+} phosphor prepared using a solution method with a water-soluble phosphate oligomer, J. Mater. Chem. C 1 (36) (2013) 5741–5746.

[104] Z. Ma, X. Yuan, Z. Zhang, D. Mei, L. Li, Z.F. Ma, J. Zhang, Novel flower-like nickel sulfide as an efficient electrocatalyst for non-aqueous lithium-air batteries, Sci. Rep. 5 (1) (2015) 1–9.

[105] Q. Hong, H. Lu, Y. Cao, Improved oxygen reduction activity and stability on N, S-enriched hierarchical carbon architectures with decorating core-shell iron group metal sulphides nanoparticles for Al-air batteries, Carbon 145 (2019) 53–60.

22 Oxide free nanomaterials for energy storage and conversion applications

[106] J. Jiang, C. Yan, X. Zhao, H. Luo, Z. Xue, T. Mu, A PEGylated deep eutectic solvent for controllable solvothermal synthesis of porous NiCo 2 S 4 for efficient oxygen evolution reaction, Green Chem. 19 (13) (2017) 3023–3031.

[107] J. Kibsgaard, Z. Chen, B.N. Reinecke, T.F. Jaramillo, Engineering the surface structure of MoS_2 to preferentially expose active edge sites for electrocatalysis, Nat. Mater. 11 (11) (2012) 963–969.

[108] D. Voiry, H. Yamaguchi, J. Li, R. Silva, D.C. Alves, T. Fujita, M. Chhowalla, Enhanced catalytic activity in strained chemically exfoliated WS_2 nanosheets for hydrogen evolution, Nat. Mater. 12 (9) (2013) 850–855.

[109] Z.F. Huang, J. Song, K. Li, M. Tahir, Y.T. Wang, L. Pan, J.J. Zou, Hollow cobalt-based bimetallic sulfide polyhedra for efficient all-pH-value electrochemical and photocatalytic hydrogen evolution, J. Am. Chem. Soc. 138 (4) (2016) 1359–1365.

[110] P.D. Tran, M. Nguyen, S.S. Pramana, A. Bhattacharjee, S.Y. Chiam, J. Fize, J. Barber, Copper molybdenum sulfide: a new efficient electrocatalyst for hydrogen production from water, Energ. Environ. Sci. 5 (10) (2012) 8912–8916.

[111] D.Y. Wang, M. Gong, H.L. Chou, C.J. Pan, H.A. Chen, Y. Wu, H. Dai, Highly active and stable hybrid catalyst of cobalt-doped FeS_2 nanosheets–carbon nanotubes for hydrogen evolution reaction, J. Am. Chem. Soc. 137 (4) (2015) 1587–1592.

[112] M. Basu, A.K. Sinha, M. Pradhan, S. Sarkar, Y. Negishi, T. Pal, Evolution of hierarchical hexagonal stacked plates of CuS from liquid − liquid interface and its photocatalytic application for oxidative degradation of different dyes under indoor lighting, Environ. Sci. Technol. 44 (16) (2010) 6313–6318.

[113] S.K. Bhar, S. Jana, A. Mondal, N. Mukherjee, Photocatalytic degradation of organic dye on porous iron sulfide film surface, J. Colloid Interface Sci. 393 (2013) 286–290.

[114] Y.R. Girish, R. Biswas, M. De, Mixed-phase 2D-MoS_2 as an effective photocatalyst for selective aerobic oxidative coupling of amines under visible-light irradiation, Chem. A Eur. J. 24 (52) (2018) 13871–13878.

[115] M. Xie, X. Dai, S. Meng, X. Fu, S. Chen, Selective oxidation of aromatic alcohols to corresponding aromatic aldehydes using In2S3 microsphere catalyst under visible light irradiation, Chem. Eng. J. 245 (2014) 107–116.

[116] T. Li, S. Zhang, S. Meng, X. Ye, X. Fu, S. Chen, Amino acid-assisted synthesis of in $2S_3$ hierarchical architectures for selective oxidation of aromatic alcohols to aromatic aldehydes, RSC Adv. 7 (11) (2017) 6457–6466.

[117] H. Hao, X. Lang, Metal sulfide photocatalysis: visible-light-induced organic transformations, ChemCatChem 11 (5) (2019) 1378–1393.

[118] M. Acerce, D. Voiry, M. Chhowalla, Metallic 1T phase MoS_2 nanosheets as supercapacitor electrode materials, Nat. Nanotechnol. 10 (4) (2015) 313–318.

[119] H. Tang, J. Wang, H. Yin, H. Zhao, D. Wang, Z. Tang, Growth of polypyrrole ultrathin films on MoS_2 monolayers as high-performance supercapacitor electrodes, Adv. Mater. 27 (6) (2015) 1117–1123.

[120] E.G. da Silveira Firmiano, A.C. Rabelo, C.J. Dalmaschio, A.N. Pinheiro, E.C. Pereira, W.H. Schreiner, E.R. Leite, Supercapacitor electrodes obtained by directly bonding 2D MoS_2 on reduced graphene oxide, Adv. Energy Mater. 4 (6) (2014) 1301380.

[121] J. Feng, X. Sun, C. Wu, L. Peng, C. Lin, S. Hu, Y. Xie, Metallic few-layered VS2 ultrathin nanosheets: high two-dimensional conductivity for in-plane supercapacitors, J. Am. Chem. Soc. 133 (44) (2011) 17832–17838.

[122] H. Huo, Y. Zhao, C. Xu, 3D Ni_3S_2 nanosheet arrays supported on Ni foam for high-performance supercapacitor and non-enzymatic glucose detection, J. Mater. Chem. A 2 (36) (2014) 15111–15117.

[123] B. Yang, L. Yu, Q. Liu, J. Liu, W. Yang, H. Zhang, J. Wang, The growth and assembly of the multidimensional hierarchical Ni_3S_2 for aqueous asymmetric supercapacitors, CrstEngComm 17 (24) (2015) 4495–4501.

[124] Y. Tang, T. Chen, S. Yu, Morphology controlled synthesis of monodispersed manganese sulfide nanocrystals and their primary application in supercapacitors with high performances, Chem. Commun. 51 (43) (2015) 9018–9021.

[125] H. Peng, G. Ma, K. Sun, J. Mu, H. Wang, Z. Lei, High-performance supercapacitor based on multi-structural CuS@ polypyrrole composites prepared by in situ oxidative polymerization, J. Mater. Chem. A 2 (10) (2014) 3303–3307.

[126] M.S. Whittingham, et al., Lithium batteries: 50 years of advances to address the next 20 years of climate issues, Nano Lett. 20 (2020) 8435–8437. https://pubs.acs.org/doi/10.1021/acs.nanolett.0c04347.

[127] J. Zhou, J. Qin, X. Zhang, C. Shi, E. Liu, J. Li, C. He, 2D space-confined synthesis of few-layer MoS_2 anchored on carbon nanosheet for lithium-ion battery anode, ACS Nano 9 (4) (2015) 3837–3848.

[128] S. Chenrayan, K.S. Chandra, S. Manickam, Ultrathin MoS_2 sheets supported on N-rich carbon nitride nanospheres with enhanced lithium storage properties, Appl. Surf. Sci. 410 (2017) 215–224.

[129] C. Senthil, S. Amutha, R. Gnanamuthu, K. Vediappan, C.W. Lee, Metallic 1T MoS_2 overlapped nitrogen-doped carbon superstructures for enhanced sodium-ion storage, Appl. Surf. Sci. 491 (2019) 180–186.

[130] V. Palomares, P. Serras, I. Villaluenga, K.B. Hueso, J. Carretero-González, T. Rojo, Na-ion batteries, recent advances and present challenges to become low cost energy storage systems, Energ. Environ. Sci. 5 (3) (2012) 5884–5901.

[131] J.A. Bragagnolo, A.M. Barnett, J.E. Phillips, R.B. Hall, A.L.L.E.N. Rothwarf, J.D. Meakin, The design and fabrication of thin-film CdS/Cu_2S cells of 9.15-percent conversion efficiency, IEEE Trans. Electron Devices 27 (4) (1980) 645–651.

[132] M.G. Panthani, V. Akhavan, B. Goodfellow, J.P. Schmidtke, L. Dunn, A. Dodabalapur, B.A. Korgel, Synthesis of $CuInS_2$, $CuInSe_2$, and Cu (In_xGa_{1-x}) Se_2 (CIGS) nanocrystal "inks" for printable photovoltaics, J. Am. Chem. Soc. 130 (49) (2008) 16770–16777.

[133] K.H. Huang, J.W. Yeh, A Study on the Multicomponent Alloy Systems Containing Equal-Mole Elements, National Tsing Hua University, Hsinchu, 1996, p. 1.

[134] Y.F. Ye, Q. Wang, J. Lu, C.T. Liu, Y. Yang, High-entropy alloy: challenges and prospects, Mater. Today 19 (6) (2016) 349–362.

[135] J.W. Yeh, S.K. Chen, S.J. Lin, J.Y. Gan, T.S. Chin, T.T. Shun, S.Y. Chang, Nanostructured high-entropy alloys with multiple principal elements: novel alloy design concepts and outcomes, Adv. Eng. Mater. 6 (5) (2004) 299–303.

[136] X. Zhao, Z. Xue, W. Chen, Y. Wang, T. Mu, Eutectic synthesis of high-entropy metal phosphides for electrocatalytic water splitting, ChemSusChem 13 (2020) 2038–2042, https://doi.org/10.1002/cssc.20200017310.1002/cssc.202000173.

[137] S. Nellaiappan, N.K. Katiyar, R. Kumar, A. Parui, K.D. Malviya, K.G. Pradeep, K. Biswas, High-entropy alloys as catalysts for the CO_2 and CO reduction reactions: experimental realization, ACS Catal. 10 (6) (2020) 3658–3663.

[138] H. Xu, Z. Zhang, J. Liu, C.L. Do-Thanh, H. Chen, S. Xu, S. Dai, Entropy-stabilized single-atom Pd catalysts via high-entropy fluorite oxide supports, Nat. Commun. 11 (1) (2020) 1–9.

[139] H.J. Qiu, G. Fang, J. Gao, Y. Wen, J. Lv, H. Li, S. Sun, Noble metal-free nanoporous high-entropy alloys as highly efficient electrocatalysts for oxygen evolution reaction, ACS Mater. Lett. 1 (5) (2019) 526–533.

24 Oxide free nanomaterials for energy storage and conversion applications

[140] D. Wu, K. Kusada, T. Yamamoto, T. Toriyama, S. Matsumura, I. Gueye, H. Kitagawa, On the electronic structure and hydrogen evolution reaction activity of platinum group metal-based high-entropy-alloy nanoparticles, Chem. Sci. 11 (47) (2020) 12731–12736.

[141] S. Gao, S. Hao, Z. Huang, Y. Yuan, S. Han, L. Lei, J. Lu, Synthesis of high-entropy alloy nanoparticles on supports by the fast moving bed pyrolysis, Nat. Commun. 11 (1) (2020) 1–11.

[142] G. Zhang, K. Ming, J. Kang, Q. Huang, Z. Zhang, X. Zheng, X. Bi, High entropy alloy as a highly active and stable electrocatalyst for hydrogen evolution reaction, Electrochim. Acta 279 (2018) 19–23.

[143] Y. Yao, Z. Huang, T. Li, H. Wang, Y. Liu, H.S. Stein, L. Hu, High-throughput, combinatorial synthesis of multimetallic nanoclusters, Proc. Natl. Acad. Sci. 117 (12) (2020) 6316–6322.

[144] X. Chen, C. Si, Y. Gao, J. Frenzel, J. Sun, G. Eggeler, Z. Zhang, Multi-component nanoporous platinum–ruthenium–copper–osmium–iridium alloy with enhanced electrocatalytic activity towards methanol oxidation and oxygen reduction, J. Power Sources 273 (2015) 324–332.

[145] T. Wang, H. Chen, Z. Yang, J. Liang, S. Dai, High-entropy perovskite fluorides: a new platform for oxygen evolution catalysis, J. Am. Chem. Soc. 142 (10) (2020) 4550–4554.

Chapter 2

Novel synthesis methods of nanostructured oxide-free materials for energy storage and conversion applications

Arun Prasad Murthy

Department of Chemistry, School of Advanced Sciences, Vellore Institute of Technology, Vellore, Tamil Nadu, India

Chapter outline

1. Introduction		**26**
1.1 Template synthesis method		26
1.2 A common strategy to synthesize CoP and CoX$_2$ (X=S, Se, and Te) nanoframes		27
1.3 Charge transfer engineering through multiheteroatom doping		29
1.4 Novel single-step synthesis of 1D nanoarchitecture		30
1.5 Synthesis of different electrocatalysts through annealing gas regulation		31
1.6 Multicomponent hybrid materials as multifunctional electrodes for energy applications		32
1.7 Strategy of direct growth of metal-organic frameworks on conducting substrates		34
1.8 Synthetic strategy of excluding toxic NH$_3$ gas in the nitridation of transition metals		34
1.9 Facile synthesis of multimetallic aerogels using a strong salting-out agent		35
1.10 General synthesis of 1D multicomponent hybrid heterostructures through modulated electronic state		36
1.11 Comparison between hydrothermal and electrochemical methods of preparing catalysts		37
1.12 Role of ultra-sonication in the synthesis of noble metal electrocatalysts		38
1.13 Strategy to achieve both high dispersibility and conductivity of polyaniline in aqueous media		39
1.14 Aerogel preparation through anion exchange method		40
1.15 Synthesis of carbon-supported Pt nanoparticle through a		

Oxide Free Nanomaterials for Energy Storage and Conversion Applications
https://doi.org/10.1016/B978-0-12-823936-0.00017-6
Copyright © 2022 Elsevier Inc. All rights reserved.

26 Oxide free nanomaterials for energy storage and conversion applications

modified borohydride method in a mixed solvent 41
1.16 Synthetic strategy to prepare a new class of advanced nanoalloy aerogel 41
1.17 Synthesis of multimetallic anode electrocatalyst for direct methanol fuel cell in high quantity 43
1.18 Exfoliant-assisted liquid phase exfoliation 44
1.19 Significance of carbon supports in the syntheses of catalysts for energy applications 45
2. Conclusions 46
References 47

1. Introduction

Several hundred million barrels of crude oil are being used daily and the depletion of fossil fuels is very imminent. The environmental effects of the combustion of fossil fuels are drastic and destructive. These dual crises of energy source depletion and environmental deterioration can be overcome by embracing renewable and sustainable energy sources such as solar energy, wind energy, biomass energy, etc. As the world is moving towards green technologies, devices, and fuels, the syntheses of various catalytic and electrocatalytic materials needed to bring about the revolution in renewable and sustainable technologies in the fields of energy storage and conversion become the need of the hour. In the past years, many innovative synthetic methods and strategies have been reported in the literature for several-energy storages and conversion applications. Because of these novel methods and strategies catalysts and electrodes of low cost, high durability, superior activity, and nontoxicity have been fabricated.

In this chapter, various recent and novel synthetic strategies designed for the preparation of oxide-free nanomaterials for many energy-related applications such as hydrogen evolution reaction, oxygen evolution reaction, oxygen reduction reaction, methanol oxidation reaction, hydrogen oxidation reaction, fuel cells, acidic, and alkaline water electrolyzers, supercapacitor, etc. have been discussed. These synthetic methods include template synthesis method, direct growth method, anion exchange method in the preparation of aerogels, exfoliant-assisted liquid-phase exfoliation method, etc. Similarly, various synthetic strategies such as multiheteroatom doping, regulation of annealing gas, exclusion of toxic gas in the synthetic process, etc. have also been discussed. As the result of these methods and strategies multicomponent materials, multifunctional catalysts, advanced nanoalloy aerogels, multimetallic anode electrocatalysts for fuel cells, etc. have been synthesized and employed in various energy storage and conversion applications.

1.1 Template synthesis method

Transition metal phosphides have attracted wide research interest because of their promising catalytic activities in energy storage and conversion applications.

In the catalytic process, metal centers (M) and phosphorous (P) both can play the active sites. Furthermore, strong M—P bonds ensure material physical strength, thermal stability along resistance to strong chemicals. Synthetic strategies for the preparation of transition metal phosphides include the topotactic conversion method, temperature-controlled synthesis method, etc. resulting in a variety of morphologies such as nanowires, nanoneedles, nanoflowers, etc. In this context, for the synthesis of one-dimensional phosphides template as well as nontemplate methods are generally employed. Phosphides prepared from nontemplate methods suffer from breakdown during catalytic operations and the poor charge transfer between the particles leading to reduced catalytic activity. In the template synthesis method metal as well as carbon-based support materials are employed for the fabrication of advanced electrocatalysts. These support materials include Ni foam, TiO_2, carbon nanotubes, graphene, nanofibers, etc. There are two significant disadvantages in the template synthesis method; (a) synthetic procedures are expensive as well as expansive involving costly methods and toxic chemicals and (b) metal-based supports increase the weight and lower the specific energy density, moreover, large-scale synthesis of carbon nanotubes and graphene-based materials are tedious. Hao et al. [1] employed micro carbon spheres (CS) as lightweight support material for the synthesis of nickel-cobalt phosphide nanoneedle arrays. Micro carbon spheres exhibit several benefits such as high mechanical strength, porous structure, high dispersion of catalyst nanoparticles, large-scale quick synthesis process, precursors are low-cost and can be obtained from bio-sources, etc.

Fig. 1A illustrates schematically the processes involved in the synthesis of Ni-Co-P@CS. Carbon spheres were obtained from the glucose source through hydrothermal method at 190°C for 7h followed by free drying and Fig. 1B shows carbon spheres of uniform spherical shape. Ni-Co-OH@CS were then prepared by hydrothermal method using $Ni(NO_3)_2 \cdot 6H_2O$, $Co(NO_3)_2 \cdot 6H_2O$, and urea at 120°C for 6h followed by freeze-drying. Fig. 1C and D show high-density vertically aligned nanoneedles arrays on the surface of carbon spheres. Phosphidization of Ni-Co-OH@CS was conducted in a quartz tube using $NaH_2PO_2 \cdot H_2O$ at 300°C under Ar atmosphere for 3h. The morphology of Ni-Co-P@CS is similar to that of Ni-Co-OH@CS as shown in Fig. 1E–G.

1.2 A common strategy to synthesize CoP and CoX_2 (X = S, Se, and Te) nanoframes

Ji et al. [2] demonstrated the design and fabrication of frame-like hollow nanostructures of CoP and CoX_2, (X = S, Se, and Te) using a generalized strategy. The advantages of hollow nanostructures are high atom utilization, large surface area, high surface area to volume ratio, etc. Besides, removing unutilized atoms result in increased mass efficiency as well as reduced catalyst mass. Synthetic reports on nano-frames have been mostly confined to noble metals-based catalysts in the literature. Prussian blue analogs (PBA) is used as precursors in the synthesis of nanomaterials for energy-related applications

28 Oxide free nanomaterials for energy storage and conversion applications

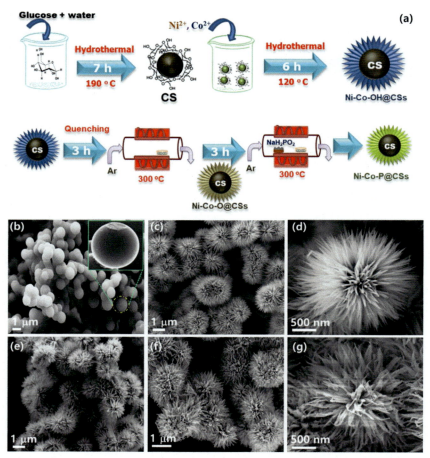

FIG. 1 (A) Illustration showing the synthesis of Ni-Co-P@CS, (B) FE-SEM image of carbon spheres with high magnification image at the inset, (C and D) FE-SEM images of Ni-Co-OH@CS at various magnifications and (E–G) FE-SEM images of Ni-Co-P@CS at various magnifications. *(Reproduced from V.H. Hoa, D.T. Tran, H.T. Le, N.H. Kim, J.H. Lee, Hierarchically porous nickel–cobalt phosphide nanoneedle arrays loaded micro-carbon spheres as an advanced electrocatalyst for overall water splitting application, Appl. Catal. B, 253 (2019) 235–245, https://doi.org/10.1016/j.apcatb.2019.04.017, copyright 2019, Science Direct.)*

such as supercapacitors, lithium-ion batteries, etc. For example, by annealing PBA precursors transition metal oxides are easily obtained. The authors developed a facile strategy involving precipitation, chemical etching, and low-temperature phosphidation to prepare CoP nanoframes. In the initial step, Co-Co PBA nanocubes were prepared by precipitation. Briefly, a solution of cobalt acetate and sodium citrate was mixed with a solution of potassium hexacyanocobaltate and the reaction was continued for 8h at room temperature. Then chemical etching was performed in ammonia solution for 10min at room temperature to obtain Co-Co PBA nanoframes. During low-temperature phosphidation at 300°C for 2 h under argon atmosphere using NaH_2PO_2 in a tube furnace, transition metal ions, as well as cyano ligands, were transformed into transition metal phosphide nanomaterials. The authors employed the same synthetic strategy to prepare cobalt dichalcogenides nanoframes (CoX_2 where $X = S$, Se, and Te). The authors used simple sulfuration, telluridation, selenylation, etc., methods in place of final phosphidation of Co-Co PBA nanoframe precursor. In the case of sulfuration, sulfur powder was used while Se and Te powders were used for selenylation and telluridation respectively in the tube furnace. The nanoframe morphology of the Co-Co PBA precursor was maintained in the final products after sulfuration, telluridation, selenylation, etc. Employing the above synthetic strategy, a series of CoP, CoS_2, $CoSe_2$, and $CoTe_2$ nanoframes were prepared and their relative catalytic activities were compared by the authors.

1.3 Charge transfer engineering through multiheteroatom doping

A common strategy to prepare hybrid catalyst materials is either via embedding or decorating active material nanoparticles on carbon-based supports such as carbon nanotubes, carbon spheres, doped graphenes, etc. In this context, co-doping heteroatoms may increase catalytic activity through maximizing active sites and synergistic interactions. However, engineering multihetero atoms within a catalyst hybrid is a challenging task. Xu et al. [3] fabricated dual carbon coupled cobalt phosphide with triple heteroatom (B, N, and S) doping (B,N, S-CoP@C@rGO) as water splitting electrocatalyst. The synthesis method for B, N, and S co-doped CoP@C@rGO involved three steps. In the initial step, ZIP-67@GO was prepared through co-precipitation in which methanol solution of cobalt nitrate was added to graphene oxide solution in which cobalt ions were anchored on graphene oxide. Subsequently, methanol solution of 2-methylimidazole was added and the suspension was maintained for 1 day under ambient conditions resulting in nucleation and formation of cobalt coordinated zeolite imidazole framework. The suspension was then centrifuged, washed and vacuum freeze-dried to obtain ZIP-67@GO. In the second step, B,N-CoP@C@rGO was obtained through one-pot thermal annealing resulting in the formation of CoP nanoparticles and the reduction of graphene oxide to reduced graphene oxide (rGO). Typically, ZIP-67@GO obtained in the first step was mixed with NaH_2PO_2 and $[N(CH_3)_4]B_3H_8$ and heated in a tube furnace at 800°C for 2 h in the presence of Ar atmosphere. Finally, B,N, S-CoP@C@rGO was obtained by heating B,N-CoP@C@rGO, and thiourea in a tube furnace at 400°C for 1 h under Ar atmosphere. Fig. 2 shows the schematic illustration of the fabrication of B,N,S-CoP@C@rGO via zeolitic imidazolate framework on graphene oxide as well as increased charge transfer with an increased number of heteroatoms. Both experimental and density functional theory investigations revealed electronic interactions among B, N, and S

30 Oxide free nanomaterials for energy storage and conversion applications

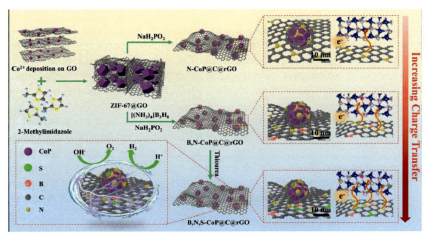

FIG. 2 Schematic illustration showing fabrication of N-CoP@C@rGO, B,N-CoP@C@rGO and B,N,S-CoP@C@rGO as well as increased charge transfer with an increased number of heteroatoms. *(Reproduced from H. Xu, H. Jia, B. Fei, Y. Ha, H. Li, Y. Guo, M. Liu, R. Wu, Charge transfer engineering via multiple heteroatom doping in dual carbon-coupled cobalt phosphides for highly efficient overall water splitting, Appl. Catal. B, 268 (2020) 118404, https://doi.org/10.1016/j.apcatb. 2019.118404, copyright 2020, Science Direct.)*

heteroatoms resulting in enhanced interfacial charge transfer and optimized energy barriers of the key intermediates. The constructed 3D architecture also helped to prevent the aggregation of cobalt phosphide nanoparticles.

1.4 Novel single-step synthesis of 1D nanoarchitecture

Materials with one-dimensional nanoarchitecture include nanorods, nanotubes, nanowires, etc. and they possess excellent physicochemical properties suitable for energy storage and conversion applications. Compared to powder-based catalyst materials, 1D nanoarchitecture offers several advantages including high aspect ratio, large surface area, abundant active sites, fast charge and mass transport, nonrequirement of binders, etc. From the synthetic point of view, 1D nanostructures can be fabricated in a facile way on substrates through chemical vapor deposition or electroplating methods. Other synthetic methods generally reported for the preparation of 1D nanostructures include a hydrothermal method, anodic aluminum oxide membrane templated electrodeposition, liquid-phase reduction, etc. Huang et al. [4] fabricated 1D nanotube arrays of NiFe metallic alloy in a single step using the bubble-releasing assisted pulse electrodeposition method. This method is simple, capable of producing open-end tubes, and highly efficient compared to two-step traditional methods for preparing nanotube arrays involving sacrificial templates. The authors used anodic aluminum oxide porous membrane with pore sizes between 200 and 300 nm as a template for the electrodeposition. To provide conductivity and

Novel synthesis methods **Chapter | 2 31**

mechanical strength 300 nm and 20 μm, thick Au and Ni thin layers respectively were deposited. In the next step, NiFe was deposited in the pore space of anodic aluminum oxide membrane in a two-electrode cell assembly in which anodic aluminum oxide membrane acted as the working electrode. $NiSO_4 \cdot 5H_2O$ and $FeSO_4 \cdot 7H_2O$ were used as Ni and Fe precursors respectively in the electrolyte. Several molar ratios of Ni/Fe were investigated to study the effect of the electrolyte composition. The pH of the electrolyte was maintained at 2 using H_2SO_4 and the temperature of the electrodeposition cell was controlled at 50°C. Pulsed current densities used for the electrodeposition ranged between -0.25 and $-2\,A\,cm^{-2}$ with an on/off cycle of 0.02/1 s respectively. The authors used shorter deposition times at higher current densities to maintain a constant deposition charge. The prepared NiFe nanotube arrays were catalytically highly active and stable and also proved to be a promising novel catalyst architectural design for energy applications.

1.5 Synthesis of different electrocatalysts through annealing gas regulation

A synthesis method wherein two electrocatalysts were prepared for two different electrochemical reactions (hydrogen and oxygen evolution reactions) using the same procedure with the same precursors. Electrocatalysts for hydrogen and oxygen evolution reactions are generally prepared by completely different methods involving different precursors. This increases the cost of making water electrolyzers especially for large-scale production of hydrogen. Bimetallic alloys are active for hydrogen evolution reaction whereas bimetallic oxides are efficient for oxygen evolution reaction. Zhang et al. [5] demonstrated the synthesis of hydrogen and oxygen evolution electrocatalysts through annealing gas regulation. 3D reduced graphene oxide architecture was formed via the hydrothermal method in the presence of thiourea and metal ions. Further pyrolysis with gas regulation led to $CoFe_2O_4$ encapsulated nitrogen, sulfur-doped reduced graphene oxide aerogel (CoFeO@N/S-rGO) and amorphous CoFe encapsulated nitrogen-doped reduced graphene oxide aerogel (CoFe@N-rGO). The authors prepared graphene oxide from graphite using a modified Hummers method. Through ultrasonication graphene oxide dispersion, thiourea, $Co(NO_3)_2 \cdot 6H_2O$ and $Fe(NO_3)_3 \cdot 9H_2O$ were mixed. In the hydrothermal process, the mixture was heated at 120°C in an autoclave for 2 h to form a hydrogel. The products were then freeze-dried to obtain aerogels and subsequently treated in a tube furnace at the temperature of 800°C for 1 h using different gas atmospheres. When nitrogen atmosphere was used the obtained product was CoFeO@N/S-rGO whereas in the presence of NH_3 CoFe@N-rGO was obtained. The authors demonstrated through the regulation of N_2 or NH_3 gas in the pyrolysis desirable products could be obtained. Fig. 3 shows a schematic illustration of the synthetic method for the preparation of CoFe@N-rGO and CoFeO@N/S-rGO through the regulation of gasses. Simplification of synthesis

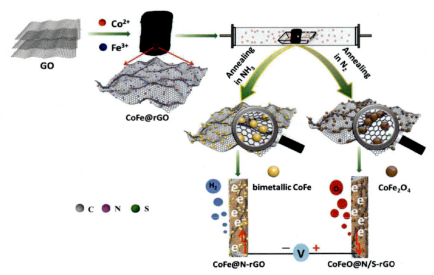

FIG. 3 Schematic illustration of the synthetic method for the preparation of CoFe@N-rGO and CoFeO@N/S-rGO through the regulation of gasses. *(Reproduced from B. Zhang, H. Wang, Z. Zuo, H. Wang, J. Zhang, Tunable CoFe-based active sites on 3D heteroatom doped graphene aerogel electrocatalysts via annealing gas regulation for efficient water splitting, J. Mater. Chem. A, 6 (2018) 15728–15737, https://doi.org/10.1039/C8TA05705B, copyright 2018, the Royal Society of Chemistry.)*

method using same procedure and precursors for different catalytic reactions would be a promising strategy to minimize manufacturing cost of large-scale productions.

1.6 Multicomponent hybrid materials as multifunctional electrodes for energy applications

Novel classes of electrodes with high conductivity, large surface area, high porosity, etc. have been reported in recent years for energy storage and conversion applications. Especially, multicomponent high-performance active materials are fabricated as multifunctional electrocatalysts. Furthermore, a recent school of thought consists in hybridizing active materials with heteroatom doped graphene. Xu et al. [6] fabricated nanocomposite architecture of an assembly of 2D WSe$_2$ and NeFe-layered double hydroxide (LDH) nanosheets on 3D nitrogen, sulfur codoped graphene framework. Compared to 2D graphene, 3D graphene is beneficial in terms of porosity, conductivity, etc. The authors successfully employed the ternary hybrid material as a multifunctional electrode for hydrogen and oxygen evolution reactions and supercapacitors. Integration of different components such as 2D nanosheets and layered double hydroxides with graphene as well as maximizing the synergistic effects is a

Novel synthesis methods **Chapter | 2** **33**

challenge. The authors used modified Hummers' method to prepare graphene oxide from graphite powder. Nanosheets of WSe$_2$ were prepared from bulk WSe$_2$ through the liquid-phase exfoliation method. This exfoliation method involved sonication followed by centrifugation. In a typical preparation of NiFe-LDH, Ni and Fe precursors were mixed with urea and trisodium citrate in distilled water and sonicated. The solution was then transferred to Teflon-lined stainless steel, sealed, and heated to 150°C for 20 h. After cooling to room temperature, the resultant product was washed and dried overnight under vacuum at 60°C. The liquid phase exfoliation method was used to obtain NiFe-LDH nanosheets through sonication in degassed formamide followed by centrifugation. The hydrothermal method followed by freeze-drying were used to prepare nitrogen, sulfur codoped graphene, and WSe$_2$ based aerogel (N,S-rGO/WSe$_2$). During this process, graphene oxide was doped with nitrogen and sulfur in the presence of L-cysteine and also was reduced to graphene. In the final step, the obtained aerogel was immersed into NiFe LDH suspension for 24 h yielding 3D highly coupled N,S-rGO/WSe$_2$/NiFe-LDH ternary hybrid material through electrostatic self-assembly. Fig. 4 shows various processes involved in the synthesis of N,S-rGO/WSe$_2$/NiFe-LDH aerogel and its application in supercapacitor and hydrogen and oxygen evolution reactions.

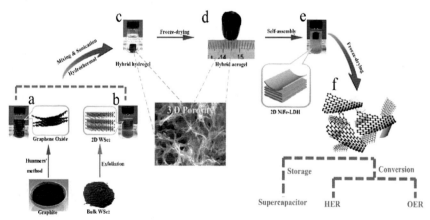

FIG. 4 Various processes involved in the synthesis of N,S-rGO/WSe$_2$/NiFe-LDH aerogel. (A) graphene oxide and graphene oxide dispersion in water, (B) WSe$_2$ dispersion, (C) hybrid hydrogel, (D) corresponding aerogel, (E) formation of N,S-rGO/WSe$_2$/NiFe-LDH, and (F) N,S-rGO/WSe$_2$/NiFe-LDH aerogel and its application in supercapacitor and hydrogen and oxygen evolution reactions. *(Reproduced from X. Xu, H. Chu, Z. Zhang, P. Dong, R. Baines, P.M. Ajayan, J. Shen, M. Ye, Integrated energy aerogel of NS-rGO/WSe$_2$/NiFe-LDH for both energy conversion and storage, ACS Appl. Mater. Interfaces, 9 (2017) 32756–32766, https://doi.org/10.1021/acsami.7b09866, copyright 2017, the American Chemical Society.)*

1.7 Strategy of direct growth of metal-organic frameworks on conducting substrates

Electrocatalysts in powder form suffer from aggregation and require low-conducting polymer binder as an additive. In recent years, the strategy of direct growth of active materials on the conducting substrates has been adopted. For example, direct growth of metal-organic frameworks (MOF) on highly porous and conducting supports is reported for preparing efficient electrocatalysts for electrochemical energy devices. MOFs are employed as pyrolysis precursors to fabricate heteroatom doped carbon nanomaterials. It may be noted that these MOF-based catalysts are in powder form and are coated on conducting substrates like nickel foam, glassy carbon electrodes, carbon paper, etc. This strategy also requires nonconducting polymer binders as an additive during the coating process. Polymer binders being poor conductors reduce the conductivity of the electrode and also mask the active sites of the catalyst. The strategy of direct growth of MOFs on graphene, stainless steel may address the issue, however, sufficient loading of the catalyst cannot be achieved on 2D conducting substrates. Hence, 3D conducting materials can be the suitable substrates for the direct growth of MOFs as demonstrated by Ming et al. [7] Recently, Yuan et al. [8] constructed MOF-derived 3D nitrogen-doped carbon nanotube frameworks on nickel foam. Doping of nanocarbons with nitrogen, boron, etc. heteroatoms can modulate the electronic structure resulting in high catalytic activity. The authors carbonized cobalt-based MOF grown on nickel foam in the presence of dicyandiamide as a nitrogen source and the overall synthesis process involved two steps. In the initial step, cobalt-MOF/nickel foam was fabricated via the solvothermal method. A clear solution of cobalt precursor, terephthalic acid, and NaOH in N,N-dimethyllformamide was transferred to Teflon-lined stainless steel autoclave and heated to 115°C for 24h in the presence of as-cleaned nickel foam. In the second step, the resulting maroon Co-MOF/nickel foam along with dicyandiamide was placed in a quartz boat and heated in a tube furnace in a vacuum at 600°C for 3h to obtain Co embedded N-doped carbon nanotube frameworks supported on 3D nickel foam.

1.8 Synthetic strategy of excluding toxic NH_3 gas in the nitridation of transition metals

Metal nitrides, especially transition metal nitrides, are widely employed as catalyst materials for various energy-related applications on account of high corrosion resistance and electrical conductivity. Moreover, interstitial nitrogens of metal nitrides modulate the electronic structure suitable for catalytic applications. Ray et al. [9] developed integrated 3D nickel-cobalt nitrides on conductive carbon cloth support (NCN/CC). The authors used a simple strategy of controlled pyrolyzation of Ni and Co precursors on polyaniline deposited carbon cloth to fabricate grass-like cobalt and nickel nitrides (CoN and

Novel synthesis methods **Chapter | 2 35**

Ni_3N) in addition to N-doped carbon. The material was designed as a self-supporting, robust, binder-free, and mechanically stable electrocatalyst. The beneficial feature of this synthetic strategy was that the use of toxic NH_3 was excluded and the carbonization of polyaniline created the NH_3 environment necessary for the nitridation of cobalt and nickel precursors. The fabrication of NCN/CC involved three steps. In the first step, aniline was polymerized in the presence of ammonium persulfate and pretreated carbon cloth. The polymerization was carried out in 1 M H_2SO_4 for 6 h at room temperature. The obtained polyaniline/carbon cloth was then deposited with metal hydroxides via the hydrothermal method. Teflon-lined stainless steel autoclave containing polyaniline coated carbon cloth, metal precursors, etc. was heated at 120°C for 12 h to obtain NiCo-Polyaniline/carbon cloth. In the final step, the above hybrid was subjected to 700°C under argon atmosphere for 2 h to obtain 3D nickel-cobalt nitrides on conductive carbon cloth support. This synthetic strategy delivered the dual advantage of polyaniline-induced synthesis of cobalt and nickel nitrides avoiding the use of common nitridation reagent toxic NH_3 gas and also the formation of N-doped carbon between catalyst and carbon support.

1.9 Facile synthesis of multimetallic aerogels using a strong salting-out agent

Aerogels are a class of porous materials characterized by ultralow density and large specific surface area. These self-supported architectures can be fabricated from a variety of building blocks like organic-inorganic hybrids, metal oxides, carbons, metal chalcogenides, polymers, ceramics, etc. Aerogels have been employed in the research fields of catalysis, energy storage, and conversion, adsorption, etc. Interconnected pores of the aerogels afford open channels for facile mass transport. 3D architectures of aerogels are suitable especially for the reduction/oxidation of small molecules like hydrogen, methanol, oxygen, ethanol, etc. In this context, noble metal aerogels can be described as relatively recent members of the aerogel family. Noble metal aerogels possess all the desirable properties of aerogels in addition to self-supported architectures, excellent catalytic activity, and electrical conductivity [10, 11]. These aerogels may thus fill the gap between nanomaterials and macro-materials. Some metal aerogels have been reported to exhibit characteristics that are superior to those of respective metal nanoparticles [12]. Noble metal aerogels have been shown to display higher performance in terms of activity and stability compared to commercial noble metal catalysts like Pd/C and Pt/C [13]. Manipulation of parameters such as ligaments sizes, spatial element distributions, etc. requires a complete understanding of structure performance correlations. Du et al. [13] synthesized Au-Pt and Au-Rh aerogels and employed them as electrocatalysts for hydrogen evolution and oxygen reduction reactions. These aerogels achieved remarkable pH-universal performance exceeding the activity of

36 Oxide free nanomaterials for energy storage and conversion applications

commercial Pt/C itself. In general, the synthetic process in these cases involves tedious procedures and also long fabrication durations. The authors, however, developed a salting-out gelation process in which the interactions between salt ions and metal nanoparticles were regulated based on specific ion effects resulting in the efficient preparation of tailored structures of noble metal aerogels. Gelation was started by NH_4F, a strong salting-out agent. For example, in the synthesis of Au-Pt aerogel, trisodium citrate dehydrates, and the solutions of $HAuCl_4 \cdot 3H_2O$, K_2PtCl_4, and $NaBH_4$ were mixed initially. Subsequently, NH_4F was added to the above-formed nanoparticle solution and the obtained Au-Pt hydrogel was solvent exchanged with t-butanol. Finally, the wet gel was flash frozen and freeze-dried to achieve the multimetallic aerogel of Au-Pt.

1.10 General synthesis of 1D multicomponent hybrid heterostructures through modulated electronic state

The electronic structure of nonnoble metals can be modulated via the introduction of foreign metal atoms. Charge transfer can be boosted between different metal atoms resulting in high catalytically active sites. Similarly, carbon nanomaterials doped with heteroatoms such as N, S, and O also exhibit unique electronic structures and in turn reduce local work function. Liu et al. [14] reported a general method for the synthesis of one-dimensional hybrid heterostructures with multiple active components of Mo_2C and Co_4S_3 embedded in nitrogen and sulfur co-doped carbon. The electrocatalyst was porous hollow structures with large specific surface areas with tunable electronic structures. The synthesis process consisted of several steps. Initially, 1D MoO_3 in the form of nanorods was obtained using ammonium molybdate tetrahydrate and nitric acid via hydrothermal process. In the next step, poly-2-aminothiazole was formed on the surface of MoO_3 nanorods (P-2AT@MoO_3) through a facile polymerization in which 2-aminothiazole was used as the monomer. The role of poly-2-aminothiazole was multipurpose, viz it acted as nitrogen and sulfur-containing precursor as well as supplied sulfur and carbon for the formation of Mo_2C and Co_4S_3. Also, nitrogen and sulfur sites could adsorb easily Co^{2+} species on the surface of the polymer. Briefly, P-2AT@MoO_3 was dispersed in water and ethanol mixture into which $Co(CH_3COO)_2$ was added subsequently. After 24 h of vigorous stirring at room temperature, the mixture was frozen and freeze-dried to obtain Co^{2+}@P-AT@MoO_3. Finally, Co_4S_3/Mo_2C-NSC was obtained through controlled calcination of Co^{2+}@P-AT@MoO_3 under nitrogen atmosphere in which Co^{2+} and MoO_3 were transformed into Co_4S_3 and Mo_2C in the carbon. It can be noted in the pyrolysis process that poly-2-aminothiazole was finally converted into sulfur and nitrogen co-doped graphitic shells. The entire synthetic process of Co_4S_3/Mo_2C-NSC involving various steps is illustrated in Fig. 5. This general, facile and controllable synthesis method presents many advantages like large specific area, controllable composition, and

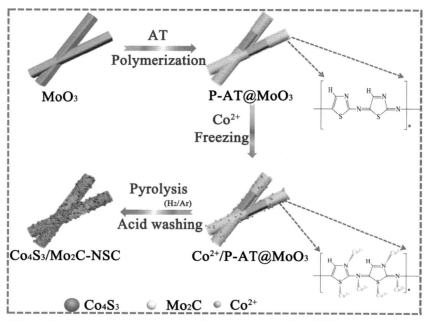

FIG. 5 Various steps involved in the synthetic process of Co$_4$S$_3$/Mo$_2$C-NSC. *(Reproduced from Y. Liu, X. Luo, C. Zhou, S. Du, D. Zhen, B. Chen, J. Li, Q. Wu, Y. Iru, D. Chen, A modulated electronic state strategy designed to integrate active HER and OER components as hybrid heterostructures for efficient overall water splitting, Appl. Catal. B, 260 (2020) 118197, https://doi.org/10.1016/j.apcatb.2019.118197, copyright 2020, Science Direct.)*

regulation of electron structures, etc. for the fabricating of advanced electrocatalysts for energy storage and conversion applications.

1.11 Comparison between hydrothermal and electrochemical methods of preparing catalysts

In many energy conversion and storage applications, efficient contact between reactants and catalysts is necessary for facile mass transport and charge transfer. A rational catalyst design exposing a large number of active sites and preventing aggregation of catalyst nanoparticles would pursue activities comparable to those of noble metal-based catalysts. In this context, synthesis methods play a very important role in determining the efficiency of a catalyst. Dang et al. [15] reported the synthesis of 3D holey cobalt phosphide ultrathin nanosheets on carbon cloth. The authors developed two synthetic routes to obtain cobalt phosphide nanostructures, viz, electrochemical and hydrothermal methods. Through electrochemical route holey and ultrathin nanosheets of cobalt phosphide on carbon cloth, CoP UNS/CC (EC), was fabricated while

38 Oxide free nanomaterials for energy storage and conversion applications

hydrothermal route afforded CoP NS/CC (HT). CoP UNS/CC (EC) was prepared via a two-step process involving potentiodynamic electrodeposition followed by low-temperature phosphidation. Initially, carbon cloth was cleaned through sonication in water, ethanol, and acetone sequentially and then immersed in $Co(NO_3)_2$ solution. The potentiodynamic deposition method was used to prepare $Co(OH)_2$/CC in a three-electrode cell assembly that consisted of carbon cloth working electrode, saturated calomel reference electrode, and graphite rod counter electrode. During electrodeposition, the potential was scanned between 0 and -1.2 V vs SCE at a scan rate of $100\,mV\,s^{-1}$ for 40 cycles. In a second step, $Co(OH)_2$/CC was converted to CoP UNS/CC (EC) using NaH_2PO_2 as a phosphidation agent at a temperature of 300°C for 2 h under argon atmosphere. In the case of CoP NS/CC (HT), the phosphidation step was identical to the case of CoP UNS/CC (EC) whereas the precursor $Co(OH)_2$/CC was obtained from the hydrothermal method. The authors observed CoP UNS/CC (EC) to be appealing compared to CoP NS/CC (HT) because of high surface area, super hydrophilicity, etc. Compared to electrochemical methods hydrothermal methods involve high energy input hence cannot be suitable for the synthesis of inexpensive catalysts on large scale. Furthermore, electrochemical methods are simple, efficient, and tunable nanostructures that can be prepared on large conductive substrates. The authors concluded that the electrochemical deposition method afforded better control over the nanostructure morphology which in turn enhanced the catalytic activity.

1.12 Role of ultra-sonication in the synthesis of noble metal electrocatalysts

Porous materials are widely employed in various energy-related applications in the past few years since these materials can have efficient contact between catalytic sites and electrolytes not only on the surface but throughout the bulk. In this context, metallic aerogels have received wide attention owing to their high porosity leading to the high specific surface area, ultra-low density along high conductivity. Kundu et al. [12] synthesized gold electrocatalyst on thin carbon nitride sheets through two synthetic routes. When the precursor $HAuCl_4$ was reduced with sodium borohydride along with ultra-sonication in the presence of carbon nitride nanosheets gold aerogel was obtained (Au-aerogel-CN_x). On the other hand, when $HAuCl_4$ was reduced only with ultrasonic treatment in the presence of carbon nitride sheets, gold nanoparticles of about 2 nm were produced (AuNPs-CN_x). Au-aerogel-CN_x was an interconnected high surface area porous network while AuNPs-CN_x was highly dispersed nanoparticles of ultra-small size on carbon nitride sheets. Graphitic carbon nitride is a 2D material similar to graphene and plays the role of efficient support for nanostructures. Graphitic carbon nitride was prepared by heating in a microwave synthesizer using formamide as a precursor. The prepared graphitic carbon

Novel synthesis methods **Chapter | 2** **39**

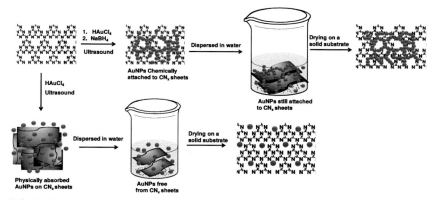

FIG. 6 Schematic description of the synthetic process of Au-aerogel-CN$_x$ and AuNPs-CN$_x$. *(Reproduced from M.K. Kundu, T. Bhowmik, S. Barman, Gold aerogel supported on graphitic carbon nitride: an efficient electrocatalyst for oxygen reduction reaction and hydrogen evolution reaction, J. Mater. Chem. A, 3 (2015) 23120–23135, https://doi.org/10.1039/C5TA06740E, copyright 2015, the Royal Society of Chemistry.)*

nitride was dispersed in water through sonication to which an aqueous solution of HAuCl$_4$ was added and finally, NaBH$_4$ was introduced into the mixture. The mixture was then subjected to 400-W ultra-sound treatment at 28 kHz for 3 h. The solid product obtained after the reaction was centrifuged at 8000 rpm, washed with water, and dried in a vacuum. In the case of the synthesis of AuNPs-CN$_x$, only ultrasonication was used without the addition of the reducing agent NaBH$_4$. Fig. 6 schematically describes the fabrication of Au-aerogel-CN$_x$ and AuNPs-CN$_x$. Based on the catalytic performances of the two gold based electrocatalysts the authors concluded that Au-aerogel-CN$_x$ was superior compared to AuNPs-CN$_x$ because of the porous network architecture of Au-aerogel-CN$_x$.

1.13 Strategy to achieve both high dispersibility and conductivity of polyaniline in aqueous media

Polyaniline (PANI) is one of the widely studied synthetic polymers and emeraldine acid forms of which has good electrical conductivity. During protonation with acid, the oxidation state of PANI does not change whereas in other conducting polymers doping leads to partial reduction or oxidation of π electrons system. When doped with small-molecule acids like HCl emeraldine PANI is highly conducting but becomes insoluble in water. The polymer's wide applications in catalysis and energy-related fields are limited because of the insolubility of emeraldine acid PANI doped with small-molecule acids. On the other hand, when doped with polymer acids like poly-(acrylic acid) and poly(styrene sulfonic acid) dispersibility of PANI can be enhanced but with concurrent loss

40 Oxide free nanomaterials for energy storage and conversion applications

in the conductivity. In this context, it was reported that PANI when doped with poly(2-acryl-amido-2-methyl-1-propanesulfonic acid) (PAAMPSA) both have high dispersibility as well as conductivity up to $1\,S\,cm^{-1}$ could be achieved [16]. This was not possible with traditional polymer acids like poly(styrene sulfonic acid) and was attributed to specific interactions involving amide groups of PAAMPSA facilitating hydrogen bonds. The aqueous dispersion of PANI-PAAMPSA is highly stable suitable for long-term storage. Murthy et al. [17] synthesized PANI-PAAMSA as well as Poly(*ortho*-toluidine)-PAAMPSA and used them as ionic-electronic conducting ionomers for the direct methanol fuel cells. Aniline and *ortho*-toluidine were used as monomers respectively for the preparation of PANI-PAAMSA and Poly(*ortho*-toluidine)-PAAMPSA. Initially, PAAMPSA was dissolved in water then the monomer was added. The resulting solution was stirred for 2 h at room temperature. Subsequently, ammonium peroxydisulfate was added to the above solution and oxidative polymerization was carried out in an ice bath for 6 h. The final product was washed with acetone and dried in a vacuum oven for 8 h. PANI-PAAMSA and Poly(ortho-toluidine)-PAAMPSA were completely dispersed in water by stirring continuously for 2 weeks.

1.14 Aerogel preparation through anion exchange method

Compared to metal oxides their sulfide counterparts possess rich redox chemistry, better selectivity of metal ions, and higher conductivity. In the anion exchange method of synthesis of metal sulfide nanostructures, the precursor oxides, hydroxides, or carbonates, etc., are mixed with sulfur sources such as thiourea, thioacetamide, etc. Anion exchange occurs below 200°C leading to metallic sulfides. Through this method, porosity may be introduced via Ostwald ripping and Kirkendall effect and also the morphology of the precursor can be maintained. Among several types of nanostructures, aerogels consist of the ultra-high surface area along with large pore volume suitable for energy-related applications. Synthetic methods of nanoparticle condensation and metathesis are followed for the preparation of metal sulfides. In this context, Gao et al. [18] introduced anion exchange for the first time into the sol-gel method and synthesized metal sulfide-based aerogel. The prepared Co_9S_8 aerogel possessed a large pore volume of $0.87\,cm^3\,g^{-1}$, the surface area of $274.2\,m^2\,g^{-1}$, and specific capacitance of $950\,F\,g^{-1}$ at $1\,A\,g^{-1}$. In this method, metal oxide wet gel can be prepared in the first step, and then sulfur anion can be exchanged during gel aging. The slow kinetics of anion exchange ensures the initial morphology of metal oxide to be maintained in the final product. Cobalt nitrate hexahydrate and citric acid were used as metal precursors and gel accelerators respectively. Both metal precursor and gel accelerator were mixed along with small amount of formamide to form a sol. The sol was maintained at 60°C in a sealed container for 4 h to form cobalt citrate wet gel. Then the anion exchange process was carried out by aging the wet gel in thioacetamide/ethanol/water solution at 80°C. The process was carried out for 3 days during which fresh thioacetamide/ethanol/water solution was exchanged several times. The wet gel was taken into an autoclave filled with ethanol and then purged with nitrogen. The sealed autoclave was heated to 260°C and after cooling to room temperature Co_9S_8 aerogel was obtained. Since metal oxide gel can be readily prepared, various compositions and microstructures of metal sulfide aerogels may be synthesized through the anion exchange method in a facile way.

1.15 Synthesis of carbon-supported Pt nanoparticle through a modified borohydride method in a mixed solvent

Like in many catalytic reactions the best catalyst for adsorption and dehydrogenation of methanol is Pt. For further oxidation of methanol, the second metal of Ru is incorporated in the catalyst to make PtRu so that adsorbed CO can be oxidized at lower potentials. However, long-term operation of PtRu results in dissolution of Ru from the anode and further contamination of cathode. This diminishes methanol oxidation activity and in turn overall performance of direct methanol fuel cells. Lee et al. [19] synthesized carbon-supported Pt nanoparticles through a modified borohydride method using water-ethylene glycol mixed solvents. Pt/C prepared in this method could oxidize the adsorbed CO systematically at lower potentials. For example, Pt nanoparticles prepared by a reduction in the composition of water-ethylene glycol mixture with ethylene glycol content of 67% could oxidize adsorbed CO at about 100 mV lower than the potential when adsorbed CO was reduced at Pt nanoparticles synthesized via a conventional reduction in water. In this reduction method, carbon black was thoroughly dispersed in a water-ethylene glycol mixture of various compositions. The precursor $H_2PtCl_6 \cdot 6H_2O$ was added to the carbon dispersion and the pH of the medium was adjusted to 2. Then excess sodium borohydride was added to the above mixture at room temperature. After completion of the reaction the product was filtered, washed, and dried in an oven. Fig. 7 shows TEM images of carbon-supported Pt nanoparticles prepared in various water-ethylene glycol mixtures. Pt/C prepared in a particular composition of 67% ethylene glycol could oxidize adsorbed CO at lower overpotential and the catalytic activity of Pt/C for methanol oxidation reaction was increased 2.5 times. The authors demonstrated that a simple variation in the composition of the reaction medium could enhance the morphology as well as the activities for adsorbed CO oxidation and overall methanol oxidation reaction.

1.16 Synthetic strategy to prepare a new class of advanced nanoalloy aerogel

Catalytic or electrocatalytic performance can be improved through effectively designing the nanostructures of catalytic materials. In this context, metallic aerogels provide ultra-large surface area, 3D architectures with high porosity,

42 Oxide free nanomaterials for energy storage and conversion applications

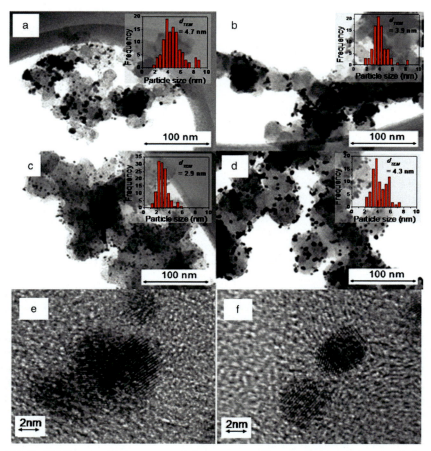

FIG. 7 TEM images of carbon-supported platinum prepared in various water-ethylene glycol mixtures with ethylene glycol percentages of (a) 0, (b) 25, (c) 67, (d) 10, (e) 0 (high resolution), and (f) 67 (high resolution). *(Reproduced from E. Lee, A. Murthy, A. Manthiram, Carbon-supported Pt nanoparticles prepared by a modified borohydride reduction method: effect on the particle morphology and catalytic activity for COad and methanol electro-oxidation, Electrochem. Commun. 13 (2011) 480–483, https://doi.org/10.1016/j.elecom.2011.02.026, copyright 2011, Science Direct.)*

low density along high conductivity. Jin et al. [20] reported platinum-rhodium advanced nanoalloy aerogel (PtRh NAA) through an easy and controllable method. PtRh NAA was a new class of aerogel with a unique lamellar nanostructure. The authors desired to prepare Pt-M atomically alloyed nanophase where M represents oxophilic metal such as Rh, Cu, Ru, etc. to weaken the hydrogen binding energy of Pt. Besides, the authors opined that quasi two-dimensional lamellar architectures would provide excellent mass transport, high atom utilization, and good contact between lamellar architecture and substrates when compared to three-dimensional regular aerogel architectures.

A synthetic strategy of the one-pot aqueous method was successfully designed at 60°C for the preparation of atomic-scale PtRh aerogel nanoalloy. A freshly prepared $NaBH_4$ aqueous solution was added to preheated ultrapure water at 60°C. Then Pt and Rh precursors (chloroplatinic acid hexahydrate and rhodium chloride) in an aqueous medium were immediately added to the above solution with strong stirring for 1 min and the reaction medium was then kept in a thermostatic water bath at 60°C without stirring. Finally, the product was formed as a thin layer at the bottom of the beaker. The carefully washed product was subjected to freeze-drying overnight to obtain PtRh NAA. The synthesized lamellar nanoalloy of PtRh aerogel possessed several advantages as an electrocatalyst like hierarchical pores, low-coordinated atoms, synergetic effects, ultralow loading of the electrocatalyst, etc.

1.17 Synthesis of multimetallic anode electrocatalyst for direct methanol fuel cell in high quantity

Direct methanol fuel cell is a well-known low-temperature fuel cell with several advantages. Methanol is a less expensive fuel with high theoretical energy density and the infrastructure for transporting and handling liquid fuel is presently available. Methanol oxidation reaction (MOR) taking place at the anode of the fuel cell is complicated producing carbon monoxide poisoning the electrode and other intermediates such as formic acid and formaldehyde. Platinum-based catalysts are investigated for MOR, especially, Pt-Ru with Ru acting as a secondary metal aiding oxidation of methanol to carbon dioxide. However, Pt-Ru suffers from Ru dissolution during long period operations. Other secondary metals that have high operational stability and negligible dissolution include Sn, Ce, etc. Hence it is prudent to incorporate other secondary metals in the synthesis of electrocatalyst for MOR with enhanced activity and stability. The research group of Manthiram synthesized a series of carbon-supported electrocatalysts, viz, Pt/C, Pt-Ru/C, Pt-Sn/C, Pt-Sn-Ce/C, and Pt-Ru-Sn-Ce/C and employed them for methanol oxidation reaction [19, 21–23]. The authors assembled a direct methanol fuel cell stack with 20 cells using novel multimetallic carbon-supported anode electrocatalyst (Pt-Ru-Sn-Ce/C) achieving a maximum power of 20 W. Fig. 8 shows photographs of membrane electrode assembly, hardware, and direct methanol fuel cell stack. Pt-Ru-Sn-Ce/C was synthesized in large quantity for the above fuel cell stack and about 400 mg of the electrocatalyst was prepared in each lot using a 2 L three-neck round bottom reaction flask. In the synthesis of Pt-Ru-Sn-Ce/C, $H_2PtCl_6 \cdot 6H_2O$, $RuCl_3 \cdot xH_2O$, $SnCl_2 \cdot 2H_2O$, and $Ce(NO_3)_3 \cdot 6H_2O$ were used as metal precursors. Initially, the metal precursors were dissolved in ethylene glycol which acted as both solvent and reducing agent. Aqueous NaOH was added to the above solution and refluxed at 178°C under an open atmosphere for 3 h. The formed colloid of metal nanoparticles was added to carbon dispersion in ethylene glycol and the pH of the mixture was adjusted to 4.5 by adding H_2SO_4.

44 Oxide free nanomaterials for energy storage and conversion applications

FIG. 8 Photographs of the membrane electrode assembly, hardware, and direct methanol fuel cell stack. *(Reproduced from X. Zhao, W. Li, A. Murthy, Z. Jiang, Z. Zuo, A. Manthiram, A DMFC stack operating with hydrocarbon blend membranes and Pt–Ru–Sn–Ce/C and Pd–Co/C electrocatalysts, Int. J. Hydrogen Energy, 38 (2013) 7448–7457, https://doi.org/10.1016/j.ijhydene.2013.04.013, copyright 2013, Science Direct.)*

Metal nanoparticles were allowed to deposit onto the carbon by vigorously stirring the dispersion for 8 h. Finally, the obtained product was washed with water and dried in a vacuum oven overnight at 80°C.

1.18 Exfoliant-assisted liquid phase exfoliation

Transition metal dichalcogenides are two-dimensional materials widely employed in energy storage and conversion applications. Especially, molybdenum disulfide thin sheets with sulfur edges show good catalytic activity. Hence by reducing the size of the sheets and the number of layers the catalytic activity of MoS_2 can be further enhanced especially for hydrogen evolution reactions. Furthermore, good electrical conductivity can be established between catalytic sites of MoS_2 and conducting substrates by incorporating conducting materials into the nanostructures. Synthesis of exfoliated MoS_2 thin sheets would afford the electrocatalyst with a large number of exposed active sulfur sites rather than inactive basal planes. Exfoliated MoS_2 thin sheets can be prepared from mechanical exfoliation or chemical vapor deposition methods. However, these methods suffer from high cost and limited scalability in the case of the former method and difficulty in controlling the stoichiometry ratio of the precursors in the latter case. A more facile method for the preparation of MoS_2 thin sheets is exfoliant-assisted liquid-phase exfoliation. In this method, MoS_2 is dispersed in a solvent in the presence of an exfoliant to mitigate the attractive forces between the layers of MoS_2. Sonication or shearing is then applied to exfoliate bulk MoS_2 into thin sheets with a large number of edge S sites. Chen et al. [24] developed 3D MoS_2 aerogel through the hydrothermal process and employed it as an efficient hydrogen evolution reaction electrocatalyst. Initially, chlorophyll was extracted from *Sapium sebiferum* leaves in acetone and then centrifuged to obtain chlorophyll extract solution. In the second step, MoS_2 and chlorophyll extract solution were mixed in acetone and sonicated at a temperature of $10°C$ to obtain exfoliated MoS_2 thin sheets. Aqueous agar solution was mixed with exfoliated MoS_2 from the above step to form a suspension which was then refrigerated at $-20°C$. The frozen mixture was subjected to a 10^{-3} Torr vacuum to obtain MoS_2 aerogel. The authors also prepared MoS_2 quantum dot aerogel from exfoliated MoS_2 through the hydrothermal method. Initially, exfoliated MoS_2 suspension was placed in an autoclave for 8 h at $120°C$. Aqueous agar solution and hydrothermal treated MoS_2 suspension from the above step were mixed and subjected to hydrothermal treatment at $120°C$ for 3 h. Finally, the product was cooled to $-20°C$ to freeze the solution, and then the frozen solution was subjected to a 10^{-3} Torr vacuum to obtain MoS_2 quantum dot aerogel.

1.19 Significance of carbon supports in the syntheses of catalysts for energy applications

Fabrication of catalysts/electrocatalysts for energy storage and conversion applications very often requires supports for the active materials. Loading of active nanoparticles on supports results in both stabilizations as well as improvement in the catalytic activity of the nanoparticles. For example, the activity of gold catalyst depends on the nature of the support as well as the size of the nanoparticles. There are several materials available for supporting active components of the catalysts such as various carbon materials, tin oxide, titanium dioxide, Ti_2O_3, Ti_4O_7, WO_3, WC, etc. Among these support materials, carbon-based supports are widely used in catalysis and energy-related applications. Carbon allotropes like carbon nanotubes, graphene, carbon nanodots, nanospheres, etc. are important carbon materials extensively investigated as supports. Various advantages of carbon supports are—(1) Carbon supports act as conducting pathways between active components and electrodes. (2) Carbon materials are available in abundance. (3) Active components of the catalysts can be directly grown on carbon supports. (4) Aggregation of nanoparticles can be prevented by depositing them on carbon supports. (5) Carbon support enhances the endurance of the catalyst in acid and basic media. (6) Catalytic activity can be increased through synergistic cooperation between active components and carbon supports. Numerous catalytic active materials such as earth-abundant elements, semiconductors, nonnoble metals, noble metals, etc. have been fabricated on carbon materials [25]. In this context, several synthetic methods are available for the incorporation of carbon supports during the fabrication of catalytic/electrocatalytic materials for energy applications. Few representative examples are given below.

Noble metals—Noble metals are prepared by reducing metal precursors using solvent as a reducing medium or using external reducing agents. For example, $H_2PtCl_6 \cdot 6H_2O$ can be reduced in ethylene glycol through refluxing

46 Oxide free nanomaterials for energy storage and conversion applications

till the dark nanoparticles are formed. Carbon material such as carbon black is dispersed and added to the above noble metal dispersion with vigorous stirring so that metal nanoparticles can be adsorbed onto the carbon surface [26]. In the case of the external reducing agent method, metal precursor and carbon support are initially dispersed in a suitable solvent and the reducing agent is subsequently added to the mixture to form noble metal nanoparticles deposited on a carbon support.

Transition metal dichalcogenides—Molybdenum disulfide can be fabricated on carbon support via forming MoO_3/C initially through pyrolysis using ammonium heptamolybdate as a metal precursor. MoO_3/C is subsequently subjected to sulfurization using sulfur powder at 600°C. MoS_2/C can also be prepared via direct vaporization of MoO_3 and sulfur powder on carbon substrates like graphene. Murthy et al. [27] synthesized various carbon-supported MoS_2 using the hydrothermal method. Carbon materials used as supports were carbon nanofibers, acetylene black, carbon nanotubes, and Vulcan carbon. Sodium molybdate dihydrate and thiourea were used as molybdenum and sulfur precursors. Carbon materials were dispersed individually in water initially to which the precursors were subsequently added under vigorous stirring. The mixture was then subjected to hydrothermal treatment at 250°C to finally obtain MoS_2/C.

Transition metal carbides—In the synthesis of molybdenum carbide, dispersion of carbon support, transition metal precursors-like ammonium molybdate, and carbon source are thoroughly mixed. The mixture is subjected to a hydrothermal process at about 200°C and the product is separated by centrifugation then washed and dried in a vacuum oven overnight. Finally, the above-obtained product is carbonized under an inert atmosphere at 900°C to form molybdenum carbide on a carbon support. Alternatively, ammonium molybdate and carbon support mixture can be dried completely and then subjected to carburization in a tube furnace at 800°C under an inert atmosphere [25].

Transition metal nitrides—In the synthesis of molybdenum nitride, carbon support and transition metal precursor-like $MoCl_5$ are dispersed separately in ethanol and then mixed by vigorous stirring. Subsequently, the required amount of urea is added to the above mixture, stirred thoroughly, and dried in an oven to remove the solvent. The dry mixture is then subjected to calcination at 750°C under a nitrogen atmosphere to finally obtain carbon-supported molybdenum nitride [28].

2. Conclusions

Recent development and evolution in synthetic methods and strategies for preparing various advanced oxide-free catalysts/electrocatalysts have been discussed in the context of energy storage and conversion applications in this chapter. It can be understood from the above discussion that efficient synthetic strategies are available for the fabrication of active materials of various sizes, shapes, morphologies, and dimensions. Especially, nano-architectures of lower dimensions find tremendous application potential in energy-related fields and devices. A variety of catalyst materials from traditional nanoparticles to hetero-structures of multidimensions have been reported and these materials afford desirable properties of large specific surface area, high electrical conductivity, good durability, etc. From the discussion, it can be noted that hetero-structured materials are more efficient than their single-component counterparts. Besides, a recent trend in the fabrication of electrodes is the incorporation of carbon and carbon-based materials as supports that can enhance both the performance and durability of the active components of the electrodes.

References

[1] V.H. Hoa, D.T. Tran, H.T. Le, N.H. Kim, J.H. Lee, Hierarchically porous nickel–cobalt phosphide nanoneedle arrays loaded micro-carbon spheres as an advanced electrocatalyst for overall water splitting application, Appl. Catal. B 253 (2019) 235–245, https://doi.org/10.1016/j.apcatb.2019.04.017.

[2] L. Ji, J. Wang, X. Teng, T.J. Meyer, Z. Chen, CoP nanoframes as bifunctional electrocatalysts for efficient overall water splitting, ACS Catal. 10 (2019) 412–419, https://doi.org/10.1021/acscatal.9b03623.

[3] H. Xu, H. Jia, B. Fei, Y. Ha, H. Li, Y. Guo, M. Liu, R. Wu, Charge transfer engineering via multiple heteroatom doping in dual carbon-coupled cobalt phosphides for highly efficient overall water splitting, Appl. Catal. B 268 (2020) 118404, https://doi.org/10.1016/j.apcatb.2019.118404.

[4] C.-L. Huang, X.-F. Chuah, C.-T. Hsieh, S.-Y. Lu, NiFe alloy nanotube arrays as highly efficient bifunctional electrocatalysts for overall water splitting at high current densities, ACS Appl. Mater. Interfaces 11 (2019) 24096–24106, https://doi.org/10.1021/acsami.9b05919.

[5] B. Zhang, H. Wang, Z. Zuo, H. Wang, J. Zhang, Tunable CoFe-based active sites on 3D heteroatom doped graphene aerogel electrocatalysts via annealing gas regulation for efficient water splitting, J. Mater. Chem. A 6 (2018) 15728–15737, https://doi.org/10.1039/C8TA05705B.

[6] X. Xu, H. Chu, Z. Zhang, P. Dong, R. Baines, P.M. Ajayan, J. Shen, M. Ye, Integrated energy aerogel of NS-rGO/WSe2/NiFe-LDH for both energy conversion and storage, ACS Appl. Mater. Interfaces 9 (2017) 32756–32766, https://doi.org/10.1021/acsami.7b09866.

[7] F. Ming, H. Liang, H. Shi, X. Xu, G. Mei, Z. Wang, MOF-derived Co-doped nickel selenide/C electrocatalysts supported on Ni foam for overall water splitting, J. Mater. Chem. A 4 (2016) 15148–15155, https://doi.org/10.1039/c6ta06496e.

[8] Q. Yuan, Q. Yu, Y. Gong, X. Bi, Three-dimensional N-doped carbon nanotube frameworks on Ni foam derived from a metal–organic framework as a bifunctional electrocatalyst for overall water splitting, ACS Appl. Mater. Interfaces 12 (2019) 3592–3602, https://doi.org/10.1021/acsami.9b18961.

[9] C. Ray, S.C. Lee, B. Jin, A. Kundu, J.H. Park, S. Chan Jun, Conceptual design of three-dimensional CoN/Ni3N-coupled nanograsses integrated on N-doped carbon to serve as efficient and robust water splitting electrocatalysts, J. Mater. Chem. A 6 (2018) 4466–4476, https://doi.org/10.1039/c7ta10933d.

48 Oxide free nanomaterials for energy storage and conversion applications

[10] A. Kumar, A. Rana, G. Sharma, S. Sharma, M. Naushad, G.T. Mola, P. Dhiman, F.J. Stadler, Aerogels and metal-organic frameworks for environmental remediation and energy production, Environ. Chem. Lett. 16 (2018) 797–820, https://doi.org/10.1007/s10311-018-0723-x.

[11] D. Kobina Sam, E. Kobina Sam, X. Lv, Application of biomass-derived nitrogen-doped carbon aerogels in electrocatalysis and supercapacitors, ChemElectroChem 7 (2020) 3695–3712, https://doi.org/10.1002/celc.202000829. Ahead of Print.

[12] M.K. Kundu, T. Bhowmik, S. Barman, Gold aerogel supported on graphitic carbon nitride: an efficient electrocatalyst for oxygen reduction reaction and hydrogen evolution reaction, J. Mater. Chem. A 3 (2015) 23120–23135, https://doi.org/10.1039/C5TA06740E.

[13] R. Du, W. Jin, R. Huebner, L. Zhou, Y. Hu, A. Eychmueller, Engineering multimetallic aerogels for pH-universal HER and ORR electrocatalysis, Adv. Energy Mater. 10 (2020) 1903857, https://doi.org/10.1002/aenm.201903857.

[14] Y. Liu, X. Luo, C. Zhou, S. Du, D. Zhen, B. Chen, J. Li, Q. Wu, Y. Iru, D. Chen, A modulated electronic state strategy designed to integrate active HER and OER components as hybrid heterostructures for efficient overall water splitting, Appl. Catal. B 260 (2020) 118197, https://doi.org/10.1016/j.apcatb.2019.118197.

[15] Y. Dang, J. He, T. Wu, L. Yu, P. Kerns, L. Wen, J. Ouyang, S.L. Suib, Constructing bifunctional 3D holey and ultrathin CoP nanosheets for efficient overall water splitting, ACS Appl. Mater. Interfaces 11 (2019) 29879–29887, https://doi.org/10.1021/acsami.9b08238.

[16] J. Tarver, J.E. Yoo, T.J. Dennes, J. Schwartz, Y.-L. Loo, Polymer acid doped polyaniline is electrochemically stable beyond pH 9, Chem. Mater. 21 (2009) 280–286, https://doi.org/10.1021/cm802314h.

[17] A. Murthy, A. Manthiram, Highly water-dispersible, mixed ionic-electronic conducting, polymer acid-doped polyanilines as ionomers for direct methanol fuel cells, Chem. Commun. (Camb.) 47 (2011) 6882–6884, https://doi.org/10.1039/c1cc11473e.

[18] Q. Gao, Z. Shi, K. Xue, Z. Ye, Z. Hong, X. Yu, M. Zhi, Cobalt sulfide aerogel prepared by anion exchange method with enhanced pseudocapacitive and water oxidation performances, Nanotechnology 29 (2018) 215601, https://doi.org/10.1088/1361-6528/aab299.

[19] E. Lee, A. Murthy, A. Manthiram, Carbon-supported Pt nanoparticles prepared by a modified borohydride reduction method: effect on the particle morphology and catalytic activity for COad and methanol electro-oxidation, Electrochem. Commun. 13 (2011) 480–483, https://doi.org/10.1016/j.elecom.2011.02.026.

[20] Y. Jin, F. Chen, J. Wang, L. Guo, T. Jin, H. Liu, Lamellar platinum-rhodium aerogels with superior electrocatalytic performance for both hydrogen oxidation and evolution reaction in alkaline environment, J. Power Sources 435 (2019) 226798, https://doi.org/10.1016/j.jpowsour.2019.226798.

[21] E. Lee, A. Murthy, A. Manthiram, Comparison of the stabilities and activities of Pt–Ru/C and Pt3–Sn/C electrocatalysts synthesized by the polyol method for methanol electro-oxidation reaction, J. Electroanal. Chem. 659 (2011) 168–175, https://doi.org/10.1016/j.jelechem.2011.05.022.

[22] A. Murthy, E. Lee, A. Manthiram, Electrooxidation of methanol on highly active and stable Pt–Sn–Ce/C catalyst for direct methanol fuel cells, Appl. Catal. B 121–122 (2012) 154–161.

[23] X. Zhao, W. Li, A. Murthy, Z. Jiang, Z. Zuo, A. Manthiram, A DMFC stack operating with hydrocarbon blend membranes and Pt–Ru–Sn–Ce/C and Pd–Co/C electrocatalysts, Int. J. Hydrogen Energy 38 (2013) 7448–7457, https://doi.org/10.1016/j.ijhydene.2013.04.013.

[24] I.W.P. Chen, C.-H. Hsiao, J.-Y. Huang, Y.-H. Peng, C.-Y. Chang, Highly efficient hydrogen evolution from seawater by biofunctionalized exfoliated MoS2 quantum dot aerogel

electrocatalysts that is superior to Pt, ACS Appl. Mater. Interfaces 11 (2019) 14159–14165, https://doi.org/10.1021/acsami.9b02582.

[25] A.P. Murthy, J. Madhavan, K. Murugan, Recent advances in hydrogen evolution reaction catalysts on carbon/carbon-based supports in acid media, J. Power Sources 398 (2018) 9–26, https://doi.org/10.1016/j.jpowsour.2018.1007.1040.

[26] A. Murthy, A. Manthiram, Application of derivative voltammetry in the analysis of methanol oxidation reaction, J. Phys. Chem. C 116 (2012) 3827–3832, https://doi.org/10.1021/jp2092829.

[27] A.P. Murthy, J. Theerthagiri, J. Madhavan, K. Murugan, Highly active MoS_2/carbon electrocatalysts for the hydrogen evolution reaction—insight into the effect of the internal resistance and roughness factor on the Tafel slope, Phys. Chem. Chem. Phys. 19 (2017) 1988–1998, https://doi.org/10.1039/c6cp07416b.

[28] D.H. Youn, S. Han, J.Y. Kim, J.Y. Kim, H. Park, S.H. Choi, J.S. Lee, Highly active and stable hydrogen evolution electrocatalysts based on molybdenum compounds on carbon nanotube-graphene hybrid support, ACS Nano 8 (2014) 5164–5173.

Chapter 3

Fundamentals, basic components and performance evaluation of energy storage and conversion devices

Durai Govindarajan[a,*] and Karthik Kumar Chinnakutti[b,*]

[a]*Centre for Nanoscience and Nanotechnology, Sathyabama Institute of Science and Technology, Chennai, India,* [b]*Department of Chemistry, Vinayaka Mission's Kirupananda Variyar Arts and Science College, Vinayaka Mission's Research Foundation (Deemed to be University), Salem, India*

Chapter outline

1. **Introduction**   **52**
2. **Batteries**   **52**
   2.1 Principles of the Li-S battery   53
   2.2 Fundamental challenges of Li-S batteries   54
3. **Supercapacitors**   **54**
   3.1 Principle of supercapacitors   55
   3.2 Characteristics of a supercapacitors   56
   3.3 Classification of supercapacitors   57
   3.4 Material selection for supercapacitors   60
   3.5 Electrolytes   60
4. **Solar energy and solar cells**   **61**
   4.1 Principle of photovoltaic   61
   4.2 Charge carrier's generation   62

   4.3 Subsequent separation of the photo-generated charge carriers in the junction   63
   4.4 Collection of the photo-generated charge carriers at the terminals of the junction   64
   4.5 Solar cells types   64
   4.6 First generation solar cells (silicone cells)   65
   4.7 Second-generation solar cells (thin-film solar cells)   66
   4.8 Third-generation solar cells   67
   4.9 Overview of thin-film solar cell   68
   4.10 Thin-film solar cell challenges   71
5. **Conclusions**   **71**
References   **72**

---

* These authors contributed equally to this work.

Oxide Free Nanomaterials for Energy Storage and Conversion Applications.
https://doi.org/10.1016/B978-0-12-823936-0.00010-3
Copyright © 2022 Elsevier Inc. All rights reserved.

52 Oxide free nanomaterials for energy storage and conversion applications

1. Introduction

There are a variety of energy storage sources based on electromagnetic, thermal, hydrogen, mechanical, and electrochemical energy that are being presently developed [1–3]. Among all the energy storage approaches, electrochemical energy storage/conversion devices such as batteries, supercapacitors, and solar cells are playing a crucial role due to their long cycle life, high energy/power density, high efficiency, versatility, and flexibility [4, 5]. Generally, electrochemical batteries are the most suitable electrical energy storage devices for many applications in day-to-day life, since batteries have a larger energy density than capacitors. The electrochemical batteries offer several disadvantages such as, (i) lower power density than the supercapacitor, (ii) a standard battery such as lead-acid battery has a shorter lifetime, and (iii) contain hazardous waste substances which impose extra difficulties for device recycling. In this chapter, the fundamentals, basic components, and performance evaluation of energy storage/conversion systems are focused on, especially batteries, supercapacitors, and solar cells are elucidated in detail.

2. Batteries

A battery is a basic electrochemical device that converts chemical energy contained within its active materials directly into electric energy using an electrochemical oxidation-reduction (redox) reaction. This type of reaction involves the transfer of electrons from one material to another via an electric circuit. Batteries usually consist of electrochemical cells connected in parallel or series or both providing the required potential and current [6]. A battery provides electrical power when connected to external devices like flashlights, smartphones, laptops, and even vehicles like electric cars. Gaston Plante was the first person to invent a secondary lead-acid battery in 1859 [7]. In a battery, electrical energy is directly converted into chemical energy using an electrochemical reaction called the redox (reduction-oxidation) process [8]. An important parameter for the battery is the standard potential $E°$ of a battery cell, which is in association with redox reactions that can be calculated using the formula:

$$E° = -\Delta G°/zF \tag{1}$$

where F is the Faraday constant, $\Delta G°$ is the standard Gibbs free energy, and z is the number of electrons. The overall theoretical cell voltage is the difference between positive electrode potential $E°^+$ and the negative electrode potential $E°^-$;

$$\Delta E° = E°^+ - E°^- \tag{2}$$

For several decades, batteries have been more prevalent in portable electronics applications due to their capability of storing high energy density [9]. But it takes a long time to charge and discharge resulting in the degradation of

Energy storage and conversion devices **Chapter | 3** 53

chemical compounds present in the battery. In addition to the material damage, the ability to deliver power is also limited even in technologies like lead-acid and lithium-ion batteries [10, 11]. Because of these shortcomings, various research works are being proposed for the hybridization of batteries along with supercapacitors to obtain improved performance. Battery technology with the application of sodium-ions is turning out to be very promising for future renewable applications but still faces certain technical limitations.

2.1 Principles of the Li-S battery

A typical Li-S cell consists of a lithium-metal anode, a separator, a sulfur-based cathode, and an electrolyte. A diagram shows the typical Li-S cell setup and both types of charging/discharge voltage profiles in Fig. 1. Lithium-metal is oxidized into lithium ions during the discharge process and travels to the sulfur cathode through the electrolyte in which Li forms conversion type Li-S compounds. Fig. 1B (left) shows typical discharge-charge files for an electrochemical solid-liquid Li-S phase reaction. S_8 has been reduced to Li_2S_4 on the first plateau by 2.3 V, with a 1/2 electron transfer by sulfur atom delivering 1/4 theoretical capacitance (418 mA h g^{-1}). Li_2S_4 then receives 3/2 electrons per sulfur atom and decreases to Li_2S on a plateau of about 2.1 V for 1254 mA h g^{-1} capacity [12]. Reverse reactions occur and the Li_2S is returned to S_8 during the charging process. The corresponding reaction equations are as follows:

Discharge;

$$Anode: 16Li \rightarrow 16Li^+ + 16e^-$$

$$Cathode: S_8 + 4Li^+ + 4e^- \rightarrow 2Li_2S_4 \quad (Step\ 1)$$

$$2Li_2S_4 + 12Li^+ + 12e^- \rightarrow 8Li_2S \quad (Step\ 2)$$

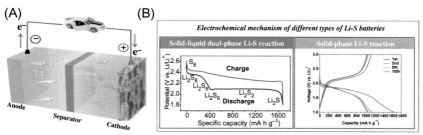

FIG. 1 schematic illustrative of a Li-S cell configuration (left) and a solid-phase Li-S reaction of the typical charge/discharge voltage profile (right). *(Reproduced with permission from Z. Ma, Z. Li, K. Hu, D. Liu, J. Huo, S. Wang, The enhancement of polysulfide absorbs ion in Li-S batteries by hierarchically porous CoS₂/carbon paper interlayer, J. Power Sources 325 (2016) 71–78, Copyright 2016, Royal Society of Chemistry. Reprinted with permission of Z. Li, L. Yin, Nitrogen-doped MOF-derived micropores carbon as immobilizer for small sulfur molecules as a cathode for lithium sulfur batteries with excellent electrochemical performance, ACS Appl. Mater. Interfaces 7 (2015) 4029–4038, Copyright 2015, American Chemical Society.)*

Charge;

$$\text{Anode}: 16Li^+ + 16e^- \rightarrow 16Li$$

$$\text{Cathode}: 8Li_2S \rightarrow S_8 + 16Li^+ + 16e^-$$

Li-S batteries electrochemical reaction above belongs to the dual-phase reaction of "solid-liquid" [13]. These types of batteries use ether-based electrolyte systems and are the most important systems in the field of research and application. However, the double-phase solid-liquid reaction is not the only route that completes the electrochemical Li-S reaction.

2.2 Fundamental challenges of Li-S batteries

Lithium-ion batteries (LIB's) have been quickly developed since the 1990s with stable electrochemical tests and are considered ideal energy supplies for mobile electronic devices like mobile telephones, computers, and electric vehicles. LIBs based on insertion-type transition metals/metal oxides are unfortunately not capable of providing enough energy density to meet the growing demands of electric vehicles with long-range power [14]. It is therefore important to seek new materials of electrodes that have low molecular/atomic weight and are in a position to transfer multiion/electron per molecular/atomic material [15]. Sulfur has a relatively low atomic weight of $32\,g\,mol^{-1}$, which is one of the largest elements in this earth's crust. It is a cost-efficient and ecological alternative to traditional LIBs. Li-S batteries (in combination with Li metal anodes) have huge power as power storage devices because of their $2600\,W\,h\,kg^{-1}$ high theoretical energy density based on the two transfers by atom S [16]. Li-S batteries have been given greater attention since 2009 and are considered to be one of the most promising candidates for rechargeable next-generation batteries [17].

Li-S batteries suffer from a variety of problems, so far limiting their development to large-scale commercial technology. The main problem is the dissolution and shuttling of the polysulfides of the active material, which leads to a wide range of effects, including increased solid electrolyte interface (SEI) resistivity, due to the deposition of isolation products on Li-metal anode, active material loss, or reduction of coulombic efficiency. Altogether this results in a low life cycle of the Li-S cell.

3. Supercapacitors

Supercapacitors are also known as "supercaps" or "ultracapacitors" or "electrochemical capacitors" are electrochemical energy sources with high power delivery and exceptional cycle life. Supercapacitors are considered as third-generation capacitive devices and can be used where high-power demands such as power buffer and power-saving units are required. Also, they are of great interest for energy recovery. It has the potential to achieve higher energy density than conventional capacitors and to deliver higher power density than other energy storage devices. Supercapacitors can store 10–100 times more energy per unit volume which is very high when compared to batteries and which fill the power gap between ordinary capacitors and batteries in a small and light-weight package. These can withstand many charges and discharge cycles when compared to rechargeable batteries.

It is reported that nearly 30% of power is wasted due to poor storage techniques. To avoid this wastage, some good storage methods have to be implemented. Consequently, there is a need to create and commercialize renewable energy sources with state-of-the-art technology in power electronics. In this particular scenario, electrochemical supercapacitors assume significance in the light of their extraordinary properties in contrast with other energy storage devices like batteries. The supercapacitor has many advantages like high power density and specific capacitance, long life cycle, eco-friendly, and flexible working temperature. Additionally, they can rapidly charge with quick power conveyance and are competent to replace conventional capacitors. Also, supercapacitors bridge the technology gap between conventional capacitors and batteries. A comparison between the battery and supercapacitor is shown in Table 1. Due to the physical charge storage, the supercapacitors have larger power densities when compared with batteries and fuel cells, but the energy densities of supercapacitors are low when compared with the other devices.

3.1 Principle of supercapacitors

The supercapacitor is a new class of electrical storage devices with characteristics intermediate between conventional capacitors and batteries [18]. In 1957, the first patent on a supercapacitor was registered by Becher and the first of this kind of device was produced by SOHIO, United States in the year 1969. Fig. 2 shows the "Ragone plot" highlighting the energy density and power density of

TABLE 1 Comparison between the battery and supercapacitor [18].

Properties	Supercapacitor	Battery
Charging time	1–10s	10–60min
Cycle life	~10,00,000 cycles	500–1000 cycles
Service life	10–15 years	5–10 years
Cell voltage	Drop	Remain same
Power density	High	Low
Energy density	Moderate	High
Storage mechanism	Static/EC	Electrochemical

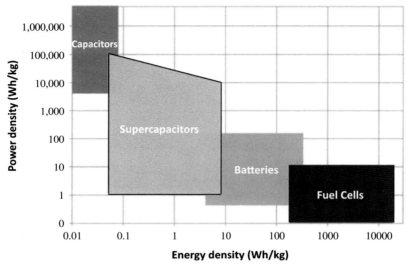

FIG. 2 Ragone plot illustrating power density vs energy density for selected energy storage devices. *(Reproduced with permission from B.E. Conway, Electrochemical Supercapacitors: Scientific Fundamentals and Technological Applications (POD), Kluwer Academic/Plenum, New York, 1999.)*

various electrochemical storage devices. From this Ragone plot, it is seen that the supercapacitor fills the gap between batteries and conventional capacitors such as electrolytic capacitors or metalized film capacitors. It can be seen from Fig. 2; the supercapacitors not only overlap with capacitors but also overlaps with batteries in terms of power density and energy density. It seems that supercapacitors can have specific power densities comparable to those of conventional capacitors and specific energy densities as high as those of some types of batteries. However, to achieve more promising power density and energy density, supercapacitors still need to be developed. The key to this lies in improving a fundamental understanding of the processes in supercapacitors and their electrode materials.

3.2 Characteristics of a supercapacitors

Supercapacitors have advantages in applications where a large amount of power is needed for a relatively short time, or where a very high number of charge/discharge cycles or a longer life is required. Some of the important characteristics to be considered when selecting a supercapacitor are the following:

- Operating voltage
- Energy capacity
- Specific energy and specific power

- Lifetime
- Capacitance
- Operating temperature voltage
- Self-discharge rate

3.3 Classification of supercapacitors

A supercapacitor is mainly classified into three types based on energy storage mechanism (Fig. 3) They are (i) electrochemical double-layer capacitors (EDLCs), (ii) pseudo capacitors, and (iii) hybrid supercapacitors.

3.3.1 Electrochemical double-layer capacitors

Electrochemical double-layer capacitors (EDLCs) are supercapacitors that employ electrostatic charge separation only. The energy storage process of EDLCs takes place at the interface between the electrode surface and the electrolyte [6]. The electrostatic charge transfer is fully reversible, which results in efficient devices with a long lifetime. EDLCs consist of at least two electrodes that are separated by a separator. The separator is ion-permeable and also prevents short circuits between the electrodes. The space between the electrodes is filled with electrolyte. By charging the device, two layers of opposite charge are formed at the interface between the electrode and the electrolyte.

These double-layer capacitors generally consist of two electrodes similar to the one in normal capacitors and these capacitors also contain a separator and an electrolyte. The electrolyte is a mixture of both positive and negative ions which are dissolved in polar solvents like water. The two electrodes get separated by a separator. The schematic diagram of the EDLC type supercapacitor is shown in Fig. 4 [6, 7]. Generally, carbon-based electrodes are called EDLC type supercapacitors. The amount of charge stored (ΔQ) in EDLC is correlated with the potential difference (ΔV) which is developed across the interface thereby producing a capacitance which is also known as double-layer capacitance, represented as C_{DL} and is shown in the following relation (3);

$$C_{\text{DL}} = \Delta Q / \Delta V \tag{3}$$

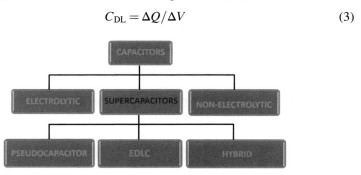

FIG. 3 Classification of supercapacitors.

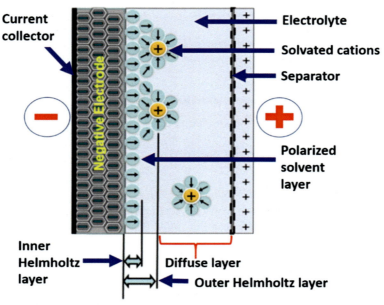

FIG. 4 Schematic of an electrochemical double-layer capacitor (EDLC). *(Reproduced with permission from O. Haasa, E.J Cairns, Electrochemical energy storage, Annu. Rep. Prog. Chem., Sect. C: Phys. Chem. 95 (1999) 163–197.)*

3.3.2 Pseudocapacitors

Pseudo-capacitance is another type of energy storage process which involves fast reversible Faradaic reactions like redox reactions, electrosorption, and intercalation/deintercalation processes at the electrode/electrolytic interface [19], and the working principle of a pseudo-capacitor is shown in Fig. 5. Primarily, the following three types of electrochemical processes are involved in the working of the pseudo-capacitors;

  (i) surface adsorption or chemisorption of ions from the electrolyte,
 (ii) redox reactions that include ions from the electrolyte and
(iii) doping and undoping of active electrodes.

Out of these, the first two processes are more surface-dependent. Due to the presence of faradaic reactions, electrochemical capacitors are similar to batteries. Thus, electrochemical capacitors are also called pseudo-capacitors, and the capacitance of electrochemical capacitors is defined as pseudo-capacitance. When a voltage is applied to the two electrodes of electrochemical capacitors, a faradaic redox reaction occurs on the surface of the electrode to transfer charges between electrodes and the electrolyte, generating a circuit in the electrochemical capacitors system. Therefore, the resulting charge transfer capacitance, $C = dQ/dV$ depends on the applied voltage. Theoretically, pseudo-capacitance takes

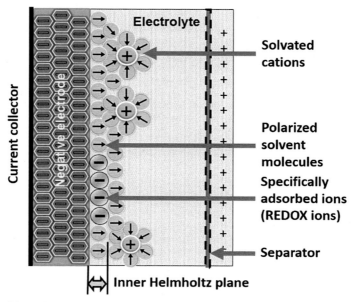

FIG. 5 Schematic diagram illustrates the working mechanism of a pseudo-capacitor. *(Reproduced with permission from A. Davies, Y.A. Ping, Material advancements in supercapacitors: from activated carbon to carbon nanotube and grapheme, Can. J. Chem. Eng. 89 (2011) 1342.)*

place due to thermodynamic reasons, where the potential (V) is a function of the integrated charge (Q) of the electrode. In this case, the extent of reaction (Q) is a continuous function of reaction potential (V), and its derivative at any point can be expressed as pseudo-capacitance $C_\Phi = dQ/dV$.

The most commonly known electrode materials for pseudo-capacitors include metal oxides [20], metal nitrides/sulfides [21], conducting polymers [22], etc. In a specifically designed supercapacitor, pseudo-capacitance can be made much higher than EDLCs. These pseudo-capacitance materials have been widely applied in supercapacitors because of their excellent properties such as high conductivity, large capacitances, and good flexibility and making them highly desirable for next-generation supercapacitors.

3.3.3 Hybrid supercapacitors

The charge storage mechanism of the hybrid supercapacitor is the combination of the properties of nonfaradaic (EDLC) and faradaic (pseudo-capacitor) processes to store charge. The hybrid supercapacitor delivers higher specific capacitance in comparison to the existing EDLCs and pseudo-capacitors. Generally, the asymmetric behavior of the hybrid supercapacitor can act as an enhancer in its respective capacitance values as well as power/energy densities. The operating voltages of the normal capacitors are very high but in the case of supercapacitors, the operating voltage lies between $\sim$2.5 and $\sim$2.7 V. This hybrid approach marks a new beginning towards the much-needed pollution-free, long-lasting, and proficient energy-storing performance [23].

3.4 Material selection for supercapacitors

The supercapacitor is a multilayered device, and several materials are required to make a supercapacitor device. Two metal oxides/nitrides-based electrodes (a thin layer of metal oxide/nitride coating on stainless steel substrate as a current collector) are required. A porous separator layer that contains the electrolyte is required to prevent short-circuiting between the electrodes. The following important parameters may be considered to select a good electrode material: (i) multiple oxidation states, (ii) superior electrical conductivity, (iii) high mechanical stability, (iv) high surface area and chemical stability, and (v) electrochemical activity.

3.5 Electrolytes

Electrolytes have been identified as one of the most important components in electrochemical supercapacitor performance. Generally, the electrolytes are classified into several categories, such as aqueous electrolytes (KOH, KCl, Na_2SO_4, H_2SO_4, etc.), organic electrolytes (TEABF$_4$/ACN, TEABF$_4$/PC, TEABF$_4$/HFIP, SBPBF$_4$/ACN, etc.), ionic liquids [EMIM][BF$_4$], [BMIM] [BF$_4$], etc. and solid-state or semisolid-state electrolytes (LiClO$_4$-Al$_2$O$_3$, PVA/KOH, PVA/H$_3$PO$_4$, etc.). Actually, it is very difficult to select a suitable electrolyte to meet all of the requirements, as each electrolyte has its own merits and demerits. In general, a good electrolyte should have the following important features: (i) high ionic conductivity, (ii) large potential window, (iii) wide operating temperature range, (iv) high chemical and electrochemical stability, (v) low volatility and flammability, (vi) environmentally friendly, and (vii) low cost.

The most commercially available electrolytes are organic electrolytes and have some advantages like large operating voltage. However, compared with aqueous electrolytes, organic electrolytes are expensive, highly flammable, and, in some cases, toxic. Even though aqueous electrolytes are examined by a cramped working voltage, the ions present in the aqueous electrolytes are capable of providing incredibly faster carrier rates than organic electrolytes and can achieve better performance of electrochemical supercapacitors. With the development of ionic liquid (IL) electrolytes, the operating cell voltage of the corresponding electrochemical supercapacitors was further increased to $\sim$4 V, although they have low ionic conductivity, high viscosity, and high cost, which can limit their practical use in supercapacitors. Additionally, the electrolyte properties such as (i) ion size, (ii) ion concentration, (iii) interaction between the ion and solvent, (iv) interaction between the electrolyte and electrode materials, and (v) potential window all influence the specific/areal capacitance, energy/power densities, and cycle-life.

4. Solar energy and solar cells

Solar energy is a remarkable sustainable energy source that can address a large number of the difficulties confronting the globe. There are many reasons for promoting its energy market share. This power source is famous because it is flexible and offers different benefits for people and nature. While most of the world's supply of electricity is produced from fossil fuels such as coal, oil, and natural gas, conventional energy sources face a range of challenges. Import dependence from several countries that have substantial unsustainable energy sources and create biological concern about the risk of natural change associated with the generation of electricity by fossil fuels. Solar energy has increased as one of the fastest-growing renewable power generation resources. The Earth is only 2% more than our yearly necessity, with a total amount of solar radiation of 1527 kW h. Presently, 19 Terawatts (TW) are the most important all-inclusive energy, and 10 TW must be increased in the next 30 years [24]. The amount of light to the surface of the earth is 120,000 TW. In addition, 20 min of sunlight are sufficient to meet the global need for electricity for a year.

The first pragmatic silicone solar cells were shown in April 1954 at the Bell research centers. A first examination of the PV impact in 1839 was carried out by the narration of solar cells. The French scientist Antoine Cesar Becquerel and his father used metallic anodes in an electrolyte setting while he saw that when metals were displayed to light, little electrical flow was provided but he could not explain the effect. Some years after that, the first selenium solar cells were prepared by the American discoverer Charles Fiedens in 1883. While Fritz trusted his solar cells to fight the Edison coal-fired energy plants, they converted sunlight to power <1% and were not viable as a result. The first genuine silicon solar cell was made by the group 13 years after the event. Researchers working in Bell labs collectively. Sunlight has changed to power with the first silicon solar cells of about 6% efficiency, a massive development of previous solar cells ended. Fig. 6 shows the graphic structure and improved efficiency of the broad range of photovoltaic innovations in recent years.

4.1 Principle of photovoltaic

A photovoltaic device (PV) is one of the renewables that turn light energy into electricity. The sun roughly transfers around 174,000 TW of the average power density of $1366 \, \mathrm{W \, m^{-2}}$ to the upper dimensions of Earth's atmosphere called a solar constant [25]. By atmospheric adsorption and dissipation, this value is reduced to $1000 \, \mathrm{W \, m^{-2}}$. By this measure, PV exceeds both now and infinitely the capacity to fulfill world energy needs. The current wafer-based silicon

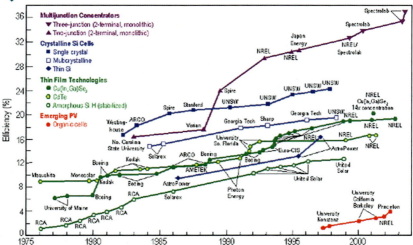

FIG. 6 The number of photovoltaic technologies and their increased efficiencies over the years.

innovation is dependent on costly assembly procedures and has high balance-sheet costs, but it may also be capable of meeting future energy requirements.

Generally, solar cells are converted into electricity directly by the internal photoelectric effect of solar radiation. So, any solar cell would like a PV absorber material which, in addition to the mobile carriers, electrons, and holes, isn't just able to absorb external light capably and which can be insulated at the terminal units without a notable energy loss. In 1905, Albert Einstein understood this impact to be clarified by expecting the light to contain characterized quantities of energy called photons all over the place. $E = h\nu$ assumes the power of the photon, where h is Planck's constant and ν is the light frequency, which is given by $\nu = c/\lambda$, c is the light speed and λ is the wavelength. In 1921 he received the Noble Prize on Physics with his innovation of photoelectric impact [26]. The PV impact can be divided into three fundamental procedures.

4.2 Charge carrier's generation

Photon absorption in material means it uses energy to lead an electron from the preliminary energy level E_i to the upper energy level E_f, as illustrated in Fig. 7. If energy levels of E_i and E_f are available, the photons should also be engaged, so photon energy, $h\nu = E_f - E_i$, is their distinction. In an ideal semiconductor, electrons can occupy power levels below the edge of the valence band, E_V, or more of the alleged edge of the conductive band, E_C. There are no allowed energy states between those two bands which could be occupied by electrons. This energy

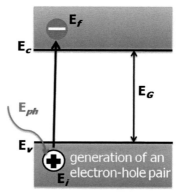

FIG. 7 Illustrating the absorption of a photon in a semiconductor with bandgap E_G. The photon with energy $E_{ph} = h\nu$ excites on an electron from E_i to E_f a hole is created.

difference is therefore referred to as the band gap, $E_g = E_C - E_V$. If a photon of energy less than E_g reaches a perfect semiconductor, it will not be consumed without contact. In a specific semiconductor, the valence and conductivity bands are not constant, but they fluctuate based on the supposed k-vector that represents energy of an electron in the semiconductor. It means that an electron's energy depends on the momentum of the semiconductor crystal because of its regular structure. On the off possibility of limiting the valence band and the conductive band base existing at a common k-vector, the electron can be energized from the valence to the conductive band without any change of momentum.

This semiconductor is classified as a direct bandgap material. We talk about an indirect bandgap that is unlikely to energize the electron without changing its momentum. The electron can only change its momentum by changing the equilibrium with the crystal, for instance, by obtaining momentum from the crystal lattice or by giving energy to them. The absorption coefficient in a material with direct band gaps is considerably greater than in an indirect bandgap substance and may be much thinner than those semiconductors that are often referred to as the absorber. When E_i to E_f excites an electron, a void will be produced at E_i. This space is called a hole and works as a constructive charge particle. The photon absorption thus helps to develop an electron-hole pair as shown in Fig. 8. The radiative energy of the photon is transferred into the chemical energy of the electron-hole pair. The optimal conversion efficiency from radiative to chemical energy is restricted by thermodynamics.

4.3 Subsequent separation of the photo-generated charge carriers in the junction

Typically, the electron-hole pair recombines, e.g., the electron drops back to the initial E_i energy level. At this point, the energy is released as the photon

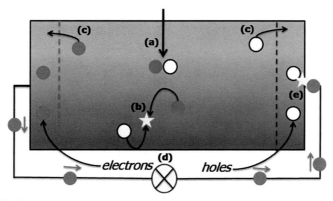

FIG. 8 Simple solar cell model. (a) The absorption of a photon leads to the generation of an electron-hole pair. (b) The electrons and holes will recombine. (c) Electrons and holes can be separated. (d) The separated electrons can be used to drive an electric circuit. (e) After the electrons passed through the circuit, they will recombine with holes.

(radiative recombination) or substituted by a variety of electrons or holes or lattice vibrations (nonradiative recombination). If the energy contained in the external circuit is to be used, the semipermeable membranes must be accessible on both sides of the absorber to allow only electrons to pass through a single membrane and only holes to pass through the other membrane, as shown in Fig. 8(c). These membranes are composed of materials of n and p-type for a bulk of thin-film solar cells (TFSC). The final goal of the solar cell is for electrons and holes to enter the membranes until they recombine, i.e., it must be less than their lifetime if the charge carriers are to hit the membranes. This condition reduces the thickness of the absorber.

4.4 Collection of the photo-generated charge carriers at the terminals of the junction

Finally, the charge carrier is collected from solar cells using electric conduct to carry out a further circuit (Fig. 8(d). The chemical energy of the electron-hole pairs is ultimately converted to electricity. After electrons are passed through the circuit, the metal absorber interface hole is recombined as set out in Fig 8(e).

4.5 Solar cells types

Solar cells can be classified into three generations ages, using their findings and applications (Fig. 9).

(1) First solar cells (Silicone cells).
(2) Second-generation solar cells (Thin-film solar cells).
(3) Third-generation solar cells.

Energy storage and conversion devices Chapter | 3 **65**

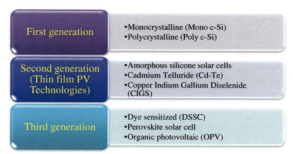

FIG. 9 Classification of solar cells.

4.6 First generation solar cells (silicone cells)

Bell Laboratories designed the primary silicon solar cell in 1954 with an efficiency of 6%. Silicon is today the most competent private solar panels available and reports that more than 80% of the total solar panel sold in the world is available. For photovoltaic single-cell systems, silicon solar cells are the most efficient, and silicon, which second only to oxygen, is the most abundant part on earth. It is a semiconductor material with a 1.1 eV energy bandgap range for use in PV applications. Crystalline silicon cells are arranged for the production of Si wafers in three different types. The silicon varieties depend on the type directly mentioned in Fig. 9.

4.6.1 Monocrystalline silicon solar cells

Silicon is the innovative material most frequently used in PV devices. Until 2004, the PV sector accounted for ~95%. In terms of physical science, Si is not ideally suited for PV because the absorption coefficient is usually low. A large Si thickness of ~125 μm is required to absorb >90% of the overband gap by capacity. Si is ready for PV because it is relatively economical to compare with various materials. A lot of advances in Si PV focus on crystalline p-n junction devices. Due to the lower cost of such a wafer compared to the n-type wafer, the absorber is normally p-type. The late growth of photovoltaic companies led to Si shortages for photovoltaics and thus increased Si expenses. Si solar modules' typical industrial efficiency performance is about 14%–17% [27]. The indistinct Si-based cells, recorded at ~22.3% and 18%–20% performance, offer the most significant accessible industrial output [27, 28]. The key mechanisms for decaying output in Si systems include mass recombination, depletion recombination, recombination of the front and back surface contact and contact loss, loss of resistance in series, loss of reflection, and others. Mehul C and collaborators with advanced productivity of around ~25% by use of high-quality Si have demonstrated the worldwide fastest Si-driven sunlight-based cell [29].

4.6.2 Polycrystalline silicon solar cells

Polycrystalline PV modules are usually made of several crystals and fastened to each other in one cell. Processing of polycrystalline Si solar cells has been increasingly productive and liquid Si graphite-filled structure is being provided. The solar cells of polycrystalline Si are the most growing solar cells today. In the middle of 2008, up to 48% of all solar cell output is permitted [30]. During the solidification of the molten silicon different crystal structures are created. Although they are slightly cheaper than single Si solar panels, they are still less efficient at about 12%–14% [31].

4.7 Second-generation solar cells (thin-film solar cells)

Thin films are expected to be second-generation photovoltaic innovations that offer less cost-intensive modules, such as glass, metal films, polymers, etc., by direct deposition of thin absorbers on the low-cost substrate. In the production of solar cells, porous silicone structures are also used. Silicone solar cells have light absorption layers of up to 350 µm of thickness while thin solar cell films usually have thin absorption layers of $\sim$1 µm of thickness [32]. Three types of thin-film solar cells are normally classified as:

(1) Amorphous silicon
(2) Cadmium telluride (CdTe)
(3) Copper indium gallium di-selenide (CIGS)

4.7.1 Amorphous silicon

Amorphous Si (a-Si) PV modules are the earliest solar cells first produced in the industrial sector. The low preparation of a-Si solar cells allows the use of various low-cost polymers and other adaptable substrates. Such substrates require less handling energy [33]. The usually low 5%–6% range of the typical output of the amorphous commercial silicon module. The Amorphous Si is incorporated with the heterojunction system or the mounting system in higher efficiency systems. Various material bandwidths increase the absorption of the cell.

4.7.2 Cadmium telluride

CdTe is one of the main vendors for the development of affordable PV systems among the thin-film solar cells and is also the first low-cost PV breakthrough [34]. The bandgap of CdTe also is $\sim$1.5 eV, and there are strong coefficients of optical and chemical absorption. These features make CdTe the most attractive product for thin-film solar cell design. CdTe was recorded in efficiency acquired 22.1% (Fig. 1) of the First Solar ZSW. In all cases, the use of Cd is easily debatable due to the toxicity of Cd in solar cell innovation and therefore risky to nature.

Energy storage and conversion devices **Chapter | 3** **67**

4.7.3 Copper indium gallium di-selenide

CIGS consist of four components: copper, indium, gallium, and selenium and is a quaternary compound semiconductor. CIGS is also direct band divide semiconductors. CIGS is $\sim10\%$–12% higher efficiency than the CDTe thin-film solar cell [35]. The solar cell technology associated with CIGS is an exciting tool to advance thin film because of its fundamental strength and economic efficiency.

4.8 Third-generation solar cells

Solar cells in the third generation are the breakthrough in the solar industry with promising technology. It shall be classified as follows:

(1) Polymer-based/Organic solar cells.
(2) Dye-sensitized solar cells.
(3) Perovskite solar cells.

4.8.1 Polymer-based/organic solar cells

The substrate used in organic solar cells to capture energy is organic material such as conjugated polymers. The phenomenon in which polymers are capable of acting as a semiconductor is a breakthrough for the 2000 Noble Prize in Chemistry for Alan J. Heeger, Alan Macdiarmid, Hideki Shirakawa. The first polymer solar cell is made of mixed poly [2-methoxy-5-(2′-ethylhexyloxy)-p-phenylene vinylene] (PPV), C60, and its numerous variants with high energy conversion efficiency [36]. This technique contributed to a further increase in the age of polymer products for the capture of solar energy. As fundamental criteria have progressed, experts have achieved a performance of more than 3.0% for the PPV type polymer solar cells [37, 38]. Such special features of polymer solar cells have paved the way for the production of extendable solar systems, such as new applications design and textures [37]. The efficiency of polymer solar cells has already been shown to be 14.2% [39]. The lifetime also has improved greatly and for some time plastic solar cells have been shown to be useful [40, 41].

4.8.2 Dye-sensitized solar cells

Dye-sensitized solar cells (DSSC) was developed by Michael Graetzel and Brain O'Regan and is often referred to as the Gräetzel Cell at the École Polytechnique Fédérale de Lausanne (EPFL), Switzerland, in the year 1991. DSSC's new development comes from photosensitizing TiO_2 nano grain coating combined with active dyes, thus making more than 10% efficiency maximum [42–44]. However, other challenges still exist, such as color loss and future stability problems [35]. This is due to the low optical absorption of the sensitizers, leading to week conversion efficiency.

4.8.3 Perovskite solar cells

The term perovskite solar cell comes from an ABX_3 crystal structure of the absorbing element known as a perovskite structure. The most commonly known absorber of perovskite with an optical bandgap between 2.3 and 1.6 eV is methylammonium lead trihalide ($CH_3NH_3PbX_3$, which includes an X halogen, for example, I-, Br-, Cl-). Formamidinium lead trihalide ($H_2NCHNH_2PbX_3$) also exists with a 2.2 and 1.5 eV bandgap. The minimum bandgap is similar to that for a single junction device for methylammonium lead trihalide, which makes them suitable for higher efficiencies [45]. The insertion of lead as part of perovskite materials is a typical problem; for example, $CH_3NH_3SnI_3$ solar cells based on tin-perovskite absorbers are also responsible for lower conversion efficiency [46]. Perovskite-based solar cells in tandem configuration have an efficiency of up to 31% [47]. According to an intriguing analysis by Volkswagen at a later date [47]. Perovskite can also perform an important job in batteries for next-generation electric cars. However, their stability and durability present problems with perovskite solar cells. After some time, the substance degrades and therefore decreases in high productivity. Therefore, more work is needed to commercialize these cells.

4.9 Overview of thin-film solar cell

A thin-film solar cell consists of two different types of semiconductors, such as p- and n-type. The n-type semiconductor is delivered by adding the impurity of the donor, e.g., arsenic, antimony, or phosphorus, to the intrinsic semiconductor and therefore the n-type semiconductor is one that has the donor dopants deposited in its crystal lattice and the majority has a negative charge. The electron concentration (n) in the n-type semiconductor is related to the position of the Fermi level (E_{Fn}) by [48];

$$n = N_c \exp\left[-\left(\frac{E_c - E_{Fn}}{k_B T}\right)\right] \tag{4}$$

Eq. (4) can be modified for an intrinsic semiconductor

$$n = N_c \exp\left[-\left(\frac{E_c - E_{Fi}}{k_B T}\right)\right] \tag{5}$$

From Eqs. (4), (5) the Fermi level position in the n type extrinsic semiconductor with respect to the intrinsic Fermi level

$$E_{Fn} - E_{Fi} = k_B T \ln\left(\frac{n}{n_i}\right) \tag{6}$$

A p-type semiconductor is delivered by adding an acceptor impurity, for example, gallium, boron, or indium to an intrinsic semiconductor. Here, the holes are called the majority and the majority carriers have a positive charge. The Fermi level position (E_{Fp}) is given below;

$$E_{Fp} - E_{Fi} = -k_B T \, \ln\left(\frac{p}{n_i}\right) \tag{7}$$

Thin-film solar cells are made from films that are much thinner than wafers that form the basis for PV production. As demonstrated by Deb et al. [32] a thin film is a layer produced by a random process of nucleation of an atomic/ionic/molecular species, independently condensing on a substrate. A large number of depositions are specifically subjected to the structural, chemical, metallurgical, and physical properties of that material. Thin-film solar cells were relied on to end up being cheaper than first-generation solar cells. Despite the present decline in the value of wafer-based solar cells, thin-film solar cells have not yet proved to be intriguing from a monetary point of view [49]. Thin-film cells typically have a low energy efficiency value compared to c-Si solar cells. The periodic guideline is a special case for GaAs [50]. Thin-film solar cells require a carrier to ensure mechanical stability rather than self-supporting silicon solar cells. In thin-film solar cells, the active semiconductor layers between a TCO layer and an electrical back contact are sandwiched. A back reflector on the back of the cell is regularly presented to limit transmissible losses. For thin-film solar cells, a wide range of semiconductors is being used. Some semiconductive devices require very rare components, such as indium (In), selenium (Se), or tellurium (Te). In the case of the terawatt PV scale, solar cells should be entirely based on abundant elements. The junction p-n comprises a mixture of semiconductor substances of p-type and n-type. The p-type has a large hole level and the n-type is built by large concentrations of electrons. The hole of p-type material diffuses in the n-type material and a different direction when the substances p and n are connected. Due to this impact, a depletion region is formed in the middle of the p-n junction where an electrical field is created. This procedure proceeds until a state of equilibrium is established. At the point where light occurs on a p-n junction electron-hole pair is created. The electron-hole pair may be created when the photon has sufficient energy than the bandgap energy. In particular, the electrons in the n-type material and the holes in the p-type material are not stable and exist only for a while in proportion to the life of the minority carrier before recombination. If they recombine, no current is produced as all the electron-hole pair is lost. P-n junction keeps this recombination, which separates the electron and the holes. The electrical field in the p-n junction isolates the carrier. Now, when the photo-generated carriers reach the depletion region, the electrical field activity pulls them over the opposite side. These carriers then flow through the external circuit, which includes the flow of electrical current. The TFSC consists of different materials in thin-film form (Fig. 10).

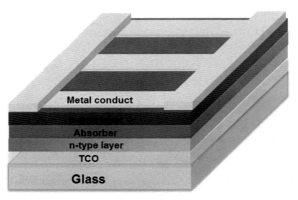

FIG. 10   Schematic diagram of thin-film solar cell.

In general, TFSC comprises of a substrate, TCO, window layer (p or n-type), absorber layer (I or p-type), and the metal contact layer. Typical TCO layers are produced using fluorine-doped tin oxide (FTO), aluminum-doped zinc oxide (ZnO:Al), boron-doped zinc oxide (ZnO:B), hydrogen doped (hydrogenated) indium oxide (In_2O_3:H) [51], and ITO, which is a blend of about 90% indium oxide (In_2O_3) and 10% tin oxide (SnO_2). These films are prepared to utilize either sputtering, low pressure-synthetic vapor statement (LP-CVD, utilized e.g., for ZnO:B), metal-organic chemical vapor deposition (MO-CVD), or atmospheric-pressure chemical vapor deposition (AP-CVD, utilized e.g., for FTO). Various reviews on TCOs have shown up in literature [52–55].

Assembling of photovoltaic modules includes the consecutive deposition of various thin films on a large area substrate. An average polycrystalline superstrate module producing process starts with cleaning of the glass substrate pursued by the TCO deposition, window layer, and absorber layer arrangement. Thin-film solar cell efficiencies are ascending and noticeable. Truth be told it was the first solar's CdTe thin-film solar panels that broke the $1/W achievement in mid-2009 [56]. The CdTe business is predominantly overwhelmed by the First solar, which has likewise as of late shown a record module of 18.6% efficiency [56] higher than that of the best commercial application multicrystalline module at any point recently recorded. PV industry shipments have become 15% in the most recent year, from 34.0-GWp in 2013 to 34.0-GWp in 2014 [57]. Inside the PV business, the development of thin-film organizations has launched, with in excess of 100 organizations entering the market somewhere in the range of 2001 and 2009 and production expanding from 14 to 2141 MW [58]. It is normal that in the long term, thin-film PV innovation will outperform crystalline advancements if the efficiency and reliability are bankable. In 2008, thin-film innovations had a 14% share, which further expanded in 2009 to 17%. The market share of thin-film PV is logically diminishing, with just a 7%–8% market share in 2014 [58, 59]. Yet, with the diminishing piece of the overall industry of thin-film, it must be understood that if the thin-film does not decisively supplant commercial crystalline silicon innovations, it has a substantial favorable position in numerous Sunbelt nations with diffuse light conditions and hot temperatures. Thin-film innovations with better temperature coefficients and perfect power conversion efficiencies in adverse environments do have points of interest over crystalline silicon.

4.10 Thin-film solar cell challenges

In recent years the market of photovoltaics has boomed enormously, with sales fully dominated by crystalline silicon, such as micro-digital electronics. More than half of the total cost of the module is covered by wafer costs. One way to eliminate this significant cost aspect is to replace wafers with thin films of semiconductors deposited on a supporting substrate (or even more commonly, glass). It offers an opportunity for thin-film products to increase market share and to place their accreditation on the market not too impressed by such devices, partly because of the apparent high reliability and durability of the wafer-based technology. Although they are the most complex of the basic physics of thin-film devices, they give two key points of interest. Since the thickness of the active layers will not be the same as that of the crystalline Si devices by two or three orders due to high optical absorption, the cost of the material will remain a small part of the total cost of the cell and the thin film deposition technique can be effectively modified for large area deposition without affecting the consistent processing of the "production line." The challenge of growing large-scale methods and building machines to produce thin-film solar cell arrays in a consistent process is daunting. When processes and technologies are built, these machines should be able to deliver enormous performance and therefore result in low unit costs. In any case, considering that every beneficial thing wants value, the ability to customize the different properties of thin films needed for an effective solar cell requires a good understanding of the material thus produced with the help of a scope test and analytic equipment. It should also be noted that the high resistance of film properties to deposition parameters will produce a large number of undesirable results; in this way, thin-film materials should be treated with due respect and understanding.

5. Conclusions

The primary focus of this chapter is on the fundamentals, basic components, and performance assessment of different energy storage and conversion systems, notably batteries, supercapacitors, and solar cells. The current literature on relevant concepts, working mechanisms, and classification of batteries, supercapacitors, and solar cells are also reviewed in this chapter, with a focus on the different basic components used in energy storage and conversion devices.

72 Oxide free nanomaterials for energy storage and conversion applications

References

[1] Y. Wang, G. Cao, Developments in nanostructured cathode materials for high-performance lithium-ion batteries, Adv. Mater. 20 (2008) 225.

[2] A.J. Salkind, J.J. Kelly, A.G. Cennone, in: D. Linden (Ed.), Hand Book of Batteries, McGraw Hill, New York, 1995.

[3] M.S. Whittingham, Lithium batteries and cathode materials, Chem. Rev. 104 (2004) 427.

[4] S.U. Falk, S.J. Salkind, Alkaline Storage Batteries, Wiley, New York, 1969.

[5] M. Winter, R.J. Brodd, What are batteries, fuel cells, and supercapacitors, Chem. Rev. 104 (2004) 4245.

[6] O. Haasa, E.J. Cairns, Electrochemical energy storage, Annu. Rep. Prog. Chem., Sect. C: Phys. Chem. 95 (1999) 163–197.

[7] G. Plante, The Storage of Electrical Energy, MT Publisher Kessinger Publishing, 2007.

[8] R. Anthony, West, Solid State Chemistry and its applications, John Wiley Sons, 2005.

[9] D. Gielen, Electricity Storage and Renewable for Island Power: A Guide for Decision Makers, International Renewable Energy Agency, 2012.

[10] G. Coppez, S. Chowdhury, S.P. Chowdhury, Review of battery storage optimisation in distributed generation, in: Power Electronics, Drives and Energy Systems (PEDES) & 2010 Power India, 2010, pp. 1–6.

[11] T. Lambert, P. Gilman, P. Lilienthal, Micropower System Modelling with HOMER: Integration of Alternative Sources of Energy, John Wiley & Sons, New York, 2006.

[12] X. Yang, Y. Yu, N. Yan, H. Zhang, X. Li, H. Zhang, 1-D oriented cross-linking hierarchical porous carbon fibers as a sulfur immobilizer for high performance lithium-sulfur batteries, J. Mater. Chem. A. 4 (2016) 5965–5972.

[13] J. Liu, X. Sun, Elegant design of electrode and electrode/electrolyte interface in lithium-ion batteries by atomic layer deposition, Nanotechnology 26 (2015) 24001.

[14] S. Zhang, K. Ueno, K. Dokko, M. Watanabe, Recent advances in electrolytes for lithium-sulfur batteries, Adv. Energy Mater. 5 (2015) 1–28.

[15] Z. Li, Y. Huang, L. Yuan, Z. Hao, Y. Huang, Status and prospects in sulfur-carbon composites as cathode materials for rechargeable lithium-sulfur batteries, Carbon 92 (2015) 41–63.

[16] P.G. Bruce, S.A. Freunberger, L.J. Hardwick, J.-M. Tarascon, Li-O_2 and Li-S batteries with high energy storage, Nat. Mater. 11 (2011) 19–29.

[17] X. Ji, K.T. Lee, L.F. Nazar, A highly ordered nanostructured carbon–sulphur cathode for lithium–sulphur batteries, Nat. Mater. 8 (2009) 500–506.

[18] B.E. Conway, Electrochemical Supercapacitors: Scientific Fundamentals and Technological Applications (POD), Kluwer Academic/Plenum, New York, 1999.

[19] A. Davies, Y.A. Ping, Material advancements in supercapacitors: from activated carbon to carbon nanotube and grapheme, Can. J. Chem. Eng 89 (2011) 1342.

[20] S. Nagamuthu, K.S. Ryu, Synthesis of Ag/NiO honeycomb structured nanoarrays as the electrode material for high performance asymmetric supercapacitor devices, Sci. Rep. 9 (2019) 1.

[21] S. Ghosh, S.M. Jeong, S.R. Polaki, A review on metal nitrides/oxynitrides as an emerging supercapacitor electrode beyond oxide, Korean J. Chem. Eng. 35 (2018) 1389–1408.

[22] Y. Wang, Y. Ding, X. Guo, Conductive polymers for stretchable supercapacitors, Nano Res. 12 (2019) 19781987.

[23] A. Muzaffar, M.B. Ahamed, K. Deshmukh, J. Thirumalai, A review on recent advances in hybrid supercapacitors: design, fabrication and applications, Renew. Sustain. Energy Rev. 101 (2019) 123–145.

[24] U.S. EIA (Org.), International Energy Outlook 2017 Overview: U.S. Energy Information Administration, 2017, p. 76.

[25] C. Fröhlich, J. Lean, Solar radiative output and its variability: evidence and mechanisms, Astron. Astrophys. Rev. 12 (2004) 273–320.

[26] A. Einstein, Über einen die Erzeugung und Verwandlung des Lichtes betreffenden heuristischen Gesichtspunkt, Ann. Phys. 322 (1905) 132–148.

[27] Y. Hishikawa, E.D. Dunlop, D.H. Levi, M.A. Green, J. Hohl, E. Masahiro, Y. Anita, W.Y.H. Baillie, Solar Cell Efficiency Tables (Version 53), vol. 2, 2019, pp. 3–12.

[28] L.C. Andreani, A. Bozzola, P. Kowalczewski, M. Liscidini, L. Redorici, Silicon solar cells: toward the efficiency limits, Adv. Phys. X. 4 (2019) 1–24.

[29] M. Drahansky, M. Paridah, A. Moradbak, A. Mohamed, F.A.T. Owolabi, M. Asniza, S.H. Abdul Khalid, Industrial Silicon Solar Cells, vol. 13, Intech, 2016, pp. 303–319.

[30] T. Saga, Advances in crystalline silicon solar cell technology for industrial mass production, NPG Asia Mater. 2 (2010) 96–102.

[31] Jayakumar, Biomass Energy: Resource Assessment Handbook Asian and Pacific Centre for Transfer of Technology, Economic and Social Commission for Asia and the Pacific: United Nation (UN), 2009.

[32] S.K. Deb, Thin-film solar cells: an overview, Renew. Energy 8 (1996) 69–92.

[33] M. Imamzai, M. Aghaei, Y.H. Thayoob, A review on comparison between traditional silicon solar cells and thin- film CdTe solar cells, Proc. Natl. Grad. Conf. 2012 (2011) 8–10.

[34] S. Luque, A. Hegedus, Handbook of Photovoltaic Science and Engineering, John Wiley & Sons, New York, 2011.

[35] A. Mohammad Bagher, Types of solar cells and application, Am. J. Opt. Photonics. 3 (2016) 94–113.

[36] F. Wudl, G. Srdanov, Conducting Polymer Formed of Poly(2-Methoxy,5-(2′-Ethyl-Hexyloxy)-p-Phenylenevinylene): US Patent, 1993.

[37] G. Li, R. Zhu, Y. Yang, Polymer solar cells, Nat. Photo-Dermatology 6 (2012) 153–161.

[38] C.J. Brabec, S.E. Shaheen, C. Winder, N.S. Sariciftci, P. Denk, Effect of LiF/metal electrodes on the performance of plastic solar cells, Appl. Phys. Lett. 80 (2002) 1288–1290.

[39] S. Li, L. Ye, W. Zhao, H. Yan, B. Yang, D. Liu, W. Li, H. Ade, J. Hou, A wide band gap polymer with a deep highest occupied molecular orbital level enables 14.2% efficiency in polymer solar cells, J. Am. Chem. Soc. 140 (2018) 7159–7167.

[40] J. Nowotny, C.C. Sorrell, L.R. Sheppard, T. Bak, Solar-hydrogen: environmentally safe fuel for the future, Int. J. Hydrog. Energy 30 (2005) 521–544.

[41] E. Coronel, A. Hultqvist, Solar cell efficiency tables: version, Prog. Photovolt. Res. Appl. 17 (2009) 115–125.

[42] B. Li, L. Wang, B. Kang, P. Wang, Y. Qiu, Review of recent progress in solid-state dye-sensitized solar cells, Sol. Energy Mater. Sol. Cells. 90 (2006) 549–573.

[43] M. Graetzel, R.A.J. Janssen, D.B. Mitzi, E.H. Sargent, Materials interface engineering for solution-processed photovoltaics, Nature 488 (2012) 304–312.

[44] M. Liang, W. Xu, F. Cai, P. Chen, New triphenylamine-based organic dyes for efficient dye-sensitized solar cells, J. Phys. Chem. C (2007) 4465–4472.

[45] G.E. Eperon, S.D. Stranks, C. Menelaou, M.B. Johnston, L.M. Herz, H.J. Snaith, Formamidinium lead trihalide: a broadly tunable perovskite for efficient planar heterojunction solar cells, Energy Environ. Sci. 7 (2014) 982–988.

[46] N.K. Noel, S.D. Stranks, A. Abate, C. Wehrenfennig, S. Guarnera, A.A. Haghighirad, A. Sadhanala, G.E. Eperon, S.K. Pathak, M.B. Johnston, A. Petrozza, L.M. Herz, H.J. Snaith,

74 Oxide free nanomaterials for energy storage and conversion applications

Lead-free organic-inorganic tin halide perovskites for photovoltaic applications, Energy Environ. Sci. 7 (2014) 3061–3068.

[47] D. Shi, Y. Zeng, W. Shen, Perovskite/c-Si tandem solar cell with inverted nanopyramids: realizing high efficiency by controllable light trapping, Sci. Rep. 5 (2015) 1–10.

[48] M.D. Sturge, Statistical and Thermal Physics Fundamentals and Applications, CRC Press, 2018.

[49] M.S. Richard, P. Feynman, R.B. Leighton, The Feynman Lectures on Physics, Basic Books, New York, 1963.

[50] M.A. Green, K. Emery, Y. Hishikawa, W. Warta, E.D. Dunlop, Solar cell efficiency tables (version 44), Prog Photovoltaics Res. Appl. 22 (2014) 701–710.

[51] T. Koida, H. Fujiwara, M. Kondo, Hydrogen-doped In_2O_3 as high-mobility transparent conductive oxide, Jpn J. Appl. Phys, Part 2 Lett. 46 (2007) L685–L687.

[52] K.L. Chopra, S. Major, D.K. Pandya, Transparent conductors-A status review, Thin Solid Films 102 (1983) 1–46.

[53] R.G. Gordon, Criteria for choosing transparent conductors materials properties relevant to transparent conductors optical and electrical performance of transparent conductors, MRS Bull. 25 (2011) 52–57.

[54] P. Löbmann, Transparent conducting oxides, MRS Bull. 25 (2000) 15–18.

[55] S. Major, A. Banerjee, K.L. Chopra, Highly transparent and conducting indium-doped zinc oxide films by spray pyrolysis, Thin Solid Films 108 (1983) 333–340.

[56] T.D. Lee, A.U. Ebong, A review of thin film solar cell technologies and challenges, Renew. Sustain. Energy Rev. 70 (2017) 1286–1297.

[57] IDTechEx, Photovoltaic Manufacturer Shipments: Capacity, Production, Prices and Revenues to 2019, 2014.

[58] G. Adiboina, Thin Film Photovoltaic (PV) Cells Market Analysis to 2020, Alterenergymag. com, 2019, pp. 1–10.

[59] IDTechEx, Global Photovoltaic Shipments Jump 15% in 2014 Photovoltaic Technology Trends : A Supply Perspective, 2019, pp. 1–9.

Chapter 4

Oxides free materials for symmetric capacitors

Govindhasamy Murugadoss[a], Sunitha Salla[b], and Prabhakarn Arunachalam[c]

[a]Centre for Nanoscience and Nanotechnology, Sathyabama Institute of Science and Technology (Deemed to be University), Chennai, Tamil Nadu, India, [b]Department of Chemistry, Sathyabama Institute of Science and Technology (Deemed to be University), Chennai, Tamil Nadu, India, [c]Electrochemical Science Research Chair, Chemistry Department, College of Science, King Saud University, Riyadh, Saudi Arabia

Chapter outline

1. Introduction 75	5.1 Types of TMS based SCs 83
1.1 Electrochemical energy storage systems 76	5.2 Other transition TMS 86
1.2 SCs and capacitor 76	**6. Graphene, CNT, and carbon-based materials for SCs** 87
1.3 Structure of SCs 77	6.1 Graphene with metal oxides and metal sulfides nanocomposites 88
1.4 Supercapacitor mechanism and its type 77	6.2 CNTs for high performance of SCs 89
2. TMO-based SCs-PCs 81	6.3 AC 89
3. TMOs merits and drawbacks 81	**7. Summary and conclusion** 90
4. Metal-based chalcogenides on electrochemical SCs 82	**References** 90
5. Sulfide based SCs 82	

1. Introduction

The climate variation and lack of availability of fossil fuels have greatly influenced the world's economy and ecology. In recent years, sustainable and renewable energy storage technologies have been significantly developed because of conserving natural resources and regulating energy consumption. The huge requirement for solving the future energy crisis and increasing environmental concerns has driven high-performance energy storage technologies. Among the various energy storage devices such as batteries, capacitors, and supercapacitors (SCs), SCs are emerging energy devices that bridge the gap between

Oxide Free Nanomaterials for Energy Storage and Conversion Applications.
https://doi.org/10.1016/B978-0-12-823936-0.00005-X
Copyright © 2022 Elsevier Inc. All rights reserved.

76 Oxide free nanomaterials for energy storage and conversion applications

dielectric capacitors and batteries. SCs are the electrochemical devices produced to find solutions for increasing energy demand, which is a rising problem in the future. There is a developing interest in using SCs in energy storage technologies owing to their high specific power, fast charge/discharge rates, and longer electrochemical durability. SCs found many applications in industries, most notably for military vehicles and automotive sectors having electric vehicles because of their ecological features, storage capabilities, etc., and considered a substitute source to batteries [1].

1.1 Electrochemical energy storage systems

Nowadays, it is mandatory to store the harvested energy for enormous applications to enhance energy storage devices. Batteries, fuel cells, capacitors, and SCs are energy-storing devices mostly utilized for several purposes. Generally, batteries comprise two or more voltaic cells associated in series to deliver a steady dc voltage at its output terminals. The voltage is created by a substance response inside the cell. Electrodes are submerged in an electrolyte that powers the separation of electrical charges into ions and free electrons. Batteries can hold a lot of power; however, they require hours to energize. Fuel cells have low energy density. Hence, in batteries and fuel cells, the circuit in a flash attachment usages a capacitor to store energy. Consequently, rather than a battery and power device, capacitors kept their energy as an electric field than the chemicals that undergo reactions; they can be recharged repeatedly. Besides, capacitors charge promptly however store only small amounts of power. So, according to demand, rapidly storing and releasing large amounts of electricity approach is necessary. Thus, energy storage devices such as electrochemical capacitors with high power density, longer life cycle, cost-effectiveness, and small size are useful. Due to its tremendously high capacitance density, electrochemical capacitors are also known as SCs.

1.2 SCs and capacitor

Electrochemical SCs are like different kinds of capacitors because they assist a similar capacity—charging and discharging (CD). What makes a difference is how much energy they can store and at what voltages they work. These assortments are achieved by the distinctions in how SCs are planned associated with other capacitor advancements. The dielectric in ceramic, aluminum and tantalum, capacitors include thousands of particles, while SCs utilize one atom to shape the dielectric. The dielectric material is created to make contact with carbon and creates an inner Helmholtz Layer. The external Helmholtz layer presented in Fig. 1 functions as the cathode of one of the capacitors on one terminal and one of the capacitor's anodes on another electrode.

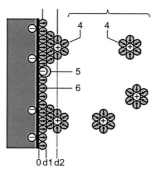

FIG. 1 Formation of Helmholtz double layer.

1.3 Structure of SCs

In the arrangement of SCs, the external Helmholtz layer might confuse both cathode and anode. The outer Helmholtz layer structures one cathode and one anode, while the other cathode and anode are formed on the different sides of the separator when the capacitor is charged. This is where the name electrochemical double-layer capacitors (EDLCs) come from. The two terminals are covered with a slight carbon layer and the electrolyte, which has a separator in the center. The formation of the double layer is presented in Fig. 2.

The higher specific capacitance (Cs) is attained when it was used highly porous carbon. This raises the specific surface area (SSA), which is directly related to the capacitance of a capacitor. Since there is just a single molecule, the electrolyte is extremely flimsy, which implies the distance between the terminals is very thin than in different systems. This allows the part to have a prevalent Cs as the distance between terminals is inversely proportionate to the capacitance as existing in the equation given below,

$$C = \varepsilon_0 \kappa A / d \qquad (1)$$

This likewise explains why the voltage rating of SCs is low. Since the dielectric is very small thick, voltages better than the appraised voltage place electrical pressure across the electrolyte molecule and tears it apart. This harm is irreversible, and the part can't self-recuperate like other capacitor dielectrics.

1.4 Supercapacitor mechanism and its type

Electrochemical SCs comprised of a couple of electrodes, one electrolyte, and a separator are the essential constituents to form fuel-cell SCs. The arrangement of these SCs may be symmetric or asymmetric. The two electrodes are separated by a separator using filter paper, cellulose membrane with better porous properties for particle transportation. SCs are categorized into, (i) EDLCs, (ii) Pseudocapacitors (PCs) and (iii) Hybrid capacitors (HCs).

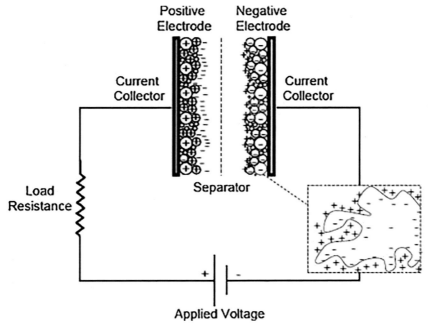

FIG. 2 Schematic sketch of an electrical double-layer capacitor. *(From A.K. Shukla, A. Banerjee, M.K. Ravikumar, A. Jalajakshi, Electrochemical capacitors: technical challenges and prognosis for future markets, Electrochimica Acta 84 (2012) 165–173, https://doi.org/10.1016/j.electacta.2012.03.059.)*

1.4.1 EDLCs

For the most part, the main sort called EDLCs uses carbon materials, for example, graphene, enacted carbon (AC), nanostructured carbon, and carbon aerogels (CA), for the collection of charge reversible adsorption/desorption of particles at the interface between the anode/electrolyte. EDLC materials have been broadly investigated owed to their more prominent SSA, great electrical conductivity, and excellent mechanical stability, yet they hurt from a low Cs. There are three sorts of EDLCs as far as the carbon content, prompting different functions in the device, like morphology, hybridization, and structural defects, and are discussed below.

Carbon aerogels, carbon foams and, carbide-derived carbon

They are the lightest materials with a high SSA, good thermal, electrical, mechanical properties, and great adsorption levels. They are synthesized by a sol-gel process containing a 3D system of CNTs with an enlarged Young's modulus. Various synthetic approaches permit controlling the pores' concentration independently, which the carbon aerogel acts as appropriate material. If the metal is included as a starting material, it affects modifying in pH, activation, etc., creating it very hard to employ the reliability of the pores [2].

Graphene and CNTs

Graphene is a one-atom-thick layer of carbon particles with great electrical and mechanical properties and is synthesized by different methods. It has various structural dimensions from 0D to 3D providing a fine SSA and high electrical and thermal properties. It would be great to embed this graphene's in SCs for long and better cycle performance. Graphene also forms nanosized composites with various groups like conducting polymers (CPs), transition metal oxides (TMO), etc. These nanosized composites permit multiple features to be realized, including high SSA, power density, and good electrical conductivity. When these graphene sheets were folded up into a cylinder, it forms CNTs which also exhibits good electrical conductivity and high SSA. There are two kinds of CNTs: single-walled carbon nanotubes (SWCNTs) and multi-walled carbon nanotubes (MWCNTs), in which SWCNTs have high flexibility and size of around 5 nm. MWCNTs comprise a large number of defects in their arrangement than the SWCNTs. These CNTs can be obtained through physical vapor deposition (PVD), chemical vapor deposition (CVD), electric arc discharge method. Even though CNTs are professed to have profoundly looked for highlights, their nanosized composite natures are chemically changed, and supported CNT materials will, in general, have a higher evaluation and elevated qualities [3].

AC

EDLC devices based on AC electrodes exhibit greater electrochemical performance because of their high oxidizing property. Such property can be defined by controlling the pore size and also C-C dangling bonds. The pore size and ion size of an electrolyte are straightforwardly related, and interaction between these two limits provides an ideal electrochemical presentation in a capacitor. The main disadvantage of the AC electrode is that it does not pay for high capacitance because of the discrepancy of ion electrolytes [4, 5].

1.4.2 Pseudocapacitors (PCs)

The next type is called pseudocapacitors (PCs), its energy is kept through a quick and reversible faradaic process over the surface of the active materials. Appropriate electrode candidates for PCs are systematically explored, as TMOs offer moderately high Cs and higher specific energy while CPs have good intrinsic conductivity, subsequently creating distinct sources for better electrochemical activities for SCs. Unfortunately, CPs experience low cycling-stability features because of the continued growing and contracting of the polymer chains during the doping and undoping measure [6]. Mostly, all electrode materials of this type of capacitor contain TMO, metal-doped carbon, and

CPs. The major disadvantage of this type is poor conductivity, and due to this, the Cs value is poor.

CPs pseudocapacitor

They possess high capacitance, high potential with good potential densities, and example materials, including polypyrrole, polyaniline (PANI), and polythiophene. They are used to enhance capacitance as a nanofiller, and these polymer-based electrodes exhibit low cyclic stability than carbon-based electrodes [7].

TMO PCs

To develop high-energy storage technologies, TMO PCs have fascinated much consideration as PC electrodes because of their exceptional specific energy output compared with EDLC materials. Different TMO based on Ru, Mn, Ni, Co, V-oxides is most normally utilized in SCs due to their greater theoretical Cs from the faradaic charge transfer process [8–12]. Among these oxides, RuO_2 is considered an essential electrode for TMO-based SCs, since it has high theoretical capacitance and better rate capability. Although it is deemed to be superior potential SCs material than EDLC and conducts polymer-based electrodes because of its agglomeration nature, its electrical performance is hampered during the CD process [13].

1.4.3 Hybrid capacitors (HCs)

Because of the various limitations for these two types of SCs, huge attempts were made to develop hybrid SCs [14], which have the advantages of both EDLCs and PCs; HCs consist of polarizable and nonpolarizable electrodes for storing charges. It employs both the Faradaic and non-Faradaic processes to obtain high energy storage, ensuring better electrochemical durability with low cost than EDLCs. It is of three important types, namely asymmetric, composite, and battery type.

Asymmetric hybrid SCs

This is a distinct type of SCs consisting of two different electrodes intended to function concurrently to state the power density and energy density requirements. One electrode works as a capacitive electrode among the two electrodes, and the other electrode works as a faradaic electrode. Carbonaceous nanomaterials work as a negative electrode, and metal or TMOs work as a positive electrode. These types of capacitors have high energy density and cycling stability. It is seen that there is a reduction in the conductivity with an increment in the internal resistance because all electrolytes, exhaustion of particles and electrodes will occur [15].

Composite hybrid SCs

Composite hybrid SCs combine both carbon and TMO properties to offer the synergistic electrochemical features sought after, including Cs, cycling durability, and greater electronic conductivity. Carbon supports charge transport, and the TMO will stock charge by redox responses. Due to these limitations, like the efficiency, it began to reduce when the layer became thicker when the TMO was layered on a carbon source [16].

2. TMO-based SCs-PCs

TMO-based SCs are exclusive from the EDLS capacitors wherein a potential is introduced to an SCs and undergoes redox reactions at the electrode substances and contain the route of charge throughout the double layer, alike to the CD processes that arise in batteries, ensuing in Faradaic current passing over the SCs cell. The oxide substances such as RuO_2, MnO_2, Co_3O_4, etc., are employed [17–28].

3. TMOs merits and drawbacks

To meet sustainable and renewable energy storage requirements, such as electronic devices, hybrid electric vehicles, and larger industrial machinery, SCs have been recognized as an ideal storage device in meeting future demand. In developing sophisticated, higher-performance energy storage technologies, TMOs have been considered an appropriate PC electrode due to admirable specific energy output associated with EDLC electrodes. SCs electrodes-based promises dramatically to endorse the capacitance of SCs for its faradic charge-storage processes. All TMOs material provides a much higher energy density than carbonaceous materials with EDLC. The electrochemical activities of electrodes are strongly affected by their crystalline nature, electrical conductivity, and storage mechanisms. However, the problem of electrode-based TMOs is a comparatively lower conductivity and slow ion transfers, which can cause damage to the capabilities of the rate and stability of the cycle. Besides, the nanohybrid consists of TMOs, and conductor material matrices such as carbon materials must be stipulated smoothly, leading to increasing capacitance, the ability of a longer cycling level, and stability. Some approaches, including doping elements, integrating surface oxygen vacancies, and functionalization, were initially advanced but unsatisfactory. More importantly, the mechanism of electrochemical reactions for many in various SCs systems is still unclear and feasible to investigate in depth. Improperly, the factors that prevent arriving at a high-explicit limit are simple agglomeration on high mass stacking [29], the idea of the cycle, and low conductivity [30].

4. Metal-based chalcogenides on electrochemical SCs

The technologically essential and systematically substantial transition metal-based chalcogenides (TMC- S, Se) have gained significant consideration in the last decades owing to their anisotropic features. Generally, transition metals of sets IV to B syndicate with VI A group elements, namely S, Se, and Te, to create crystal structure with double-layer [31, 32]. Further, TMC is mainly built on the transition metal grouping, the number of layers, and the attendance or nonexistence of adopting atoms. Therefore, its bandgaps are varied between 0 and 2 eV. TMC's optical features are tunable and have become commercially viable electrodes for electrochemical SCs [33].

In this section, we mainly discuss the importance of TMC in symmetric SCs. They have attracted significant devotion owing to their greater power densities and higher durability natures, and they provide significantly better tolerance than batteries in various kinds of uses in consumer electronics, hybrid electronic vehicles, buffer powers, and so forth [32]. Further, the chalcogenides have been widely recognized in the arena of fuel cells, batteries, sensors, electrocatalytic materials, thermoelectric devices, and memory devices and are commonly applied in electrochemical SCs. Moreover, we define the promising future areas of TMC, covering both their electrochemical features and their applications in SCs. Definitely, TMS reveals significantly enhanced electrochemical features, which mostly came from its greater electrical conductive natures, and mechanical and thermal stability. Besides, it is recognized that energy storage technologies' performance is mainly built on the crystalline nature, nanosized electrocatalytic materials, morphological and textural features, composition, and electrodes' design [32]. Transition metal selenides (TMSe) have fascinated significant consideration as favorable electrode candidates, retaining rich redox chemistry and superior electrical features as a new kind of battery-like electrode material. However, compared to TMS, the TMSe were less reported than TMS.

5. Sulfide based SCs

In recent years, nanostructured has played a substantial part in electronics, mainly optoelectronic devices, owing to its distinctive, outstanding physicochemical performances, like MoS, CoS, NiS, MnS, FeS, etc., signify attractive candidates for energy-related systems due to the exceptional electrochemical characteristics they showed [34]. The electrochemical feature is much superior to the electrochemical features. This can be clarified by the existence of sulfur atoms in their place of oxygen atoms. Has fascinated much consideration in numerous fields, including SCs, solar cells, and LIBs, because of their characteristic optical and electrical features [35]. These greater electrical features are primarily associated with their precise forms and crystal structures with unique surface morphology in terms of having distinctive shapes [36].

5.1 Types of TMS based SCs

5.1.1 Nickel sulfide

The Nickel Sulfide NiS have been extensively explored to examine their ability to be used as SCs. NiS is a semiconductor and can be synthesized by various methods in numerous compositions. Xu et al. [37] recently demonstrated nano-sized electrode composites based on NiS and $NiCo_2S_4$ hydrothermally, using AC as a negative electrode and $NiCo_2S_4$/NiS as a positive one [37]. They obtained enhanced electrochemical features such as a power density of $160 \, W \, kg^{-1}$ and an energy density of $43.7 \, W \, h \, kg^{-1}$. Further, Zhu et al. demonstrated the Ni_3S_2 with nanosheet structure on a CNT surface with a Cs of $514 \, F \, g^{-1}$ at $4 \, A \, g^{-1}$ and exceptional electrochemical durability [38]. To improve the Cs of nanostructured Ni_3S_2/graphene materials, a facile method directed by changing the degree of sulfidation was accounted for by Ou et al. [39].

5.1.2 Copper sulfide

CuS has been extensively studied for energy-related systems, gas sensors, and photocatalysts. Recently, Peng et al. prepared CuS with dissimilar morphological features through a low-temperature solvothermal approach and applied it for SCs [40] with superior SSA flower-like CuS delivered an exceptional Cs of $597 \, F \, g^{-1}$. Particularly, the CuS-based chalcogenides offer an electronic conductivity of $10^{-3} \, S \, cm^{-1}$, and it tends to possess the theoretical specific capacity of $561 \, mA \, h \, g^{-1}$. Notable, the examined electrode materials are not appropriate for SCs because bare CuS is a semiconductor with comparatively lower electronic conductivity related to carbonaceous materials and CPs, and its volume variation through cycling endorses poor cycling durability. Therefore, it desires to geometrically regulate the fabrication of CuS nanosized materials and associate with electronically conductive substances to significantly promote SCs features.

5.1.3 Cobalt sulfide

In the past decades, numerous nanosized electrode materials built on CoS have been demonstrated to exploit energy storage and SCs. Govindasamy et al. synthesized d nanostructured nickel cobaltite sulfide/CoS, which reveals an exceptional Cs of $1565 \, F \, g^{-1}$ at $1 \, A \, g^{-1}$ and maintained 91% of its initial SC after several 8000 cycles at $1 \, A \, g^{-1}$ [41]. Chen et al. demonstrated a highly active SC by employing Co_3S_4 NS arrays on NF, and the Cs and electrochemical durability of Co_3S_4 NS arrays electrodes were fourfold enhancements in electrochemical features of Co_3O_4 NS (Fig. 3) [42].

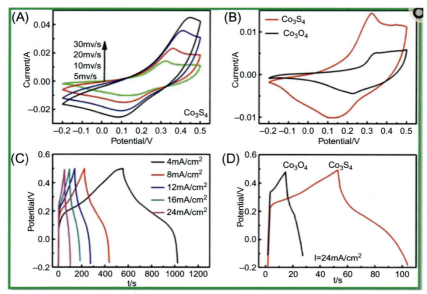

FIG. 3 (A) The cyclic voltammetric plots of Co$_3$S$_4$ NS arrays on NF at numerous sweep rates; (B) comparative cyclic voltammetric plots of the Co$_3$S$_4$ and Co$_3$O$_4$ on NF electrodes at 5 mV s^{-1}; (C) The galvanostatic charge-discharge (GCD) properties of the Co$_3$S$_4$ NS arrays at various current densities; (D) comparative GCD plots of the Co$_3$S$_4$ NS and Co$_3$O$_4$ nanowire arrays on NF.

5.1.4 Iron sulfide

Balakrishnan et al. prepared a hybrid SCs based on iron sulfide (FeS$_2$) and reduced graphene oxide (rGO) hydrothermally with a more excellent value of Cs than pure FeS$_2$ a current density of 0.3 mA cm^{-2}, it maintained 90% of its initial SC after 10,000 cycles [43]. Afterward, FeS$_2$ has fascinated worldwide researchers for its potential usage in energy-related devices (Fig. 4).

5.1.5 Molybdenum disulfide

Molybdenum disulfide (MoS$_2$) and its nanostructured composites have been explored broadly in different fields such as electrocatalysis, energy-related systems, SCs, and batteries [44]. Recently, hydrothermal synthesis of new kinds of SCs based on MoS$_2$ and graphitic carbon nitrides materials was reported, and it reveals a Cs of 532.7 F g^{-1} at 1 A g^{-1}. Moreover, the fabricated composites maintained their 86% initial capacitance after 1000 lifecycles [45]. Similarly, Manuraj et al. demonstrated a nanocomposite hetero-structured solid substance comprising MoS$_2$, nanowires, and RuO$_2$ nanoparticles via hydrothermal routes. The obtained MoS$_2$-RuO$_2$ composite electrode reveals Cs maintained 972 F g^{-1} at 1 A g^{-1}. Moreover, MoS$_2$-RuO$_2$ hybrid electrode displays a greater energy density value of 35.92 W h kg^{-1} at 0.6 kW kg^{-1} [46].

Oxides free materials for symmetric capacitors **Chapter | 4** **85**

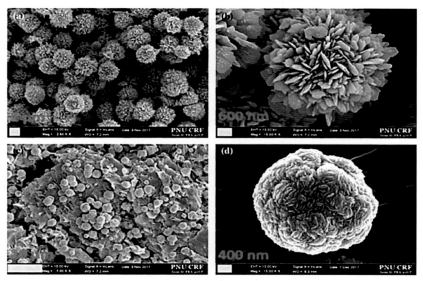

FIG. 4 Scanning electron microscopy (SEM) photographs of (A, B) micro flowers of FeS$_2$ and (C, D) microspheres of rGO/FeS$_2$ hybrid. *(Reproduced from B. Balakrishnan, S.K. Balasingam, K.S., Nallathambi, A. Ramadoss, M. Kundu, J.S. Bak, …, H.J. Kim, Facile synthesis of pristine FeS$_2$ microflowers and hybrid rGO-FeS$_2$ microsphere electrode materials for high performance symmetric capacitors, J. Ind. Eng. Chem. 71 (2019) 191–200. Copyright (2019) Elsevier.)*

5.1.6 Binary TMS

Although many metallic TMS are explored as electrode materials for SCs, binary-based TMS is quite impressive owing to its highest active redox sites and mechanical and thermal stability than its equivalent counterparts. Most binary-based nanostructured TMS is obtained by introducing the Kirkendall effect [47], and nowadays, several binary TMS is prepared in keeping with them [48]. The Kirkendall effect depends on the two-metal mutual diffusion process over an interface, so the diffusion of vacancies occurs to complete the inequality of the flow of material and in which the initial interface moves. However, significant research efforts on binary TMS, since the SCs materials remain narrow.

5.1.7 NiCo$_2$S$_4$

NiCo$_2$S$_4$ nanotubes were synthesized and used before as PC electrodes with good electrochemical performance. NiCo$_2$S$_4$ nanoplates hexagonal hollow were reported by Pu et al. [48], which reveals a high Cs of $437 \, \text{F} \, \text{g}^{-1}$ at $1 \, \text{A} \, \text{g}^{-1}$ in the electrolyte solution of 3M KOH. NiCo$_2$S$_4$ hollow spheres by Self-templating synthesis show the best electrochemical properties where intrinsic electronic conductivity is found 100 times higher than the appropriate binary TMOs [49]. While the NiCo$_2$S$_4$ nanostructure is prepared by polyol

86 Oxide free nanomaterials for energy storage and conversion applications

method exhibited high Cs with extraordinary capacitance retention and is demonstrated as potential PCs materials for the SCs [50]. $NiCo_2S_4$ based TMS loaded over carbon fiber cloth and paper was examined, and $NiCo_2S_4$ on carbon fiber cloth showed a favorable charge-transfer kinetics and rapid electron transportation related to $NiCo_2S_4$ carbon fiber cloth and therefore had greater electrochemical features [51].

5.1.8 Manganese cobalt sulfides (MCS)

The composite of MCS is very restricted even though they are favorable electrode materials for the SCs. In recent years—based electrodes have received attention because of their environmentally friendly nature and good redox properties. Ultrathin mesoporous NS was developed on Ni foam (NF) by electrodeposition approaches for SCs [52]. Zhao et al. was synthesized with a pinecone-like and hierarchical porous nanosphere structure which reveals Cs of g^{-1} and g^{-1} at $1\,A\,g^{-1}$, correspondingly [53]. Synergistic effects between hierarchical CoS and pinecone-like MnS nanospheres contribute to increased Cs. NS array sticks onto rGO/NF (/rGO/NF) by a wisely regulated sulfurization time process is prepared and revealed a greater specific capacity of g^{-1} at $1\,A\,g^{-1}$, and longstanding cyclic durability of 92.9% at a current density of g^{-1} over 3000 cycles [53].

5.2 Other transition TMS

5.2.1 Bi_2S_3

More attention has been paid in the last decades to Bi_2S_3 owing to its specific electrical and optical properties and applies in the SCs field [54], and it is a layered semiconductor material, and mainly in the orthorhombic form. Rod-like Bi_2S_3 micro flowers were synthesized and provided a higher Cs of $185.7\,F\,g^{-1}$ at $1\,A\,g^{-1}$ [55]. Hetero-chemically structured Bi_2S_3 nanorod/MoS_2 NS composite electrode revealed a Cs of $1258\,F\,g^{-1}$ at $10\,A\,g^{-1}$ with good retention in capacitance [56]. Afterward, Raut et al. demonstrated Bi_2S_3 electrodes on stainless steel by employing succeeding ionic layer adsorption and reaction methods, which enhanced capacitance features with longstanding durability [57].

5.2.2 La_2S_3

Lanthanum-based rare earth element TMC is treated as a favorable SC electrode due to a stable transition state. Lanthanum sulfides have three forms: LaS, La_2S_3, and La_3S_4, with exceptional pseudocapacitive performance and greater electronic conductivity than TMS [58]. Most researchers reported on La_2S_3 with the successive ionic layer adsorption and reaction method (SILAR). Patil et al. synthesized La_2S_3 using the SILAR method that shows Cs of $256\,F\,g^{-1}$ by $LiClO_4$/PC electrolyte, while the obtained electrode provided a maximum Cs of $358\,F\,g^{-1}$ at $5\,mV\,s^{-1}$ in KOH and Na_2SO_4 electrolyte [59]. The study of the effects of annealing on La_2S_3 electrode is extended later and prepared with

Oxides free materials for symmetric capacitors **Chapter | 4** **87**

a chemical bath approach, which offers increased Cs of this electrode drastically, and their air annealed materials exhibited a supreme of $294\,F\,g^{-1}$ at $0.5\,mV\,s^{-1}$, which was superior to bare La_2S_3 electrodes [60].

5.2.3 WS₂

Due to their great specific SSA and flexible electronic structures, WS_2-based materials are gaining consideration for SCs. Naturally, WS_2 has a hexagonal crystal structure, and each WS_2 monolayer comprises individual W atoms layers, which are then hexagonally packed between two trigonal atomic layers of S atoms [61]. Despite its advantages, it possesses poor electrical conductivity, which hinders its performance in SCs applications. Hence, the hydrothermal method's fabrication of WS_2/rGO electrodes exhibited an electrical capacitance of $350\,F\,g^{-1}$ at $2\,mV\,S^{-1}$ [61]. Nanoparticle-featured WS_2 was obtained through one-pot hydrothermal approaches and delivered a high *capacitance* of $1439.5\,F\,g^{-1}$ at $5\,mA\,cm^{-2}$ and exceptional cycling durability with 77.4% retention after 3000 cycles [62].

6. Graphene, CNT, and carbon-based materials for SCs

Carbon is considered to be the highly abundant electrode material in nature. Many different porous carbons have been prepared and used in various practical applications. The Porous carbon is also an ideal electrode material for efficient energy-related systems due to its large SSA, capacious pore space, and superior chemical stability compared to other porous materials. Because electrochemical capacitive activities are primarily based on EDLC and subsequently enhanced with pseudo-capacitance, high-surface carbon is desirable for this application. Carbon's porosity plays a vital role in improving performance too. Carbonaceous materials have many benefits, such as large quantities of raw materials, thermal stability, ease of processing, and modification. As a result, they have shown a lot of consideration and high potential in numerous energy storage systems [63]. Carbon materials have extraordinary features with different particle structures through several reaction lines with several advantages and disadvantages [64]. Several physical arrangements for carbon mesopores contain nanoparticles, nanotubes, and nanofibers, which can adapt to several industrial application categories. Different pore sizes in the mesoporous carbon nanostructure include micropores, mesopores, and macropores, an essential advantage for their SCs application. This book chapter discussed the present status of carbon materials for SCs application by debating the research literature on AC, graphene, and CNTs and offering prospective research in SCs technologies. The schematic representation of the formation mechanism of $NiMn_2O_4$/rGO/PANI nanocomposite materials was presented in Fig. 5.

88 Oxide free nanomaterials for energy storage and conversion applications

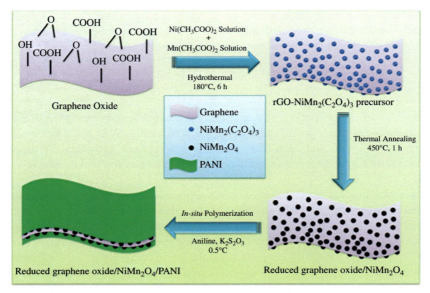

FIG. 5 Preparation of NiMn$_2$O$_4$/rGO/PANI shows the synthetic mechanism of the ternary nanocomposite. Initially, the hydrothermal circumstances made the creation of NiMn$_2$O$_4$ over the surface of graphene. Finally, an in situ polymerization method was directed to obtain PANI on the binary composite. *(Reproduced with permission from S. Sahoo, S. Zhang, J.-J. Shim, Porous ternary high performance supercapacitor electrode based on reduced graphene oxide, NiMn$_2$O$_4$, and polyaniline, Electrochim. Acta 216 (2016) 386–396, https://doi.org/10.1016/j.electacta.2016.09.030. Copyright (2016) Elsevier.)*

6.1 Graphene with metal oxides and metal sulfides nanocomposites

Graphene is a single 2d layer of sp^2 hybridized carbonaceous atoms with a hexagonal structure. Graphene has expected excellent research consideration because of its extraordinary electrochemical performances. Graphene-based composites are extensively used for the advancement of SCs. One graphene layer conducts extraordinary capacitance of around ~20.0 μF cm^{-1}, which is significantly superior to the nanomaterials with other carbon. As an electrode material, Graphene has a critical enhancement to the presentation of the SCs [65]. Graphene has many clear forms in all four dimensions as quantum dots (0D), wires and rods (1D), films (2D), and monoliths (3D). GO material and the rGO types are explored as possible electrode materials for SCs because of their excellent SSA, superior electrical conductive features [66]. Zhang et al. demonstrated an rGO/NF electrode through flame-induced reduction of dry GO onto NF and is offered Cs, which obtains 228.6 F g^{-1} at 1 A g^{-1} and maintains high cycling of 94.7% after 10,000 cycles [67]. The outstanding electrochemical highlights are credited to the cross-connecting disordered network and the irregular scattering of the caused pores that permit the rapid transport of particles to the active sites. Recently, Xu et al. demonstrated a NiS/MoS$_2$@N-rGO nanocomposite through the hydrothermal approach and applied it as an electrode revealing an extraordinary Cs of $1347.3\,Fg^{-1}$ at $10\,Ag^{-1}$. Moreover, it provides a greater $35.69\,Whkg^{-1}$ and a power density of $601.8\,Wkg^{-1}$ [68].

6.2 CNTs for high performance of SCs

Because of the more cost of graphene and graphene-based derivatives, CNTs have been widely studied and fascinated with widespread consideration due to curious and possibly beneficial structural, electrical and mechanical features. CNTs are developed when a thin graphite sheet is folded into cylinders, comprising SWCNT and MWCNT. Both CNTs have been considered as electrodes for SCs due to their uniform size distribution, greatly accessible SSA, low resistivity, and greater durability. Notably, Niu et al. [69] suggested first that CNTs could be used in SCs. To functionalize the MWCNTs surface, it was dissolved in nitric acid. These surface-functionalized MWCNTs had an SSA of $430\,m^2\,g^{-1}$, a gravimetric capacitance of $102\,Fg^{-1}$, and an energy density of $0.5\,Whkg^{-1}$ using 38 wt% sulfuric acids as the electrolyte. Pseudo-capacitance can be encouraged by the remaining functional groups and the catalyst. Thus, both Faradaic and non-Faradaic processes are involved in the SC by the CNTs. The effects of diameter and structure of CNTs, and micro-texture, and elemental composition of the materials on the capacitance were methodically examined by Frackowiak et al. [70]. Anodic aluminum oxide (AAO) template-based MWCNTs is especially appropriate for examining the size impact on super capacitance because of the uniform diameter and length. The Cs of the template-based CNT electrodes were about $50\,Fg^{-1}$. Increased capacitance is caused by CNTs uniformity-based templates compared to CNT, which is not uniform. Heating is one significant way to increase CNT graphitization and eliminate amorphous carbon. Li et al. [71] demonstrated that the Cs were enhanced by oxidization up to $650^{\circ}C$ because of increased SSA and dispersity. Pyrrole functionalized SWCNTs have high capacitance values ($350\,Fg^{-1}$), power density ($4.8\,kWkg^{-1}$), and energy density ($3.3\,kJkg^{-1}$) [72]. Plainly the SSA is anything but an exclusively prevailing variable for execution. CNT capacitance is influenced by numerous features, including SSA, pore size distribution, conductivity.

6.3 AC

The basic properties of AC are minimal effort, high electrical steadiness, and enormous SSA, which makes them helpful as a reasonable material for SCs application. ACs are created by a physical or chemical method by activating raw materials such as coal, wood, etc. The physical activation is studied by heating around 700–1200°C in the absence of atmospheric air, whereas chemical stimulation needs heating of carbon source at low temperatures in the presence

90 Oxide free nanomaterials for energy storage and conversion applications

of an activating agent [73]. Due to the above two activation methods, the SSA was increased tremendously $(3000\,m^2\,g^{-1})$ with different pore size distributions [74]. Fernandez et al. prepared mesoporous carbon by carbonizing Polyvinyl alcohol and inorganic salt, which exhibited a Cs of $180\,Fg^{-1}$ in aqueous H_2SO_4 electrolyte [75], and its performance was enhanced by introducing micropores. Thus, the electrochemical activity of mesoporous carbon can be enhanced by its functionalization and act as a proficient PCs electrode apart from EDLC. Different materials, techniques, and technologies were utilized to discover a solution for an energy-storing away device with high capacitance. There is no doubt about utilizing carbon materials as potential materials for SC electrode materials because of their enhanced performance.

7. Summary and conclusion

Undoubtedly, energy-related systems are the current research hotspot and have stimulated considerable attention in fabricating and assembling more efficient electrode materials. Therefore, in-depth knowledge of oxide-free materials and their energy-level mechanisms and the latest nanotechnology-based applications in energy-related applications will help the scientific community understand the selection of oxide-free nanomaterials and design new kinds of innovative materials. In this chapter, we have extensively discussed the oxide-free materials for the symmetric capacitor in detail. In this chapter, different kinds of SCs are picked to be dissected regarding their electrode materials, the design, electrolyte, and fabricated techniques employed, along with their Cs for compact depiction, which are reasonably built up with valuable examples. The mechanism of storage principles and the categorization of SCs oxides-free materials have been summarized in this chapter. However, the lower conductivity, low cycling stability, and changes in volume during charge/release cycles of TMOs make them reducing the performance. In order to replace this, hybrid nanostructures made out of TMOs and a conducting material matrix, for example, carbon materials, are utilized to improve electrical conductivity and upgrade morphological durability with enhanced capacitance and longer cycling solidness.

References

[1] R. Genc, M.O. Alas, E. Harputlu, S. Repp, N. Kremer, M. Castellano, E. Erdem, High-capacitance hybrid supercapacitor based on multi-colored fluorescent carbon-dots, Sci. Rep. 7 (1) (2017) 1–13.

[2] C. Moreno-Castilla, F.J. Maldonado-Hódar, Carbon aerogels for catalysis applications: an overview, Carbon 43 (3) (2005) 455–465.

[3] S.J. Lee, J. Theerthagiri, P. Nithyadharseni, P. Arunachalam, D. Balaji, A.M. Kumar, M.Y. Choi, Heteroatom-doped graphene-based materials for sustainable energy applications: a review, Renew. Sustain. Energy Rev. 143 (2021) 110849.

Oxides free materials for symmetric capacitors **Chapter | 4 91**

[4] M.S. Halper, J.C. Ellenbogen, Supercapacitors: A Brief Overview, The MITRE Corporation, McLean, VA, 2006, pp. 1–34.

[5] C. Masarapu, L.P. Wang, X. Li, B. Wei, Tailoring electrode/electrolyte interfacial properties in flexible supercapacitors by applying pressure, Adv. Energy Mater. 2 (5) (2012) 546–552.

[6] M.D. Stoller, S. Park, Y. Zhu, J. An, R.S. Ruoff, Graphene-based ultracapacitors, Nano Lett. 8 (10) (2008) 3498–3502.

[7] A. Kausar, Overview on conducting polymer in energy storage and energy conversion system, J. Macromol. Sci. A 54 (9) (2017) 640–653.

[8] J.P. Zheng, P.J. Cygan, T.R. Jow, Hydrous ruthenium oxide as an electrode material for electrochemical capacitors, J. Electrochem. Soc. 142 (8) (1995) 2699.

[9] S. Mothkuri, S. Chakrabarti, H. Gupta, B. Padya, T.N. Rao, P.K. Jain, Synthesis of MnO_2 nanoflakes for high performance supercapacitor application, Mater. Today 26 (2020) 142–147.

[10] T.F. Yi, J. Mei, B. Guan, P. Cui, S. Luo, Y. Xie, Y. Liu, Construction of spherical NiO@ MnO_2 with core-shell structure obtained by depositing MnO_2 nanoparticles on NiO nanosheets for high-performance supercapacitor, Ceram. Int. 46 (1) (2020) 421–429.

[11] P. Wang, H. Zhou, C. Meng, Z. Wang, K. Akhtar, A. Yuan, Cyanometallic framework-derived hierarchical Co_3O_4-NiO/graphene foam as high-performance binder-free electrodes for supercapacitors, Chem. Eng. J. 369 (2019) 57–63.

[12] Y. Li, S. Ji, Y. Gao, H. Luo, M. Kanehira, Core-shell VO_2@TiO_2 nanorods that combine thermochromic and photocatalytic properties for application as energy-saving smart coatings, Sci. Rep. 3 (1) (2013) 1–13.

[13] L.Y. Chen, Y. Hou, J.L. Kang, A. Hirata, T. Fujita, M.W. Chen, Toward the theoretical capacitance of RuO_2 reinforced by highly conductive nanoporous gold, Adv. Energy Mater. 3 (7) (2013) 851–856.

[14] M.A.A.M. Abdah, N.S.M. Razali, P.T. Lim, S. Kulandaivalu, Y. Sulaiman, One-step potentiostatic electrodeposition of polypyrrole/graphene oxide/multi-walled carbon nanotubes ternary nanocomposite for supercapacitor, Mater. Chem. Phys. 219 (2018) 120–128.

[15] H.D. Yoo, S.D. Han, R.D. Bayliss, A.A. Gewirth, B. Genorio, N.N. Rajput, J. Cabana, "Rocking-chair"-type metal hybrid supercapacitors, ACS Appl. Mater. Interfaces 8 (45) (2016) 30853–30862.

[16] A. Ghosh, E.J. Ra, M. Jin, H.K. Jeong, T.H. Kim, C. Biswas, Y.H. Lee, High pseudocapacitance from ultrathin V_2O_5 films electrodeposited on self-standing carbon-nanofiber paper, Adv. Funct. Mater. 21 (13) (2011) 2541–2547.

[17] H.S. Huang, K.H. Chang, N. Suzuki, Y. Yamauchi, C.C. Hu, K.C.W. Wu, Evaporation induced coating of hydrous ruthenium oxide on mesoporous silica nanoparticles to develop high-performance supercapacitors, Small 9 (15) (2013) 2520–2526.

[18] M. Naushad, T. Ahamad, M. Ubaidullah, J. Ahmed, A.A. Ghafar, K.M. Al-Sheetan, P. Arunachalam, Nitrogen-doped carbon quantum dots (N-CQDs)/Co_3O_4 nanocomposite for high performance supercapacitor, J. King Saud Univ. Sci. 33 (1) (2021) 101252.

[19] K. Zhang, X. Han, Z. Hu, X. Zhang, Z. Tao, J. Chen, Nanostructured Mn-based oxides for electrochemical energy storage and conversion, Chem. Soc. Rev. 44 (3) (2015) 699–728.

[20] J. Zhang, G. Zhang, W. Luo, Y. Sun, C. Jin, W. Zheng, Graphitic carbon coated CuO hollow nanospheres with penetrated mesochannels for high-performance asymmetric supercapacitors, ACS Sustain. Chem. Eng. 5 (1) (2017) 105–111.

[21] J. Theerthagiri, K. Thiagarajan, B. Senthilkumar, Z. Khan, R.A. Senthil, P. Arunachalam, M. Ashokkumar, Synthesis of hierarchical cobalt phosphate nanoflakes and their enhanced electrochemical performances for supercapacitor applications, ChemistrySelect 2 (1) (2017) 201–210.

92 Oxide free nanomaterials for energy storage and conversion applications

[22] R.J. Ramalingam, P. Arunachalam, J.N. Appaturai, P. Thiruchelvi, Z.A. Al-othman, A.G. Alanazi, H.A. Al-lohedan, Facile sonochemical synthesis of nanoparticle modified Bi-MnOx and Fe_3O_4 deposited Bi-MnOx nanocomposites for sensor and pollutant degradation application, J. Alloys Compd. 859 (2021) 158263.

[23] K. Thiagarajan, J. Theerthagiri, R.A. Senthil, P. Arunachalam, J. Madhavan, M.A. Ghanem, Synthesis of $Ni_3V_2O_8$@graphene oxide nanocomposite as an efficient electrode material for supercapacitor applications, J. Solid State Electrochem. 22 (2) (2018) 527–536.

[24] A.K. Das, S. Sahoo, P. Arunachalam, S. Zhang, J.J. Shim, Facile synthesis of Fe_3O_4 nanorod decorated reduced graphene oxide (RGO) for supercapacitor application, RSC Adv. 6 (108) (2016) 107057–107064.

[25] K. Thiagarajan, T. Bavani, P. Arunachalam, S.J. Lee, J. Theerthagiri, J. Madhavan, M.Y. Choi, Nanofiber $NiMoO_4$/g-C_3N_4 composite electrode materials for redox supercapacitor applications, Nanomaterials 10 (2) (2020) 392.

[26] X. Hu, W. Zhang, X. Liu, Y. Mei, Y. Huang, Nanostructured Mo-based electrode materials for electrochemical energy storage, Chem. Soc. Rev. 44 (8) (2015) 2376–2404.

[27] I. Shakir, M. Shahid, H.W. Yang, D.J. Kang, Structural and electrochemical characterization of α-MoO_3 nanorod-based electrochemical energy storage devices, Electrochim. Acta 56 (1) (2010) 376–380.

[28] W. Tang, L. Liu, S. Tian, L. Li, Y. Yue, Y. Wu, K. Zhu, Aqueous supercapacitors of high energy density based on MoO_3 nanoplates as anode material, Chem. Commun. 47 (36) (2011) 10058–10060.

[29] N. Agnihotri, P. Sen, A. De, M. Mukherjee, Hierarchically designed PEDOT encapsulated graphene-MnO_2 nanocomposite as supercapacitors, Mater. Res. Bull. 88 (2017) 218–225.

[30] T. Wang, H.C. Chen, F. Yu, X.S. Zhao, H. Wang, Boosting the cycling stability of transition metal compounds-based supercapacitors, Energy Storage Mater. 16 (2019) 545–573.

[31] P. Simon, Y. Gogotsi, Materials for electrochemical capacitors, in: Nanoscience and Technology: A Collection of Reviews from Nature Journals, World Scientific, 2010, pp. 320–329.

[32] J. Theerthagiri, G. Durai, K. Karuppasamy, P. Arunachalam, V. Elakkiya, P. Kuppusami, H.S. Kim, Recent advances in 2-D nanostructured metal nitrides, carbides, and phosphides electrodes for electrochemical supercapacitors—a brief review, J. Ind. Eng. Chem. 67 (2018) 12–27.

[33] R. Kötz, M.J.E.A. Carlen, Principles and applications of electrochemical capacitors, Electrochim. Acta 45 (15–16) (2000) 2483–2498.

[34] J. Theerthagiri, K. Karuppasamy, G. Durai, A.U.H.S. Rana, P. Arunachalam, K. Sangeetha, H.S. Kim, Recent advances in metal chalcogenides (MX; X= S, Se) nanostructures for electrochemical supercapacitor applications: a brief review, Nanomaterials 8 (4) (2018) 256.

[35] S.S. Rao, Synthesis of CNTs on ZnO/NiS composite as an advanced electrode material for high-performance supercapacitors, J. Energy Storage 28 (2020) 101199.

[36] T. Li, Y. Bai, Y. Wang, H. Xu, H. Jin, Advances in transition-metal (Zn, Mn, Cu)-based MOFs and their derivatives for anode of lithium-ion batteries, Coord. Chem. Rev. 410 (2020) 213221.

[37] R. Xu, J. Lin, J. Wu, M. Huang, L. Fan, X. He, Z. Xu, A two-step hydrothermal synthesis approach to synthesize $NiCo_2S_4$/NiS hollow nanospheres for high-performance asymmetric supercapacitors, Appl. Surf. Sci. 422 (2017) 597–606.

[38] T. Zhu, H.B. Wu, Y. Wang, R. Xu, X.W. Lou, Formation of 1D hierarchical structures composed of Ni_3S_2 nanosheets on CNTs backbone for supercapacitors and photocatalytic H_2 production, Adv. Energy Mater. 2 (12) (2012) 1497–1502.

[39] X. Ou, L. Gan, Z. Luo, Graphene-templated growth of hollow Ni_3S_2 nanoparticles with enhanced pseudocapacitive performance, J. Mater. Chem. A 2 (45) (2014) 19214–19220.

[40] H. Peng, G. Ma, J. Mu, K. Sun, Z. Lei, Controllable synthesis of CuS with hierarchical structures via a surfactant-free method for high-performance supercapacitors, Mater. Lett. 122 (2014) 25–28.

[41] M. Govindasamy, S. Shanthi, E. Elaiyappillai, S.F. Wang, P.M. Johnson, H. Ikeda, C. Muthamizhchelvan, Fabrication of hierarchical $NiCo_2S_4$@CoS_2 nanostructures on highly conductive flexible carbon cloth substrate as a hybrid electrode material for supercapacitors with enhanced electrochemical performance, Electrochim. Acta 293 (2019) 328–337.

[42] Q. Chen, H. Li, C. Cai, S. Yang, K. Huang, X. Wei, J. Zhong, In situ shape and phase transformation synthesis of Co_3S_4 nanosheet arrays for high-performance electrochemical supercapacitors, RSC Adv. 3 (45) (2013) 22922–22926.

[43] B. Balakrishnan, S.K. Balasingam, K.S. Nallathambi, A. Ramadoss, M. Kundu, J.S. Bak, H.J. Kim, Facile synthesis of pristine FeS_2 microflowers and hybrid rGO-FeS_2 microsphere electrode materials for high performance symmetric capacitors, J. Ind. Eng. Chem. 71 (2019) 191–200.

[44] A.I. Osman, J.K. Abu-Dahrieh, N. Cherkasov, J. Fernandez-Garcia, D. Walker, R.I. Walton, E. Rebrov, A highly active and synergistic $Pt/Mo_2C/Al_2O_3$ catalyst for water-gas shift reaction, Mol. Catal. 455 (2018) 38–47.

[45] X. Yang, L. Zhao, J. Lian, Arrays of hierarchical nickel sulfides/MoS_2 nanosheets supported on carbon nanotubes backbone as advanced anode materials for asymmetric supercapacitor, J. Power Sources 343 (2017) 373–382.

[46] M. Manuraj, J. Chacko, K.N. Unni, R.B. Rakhi, Heterostructured MoS_2-RuO_2 nanocomposite: a promising electrode material for supercapacitors, J. Alloys Compd. 836 (2020) 155420.

[47] Z. Wang, L. Pan, H. Hu, S. Zhao, Co_9S_8 nanotubes synthesized on the basis of nanoscale Kirkendall effect and their magnetic and electrochemical properties, CrstEngComm 12 (6) (2010) 1899–1904.

[48] J. Pu, F. Cui, S. Chu, T. Wang, E. Sheng, Z. Wang, Preparation and electrochemical characterization of hollow hexagonal $NiCo_2S_4$ nanoplates as pseudocapacitor materials, ACS Sustain. Chem. Eng. 2 (4) (0214) 809–815.

[49] C. Xia, H.N. Alshareef, Self-templating scheme for the synthesis of nanostructured transition-metal chalcogenide electrodes for capacitive energy storage, Chem. Mater. 27 (13) (2015) 4661–4668.

[50] Z. Wu, X. Pu, X. Ji, Y. Zhu, M. Jing, Q. Chen, F. Jiao, High energy density asymmetric supercapacitors from mesoporous $NiCo_2S_4$ nanosheets, Electrochim. Acta 174 (2015) 238–245.

[51] M. Sun, J. Tie, G. Cheng, T. Lin, S. Peng, F. Deng, L. Yu, In situ growth of burl-like nickel cobalt sulfide on carbon fibers as high-performance supercapacitors, J. Mater. Chem. A 3 (4) (2015) 1730–1736.

[52] S. Sahoo, C.S. Rout, Facile electrochemical synthesis of porous manganese-cobalt-sulfide based ternary transition metal sulfide nanosheets architectures for high performance energy storage applications, Electrochim. Acta 220 (2016) 57–66.

[53] Y. Zhao, Z. Shi, H. Li, C.A. Wang, Designing pinecone-like and hierarchical manganese cobalt sulfides for advanced supercapacitor electrodes, J. Mater. Chem. A 6 (26) (2018) 12782–12793.

[54] H. Zhao, F. Tian, R. Wang, R. Chen, A review on bismuth-related nanomaterials for photocatalysis, Rev. Adv. Sci. Eng. 3 (2014) 3–27.

[55] L. Ma, Q. Zhao, Q. Zhang, M. Ding, J. Huang, X. Liu, X. Xu, Controlled assembly of Bi_2S_3 architectures as Schottky diode, supercapacitor electrodes and highly efficient photocatalysts, RSC Adv. 4 (78) (2014) 41636–41641.

[56] L. Fang, Y. Qiu, T. Zhai, F. Wang, M. Lan, K. Huang, Q. Jing, Flower-like nanoarchitecture assembled from Bi_2S_3 nanorod/MoS_2 nanosheet heterostructures for high-performance supercapacitor electrodes, Colloids Surf. A Physicochem. Eng. Asp. 535 (2017) 41–48.

94 Oxide free nanomaterials for energy storage and conversion applications

[57] S.S. Raut, J.A. Dhobale, B.R. Sankapal, SILAR deposited Bi_2S_3 thin film towards electrochemical supercapacitor, Physica E 87 (2017) 209–212.

[58] S.J. Patil, C.D. Lokhande, Fabrication and performance evaluation of rare earth lanthanum sulfide film for supercapacitor application: effect of air annealing, Mater. Des. 87 (2015) 939–948.

[59] S.J. Patil, V.S. Kumbhar, B.H. Patil, R.N. Bulakhe, C.D. Lokhande, Chemical synthesis of α-La_2S_3 thin film as an advanced electrode material for supercapacitor application, J. Alloys Compd. 611 (2014) 191–196.

[60] S.J. Patil, A.C. Lokhande, C.D. Lokhande, Effect of aqueous electrolyte on pseudocapacitive behavior of chemically synthesized La2S3 electrode, Mater. Sci. Semicond. Process. 41 (2016) 132–136.

[61] S. Ratha, C.S. Rout, Supercapacitor electrodes based on layered tungsten disulfide-reduced graphene oxide hybrids synthesized by a facile hydrothermal method, ACS Appl. Mater. Interfaces 5 (21) (2013) 11427–11433.

[62] C. Nagaraju, C.V.M. Gopi, J.W. Ahn, H.J. Kim, Hydrothermal synthesis of MoS_2 and WS_2 nanoparticles for high-performance supercapacitor applications, New J. Chem. 42 (15) (2018) 12357–12360.

[63] T.N.J.I. Edison, R. Atchudan, N. Karthik, P. Chandrasekaran, S. Perumal, P. Arunachalam, Y.R. Lee, Electrochemically exfoliated graphene sheets as electrode material for aqueous symmetric supercapacitors, Surf. Coat. Technol. 416 (2021) 127150.

[64] P. Ramesh, S. Amalraj, P. Arunachalam, M. Gopiraman, A.M. Al-Mayouf, S. Vasanthkumar, Covalent intercalation of hydrazine derived graphene oxide as an efficient 2D material for supercapacitor application, Synth. Met. 272 (2021) 116656.

[65] M.F. El-Kady, Y. Shao, R.B. Kaner, Graphene for batteries, supercapacitors and beyond, Nat. Rev. Mater. 1 (7) (2016) 1–14.

[66] Y. Wang, Z. Shi, Y. Huang, Y. Ma, C. Wang, M. Chen, Y. Chen, Supercapacitor devices based on graphene materials, J. Phys. Chem. C 113 (30) (2009) 13103–13107.

[67] J. Zhang, W. Yang, J. Liu, Facile fabrication of supercapacitors with high rate capability using graphene/nickel foam electrode, Electrochim. Acta 209 (2016) 85–94.

[68] X. Xu, W. Zhong, X. Zhang, J. Dou, Z. Xiong, Y. Sun, Y. Du, Flexible symmetric supercapacitor with ultrahigh energy density based on NiS/MoS_2@ N-rGO hybrids electrode, J. Colloid Interface Sci. 543 (2019) 147–155.

[69] C. Niu, E.K. Sichel, R. Hoch, D. Moy, H. Tennent, High power electrochemical capacitors based on carbon nanotube electrodes, Appl. Phys. Lett. 70 (11) (1997) 1480–1482.

[70] E. Frackowiak, K. Metenier, V. Bertagna, F. Beguin, Supercapacitor electrodes from multi-walled carbon nanotubes, Appl. Phys. Lett. 77 (15) (2000) 2421–2423.

[71] C. Li, D. Wang, T. Liang, X. Wang, J. Wu, X. Hu, J. Liang, Oxidation of multiwalled carbon nanotubes by air: benefits for electric double layer capacitors, Powder Technol. 142 (2–3) (2004) 175–179.

[72] C. Zhou, S. Kumar, C.D. Doyle, J.M. Tour, Functionalized single wall carbon nanotubes treated with pyrrole for electrochemical supercapacitor membranes, Chem. Mater. 17 (8) (2005) 1997–2002.

[73] L.L. Zhang, X.S. Zhao, Carbon-based materials as supercapacitor electrodes, Chem. Soc. Rev. 38 (9) (2009) 2520–2531.

[74] D. Qu, H. Shi, Studies of activated carbons used in double-layer capacitors, J. Power Sources 74 (1) (1998) 99–107.

[75] J.A. Fernández, T. Morishita, M. Toyoda, M. Inagaki, F. Stoeckli, T.A. Centeno, Performance of mesoporous carbons derived from poly (vinyl alcohol) in electrochemical capacitors, J. Power Sources 175 (1) (2008) 675–679.

Chapter 5

Oxides free materials for asymmetric capacitor

Prabhakarn Arunachalam[a], Mabrook S. Amer[a], Govindhasamy Murugadoss[b], and Pugalenthi Ramesh[c]

[a]*Electrochemical Science Research Chair, Chemistry Department, College of Science, King Saud University, Riyadh, Saudi Arabia,* [b]*Centre for Nanoscience and Nanotechnology, Sathyabama Institute of Science and Technology (Deemed to be University), Chennai, Tamil Nadu, India,* [c]*Department of Chemistry, Vivekanandha College of Arts & Sciences for Women, Tirchengode, Tamil Nadu, India*

Chapter outline

1. Introduction	95	3. Transition metal nitrides for electrochemical SCs	102
2. Transition metal chalcogenides for asymmetric supercapacitors	96	3.1 Transition metal carbides for asymmetric supercapacitors	103
2.1 Transition metal sulfides for asymmetric supercapacitors	97	3.2 Transition metal phosphides for asymmetric supercapacitors	105
2.2 Transition metal selenides for asymmetric supercapacitors	98	4. Conclusions	107
		References	107

1. Introduction

An extensive global increase in the energy demand from the worldwide economy's continuous growth has caused the following issues: the major one being the depletion of fossil-fuel stocks and the other is related to an upsurge in greenhouse gas emissions in specific and ecological problems in general. The future of energy sources is determined by advanced breakthroughs in designing a cost-effective, sustainable, and efficient system for converting and storing renewable energy sources [1–4]. It is expected that long-term struggles in this area will be resulting in more fascinating approaches toward developing stable and high-efficiency oxide-free electrode candidates for energy storage systems from elective energy supplies [5, 6]. In the presently developing effort for sustainable energy development, supercapacitors are fascinating energy storage approaches owing to their larger power densities incomparable to other existing storage methods such as fuel cells and batteries, etc. [7, 8]. This is credited to their great specific capacitance (SC), higher power density, rapid charging ability, ease of

Oxide Free Nanomaterials for Energy Storage and Conversion Applications.
https://doi.org/10.1016/B978-0-12-823936-0.00004-8
Copyright © 2022 Elsevier Inc. All rights reserved.

96 Oxide free nanomaterials for energy storage and conversion applications

handling, longer galvanostatic charge/discharge cycles, and wide-ranging working temperature ranges [3, 9]. Generally, supercapacitors are classified based on the mechanism of charge storage, namely electrical double-layer capacitors (EDLCs), whereas storing charge non-Faradically at the electrode-electrolyte interface, and other kinds is pseudo-capacitors, whereas the charge is indirectly stored through a faradaic chemical process [10]. Carbonaceous materials (graphene, activated carbon (AC), CNTs) are usually employed as electrode material for EDLCs. At the same time, conventional transition-metal oxides, transition metal chalcogenides (TM chalcogenides: S, Se), nanosized transition-metal nitrides (TMNs), carbides (TMCs), and phosphides (TMPs) are the main electrode candidates that hold the pseudo-capacitance performance [11]. These appropriate electrodes with coherent nanostructured outlines have revealed overall electrochemical performances for developing efficient electrochemical supercapacitors.

Nanosized electrode materials can efficiently enhance electrochemical performances and deplete rational materials with greater energy and power densities [12–14]. Moreover, exceptional cycling rates and longer electrochemical durability are all completely mandatory for future applications [15, 16]. Significant consideration would be made to build new kinds of electrode materials for supercapacitors, and innovative concepts and designing approaches are essential for this field. Therefore, the researcher should be concerned that the designing and developing electrode should be abundant accessible active sites, low-cost, and sustainable for clean invention and credibly be used in a wide choice of utilization [13, 15]. Finally, the benefits of developing oxide-free nanosized electrode candidates in the drawing strategies for energy-related technologies are systematically provided. With these issues, a significant development in the electrodes' material examination has been executed by numerous researchers regarding low-cost and higher electrochemical capacitance [17–21].

Herein, we examine and discuss the improvement in the development of asymmetric supercapacitors that involves the nanostructured oxide-free nanomaterials of TMNs, TMCs, and TMPs are elucidated in detail. Finally, we advise our outlook on these oxide-free nanomaterials to pave the approach for further studies.

2. Transition metal chalcogenides for asymmetric supercapacitors

The technologically vital and systematically substantial TM chalcogenides (S, Se) have gained significant consideration in the last decades owing to their anisotropic features. Generally, transition metals of sets IV to B syndicate with VI A group elements, namely S, Se, and Te, to create a double-layered crystalline structure [14]. These double-layered TM chalcogenides own the standard formulation of MX_2, where M is a TM, and X is a chalcogen atom (S, Se). Most of the TM chalcogenides' structure and properties nearly resemble semimetal pristine graphene, excluding the bandgap [12], which is approximately zero. In contrast, TM chalcogenides it is mainly built on the transition metal grouping, the number of layers, and the attendance or nonexistence of adopting atoms. Therefore, their bandgaps are varied between 0 and 2 eV. Owing to the bandgap variation, a wide range of TM chalcogenides features are tunable and have turn out to be commercially viable electrode candidates for electrochemical supercapacitors [22].

In this section, we mainly discuss the application of TM chalcogenides in asymmetric supercapacitors. They have attracted significant devotion owing to their greater power densities and higher durability natures, and they provide significantly better tolerance than batteries in various kinds of uses in consumer electronics, hybrid electronic vehicles, buffer powers, and so forth [23]. Further, TM chalcogenides have been widely recognized in the arena of fuel cells, batteries, sensors, electrocatalytic materials, thermoelectric devices, and memory devices and are commonly applied in supercapacitors. Moreover, we define the promising future areas of transition metal chalcogenides, covering both their electrochemical features and their applications in supercapacitors. Definitely, TMS reveals significantly enhanced electrochemical features, which mostly came from its greater electrical conductive natures, and mechanical and thermal stability. Besides, it is recognized that energy storage technologies' performance is mainly built on the crystalline nature, nanosized electrocatalytic materials, morphological and textural features, composition, and electrodes' design [24]. Transition metal selenides (TMSe) have fascinated considerable attention as favorable electrode materials, retaining rich redox chemistry and superior electrical features as a new kind of battery-like electrode material. However, compared to TMS, TMSe is less reported than metal sulfides.

2.1 Transition metal sulfides for asymmetric supercapacitors

During the past decades, nanosized have played a substantial part in electronics, mainly optical and optoelectronic devices, owing to its distinctive excellent physicochemical performances. Definitely, nickel sulfide (NiS) is of specific consideration owing to their numerous existing phases, namely NiS, NiS_2, Ni_3S_2, Ni_3S_4, Ni_7S_{10}, and Ni_9S_8 dissimilar morphological features [25]. Though, the numerous phases and morphologies of NiS occasionally exist as a grouping of two dissimilar phases [26]. In this regard, fabricating even morphological features with highly pure NiS is considered a major hurdle that has fascinated considerable attention.

Although numerous TMS have been investigated as electrode candidates for asymmetric supercapacitors, binary TMS are pretty exciting owing to their highly active redox sites and thermomechanical stability than their equivalent single-component counterparts. In this regard, numerous electrode materials, the TMS (CoS_x, CuS, $CuCo_2S_4$, $NiCoS_x$), have been widely recognized as major TMS-based electrode candidates owing to their greater theoretical capacitance

98 Oxide free nanomaterials for energy storage and conversion applications

and multiple redox reactions, mainly the grouping of different TMS [27–29]. For instance, Xu et al. demonstrated the $NiCo_2S_4/NiS$ composite electrodes are obtained through a hydrothermal approach and obtained the energy and power density of $60.3\,mWh\,cm^{-3}$, $375\,W\,kg^{-1}$. The electrochemical asymmetric supercapacitor performances of different TMSs are tabulated in Table 1.

2.2 Transition metal selenides for asymmetric supercapacitors

The electrochemical features of TMS for asymmetric supercapacitors were discussed earlier. This part is mainly dedicated to a debate on TMSe-based TM chalcogenides for asymmetric supercapacitor applications. However, TMSe' chemical and electrochemical performance almost resembles a TMS, which demonstrated that the TMSe might have favorable asymmetric supercapacitor applications [41]. Some of the significant TMSe are debated herein.

Peng et al. reported the cobalt selenide $(Co_{0.85}Se)$ nanosheets (NS) and N_2-incorporated porous carbon networks (N-PCNs) as positive and negative electrodes, respectively [42]. The $Co_{0.85}Se$ NS are fabricated through low-temperature solvothermal techniques without template/surfactant, and the N-PCNs (Fig. 1). The authors found that the developed asymmetric device comprised of $Co_{0.85}Se//N$-PCNs composite materials keeps an extended working voltage range of $1.6\,V$. They obtained a greater energy density of $21.1\,mWh\,cm^{-3}$ at $400\,W\,kg^{-1}$ and exceptional electrochemical durability (93.8% capacitive retention) in aqueous mediums.

Zhao et al. reported the solvothermal approach to developing $Co_{0.85}Se$ NS via ethylenediamine (EN) to regulate the morphology [43]. The authors found that the EN applied as a strong alkalinity supplier, assisting the creation of $Co_{0.85}Se$ NS product (Fig. 2). They used the AC as the negative electrode instead of the N-PCN. They demonstrated an energy density of $17.8\,mWh\,cm^{-3}$ at $3.57\,kW\,kg^{-1}$ and capacitance retention of 93% after 2000 cycles for asymmetric supercapacitors.

Meanwhile, Gong et al. [44] fabricated mesostructured $Co_{0.85}Se$ NS on nickel foam (NF) obtained through a one-pot hydrothermal route to fabricate high-performance asymmetric supercapacitors applications. They substituted $Co_{0.85}Se$-NS positive electrodes with $Co_{0.85}Se$-NS/NF, which delivered greater pseudocapacitive activities with great SC and exceptional electrochemical durability. The asymmetric device configuration can spread the potential working window to $1.6\,V$ and it leads to a substantial enhancement in the energy density to $39.7\,mWh\,cm^{-3}$ at $789.6\,W\,kg^{-1}$, combines with exceptional cycle durability.

The binary TMSe are presently highly encouraging, and only limited research efforts are available in electrochemical supercapacitor applications, and their electrochemical features have not yet been completely explored. Notably, Xia and co-workers demonstrated the $(Ni, Co)_{0.85}Se$ electrode materials and it provides the maximum areal capacitance of $2.33\,Fcm^{-2}$ at $4\,mA\,cm^{-2}$. They also developed asymmetric devices, which deliver a greater energy

TABLE 1 Recent advances on TMS for asymmetric electrochemical capacitors.

S.No.	Electrode material	Electrolyte	Specific capacitance (Fg^{-1})	Cycling stability	Energy ($mWh\,cm^{-3}$), power density ($W\,kg^{-1}$)	Device	References
1	3D HAC/Ni_3S_2	6M KOH	2797.43, $1\,Ag^{-1}$	8000, 82.13%	$55.32\,mWh\,cm^{-3}$, $1053.71\,W\,kg^{-1}$	3D HAC//3D HAC-Ni_3S_2	[30]
2	NiMn-G-LDH@$NiCo_2S_4$	6M KOH	$1018\,Cg^{-1}$, $1\,Ag^{-1}$	10,000, 86.4%	$60.3\,mWh\,cm^{-3}$, $375\,W\,kg^{-1}$	NiMn-GLDH@$NiCo_2S_4$@CFC	[31]
3	$NiCo_2S_4$/NiS	3M KOH	$123\,Fg^{-1}$, $1\,mA\,cm^{-2}$	1000, 90.3%	$43.7\,mWh\,cm^{-3}$, $160\,W\,kg^{-1}$	$NiCo_2S_4$/NiS//AC	[32]
4	$CoMoS_4$@Ni-Co-S nanotubes	3M KOH	2208.5, $1\,Ag^{-1}$	5000, 91.3%	$49.1\,mWh\,cm^{-3}$, $800\,W\,kg^{-1}$	$CoMoS_4$@Ni-Co-S//AC	[33]
5	NiS-SnS (NTS) heterostructure	PVA/KOH	1653, $1\,Ag^{-1}$	4000, 98%	$83\,mWh\,cm^{-3}$, $117\,W\,kg^{-1}$	NiS-SnS	[34]
6	MoS_2/NiS yolk-shell	1M KOH	1493, $0.2\,Ag^{-1}$	10,000, 100%	$31\,mWh\,cm^{-3}$ $155.7\,W\,kg^{-1}$	MoS_2/NiS//AC	[35]
7	MXene-$NiCo_2S_4$@NF	3M KOH	$596.69\,Cg^{-1}$	3000, 80.4%	$27.24\,mWh\,cm^{-3}$, $0.48\,kW\,kg^{-1}$	MXene-$NiCo_2S_4$//AC	[36]
8	Ni@CNTs@Ni-Co-S	2M KOH	$222\,mAh\,g^{-1}$ at $4\,Ag^{-1}$	2000, 90.6%	$46.5\,mWh\,cm^{-3}$, $800\,W\,kg^{-1}$	Ni@CNTs@Ni-Co-S//CC @CNTs	[37]
9	$CoNi_2S_4$-μflower	2M KOH	2098.95	2000, 91%	$82.98\,mWh\,cm^{-3}$, $9.63\,kW\,kg^{-1}$	Ni_3S_2- nanorod//$CoNi_2S_4$-μflower	[38]
10	Ni@NiS@$NiCo_2S_4$	6M KOH	$14.32\,F\,cm^{-2}$, $5\,mA\,cm^{-2}$	5000, 78.7%	$38.96\,mWh\,cm^{-3}$, $339.62\,W\,kg^{-1}$	Ni@NiS@$NiCo_2S_4$//AC	[39]
11	C@NS@NF	2M KOH	$6.086\,F\,cm^{-2}$ at $10\,mA\,cm^{-2}$	5000, 94.72%	$27.5\,mWh\,cm^{-3}$, $532\,W\,kg^{-1}$	C@NS@NF-10//AC	[40]

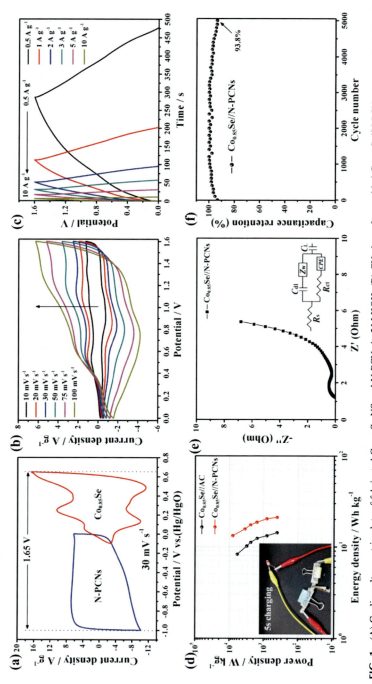

FIG. 1 (A) Cyclic voltammetric plots of fabricated $Co_{0.85}Se$ NS and N-PCNs in 2M KOH; (B) CV plots of assembled $Co_{0.85}Se$/N-PCNs asymmetric device at numerous sweep rates; (C) galvanostatic charge/discharge plots of assembled devices at numerous current densities; (D) its corresponding Ragone plots of asymmetric devices; (E) electrochemical impedance analysis of the devices (F) electrochemical durability of the fabricated devices. (Reproduced from H. Peng, G. Ma, K. Sun, Z. Zhang, J. Li, X. Zhou, Z. Lei, A novel aqueous asymmetric supercapacitor based on petal-like cobalt selenide nanosheets and nitrogen-doped porous carbon networks electrodes, J. Power Sources 297 (2015) 351–358. Copyright, Elsevier.)

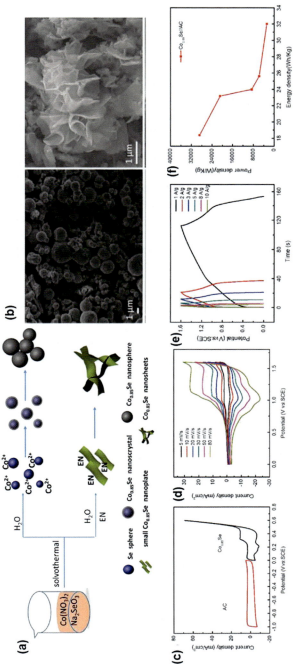

FIG. 2 (A) Schematic representation of the synthetic routes of Co$_{0.85}$Se nanostructured composites, (B) FE-SEM photographs of Co$_{0.85}$Se nanospheres, and Co$_{0.85}$Se NS, (C) Comparative cyclic voltammetric plots of an asymmetric device by involving Co$_{0.85}$Se NS and AC, in 2.0 M KOH at 10 mV s^{-1}; (D) GCD plots of the Co$_{0.85}$Se//AC at numerous current densities; (E) Ragone plot of asymmetric supercapacitors Co$_{0.85}$Se//AC. (F) Electrochemical durability of the Co$_{0.85}$Se//AC electrode up to 2000 cycles at 1 A g^{-1}. *(Reproduced from X. Zhao, X. Li, Y. Zhao, Z. Su, R. Wang, Facile synthesis of Tremelliform Co$_{0.85}$Se nanosheets for supercapacitor, J. Alloys Compd. 697 (2017) 124–131. Copyright, Elsevier.)*

102 Oxide free nanomaterials for energy storage and conversion applications

density of $2.85\,\mathrm{mW\,h\,cm^{-3}}$ at $10.76\,\mathrm{mW\,cm^{-3}}$. Further, Peng and co-workers demonstrated the different selenide composite electrodes involving $NiSe@MoSe_2$ and $Ni_{0.85}Se@MoSe_2$, through hydrothermal techniques and investigated their usage in asymmetric supercapacitors [45, 46].

3. Transition metal nitrides for electrochemical SCs

Nitrogen is the maximum abundant constituent existing in the earth's atmosphere. Identical to TMCs, nitrogen can create TMNs with practically all transition metals. With the insertion of nitrogen, the distance between metal atoms is expanded. Also, TMNs are specific because of their benefits of exceptional electronic conductivity, long ecological life, and excellent reaction selectivity. Notably, various research efforts have been dedicated to various TMNs for different kinds of applications- particularly energy-related systems, electronic storage devices, biological imaging, reforming catalysts, and electrical applications [47–49]. For instance, the TMNs are favorable electrode candidates in electrochemical supercapacitors owed to their physicochemical and mechanical features [50]. Among the transition metal components, TMNs specifically, vanadium nitride (VN) [51–53], Mo_xN [54, 55], CrN [56], RuN [57], TiN [58], and Ni_3N [59] are exciting electrode candidates for electrochemical supercapacitors owing to their lower production cost, greater durability at high temperature, exceptional electrical conductive features, and tremendous chemical resistance. Significant considerable routes were used to synthesize nitrides over many years, involving direct nitridation of the TM with N_2 or NH_3 or ammonolysis of conventional metal oxides/chlorides. Herein, we describe the recent development in the fabrication of nanosized TMNs electrodes for asymmetric supercapacitors is clarified in detail.

The real-life use of TMNs is mainly restricted by their inadequate oxidation and lower operating potential window ($\approx 1\,\mathrm{V}$) in an aqueous medium [60]. In order to overcome this, remarkable electrically conductive carbonaceous materials, like carbon nanoparticles [61], CNTs [62], and graphene NS have been hybridized with TMNs as supporting materials [63]. Generally, compared with conventional TMNs, the electrochemical features of binary or tertiary-TMNs are easily optimized by regulating the valence and electronic states of the employed TMs, causing improved capacitance features [64]. Metallic layered double hydroxide (Me-LDH) is considered a kind of 2D brucite-like anionic lamellar compounds comprising electrochemically active binary or tertiary-metal cations, e.g., Ni, Co, Al, and Fe, among others [65]. Due to greater advantages in its composition and morphological features, the thermal nitrogenization of Me-LDH NS might lead to new kinds of nanosized integrated TMNs with several electroactive sites for supercapacitors applications [66]. For instance, Peng and co-workers demonstrated supercapacitors electrode materials comprised of fluorinated graphene (FG)/LDH (CoAl-LDH/FG) electrode materials, revealing superior electrochemical features than the non-FG composites owing to the semi-ionic C—F bonds in FG assisted the ion transportation, enhanced the electrically conductive features and durability [67]. Recently, Ishaq et al. [68] reported supercapacitor electrode materials comprised of FG-incorporated Ni-Co-Fe nitride (NCF-N) on NF. They fabricated the asymmetric supercapacitor with hybrid electrode materials of NCF-N@FG/NF and it delivers superior energy (56.3 mWh cm^{-3} at 374.6 W kg^{-1}) and a power density of 7484.2 W kg^{-1} at 39.5 mWh cm^{-3}. The obtained higher electrochemical performance is credited to the enhanced carrier transfer due to its direct electrical contact among metallic NCF-N and FG/NF and developed without any polymeric binders. The electrochemical supercapacitive performances of selective TMNs are tabulated in Table 2.

3.1 Transition metal carbides for asymmetric supercapacitors

TMCs also greatly explored energy system devices, which benefited from good electrically conductive and hydrophilic surfaces [78]. Particularly, TMCs have been explored as probable electrode candidates for electrochemical supercapacitors owing to their high reactivity benefits [79]. Remarkably, enriched higher rate capability and exceptional electrochemical durability have succeeded in these carbides compared to the conventional transition-metal oxides. For instance, Pande et al. reported the mixture of nanostructured TMNs and TMCs for electrochemical supercapacitors in aq. KOH and H$_2$SO$_4$ mediums [80]. The fabricated γ-Mo$_2$N and β-W$_2$C demonstrated a decent areal capacitance, but the electrochemical activity of VC and β-W$_2$N was not very good. Afterward, two-dimensional titanium carbide (Ti$_3$C$_2$) is obtained by etching aluminum from titanium aluminum carbide (Ti$_3$AlC$_2$, a "MAX" phase) in concentrated hydrofluoric acid. The rolled films of conductive 2D TMC "clay" demonstrated by the Gogotsi team deliver a volumetric capacitance up to 245 F g^{-1} [10].

Silicon carbide (SiC) has fascinated much recognized TMCs based electrode candidates for electrochemical supercapacitors due to its high thermal and electrochemical durability, greater lifetime, and high-temperature procedures [81, 82]. Notably, Kim and co-workers reported asymmetric electrodes built on SiC/N-MnO$_2$ nanoneedles and AC that revealed high capacitance and exceptional rate activities ascribed to the synergetic action between the assembled electrode materials [83]. After optimization, it developed an SC of 59.9 F g^{-1} at 2 mV s^{-1} and greater energy and power densities (30.06 mW h cm^{-3}and 113.92 W kg^{-1}, correspondingly). Afterward, Sarno and co-workers demonstrated an AC-derived SiC nanoparticle-electrodes attained an SC of 114.7 F g^{-1} at 0.12 A g^{-1} [84]. Likewise, the Zhuang group developed SiC-based electrodes through the chemical vapor deposition (VD) method, and it revealed an aerial capacitance of 72.7 F cm^{-2} at 100 mV s^{-1} [81]. Further, the Sanger group directly fabricates SiC nano cauliflower-based electrodes via a physical VD route [85]. The obtained SiC nano cauliflower-based supercapacitors device demonstrated a greater SC of 188 F g^{-1}

TABLE 2 Recent advances on 2D nanostructured TMNs for asymmetric electrochemical capacitors.

S.No.	Electrode material	Electrolyte	Specific capacitance (Fg^{-1})	Cycling stability	Energy ($mWh\,cm^{-3}$), power density (Wkg^{-1})	Device	References
1	TiNbN//VN	0.5 M H_2SO_4	59.3 mF cm^{-2} at 1.0 mA cm^{-2}	20,000, 87.9%	74.9 mWh cm^{-3} and 8.8 W cm^{-3}	TiNbN//VN	[69]
2	m_K-BN/CNT/PANI	1 M KCl, 1 M TEABF$_4$/DMSO	515, 111, 1 Ag^{-1}	10,000, 98%	62 mWh cm^{-3}	m_K-BN/CNT/PANI	[70]
3	NiCo$_2$N@NG//NiFeN@NG	3 M KOH	262 mAh g^{-1}, 1 Ag^{-1}	20,000, 97.69%	94.93 mWh cm^{-3} at 0.79 kW kg^{-1}	NiCo$_2$N@NG//NiFeN@NG	[71]
4	ZCS/BN/CNT/PPY	1 M KCl, 1 M TEABF$_4$/ACN	534, 785	106% after 10,000	49.6 mWh cm^{-3}	ZCS/BN/CNT/PPY//CB	[72]
5	Fe-ZIF/CN//porous C/CN	6 M KOH	1096	2000, 9.3% loss	153 mWh cm^{-3}, 40,000 W kg^{-1}	Fe-ZIF/CN//porous C/CN	[73]
6	Ni$_3$Mo$_3$N	6 M KOH	264 Cg^{-1} at 0.5 Ag^{-1}	1000, 81.4%	34.89 mWh cm^{-3}, 16,000 W kg^{-1}	Ni$_3$Mo$_3$N//AC	[74]
7	C/Mo$_x$N	1 M H_2SO_4	251 at 0.5 Ag^{-1}	15,000, 78.6%	14.1 mWh cm^{-3}, 312 W kg^{-1}	PANI‖C/Mo$_x$N	[75]
8	S-gC$_3$N$_4$/CoS$_2$-II	3.5 M KOH	180 Cg^{-1} at 1 Ag^{-1}	100,000, 89%	26.7 mWh cm^{-3}, 19.8 kW kg^{-1}	S-gC$_3$N$_4$/CoS$_2$-II	[76]
9	PBA nano-box/g-C$_3$N$_4$	PVP and KOH	201.6 mAh g^{-1}	5000, 90.1%	46.9 mWh cm^{-3}, 808.2 W kg^{-1}	H Rb-NiHCF/g-C$_3$N$_4$//AC	[77]

at $5\,mV\,s^{-1}$, and reasonably higher energy density ($\sim31.43\,mWh\,cm^{-3}$) and power density ($\sim18.8\,kW\,kg^{-1}$ at $17.76\,mWh\,cm^{-3}$).

3.2 Transition metal phosphides for asymmetric supercapacitors

Like TMCs, TMPs are fascinating electrode candidates owing to their inherent material features such as greater melting points and hardness, high thermal conductivity, and electrochemical durability [86, 87]. In particular, TMPs enjoy metalloid features with greater electrical features, creating potential electrode candidates for asymmetric supercapacitors. Compared with TM-based compounds (TM oxides/hydroxides/TMS, TMPs are kinetically satisfactory for fast electron transport for attaining superior power densities of supercapacitors [88, 89]. Generally, TMPs are hard to fabricate through conventional techniques, limiting their commercial applications [90, 91]. TMPs with different stoichiometric metal/phosphorous (M/P) ratios can be obtained, containing monophosphides ($M/P = 1$), phosphorus-rich ($M/P < 1$) TMPs, and metal-rich TMPs ($M/P > 1$) [92]. Particularly, metal-rich TMPs were considered as greater electrical and heat conductors with exceptional stability [93]. Thus, different kinds of TMPs (M = Fe, Fe, Cu, Ni, etc.) were demonstrated as high-abundant electrochemically active materials [94]. Till now, numerous reports and reviews on the fabrication of nanostructured TMPs [90] and their usage in electrochemical hydrogen evolution reactions, magnetic and catalytic fields [95].

In recent years, the Ni_2P NS/NF electrode was examined for commercial uses and was assembled the Ni_2P NS/NF//AC asymmetric devices electrode in 6.0 M KOH [96]. Generally, the anode surface was covered with a PVA/KOH gel electrolyte to upsurge the Ni_2P NS/NF electrode's electrochemical durability [97, 98]. The authors demonstrated asymmetric supercapacitors systems employing Ni_2P NS/NF hybrid, and AC electrode achieved an SC of $96\,F\,g^{-1}$ at $5\,mV\,s^{-1}$ with a steady operating voltage of 1.4 V and a supreme energy density of $26\,mWh\,cm^{-3}$.

For oxide-free materials, it is profiting from the improved electron/ion transfer, enhanced accessibility of active sites/ interfaces, rich mixed valences of TM and P, and stronger exponential synergy. The binary TMPs have the potential for superior electrochemical performances. In recent years, Liang et al. reported the highly active asymmetric supercapacitors built on a plasma-assisted synthesis of NiCoP nanoplates and graphene sheets [99]. The author's group developed the devices with NiCoP nanoplates//grapheme films (Fig. 3), which offers a superior energy density of $32.9\,mWh\,cm^{-3}$ at $1301\,W\,kg^{-1}$ and exceptional cycling performance (83% capacity retention). This investigation validates that the binary TMPs based NiCoP materials shown greater activity than other NiCo-based materials, which delivers the NiCoP nanoplates promising electrode materials for energy storage devices. Similarly, Song et al. reported the hieratically porous NiCoP-CoP composite electrode materials for electrochemical supercapacitors [100]. The fabricated composite electrodes

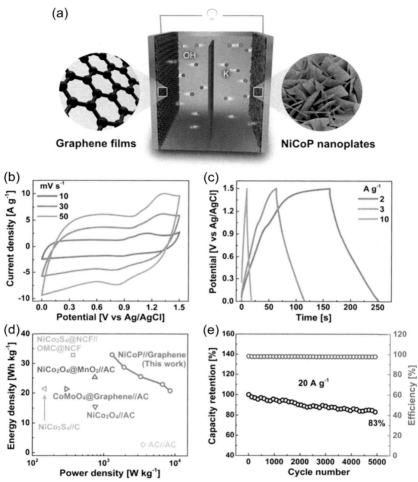

FIG. 3 Electrochemical activities of binary TMPs assembled with NiCoP//graphene asymmetric supercapacitors. (A) Representation of the assembled devices. (B) Cyclic voltammogram plots at altered sweep rates. (C) Galvanostatic charge-discharge profiles at numerous current densities. (D) Ragone plot is related to energy and power densities. (E) Electrochemical cycling performance at 20 A g^{-1}. *(Reproduced from H. Liang, C. Xia, Q. Jiang, A.N. Gandi, U. Schwingenschlögl, H.N. Alshareef, Low temperature synthesis of ternary metal phosphides using plasma for asymmetric supercapacitors, Nano Energy 35 (2017) 331–340. Copyright, Elsevier.)*

demonstrated an SC reaching 1969 F g^{-1}, much superior to its equivalent of Ni-Co-based TMS and other phosphides. Further, the author assembled asymmetric supercapacitors, the NiCoP-CoP composite as the positive electrode revealed incredible electrochemical durability with 93% retention after 5000 cycles at 8 A g^{-1}. Through structural engineering, a 4-stacked assembled device provides a superior energy density of $639\,mWh\,cm^{-2}$ at $48\,W\,cm^{-2}$. By considering the benefit of state-of-the-art electrodes and structural engineering, the assembled devices achieved greater electrochemical activities and can apply in different applications for future miniaturized electronics.

4. Conclusions

Undoubtedly, energy-related technologies and materials are the current research hotspot and have inspired considerable attention in fabricating and assembling more efficient electrode materials. The present book chapters are titled to give an overview of recent oxide-free nanostructured materials beyond the realm of science fiction and their role in the advancement of energy-related devices. Therefore, in-depth knowledge of oxide-free materials and their energy-level mechanisms and the latest nanotechnology-based applications in energy-related applications will help the scientific community understand the selection of oxide-free nanomaterials and design new kinds of innovative materials.

Presently, the fabrication of asymmetric supercapacitors as energy storage systems is of foremost significance, emphasizing their great power density. Current major research efforts are built on asymmetric supercapacitors that enhance the electrochemical features and their electrochemical durability to reduce electrode materials' production cost. The choice of appropriate electrode materials in nanosized form has greatly improved the electrochemical performances for great performance and cost reduction of developing asymmetric supercapacitors. This chapter conferred recent development in advancing oxide-free materials, and TMSe, particularly for asymmetric supercapacitors. Lastly, significant inspection should be carried out to develop the new kinds of electrode materials for asymmetric supercapacitors, and innovative plans and drawing schemes are essential for this field. Also, while selecting the materials for asymmetric supercapacitors, the worldwide researcher should be concerned that they should be highly abundant in nature, cost-effective and eco-friendly for clean invention, and possible usage in an expansive choice of utilization.

References

[1] J. Theerthagiri, K. Karuppasamy, G. Durai, A.U.H.S. Rana, P. Arunachalam, K. Sangeetha, et al., Recent advances in metal chalcogenides (MX; X = S, Se) nanostructures for electrochemical supercapacitor applications: a brief review, Nanomaterials 8 (4) (2018) 256.

[2] C. Karthikeyan, P. Arunachalam, K. Ramachandran, A.M. Al-Mayouf, S. Karuppuchamy, Recent advances in semiconductor metal oxides with enhanced methods for solar photocatalytic applications, J. Alloys Compd. 828 (2020) 154281.

[3] K. Thiagarajan, J. Theerthagiri, R.A. Senthil, P. Arunachalam, J. Madhavan, M.A. Ghanem, Synthesis of $Ni_3V_2O_8$ @ graphene oxide nanocomposite as an efficient electrode material for supercapacitor applications, J. Solid State Electrochem. 22 (2) (2018) 527–536.

108 Oxide free nanomaterials for energy storage and conversion applications

[4] A.K. Das, S. Sahoo, P. Arunachalam, S. Zhang, J.J. Shim, Facile synthesis of Fe_3O_4 nanorod decorated reduced graphene oxide (RGO) for supercapacitor application, RSC Adv. 6 (108) (2016) 107057–107064.

[5] P. Arunachalam, M.N. Shaddad, A.S. Alamoudi, M.A. Ghanem, A.M. Al-Mayouf, Microwave-assisted synthesis of Co_3 $(PO_4)_2$ nanospheres for electrocatalytic oxidation of methanol in alkaline media, Catalysts 7 (2017) 119.

[6] P. Arunachalam, M.A. Ghanem, A.M. Al-Mayouf, M. Al-shalwi, Enhanced electrocatalytic performance of mesoporous nickel-cobalt oxide electrode for methanol oxidation in alkaline solution, Mater. Lett. 196 (2017) 365–368.

[7] X. Zhao, B.M. Sánchez, P.J. Dobson, P.S. Grant, The role of nanomaterials in redox-based supercapacitors for next generation energy storage devices, Nanoscale 3 (3) (2011) 839–855.

[8] K. Karuppasamy, D. Kim, Y.H. Kang, K. Prasanna, H.W. Rhee, Improved electrochemical, mechanical and transport properties of novel lithium bisnonafluoro-1-butanesulfonimidate (LiBNFSI) based solid polymer electrolytes for rechargeable lithium ion batteries, J. Ind. Eng. Chem. 52 (2017) 224–234.

[9] J. Theerthagiri, K. Thiagarajan, B. Senthilkumar, Z. Khan, R.A. Senthil, P. Arunachalam, et al., Synthesis of hierarchical cobalt phosphate nanoflakes and their enhanced electrochemical performances for supercapacitor applications, ChemistrySelect 2 (1) (2017) 201–210.

[10] Y. Dall'Agnese, M.R. Lukatskaya, K.M. Cook, P.L. Taberna, Y. Gogotsi, P. Simon, High capacitance of surface-modified 2D titanium carbide in acidic electrolyte, Electrochem. Commun. 48 (2014) 118–122.

[11] J. Theerthagiri, R. Sudha, K. Premnath, P. Arunachalam, J. Madhavan, A.M. Al-Mayouf, Growth of iron diselenide nanorods on graphene oxide nanosheets as advanced electrocatalyst for hydrogen evolution reaction, Int. J. Hydrogen Energy 42 (18) (2017) 13020–13030.

[12] G. Wang, L. Zhang, J. Zhang, A review of electrode materials for electrochemical supercapacitors, Chem. Soc. Rev. 41 (2) (2012) 797–828.

[13] X.H. Xia, J.P. Tu, Y.Q. Zhang, Y.J. Mai, X.L. Wang, C.D. Gu, X.B. Zhao, Three-dimentional porous nano-Ni/Co $(OH)_2$ nanoflake composite film: a pseudocapacitive material with superior performance, J. Phys. Chem. C 115 (45) (2011) 22662–22668.

[14] P. Simon, Y. Gogotsi, Materials for electrochemical capacitors, Nanosci. Technol. (2010) 320–329.

[15] Y. Lu, J.K. Liu, X.Y. Liu, S. Huang, T.Q. Wang, X.L. Wang, et al., Facile synthesis of Ni-coated Ni_2P for supercapacitor applications, CrstEngComm 15 (35) (2013) 7071–7079.

[16] Z. Yu, L. Tetard, L. Zhai, J. Thomas, Supercapacitor electrode materials: nanostructures from 0 to 3 dimensions, Energy Environ. Sci. 8 (3) (2015) 702–730.

[17] J. Theerthagiri, G. Durai, K. Karuppasamy, P. Arunachalam, V. Elakkiya, P. Kuppusami, et al., Recent advances in 2-D nanostructured metal nitrides, carbides, and phosphides electrodes for electrochemical supercapacitors–a brief review, J. Ind. Eng. Chem. 67 (2018) 12–27.

[18] P. Ramesh, S. Amalraj, P. Arunachalam, M. Gopiraman, A.M. Al-Mayouf, S. Vasanthkumar, Covalent intercalation of hydrazine derived graphene oxide as an efficient 2D material for supercapacitor application, Synth. Met. 272 (2021) 116656.

[19] J.R. Rajabathar, H.A. Al-lohedan, P. Arunachalam, Z.A. Issa, M.K. Gnanamani, J.N. Appaturi, et al., Synthesis and characterization of metal chalcogenide modified graphene oxide sandwiched manganese oxide nanofibers on nickel foam electrodes for high performance supercapacitor applications, J. Alloys Compd. 850 (2021) 156346.

[20] M. Naushad, T. Ahamad, M. Ubaidullah, J. Ahmed, A.A. Ghafar, K.M. Al-Sheetan, P. Arunachalam, Nitrogen-doped carbon quantum dots (N-CQDs)/Co_3O_4 nanocomposite for high performance supercapacitor, J. King Saud Univ. Sci. 33 (1) (2021) 101252.

[21] J.R. Rajabathar, P. Arunachalam, Z.A. Issa, Synthesis and characterization of novel metal chalcogenide modified Ni-Co-MnO$_2$ nanofibers rolled with graphene based visible light active catalyst for nitro phenol degradation, Optik 224 (2020) 165538.

[22] R. Kötz, M. Carlen, Principles and applications of electrochemical capacitors, Electrochim. Acta 45 (2000) 2483–2498.

[23] X. Lu, G. Li, Y. Tong, A review of negative electrode materials for electrochemical supercapacitors, Sci. China Technol. Sci. 58 (11) (2015) 1799–1808.

[24] C. Zhao, W. Zheng, A review for aqueous electrochemical supercapacitors, Front. Energy Res. 3 (2015) 23.

[25] H. Pang, C. Wei, X. Li, G. Li, Y. Ma, S. Li, J. Chen, J. Zhang, Microwave-assisted synthesis of NiS$_2$ nanostructures for supercapacitors and cocatalytic enhancing photocatalytic H$_2$ production, Sci. Rep. 4 (2014) 3577.

[26] S. Peng, L. Li, H. Tan, R. Cai, W. Shi, C. Li, S.G. Mhaisalkar, M. Srinivasan, S. Ramakrishna, Q. Yan, MS$_2$ (M = Co and Ni) hollow spheres with tunable interiors for high-performance supercapacitors and photovoltaics, Adv. Funct. Mater. 24 (2014) 2155–2162.

[27] Y. Liu, J. Zhang, S. Wang, K. Wang, Z. Chen, Q. Xu, Facilely constructing 3D porous NiCo$_2$S$_4$ nanonetworks for high-performance supercapacitors, New J. Chem. 38 (9) (2014) 4045–4048.

[28] T. Zhu, B. Xia, L. Zhou, X.W.D. Lou, Arrays of ultrafine CuS nanoneedles supported on a CNT backbone for application in supercapacitors, J. Mater. Chem. 22 (16) (2012) 7851–7855.

[29] S.E. Moosavifard, S. Fani, M. Rahmanian, Hierarchical CuCo$_2$S$_4$ hollow nanoneedle arrays as novel binder-free electrodes for high-performance asymmetric supercapacitors, Chem. Commun. 52 (24) (2016) 4517–4520.

[30] Z. Shi, L. Yue, X. Wang, X. Lei, T. Sun, Q. Li, et al., 3D mesoporous hemp-activated carbon/Ni$_3$S$_2$ in preparation of a binder-free Ni foam for a high performance all-solid-state asymmetric supercapacitor, J. Alloys Compd. 791 (2019) 665–673.

[31] S. Chen, C. Lu, L. Liu, M. Xu, J. Wang, Q. Deng, et al., A hierarchical glucose-intercalated NiMn-G-LDH@ NiCo$_2$S$_4$ core–shell structure as a binder-free electrode for flexible all-solid-state asymmetric supercapacitors, Nanoscale 12 (3) (2020) 1852–1863.

[32] R. Xu, J. Lin, J. Wu, M. Huang, L. Fan, X. He, et al., A two-step hydrothermal synthesis approach to synthesize NiCo$_2$S$_4$/NiS hollow nanospheres for high-performance asymmetric supercapacitors, Appl. Surf. Sci. 422 (2017) 597–606.

[33] F. Ma, X. Dai, J. Jin, N. Tie, Y. Dai, Hierarchical core-shell hollow CoMoS$_4$@ Ni–Co–S nanotubes hybrid arrays as advanced electrode material for supercapacitors, Electrochim. Acta 331 (2020) 135459.

[34] N. Kumar, D. Mishra, S.Y. Kim, Self-assembled NiS-SnS heterostructure via facile successive adsorption and reaction method for high-performance solid-state asymmetric supercapacitors, Thin Solid Films 709 (2020) 138138.

[35] Q. Qin, L. Chen, T. Wei, X. Liu, MoS$_2$/NiS yolk–shell microsphere-based electrodes for overall water splitting and asymmetric supercapacitor, Small 15 (29) (2019) 1803639.

[36] H. Li, X. Chen, E. Zalnezhad, K.N. Hui, K.S. Hui, M.J. Ko, 3D hierarchical transition-metal sulfides deposited on MXene as binder-free electrode for high-performance supercapacitors, J. Ind. Eng. Chem. 82 (2020) 309–316.

[37] T. Peng, H. Yi, P. Sun, Y. Jing, R. Wang, H. Wang, X. Wang, In situ growth of binder-free CNTs@ Ni–Co–S nanosheets core/shell hybrids on Ni mesh for high energy density asymmetric supercapacitors, J. Mater. Chem. A 4 (22) (2016) 8888–8897.

[38] S.J. Patil, J.H. Kim, D.W. Lee, Self-assembled Ni$_3$S$_2$//CoNi$_2$S$_4$ nanoarrays for ultra high-performance supercapacitor, Chem. Eng. J. 322 (2017) 498–509.

110 Oxide free nanomaterials for energy storage and conversion applications

[39] P. Yang, L. Feng, J. Hu, W. Ling, S. Wang, J. Shi, et al., Synthesis of the urchin-like NiS@ $NiCo_2S_4$ composites on nickel foam for high-performance supercapacitors, ChemElectro-Chem 7 (1) (2020) 175–182.

[40] X. Nie, X. Kong, D. Selvakumaran, L. Lou, J. Shi, T. Zhu, et al., Three-dimensional carbon-coated treelike Ni_3S_2 superstructures on a nickel foam as binder-free bifunctional electrodes, ACS Appl. Mater. Interfaces 10 (42) (2018) 36018–36027.

[41] M.Z. Xue, Z.W. Fu, Lithium electrochemistry of $NiSe_2$: a new kind of storage energy material, Electrochem. Commun. 8 (12) (2006) 1855–1862.

[42] H. Peng, G. Ma, K. Sun, Z. Zhang, J. Li, X. Zhou, Z. Lei, A novel aqueous asymmetric super-capacitor based on petal-like cobalt selenide nanosheets and nitrogen-doped porous carbon networks electrodes, J. Power Sources 297 (2015) 351–358.

[43] X. Zhao, X. Li, Y. Zhao, Z. Su, R. Wang, Facile synthesis of Tremelliform $Co_{0.85}Se$ nanosheets for supercapacitor, J. Alloys Compd. 697 (2017) 124–131.

[44] C. Gong, M. Huang, P. Zhou, Z. Sun, L. Fan, J. Lin, J. Wu, Mesoporous $Co_{0.85}Se$ nanosheets supported on Ni foam as a positive electrode material for asymmetric supercapacitor, Appl. Surf. Sci. 362 (2016) 469–476.

[45] H. Peng, J. Zhou, K. Sun, G. Ma, Z. Zhang, E. Feng, Z. Lei, High-performance asymmetric supercapacitor designed with a novel NiSe@ $MoSe_2$ nanosheet array and nitrogen-doped car-bon nanosheet, ACS Sustain. Chem. Eng. 5 (7) (2017) 5951–5963.

[46] H. Peng, C. Wei, K. Wang, T. Meng, G. Ma, Z. Lei, X. Gong, $Ni_{0.85}Se$@ $MoSe_2$ nanosheet arrays as the electrode for high-performance supercapacitors, ACS Appl. Mater. Interfaces 9 (20) (2017) 17067–17075.

[47] C. Chen, D. Zhao, X. Wang, Influence of addition of tantalum oxide on electrochemical capac-itor performance of molybdenum nitride, Mater. Chem. Phys. 97 (1) (2006) 156–161.

[48] A. Guerrero-Ruiz, Q. Xin, Y.J. Zhang, A. Maroto-Valiente, I. Rodriguez-Ramos, Microcalo-rimetric study of H adsorption on molybdenum nitride catalysts, Langmuir 15 (14) (1999).

[49] K.H. Lee, Y.W. Lee, A.R. Ko, G. Cao, K.W. Park, Single-crystalline mesoporous molybdenum nitride nanowires with improved electrochemical properties, J. Am. Ceram. Soc. 96 (1) (2013) 37–39.

[50] L.F. Chen, X.D. Zhang, H.W. Liang, M. Kong, Q.F. Guan, P. Chen, et al., Synthesis of nitrogen-doped porous carbon nanofibers as an efficient electrode material for supercapacitors, ACS Nano 6 (8) (2012) 7092–7102.

[51] R. Lucio-Porto, S. Bouhtiyya, J.F. Pierson, A. Morel, F. Capon, P. Boulet, T. Brousse, VN thin films as electrode materials for electrochemical capacitors, Electrochim. Acta 141 (2014) 203–211.

[52] E. Eustache, R. Frappier, R.L. Porto, S. Bouhtiyya, J.F. Pierson, T. Brousse, Asymmetric elec-trochemical capacitor microdevice designed with vanadium nitride and nickel oxide thin film electrodes, Electrochem. Commun. 28 (2013) 104–106.

[53] D. Choi, G.E. Blomgren, P.N. Kumta, Fast and reversible surface redox reaction in nanocrys-talline vanadium nitride supercapacitors, Adv. Mater. 18 (9) (2006) 1178–1182.

[54] S.I.U. Shah, A.L. Hector, J.R. Owen, Redox supercapacitor performance of nanocrystalline molybdenum nitrides obtained by ammonolysis of chloride-and amide-derived precursors, J. Power Sources 266 (2014) 456–463.

[55] T.C. Liu, W.G. Pell, B.E. Conway, S.L. Roberson, Behavior of molybdenum nitrides as mate-rials for electrochemical capacitors: comparison with ruthenium oxide, J. Electrochem. Soc. 145 (6) (1998) 1882.

[56] O. Banakh, P.E. Schmid, R. Sanjines, F. Levy, High-temperature oxidation resistance of $Cr_{1-x}Al_xN$ thin films deposited by reactive magnetron sputtering, Surf. Coat. Technol. 163 (2003) 57–61.

[57] S. Bouhtiyya, R.L. Porto, B. Laïk, P. Boulet, F. Capon, J.P. Pereira-Ramos, et al., Application of sputtered ruthenium nitride thin films as electrode material for energy-storage devices, Scr. Mater. 68 (9) (2013) 659–662.

[58] A. Achour, R.L. Porto, M.A. Soussou, M. Islam, M. Boujtita, K.A. Aissa, et al., Titanium nitride films for micro-supercapacitors: effect of surface chemistry and film morphology on the capacitance, J. Power Sources 300 (2015) 525–532.

[59] M.S. Balogun, Y. Zeng, W. Qiu, Y. Luo, A. Onasanya, T.K. Olaniyi, Y. Tong, Three-dimensional nickel nitride (Ni_3N) nanosheets: free standing and flexible electrodes for lithium ion batteries and supercapacitors, J. Mater. Chem. A 4 (25) (2016) 9844–9849.

[60] M.S. Balogun, W. Qiu, W. Wang, P. Fang, X. Lu, Y. Tong, Recent advances in metal nitrides as high-performance electrode materials for energy storage devices, J. Mater. Chem. A 3 (4) (2015) 1364–1387.

[61] F. Ran, Z. Wang, Y. Yang, Z. Liu, L. Kong, L. Kang, Nano vanadium nitride incorporated onto interconnected porous carbon via the method of surface-initiated electrochemical mediated ATRP and heat-treatment approach for supercapacitors, Electrochim. Acta 258 (2017) 405–413.

[62] Q. Zhang, X. Wang, Z. Pan, J. Sun, J. Zhao, J. Zhang, et al., Wrapping aligned carbon nanotube composite sheets around vanadium nitride nanowire arrays for asymmetric coaxial fiber-shaped supercapacitors with ultrahigh energy density, Nano Lett. 17 (4) (2017) 2719–2726.

[63] Z.G. Chen, J. Zou, G. Liu, F. Li, Y. Wang, L. Wang, et al., Novel boron nitride hollow nanoribbons, ACS Nano 2 (10) (2008) 2183–2191.

[64] X. Jia, Y. Zhao, G. Chen, L. Shang, R. Shi, X. Kang, et al., Water splitting: Ni_3FeN nanoparticles derived from ultrathin NiFe-layered double hydroxide nanosheets: an efficient overall water splitting electrocatalyst (Adv. Energy Mater. 10/2016), Adv. Energy Mater. 6 (10) (2016).

[65] H. Lu, J. Chen, Q. Tian, Wearable high-performance supercapacitors based on Ni-coated cotton textile with low-crystalline Ni-Al layered double hydroxide nanoparticles, J. Colloid Interface Sci. 513 (2018) 342–348.

[66] Q. Wang, D. O'Hare, Recent advances in the synthesis and application of layered double hydroxide (LDH) nanosheets, Chem. Rev. 112 (7) (2012) 4124–4155.

[67] W. Peng, H. Li, S. Song, Synthesis of fluorinated graphene/CoAl-layered double hydroxide composites as electrode materials for supercapacitors, ACS Appl. Mater. Interfaces 9 (6) (2017) 5204–5212.

[68] M. Ishaq, M. Jabeen, W. Song, L. Xu, W. Li, Q. Deng, Fluorinated graphene-supported nickel-cobalt-iron nitride nanoparticles as a promising hybrid electrode for supercapacitor applications, Electrochim. Acta 282 (2018) 913–922.

[69] B. Wei, F. Ming, H. Liang, Z. Qi, W. Hu, Z. Wang, All nitride asymmetric supercapacitors of niobium titanium nitride-vanadium nitride, J. Power Sources 481 (2021) 228842.

[70] C.K. Maity, S. Sahoo, K. Verma, A.K. Behera, G.C. Nayak, Facile functionalization of boron nitride (BN) for the development of high-performance asymmetric supercapacitors, New J. Chem. 44 (19) (2020) 8106–8119.

[71] J. Balamurugan, T.T. Nguyen, V. Aravindan, N.H. Kim, J.H. Lee, Flexible solid-state asymmetric supercapacitors based on nitrogen-doped graphene encapsulated ternary metal-nitrides with ultralong cycle life, Adv. Funct. Mater. 28 (44) (2018) 1804663.

[72] C.K. Maity, N. Goswami, K. Verma, S. Sahoo, G.C. Nayak, A facile synthesis of boron nitride supported zinc cobalt sulfide nano hybrid as high-performance pseudocapacitive electrode material for asymmetric supercapacitors, J. Energy Storage 32 (2020) 101993.

[73] L. Ma, H. Fan, K. Fu, Y. Zhao, Metal-organic framework/layered carbon nitride nano–sandwiches for superior asymmetric supercapacitor, ChemistrySelect 1 (13) (2016) 3730–3738.

112 Oxide free nanomaterials for energy storage and conversion applications

[74] R. Kumar, T. Bhuvana, A. Sharma, Ammonolysis synthesis of nickel molybdenum nitride nanostructures for high-performance asymmetric supercapacitors, New J. Chem. 44 (33) (2020) 14067–14074.

[75] Y. Tan, L. Meng, Y. Wang, W. Dong, L. Kong, L. Kang, F. Ran, Negative electrode materials of molybdenum nitride/N-doped carbon nano-fiber via electrospinning method for high-performance supercapacitors, Electrochim. Acta 277 (2018) 41–49.

[76] S. Vinoth, K. Subramani, W.J. Ong, M. Sathish, A. Pandikumar, CoS_2 engulfed ultra-thin S-doped g-C_3N_4 and its enhanced electrochemical performance in hybrid asymmetric supercapacitor, J. Colloid Interface Sci. 584 (2021) 204–215.

[77] Y. Shi, P. Chen, J. Chen, D. Chen, H. Shu, H. Jiang, X. Luo, Hollow Prussian blue analogue/g-C_3N_4 nanobox for all-solid-state asymmetric supercapacitor, Chem. Eng. J. 404 (2021) 126284.

[78] S.L. Brock, K. Senevirathne, Recent developments in synthetic approaches to transition metal phosphide nanoparticles for magnetic and catalytic applications, J. Solid State Chem. 181 (7) (2008) 1552–1559.

[79] M.R. Wixom, D.J. Tarnowski, J.M. Parker, J.Q. Lee, P.L. Chen, I. Song, L.T. Thompson, High surface area metal carbide and metal nitride electrodes, MRS Online Proc. Libr. 496 (1) (1997) 643–653.

[80] P. Pande, P.G. Rasmussen, L.T. Thompson, Charge storage on nanostructured early transition metal nitrides and carbides, J. Power Sources 207 (2012) 212–215.

[81] H. Zhuang, N. Yang, L. Zhang, R. Fuchs, X. Jiang, Electrochemical properties and applications of nanocrystalline, microcrystalline, and epitaxial cubic silicon carbide films, ACS Appl. Mater. Interfaces 7 (20) (2015) 10886–10895.

[82] C.H. Chang, B. Hsia, J.P. Alper, S. Wang, L.E. Luna, C. Carraro, et al., High-temperature all solid-state microsupercapacitors based on SiC nanowire electrode and YSZ electrolyte, ACS Appl. Mater. Interfaces 7 (48) (2015) 26658–26665.

[83] M. Kim, J. Kim, Development of high power and energy density microsphere silicon carbide–MnO_2 nanoneedles and thermally oxidized activated carbon asymmetric electrochemical supercapacitors, Phys. Chem. Chem. Phys. 16 (23) (2014) 11323–11336.

[84] M. Sarno, S. Galvagno, R. Piscitelli, S. Portofino, P. Ciambelli, Supercapacitor electrodes made of exhausted activated carbon-derived SiC nanoparticles coated by graphene, Ind. Eng. Chem. Res. 55 (20) (2016) 6025–6035.

[85] A. Sanger, A. Kumar, A. Kumar, P.K. Jain, Y.K. Mishra, R. Chandra, Silicon carbide nano-cauliflowers for symmetric supercapacitor devices, Ind. Eng. Chem. Res. 55 (35) (2016) 9452–9458.

[86] H.O. Pierson, Handbook of Refractory Carbides & Nitrides: Properties, Characteristics, Processing and Applications, William Andrew, 1996.

[87] S.T. Oyama, Introduction to the chemistry of transition metal carbides and nitrides, in: The Chemistry of Transition Metal Carbides and Nitrides, Springer, Dordrecht, 1996, pp. 1–27.

[88] M. Cui, X. Meng, Overview of transition metal-based composite materials for supercapacitor electrodes, Nanoscale Adv. 2 (2020) 5516–5528.

[89] Y. Lan, H. Zhao, Y. Zong, X. Li, Y. Sun, J. Feng, et al., Phosphorization boosts the capacitance of mixed metal nanosheet arrays for high performance supercapacitor electrodes, Nanoscale 10 (25) (2018) 11775–11781.

[90] S. Carenco, D. Portehault, C. Boissiere, N. Mezailles, C. Sanchez, Nanoscaled metal borides and phosphides: recent developments and perspectives, Chem. Rev. 113 (10) (2013) 7981–8065.

[91] M. Cheng, H. Fan, Y. Xu, R. Wang, X. Zhang, Hollow Co_2P nanoflowers assembled from nanorods for ultralong cycle-life supercapacitors, Nanoscale 9 (37) (2017) 14162–14171.

[92] E. Farahi, N. Memarian, Nanostructured nickel phosphide as an efficient photocatalyst: effect of phase on physical properties and dye degradation, Chem. Phys. Lett. 730 (2019) 478–484.

[93] Y. Ni, K. Liao, J. Li, In situ template route for synthesis of porous $Ni_{12}P_5$ superstructures and their applications in environmental treatments, CrstEngComm 12 (5) (2010) 1568–1575.

[94] L. Yu, J. Zhang, Y. Dang, J. He, Z. Tobin, P. Kerns, et al., In situ growth of $Ni_2P–Cu_3P$ bimetallic phosphide with bicontinuous structure on self-supported NiCuC substrate as an efficient hydrogen evolution reaction electrocatalyst, ACS Catal. 9 (8) (2019) 6919–6928.

[95] J.F. Callejas, C.G. Read, C.W. Roske, N.S. Lewis, R.E. Schaak, Synthesis, characterization, and properties of metal phosphide catalysts for the hydrogen-evolution reaction, Chem. Mater. 28 (17) (2016) 6017–6044.

[96] K. Zhou, W. Zhou, L. Yang, J. Lu, S. Cheng, W. Mai, et al., Ultrahigh-performance pseudo-capacitor electrodes based on transition metal phosphide nanosheets array via phosphorization: a general and effective approach, Adv. Funct. Mater. 25 (48) (2015) 7530–7538.

[97] X. Lu, G. Wang, T. Zhai, M. Yu, S. Xie, Y. Ling, et al., Stabilized TiN nanowire arrays for high-performance and flexible supercapacitors, Nano Lett. 12 (10) (2012) 5376–5381.

[98] X. Lu, M. Yu, T. Zhai, G. Wang, S. Xie, T. Liu, et al., High energy density asymmetric quasi-solid-state supercapacitor based on porous vanadium nitride nanowire anode, Nano Lett. 13 (6) (2013) 2628–2633.

[99] H. Liang, C. Xia, Q. Jiang, A.N. Gandi, U. Schwingenschlögl, H.N. Alshareef, Low temperature synthesis of ternary metal phosphides using plasma for asymmetric supercapacitors, Nano Energy 35 (2017) 331–340.

[100] W. Song, J. Wu, G. Wang, S. Tang, G. Chen, M. Cui, X. Meng, Rich-mixed-valence $Ni_xCo_{3−x}P_y$ porous nanowires interwelded junction-free 3D network architectures for ultrahigh areal energy density supercapacitors, Adv. Funct. Mater. 28 (46) (2018) 1804620.

Chapter 6

Oxides free materials for flexible and paper-based supercapacitors

C. Justin Raj[a], Hyun Jung[a], and Byung Chul Kim[b]

[a]Department of Chemistry, Dongguk University, Seoul, Republic of Korea, [b]Department of Printed Electronics Engineering, Sunchon National University, Suncheon-si, Jellanamdo, Republic of Korea

Chapter outline

1. **Introduction**	**115**	3.1 Metal carbide	122
2. **Supercapacitor designs**	**117**	3.2 Metal nitride	127
2.1 Flexible substrates	117	3.3 Metal chalcogenides	128
2.2 Flexible solid-state electrolytes	119	3.4 Metal phosphides	137
		4. **Conclusion and future prospects**	**137**
2.3 Device architectures	120	**Acknowledgments**	**140**
3. **Electrode materials for flexible supercapacitors**	**122**	**References**	**140**

1. Introduction

The recent advances in electronics have offered various types of flexible, implantable, wearable, and even stretchable electronic appliances like flexible/rollup displays, health monitors, sensors, smart cards, foldable mobile phones, etc. To operate these flexible electronic devices, it requires an appropriate power source that has some specified futures like lightweight, ultrathin, flexible, and high electrochemical performance [1, 2]. A supercapacitor (SCs) or ultracapacitor is an energy storage device that can design into a flexible form and can fulfill the above requirements of flexible electronic appliances. The promising mechanical elements like bending, folding, twisting, and even stretching quality of the flexible SCs provide an added advantage to design the electronics devices in their desired architecture. Furthermore, flexible SCs exhibited a greater power density, safety, and extreme life cycles than

Oxide Free Nanomaterials for Energy Storage and Conversion Applications.
https://doi.org/10.1016/B978-0-12-823936-0.00007-3
Copyright © 2022 Elsevier Inc. All rights reserved.

116 Oxide free nanomaterials for energy storage and conversion applications

the conventional flexible rechargeable batteries [3]. However, the low energy density of the SCs than the secondary batteries hinder the practical usage of SCs in many applications [4]. So, an immense interest has been devoted to developing the SCs with comparable or even higher energy density than batteries, without much affecting their power density and cyclic stability.

Generally, the flexible substrate, electroactive materials (electrode materials), solid-state flexible electrolyte, and flexible cases (packing material) are the typical components of the flexible SCs [5]. Among these, the electrode materials play an important role to determine the electrochemical performances and enhance the energy density of SCs. A potential SC electrode material possesses good electrical conductivity, a large surface area, high catalytic activity, better mechanical stability, and low development cost [6]. Moreover, the recent invention of hybrid-supercapacitors (HSCs) by integrating a battery type Faradic electrode as energy source and a capacitor-type electrode as power sources substantially improved the energy density of the SCs [7]. This asymmetric device configuration has drawn special attention among the research communities in search of many novel electrode materials. In general, the carbon-based electrical double layer capacitive (EDLC) materials have been used as a negative electrode due to the ultrahigh power density and cycling stability. On the other hand, the conducting polymers or metal compounds (like metal hydroxides, metal oxides, metal nitrides, metal sulfides, etc.) have been widely used as a positive electrode due to the battery type Faradaic energy storage behavior [8].

Since the last decade, several electrode materials like carbonaceous materials, conductive polymers, and transition metal oxides have been extensively utilized and studied for the fabrication and performances of flexible SCs [9, 10]. There are a large number of reviews on the electrode materials to provide better knowledge about the materials and their role in the fabrications of flexible SCs [11]. But, from the last decade, various state-of-the-art approaches were arising to introduce new emerging electrode materials like metal nitrides, metal carbides, and metal chalcogenides nanostructures in flexible energy storage devices. These metal complexes have appealed to great attention in the flexible supercapacitors due to their peculiar physicochemical properties such as high electrical conductivity, excellent stability, favorable surface morphology, and outstanding mechanical strength.

The previous chapters (Chapters 4 and 5) described more detail regarding the working principle, classifications, and energy storage mechanism of the SCs, so this chapter focused on the crucial components, design, and working of flexible SCs. Further, it is mainly devoted to the comprehensive summary of the electrode materials and their recent developments especially oxide-free metal compounds like carbides, nitrides, chalcogenides (S, Se, Te), and phosphides for flexible SCs and their challenges towards the future direction. Fig. 1 schematically depicts the designs, various types of electrode materials, and few commercial applications of the flexible SCs.

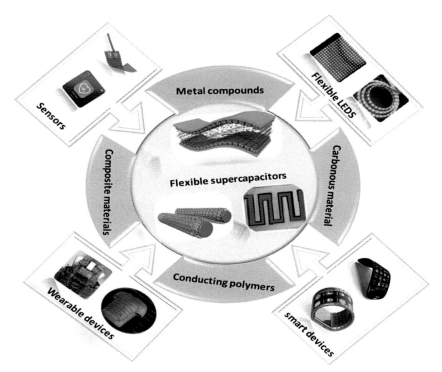

FIG. 1 Scheme of the flexible SC designs, various types of electrode materials, and applications.

2. Supercapacitor designs

As mentioned in the introduction, the substrate (current collector), electrode materials, and flexible solid-state electrolyte are the three major components of the flexible supercapacitor. If the three components match and exhibit all the relevant properties of flexible devices, one can develop a high-performance flexible SC for practical applications. A brief account of various types of substrate and electrolytes used in flexible SCs has been discussed in this section.

2.1 Flexible substrates

In flexible SCs, the substrate (current collector) used for the fabrication of electrodes must possess high electrical conductivity to guarantee the charge/discharge process of the device. Moreover, it requires some special qualities to withstand any types of mechanical deformations like bending, folding, twisting, and stretching to assemble flexible SCs relevant to flexible/wearable electronics. Thin metal sheets, polymer, carbon cloth, conducting papers, etc., are the most commonly used substrates for the fabrications of flexible SCs. Each type of substrate has its pros and cons for the development of flexible devices. The metal sheets like copper (Cu), titanium (Ti), stainless steel (SS), nickel (Ni), etc., are extensively used flexible substrates due to their extraordinary electrical conductivity and mechanical strength [12–15]. However, the high density and thickness of the metal substrate led to an increase in the overall weight of the device. And the thick active material coated over the two-dimensional (2D) metal surface restricts the charge transport path and decreases the performance of the device. So, the suggestion of utilizing three-dimensional (3D) metal substrates like flexible foam and meshes can improve the performance of the flexible SCs [16, 17]. But, the low corrosion resistance of the metallic substrate and considerable increase in device weight cannot be ruled out in this type of flexible device fabrication.

The polymer substrates are the extensively used substrate for the fabrication of flexible supercapacitors owing to their lightweight and excellent flexibility. The polymer substrates such as polyethylene terephthalate (PET), polyurethane (PU), polycarbonates (PC), latex, poly vinylidene fluoride (PVDF), etc., [9, 18–21] are commonly used substrates. But, the insulating nature of polymer substrates requires a thin conducting layer (metallic, graphene, or conducting polymers) to ensure the conductivity of the substrate for better electrochemical performances of the flexible SCs. The recent development in textile and wearable electronics leads to the utilization of flexible carbon fiber cloths and yarn-like 1D fiber substrates for the construction of flexible SCs [2, 10]. These substrates are promising in flexible SCs due to their excellent flexibility, stretchability, and easy fabrication into any desire and appropriate shapes and design.

On the other hand, the paper-based carbon-supported substrates are potential for the development of flexible supercapacitors owing to the high electrical conductivity, lightweight, considerable flexibility, and low corrosion resistance. Moreover, the 3D arrangement of well-interconnected carbon fiber networks offers a large surface area and porous architecture to accommodate a reasonable number of active materials. This 3D architecture and nanoscale pores effectively enhance the charge transportation path and improve the affinity or accessibility of electrolyte ions with the electroactive materials [22].

Generally, the active material in SC electrodes is coated by conventional slurry (paste) or drop-casting techniques, it contains a mixture of active material, binder, and conductive additives [23]. This electrode can restrict the ions' mobility and the penetration of electrolyte into the porous structure, which often leads to an increase in the internal resistance of the device and subsequently decreases the electrochemical performances. Moreover, the incorporation of additional additives like a binder and conductive materials can increase the weight of flexible SCs. Therefore, the new strategy like direct deposition of active materials over flexible 3D substrates (foam, mesh, textiles, etc.) and free-standing or self-standing of active-material electrodes can considerably solve the above issues of slurry-based electrodes [24]. Commonly, the binder-free materials are directly deposited over substrates by different methods like hydrothermal, thermal evaporation, chemical bath deposition, electrodeposition, etc. [25–27]. In freestanding electrodes, the entire electrode volume will use for the energy storage process. Because in this approach, the total flexible substrate has converted to conducting/electroactive media using various conducting carbon nanostructures or carbon-based composites. Moreover, these freestanding electrodes are lightweight, compatible, and easily processable by a simple solution-based vacuum filtration technique on paper or nanocellulose substrates [1, 28, 29].

2.2 Flexible solid-state electrolytes

For designing flexible SCs, a flexible solid-state electrolyte is necessary with essential properties like high ionic conductivity, good mechanical durability, high thermal and chemical stability, etc. Moreover, these solid-state electrolytes have to act as both ionic conducting media and the separator. The aqueous gel polymer electrolyte (AGPEs), organic gel polymer electrolytes (OGPEs), and inorganic solid electrolytes (ceramic electrolytes; CEs) are the most commonly used electrolyte systems in the flexible SCs [30]. The GPEs are solid-state electrolytes with a liquid phase and widely used electrolytes for the development of flexible SCs. In this electrolyte, the polymer act as the host matrix, the solvent (water or organic) act as a plasticizer and an electrolytic salt or acid or base determine its ionic conductivity [31]. Polyvinyl alcohol (PVA), poly(polyacrylate) (PAA), poly(methyl methacrylate) (PMMA), polyethylene oxide (PEO), polyacrylonitrile (PAN), etc. [32–35], are some highly explored polymers for AGPEs.

Among these, PVA has been widely studied proton-conducting polymeric host owing to its feasible-preparation, high hydrophilicity, good film-forming properties, nontoxic characteristics, and low cost. Further, the PVA-based AGPEs have relatively high ionic conductivity in the range 10^{-5} to 10^{-3} S cm^{-1}. Depend upon the nature of electrode materials, different types of PVA AGPEs combinations have been examined and reported [31]. For example, the carbon or some conducting polymer-based electrodes utilize strong acid compositions like (PVA)/H_3PO_4 and PVA/H_2SO_4 [18, 36]. The neutral salt-based AGPE systems such as PVA/Na_2SO_4, PVA/LiCl, PVA/$NaNO_3$, etc., were mainly employed in the flexible SCs with pseudo-capacitance electrode materials (MnO_2, V_2O_5, polypyrrole, etc.) [10, 37]. Similarly, the strong alkali PVA/KOH or PVA/NaOH gel polymer electrolyte is primarily used for the construction of hybrid-flexible supercapacitors having a pseudo-capacitance metal oxide or battery type positive electrode [4].

The organic gel polymer electrolytes (OGPEs) are another strategy of flexible solid-state electrolytes. In OGPEs, the polymer/electrolytic salt complexes are plasticizing with an organic solvent like propylene carbonate (PC), ethylene carbonate (EC), dimethylformamide (DMF), etc. Commonly, some high molecular weight polymer (PMMA and PVDF) or blended polymer matrix

120 Oxide free nanomaterials for energy storage and conversion applications

(PEO-PMMA, PEO-PAN, and PVDF-HEP) has been used as a host in OGPEs. The main advantage of a flexible device using OGPEs, it can operate at comparatively wide cell voltage (2.5–4 V) with high energy density than the AGPEs based flexible SCs [35, 38, 39]. Furthermore, inorganic composites like $Li_2S-P_2S_5$, $Li_{2.94}PO_{2.37}N_{0.75}$ (Lipon), phosphotungstic acid/$Al_2(SO_4)_3$.$18H_2O$ $Li_{1.3}Al_{0.3}Ti_{1.7}P_3O_{12}$ (LATP), etc., are some of the ceramic electrolytes utilized for the fabrications of flexible SCs [40–42]. However, these systems still hold a little amount of liquid electrolyte to maintain their ionic conductivity. Moreover, the electrode/electrolyte interfacial resistance is crucial to consider for the better performances of the CE electrolyte-based flexible SCs.

Besides these electrolytes, various alterations have been made in GPEs to improve the ionic conductivity, the operating voltage, and electrochemical performance of the flexible SCs. For instance, ionic liquids (ILs), are used to replace the electrolytic salt from the GPEs system. Since, the ILs have comparable ionic conductivities, wide potential window ($\sim$3.5 V), nonvolatile, and safety. The ILs like 1-butyl-3-methylimidazolium bis(trifluoro methylsulfonyl)imide ([BMIM][TFSI]), 1-ethyl-3-methylimidazolium tetrafluoroborate (EMIMBF4), 1-ethyl-3-methylimidazolium bis(trifluoromethylsulfonyl)-imide ([EMIM][TFSI]), 1-methylimidazolium hydrogen sulfate (MIHSO$_4$) 1-butyl-3-methylimidazolium chloride (BMIMCl), etc., [13, 43, 44] were potentially used with various host polymers as a flexible solid-state electrolyte for SCs. Further, the redox mediators like potassium iodide (KI), $K_3Fe(CN)_6$, hydroquinone, *P*-phenylenediamine, etc. have been used with the gel polymer electrolyte and investigated their improved electrochemical performances. These redox-active species in the GPEs have considerably enhanced the electrochemical capacitance and consequently improved the energy density of the flexible SCs [45, 46].

2.3 Device architectures

The performances of the flexible supercapacitors are also highly dependent upon the device configuration. The supercapacitors are either configured into symmetric or asymmetric device architecture, where the symmetric device has assembled with the positive and negative electrodes of similar type of active material having the same mass. For example, the carbon//carbon, polymer//polymer, metal compounds//metal compound-based devices [19, 22, 47]. But the asymmetric supercapacitors are assembled with the electrodes having different charge storage nature. For example, the device was designed using a pseudocapacitive (faradic) positive electrode and a capacitor-type (nonfaradaic) negative electrode. Generally, the symmetric SC has a lower working voltage due to the usage of similar electrode materials, but the voltage of asymmetric SC can be extendable than that of symmetric devices. Since the different work functions of the positive and negative electrode in the asymmetric device determine its working potential [48]. Moreover, asymmetric SCs with an EDLC material as a negative electrode (high

Oxides free materials Chapter | 6 **121**

power source) and a battery-type faradic positive electrode (high energy source), are termed as the hybrid-supercapacitors. These types of hybrid SCs exhibited excellent electrochemical performances with high energy/power density and even better long-term stability [49].

Recently, the flexible SCs have been further designed into various device architectures like conventional stacked supercapacitor (sandwich type), planar or on-chip (micro-supercapacitors) type, and yarn/fiber or wire type flexible SCs. Fig. 2 depicts the scheme of three different types of flexible SC designs. These types of flexible device designs are highly appreciable for powering any appropriate form of flexible or wearable electronic appliances or components. The stacked type device is a widely used and easily processable design for both flexible symmetric and asymmetric SCs. As shown in Fig. 2A, a pair of activated material deposited flexible electrodes are assembled face to face by sandwiching a gel polymer layer, which acts as both an electrolyte as well as a separator. This device can be bendable, foldable, twistable, or even stretchable depending upon the mechanical properties of the electrode substrates.

The micro-type and in-plane patterned flexible SCs have considerable advantages in the integrated on-chip components, wearable, and printed electronics due to their thinner, lighter, more flexible, and compatible design. In a planar device, the electrodes are constructed on the horizontal substrate surface (Fig. 2B), so no additional separator is required between the electrodes. Moreover, this device configuration offers a fast 2D transport of electrolytic ions between the electrodes and the ions transport cannot be restricted under any mechanical deformation of the device [18, 50, 51].

Recently the 1D yarn/fiber-based flexible SCs had a significant interest in the development of wearable and textile electronics due to their small volume and excellent mechanical property to integrate into any desired structural form. Commonly, the 1D flexible SCs are assembled by two yarn/fiber electrodes packaged in parallel (Fig. 2C) or twisted and also designed as a core-shell structure. These flexible yarn-based SCs are feasible power sources for portable and

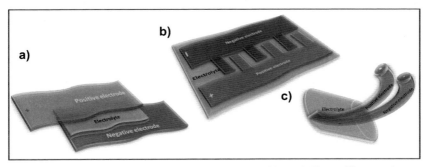

FIG. 2 Scheme of the (A) stacked type, (B) planar or on-chip, and (C) yarn/fiber or wire type flexible SCs.

122 Oxide free nanomaterials for energy storage and conversion applications

flexible electronic devices, such as wearable electronic textiles, smart medical devices and sensors, etc. [6, 52].

3. Electrode materials for flexible supercapacitors

The electrode materials are the most crucial component of the supercapacitors. Generally, the performances of the flexible SCs highly depend upon the electrode materials, especially their electrochemical affinity with the electrolytes. Since the development of supercapacitors, a variety of electrode materials have been explored and even commercialized for practical applications. However, the growing demands of the power and energy sources in modern appliances have stimulated an immunes interest among the researcher to explore the novel electrode materials to fulfill their needs. In general, the electrode materials are categorized based on the nonfaradic capacitive and faradic pseudocapacitive charge storage mechanism of SCs. For example, carbon nanostructures like carbon nanotubes, graphene, activated carbon, etc. as electric double layer capacitance (nonfaradic capacitive) materials. Meanwhile, various conducting polymers and metal compounds (metal-oxide, hydroxides, nitrides, carbides, etc.) have been termed pseudo-capacitor materials. Among these materials, carbon nanostructures, metal oxides, metal hydroxides, and conducting polymers are well explored and widely investigated materials for flexible supercapacitor electrode materials.

But, for the past decade, there is fast growth in metal compounds like metal chalcogenides, metal carbide, metal nitride, and other related composite nanostructures for flexible supercapacitor electrode materials. Since, these materials having some attractive and peculiar physical, structural, electrical, electrochemical properties satisfying some required demands of flexible SCs. Thus, this section briefly summarizes the most recent development and progress of these materials in the fabrication and performances of flexible and paper-based SCs.

3.1 Metal carbide

The transition metal carbides (TMCs), such as, molybdenum carbide (Mo_2C), vanadium carbide (VC), titanium carbide (Ti_3C_2), boron carbide (B_4C), etc., are some promising classes of materials that have recently been used for the fabrications of flexible SC electrodes [53–56]. These metal carbides exhibit ultrahigh conductivity, good electrochemical properties, mechanical stiffness, high thermal and chemical stability, which are some viable properties of flexible SC electrode materials. Boron carbide is a well-known hard ceramic with high thermal, chemical, and mechanical stabilities. The core/shell structured $B_4C@C$ synthesized by vacuum heating process was employed for the construction of flexible planar SCs (micro-SCs) using mask-assisted vacuum filtration technique as shown in Fig. 3. The device achieved a maximum areal

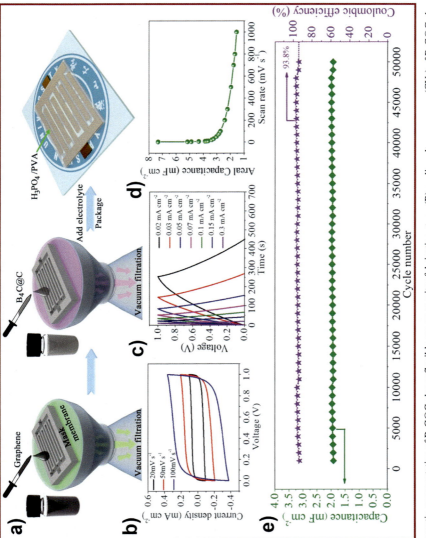

FIG. 3 (A) Schematic representation of B$_4$C@C planar flexible supercapacitor fabrication steps, (B) cyclic voltammograms (CVs) of B$_4$C@C planar flexible SC at various scan rates, (C) galvanostatic charge/discharge (GCD) curves of the flexible device at different current densities, (D) variation of areal capacitance vs scan rates and (E) cyclic stability and coulombic efficiency of B$_4$C@C planar flexible SC for 50,000 cycles. (Reprinted with permission from Y. Chang, X. Sun, M. Ma, C. Mu, P. Li, L. Li, M. Li, A. Nie, J. Xiang, Z. Zhao, J. He, F. Wen, Z. Liu, Y. Tian, Application of hard ceramic materials B4C in energy storage: design B4C@C core-shell nanoparticles as electrodes for flexible all-solid-state micro-supercapacitors with ultrahigh cyclability, Nano Energy 75 (2020) 104947. Copyright 2020, Elsevier.)

124 Oxide free nanomaterials for energy storage and conversion applications

capacitance of $7.33 \, mF \, cm^{-2}$ in H_3PO_4/PVA gel electrolyte and showed excellent cyclic stability for 50,000 charges/discharge cycles. Moreover, the device operated in a wide range of temperatures ($-25°C$ to $75°C$) and displayed excellent mechanical and environmental stability [56].

The metals carbides like Ti_2C, Ti_3C_2, Nb_4C_3, Ta_4C_3, V_2C, etc., have a unique two-dimensional structure and are commonly termed as MXenes [57]. These 2D materials have some diverse properties than those of other-dimensional, and structural materials. The morphological anisotropy, large surface area, mechanical stability and nearly atomic layer thickness are the promising properties of 2D materials, more applicable for energy storage devices [58]. Generally, the $Ti_3C_2T_X$ is the extremely researched MXene for flexible supercapacitors due to its high volumetric capacitance and electrical conductivity. Moreover, the large-surface-to-volume ratio, rich surface chemistry, hydrophobicity, and easy intercalation of various electrolytic ions between its layered structure added more advantages to these materials. In a recent study, a transparent flexible SC was fabricated utilizing spin-casted conductive $Ti_3C_2T_X$ nanosheet/PET electrodes with PVA/H_2SO_4 gel polymer electrolyte. The developed stacked type flexible SC exhibited nearly 72% of optical transparency and showed maximum areal capacitance of $1.6 \, mF \, cm^{-2}$ with $0.05 \, \mu W \, h \, cm^{-2}$ energy density and 100% stability over 20,000 charges/discharge cycles [59].

Similarly, in another study, the micro-supercapacitor was developed using ionic liquid preintercalated $Ti_3C_2T_X$ MXene films. The planar (micro) SC was designed by interdigital mask assisted pattern technique on PET substrate and $EMIMBF_4/PVDF$-HFP ionogel polymer was used as an electrolyte (Fig. 4). This flexible ionogel gel-based micro-SC operated in a wide voltage range (0–3 V) and displayed an excellent volumetric capacitance ($133 \, F \, cm^{-3}$) and energy density ($43.7 \, mW \, h \, cm^{-3}$) with excellent mechanical and cyclic stabilities [60]. The molybdenum carbide is a potential anode material in lithium-ion batteries [61] due to its ultra-high electrical conductivity, excellent mechanical, thermal, and chemical stabilities. So, the $Mo_{1.33}C$ MXene had used for the fabrication of a flexible supercapacitor. The flexible SC assembled with $Mo_{1.33}C$/poly(3,4-ethylenedioxythiophene):poly(styrenesulfonic acid) (PEDOT:PSS) and PVA/H_2SO_4 gel polymer electrolyte showed excellent electrochemical performances. The assembled $Mo_{1.33}C$/PEDOT: PSS flexible SC demonstrated the maximum volumetric capacitance and energy density of $568 \, F \, cm^{-3}$ and $33.2 \, mW \, h \, cm^{-3}$. And it exhibited an excellent mechanical property ($\sim100\%$ retention for different bending conditions) and good cyclic stability ($\sim90\%$) [62]. Further, the various advances in the fabrication and supercapacitor performances of metal carbide-based flexible supercapacitors are presented in Table 1.

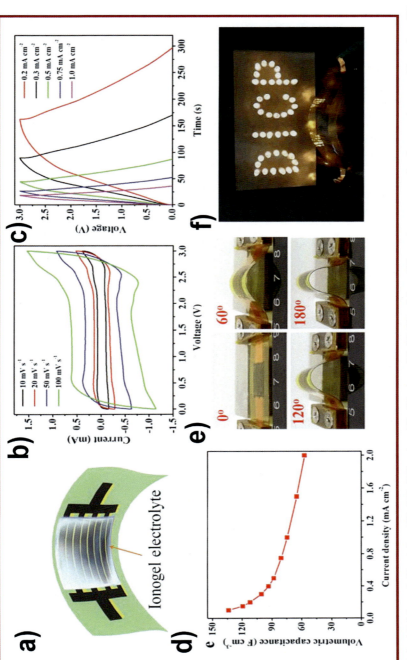

FIG. 4 (A) Schematic illustration of Ti₃C₂Tₓ MXene based flexible planar SC, (B) CVs of planar SC at various scan rates showing a maximum working voltage 3 V, (C) GCD curves of the device at different current densities, (D) plot of volumetric capacitances vs current density, (E) photographic image of the device at various bending angles, and (F) photograph of DCIP LED pattern powered by three serially connected flexible planar devices in its bending state. *(Reprinted with permission from S. Zheng, C. Zhang, F. Zhou, Y. Dong, X. Shi, V. Nicolosi, Z.-S. Wu, X. Bao, Ionic liquid pre-intercalated MXene films for ionogel-based flexible microsupercapacitors with high volumetric energy density, J. Mater. Chem. A 7 (2019) 9478–9485. Copyright 2019, Royal Society of Chemistry.)*

TABLE 1 Metal carbide-based flexible SCs symmetric (SSC) and asymmetric (ASC) and their performances.

Electrode materials	Design	Electrolyte	Voltage (V)	Capacitance	Energy density	Cyclic stability %/cycles	Ref.
rGO//Ti$_3$C$_2$T$_X$	ASC-planar	PVA/H$_2$SO$_4$	1.0	–	8.6 mWh cm^{-3}	97/10000	[63]
Ti$_3$C$_2$T$_X$@rGO	SSC-stacked	PVA/H$_2$SO$_4$	0.6	54 mF cm^{-2}	–	100/1000	[64]
PPy-Ti$_3$C$_2$T$_X$	SSC-textile	PVA/H$_2$SO$_4$	0.6	343 F g^{-1}	1.3 mWh g^{-1}	–	[65]
Ti$_3$C$_2$T$_X$	SSC-planar	PVA/H$_2$SO$_4$	0.6	61 mF cm^{-2}	0.76 μWh cm^{-2}	93.7/10000	[66]
Ti$_2$C	SSC-wire	PVA/KOH	0.7	3.1 mF cm^{-1}	210 nWh cm^{-1}	92.62/1000	[57]
Graphene/Ti$_3$C$_2$T$_X$	SSC-stacked	PVA/H$_3$PO$_4$	0.8	216 F cm^{-3}	3.4 mWh cm^{-3}	85.2/2500	[67]
PPy/L-Ti$_3$C$_2$	SSC-stacked	PVA/H$_2$SO$_4$	0.5	35 mF cm^{-2}	–	100/10000	[68]
PET/Ti$_3$C$_2$T$_X$	SSC-planar	PVA/H$_2$SO$_4$	1.0	1.44 F cm^{-3}	0.2 mWh cm^{-3}	87/1000	[69]

3.2 Metal nitride

In recent years, metal nitrides are an emerging class of SC electrode materials due to their high electrical conductivity (4000–$55,500\,S\,cm^{-1}$) than metal oxides. Since the electronic structure of the metal host is highly affected by the nitrogen atom. For instance, the titanium in its oxide form (TiO_2) has an electrical conductivity of $\sim10^{-8}\,S\,m^{-1}$, nevertheless, the titanium nitride (TiN) exhibits $\sim10^4\,S\,m^{-1}$ [70]. Moreover, the metal nitride exhibits a wide voltage range (then the metal carbide), excellent structure and chemical stabilities, better electrochemical property, and high corrosion resistance. The series of metal nitrides such as vanadium nitride, molybdenum nitride, titanium nitride, gallium nitride, nickel nitride, iron nitride, etc., are explored as some potential electrode materials for the development of flexible SCs [71]. These metal nitrides are mostly produced from their metal oxide by calcination under anhydrous ammonia (NH_3) atmosphere at the temperature ranges between 600 and 900°C.

The mesocrystal vanadium nitride (VN) nanosheets were obtained by the calcination of $Na_2V_6O_{16}$ nanosheets under the NH_3 atmosphere [72]. These VN mesocrystal nanosheets have an impressive metallic behavior and showed an electrical conductivity of $1.44 \times 10^4\,S\,m^{-1}$. The flexible asymmetric SC had assembled using vanadium pentoxide-reduce graphene oxide/Au-coated PET as a negative (V_2O_4-rGO) electrode and VN coated cellulose membrane as a positive electrode with PVA/LiCl electrolyte. The flexible device delivered a considerable volumetric capacitance of $1937\,mF\,cm^{-3}$. In another work, an ultrahigh rate flexible SC was fabricated using Ti-sheet coated molybdenum nitride (Mo_xN) and silicotungstic acid immobilized PVA-H_3PO_4 ($SiWA$-H_3PO_4-PVA) gel polymer electrolyte [73]. The flexible device showed a maximum areal capacitance of $2\,mF\,cm^{-2}$ at $10\,V\,s^{-1}$ scan rate and the SC even maintained its capacitance of $1\,mF\,cm^{-2}$ at $100\,V\,s^{-1}$ scan rate represents a high-rate capability of the device.

A group of researchers has developed titanium nitride (TN) nanowire arrays on carbon cloth for the flexible SCs with PVA/KOH gel polymer electrolytes [74]. The flexible electrode had optimized by varying the nitridation temperature from 700 to 1000°C and found that the TiN electrode prepared at 800°C exhibited better electrochemical properties. The flexible SC designed using this electrode (TiN-800) delivered a maximum volumetric capacitance of $0.33\,F\,cm^{-3}$ and $\sim0.05\,mW\,h\,cm^{-3}$ energy density with $\sim83\%$ capacitance retention for 15,000 charges/discharge cycles. But commonly, the TiN possesses poor electrochemical stability due to the formation of metal oxide layers in electrolytes, so the specific capacitance of the SC continually declines for prolonged charge/discharge cycles. So, Lie and coworkers designed carbon-coated TiN nanotube arrays, in which the carbon layer acts as a protecting layer and considerably enhances the stability of the electrode 4.5 times higher than the uncoated TiN electrode [75].

128 Oxide free nanomaterials for energy storage and conversion applications

A similar strategy was adopted to design a carbon-covered titanium nitride nanotubes (TiN@C) for fiber-based flexible SC [76]. The symmetrical and stacked type flexible fiber SCs assembled using PVA/KOH gel polymer electrolyte displayed excellent mechanical stability and electrochemical performances. The resultant flexible SC delivered a maximum areal capacitance of $19.4\,mF\,cm^{-2}$ and energy density of $2.69\,\mu W\,h\,cm^{-2}$ with considerable cyclic stability (80%) over 10,000 cycles. In addition, this flexible fiber SC demonstrated tailorable and bi-function properties represents a viable energy storage device for wearable and textile electronics. In another study, Li et al., reported a special design of vanadium nitride nanodots intercalated carbon nanosheets (VNNDs/CNSs) for the flexible SC fabrication (Fig. 5) [77]. The flexible SC demonstrated high mechanical durability and better electrochemical stability over 10,000 cycles. Moreover, the device reported a maximum volumetric capacitance of $261.5\,F\,cm^{-3}$ and an energy density of $30.9\,W\,h\,L^{-1}$ that is approximately four times that of the commercial high-energy lithium thin-film battery (Fig. 5E).

Recently, the all-metal nitride flexible asymmetric SC was reported by Fan and coworkers using the electrodes based on titanium nitride (TiN)/iron nitride (Fe_2N) coated vertically aligned graphene nanosheets (GNS) substrate [78]. The flexible asymmetric device was assembled utilizing a TiN@GNS positive electrode and a Fe_2N@GNS negative electrode with PVA/LiCl electrolyte. The hybrid flexible device operated at a wide voltage window of 1.6 V with an excellent specific capacitance $\sim 58\,F\,g^{-1}$ and comparable specific energy of $\sim 15.4\,W\,h\,kg^{-1}$. Moreover, the hybrid flexible SC exhibited outstanding cyclic stability of $\sim 98\%$ for 20,000 cycles with excellent mechanical stability. The cobalt oxynitride (CoON) based hybrid flexible supercapacitor was developed by Wu and coworkers, it exhibited excellent mechanical and cyclic stabilities. In this device, the CoON deposited carbon cloth (CC) acts as a positive electrode, and the activated carbon (AC-CC) electrode is a negative electrode. The flexible SC showed an excellent areal capacitance of $\sim 2.12\,F\,cm^{-2}$ with an energy density of $1.17\,W\,h\,cm^{-2}$ [79]. Furthermore, various recent progress in other metal nitrides and their composite-based flexible SCs are summarized with their electrochemical performances in Table 2.

3.3 Metal chalcogenides

Transition metal chalcogenides (MC; M = transition metals (Cu, Ni, Fe, Mn, Co, Mo, etc.), C = S, Se, and Te) are another potential category of SC electrode materials. These metal chalcogenides exhibit promising properties like superior electrical conductivity, cost-effectiveness, high theoretical capacitance, multiple oxidation states, considerable structural stability, etc. [87, 88]. The MCs exist in various structures, and the electrochemical and catalytic properties strongly depend upon their unique structural aspects. Based on this, the MCs have been classified into two major types layered (2D) and nonlayered

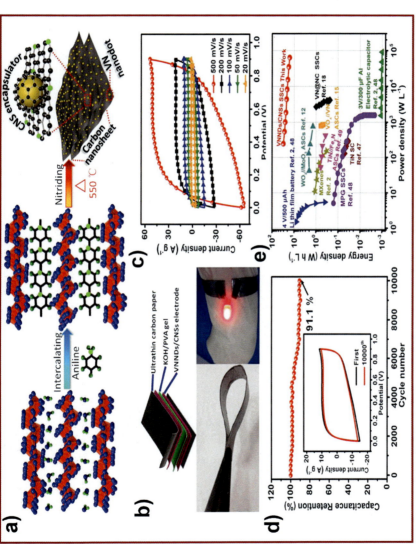

FIG. 5 (A) Scheme showing the fabrication steps of VN nanodots intercalated carbon nanosheets, (B) schematic illustration of the VNNDs/CNS flexible SC and photographic image of the device demonstrating the flexibility and wearable applications (powering a red commercial LED), (C) CVs of the flexible SC at different scan rates, (D) cyclic stability of the device for 10,000 cycles (inset shows the CVs before and after 10,000 cycles) and (E) Ragone plots of the VNNDs/CNSs flexible SC and other flexible, commercial high energy lithium thin-film battery (4V/500 μAh), high-power aluminum electrolytic capacitor (3V/300 μF). *(Reprinted with permission from Q. Li, Y. Chen, J. Zhang, W. Tian, L. Wang, Z. Ren, X. Ren, X. Li, B. Gao, X. Peng, P.K. Chu, K. Huo, Spatially confined synthesis of vanadium nitride nanodots intercalated carbon nanosheets with ultrahigh volumetric capacitance and long life for flexible supercapacitors, Nano Energy 51 (2018) 128–136. Copyright 2018, Elsevier.)*

TABLE 2 The electrochemical performance of metal nitride-based symmetric (SSC) and asymmetric (ASC) flexible SCs.

Electrode materials	Design	Electrolyte	Voltage (V)	Capacitance	Energy density	Cyclic stability %/cycles	Ref.
H-TiN NPs	SSC-stacked	PVA/H_2SO_4	0.8	$5.9\,F\,cm^{-3}$	$0.53\,mWh\,cm^{-3}$	99/3000	[47]
ZnO/Ti@TiN	SSC-Planar	PVA/LiCl	0.8	$1.24\,mF\,cm^{-2}$	$0.24\,W\,kg^{-1}$	98/5000	[80]
$NiCo_2N@NG//$ NiFeN@NG	ASC-stacked	PVA/KOH	1.6	$119\,mAh\,g^{-1}$ (capacity)	$94.93\,Wh\,kg^{-1}$	94.9/25000	[4]
CL-TiN	SSC-stacked	Na_2SO_4/carboxymethyl cellulose gel	0.6	$7.48\,F\,cm^{-3}$	$0.34\,mWh\,cm^{-3}$	<100/20000	[81]
$TiN@MnO_2//TiN$	ASC-stacked	PVA/LiCl	1.8	$1.1\,F\,cm^{-3}$	$0.55\,mWh\,cm^{-3}$	98.3/5000	[82]
$Fe_2N@Ti_2N$	SSC-stacked	PVA/LiCl	2.0	$87.21\,F\,g^{-1}$	$48.45\,Wh\,kg^{-1}$	99/20000	[83]
TiN-MON// TiN-MnO_2	ASC-Fiber	PVA/KOH	1.6	$75.1\,mF\,cm^{-2}$	$23.7\,\mu Wh\,m^{-2}$	84.5/8000	[84]
WON NW//MnO_2	ASC-stacked	PVA/LiCl	1.8	$2.73\,F\,cm^{-3}$	$1.27\,mWh\,cm^{-3}$	95.2%/ 10,000	[85]
VN/CNT	SSC-stacked	PVA/H_3PO_4	0.7	$7.9\,F\,cm^{-3}$	$0.54\,mWh\,cm^{-3}$	82%/10,000	[86]

(1D or 3D) nanostructures. The metal chalcogenides like MoC_2, VC_2, SnC_2 (C is S or Se), etc., are layered structures in which the metal layer was sandwiched between two sulfur or selenium layers (S-M-S) stacked together by a van der Waals forces. On the other hand, the nonlayered structures like FeS_2, NiS_2, $NiSe_2$, CoS_2, etc., are held by chemical covalent bonds in all 3D space [89, 90]. Recently, 2D materials have become an emerging topic in the field of flexible SCs, due to their peculiar structure which enhances the contact area with electrolyte for better electrochemical reaction (redox) and more active sites for the accumulation of electrolytic ions (EDLC). Moreover, the 2D structure offers high mechanical stability and its large interlayer distance offers easy migration of ions for more insertion/desertion process [91].

In 2011, Xie and coworkers developed an in-planar flexible SC utilizing few-layer vanadium disulfide (VS_2) ultra-thin nanosheets alternate to the graphene-based in-planar SC configuration. The VS_2 flexible planar SC (thickness $\sim$150 nm) demonstrated excellent areal capacitance of $4760 \, \mu F \, cm^{-2}$ with 100% stability over 1000 charge/discharge in $PVA/BMIMBF_4$ gel electrolyte [92]. Followed by their work, an immense interest have been devoted to design various types of flexible SCs using 2D metal chalcogenides and even with different alterations in the surface of electrode materials. For instance, a flexible fiber-based SC was developed utilizing Ti/TiO_2 deposited MoS_2 nanosheets coaxial fiber electrodes [93]. The assembled flexible fiber type SC delivered a maximum energy density $\sim$2.70 W h kg^{-1} and specific capacitance of $230 \, F \, g^{-1}$ with excellent mechanical stability and processability. In another research work, a flexible paper-based SC was designed using MoS_2 nanosheets/CNT hybrid ink on paper electrodes with PVA/H_2SO_4 electrolyte [94]. The paper SC demonstrated the maximum areal and volumetric capacitance of $\sim$16.3 mF cm^{-2} and ($\sim$13.6 F cm^{-3}) respectively. Moreover, the device achieved high energy of $0.92 \, mW \, h \, cm^{-3}$, and a high and power density of $\sim$2.1 W cm^{-3} represents a potential device for paper-based or wearable electronics.

The 2D tin selenide ($SnSe_2$) is also an excellent electrode material for the construction of flexible SCs. The $SnSe_2$ nanosheets based on flexible SC in $PVA-H_2SO_4$ electrolyte showed a maximum areal capacitance of $1176 \, \mu F \, cm^{-2}$ and no capacitance fading was observed even after 1000 charge/discharge cycles [95]. Further, a novel atomic-scale thin and semimetal WTe_2 was used for the fabrication of flexible SCs, since the Te-based material has a nearly similar catalytic character to that of Se compounds and even higher conductivity. Here, the WTe_2 was exfoliated into 2–7 layer nanosheets by liquid-phase exfoliation, and it was drop cast over the appropriate area of Au/Ti coated PET substrate. The symmetric flexible SC was fabricated using PVA/H_3PO_4 electrolyte and the as-prepared WTe_2 flexible electrodes. The resultant flexible SCs showed a maximum volumetric capacitance of $\sim$74 F cm^{-3} with high energy and power densities ($\sim$0.01 W h cm^{-3}, and $\sim$83.6 W cm^{-3}), which were significantly higher than some commercial Li thin-film batteries and even Al-based electrolytic capacitors [96].

132 Oxide free nanomaterials for energy storage and conversion applications

Besides the 2D materials, the nonlayered metal chalcogenides also have considerable interest in the fabrications of flexible SCs due to their intrinsic physical and chemical properties. Further, their high electrical conductivity, multiple oxidation states natural abundance, high theoretical capacity than metal oxides, tunable electronic and morphologies (nanorods, nanoflakes, nanospheres, etc.) properties added additional interest for SC electrodes. The flexible asymmetric SC reported by Wen and coworkers has a coaxial fiber-type design with Ni_3S_2 nanorods array coated Ni wire acts as a positive electrode and the pen ink as a negative electrode (Fig. 6). This flexible fiber-type ASC operates in a stable voltage window of 1.4 V and showed a maximum specific capacitance $\sim 35\,Fg^{-1}$ with a considerable energy density of $8.2\,Whkg^{-1}$ [97]. Similarly, a flexible asymmetric SC had designed using $NiSe_2$ nanoarrays as a positive electrode, activated carbon as a negative electrode, and the PVA/KOH as electrolyte [98]. The battery type behavior of the $NiSe_2$ nanoarrays electrode offered a hybrid flexible SC with wide operating voltage (1.6 V) and a specific capacity of $42.7\,mAhg^{-1}$. Moreover, the device achieved maximum specific energy of $33\,Whkg^{-1}$ at $215.9\,Wkg^{-1}$ specific power with $\sim 90\%$ capacity retention after 5000 cycles of charge/discharges. In a report, the hierarchical CuS deposited 3D graphene was utilized as the active materials for the construction of a flexible SC [99]. This CuS hierarchical structure enhances the rapid ions transport to the conductive 3D graphene matrix and also controls the stress caused by the volumetric change during continuous charge/discharge process. The resultant flexible SC showed a device-specific capacitance of $32\,Fg^{-1}$ with energy density $\sim 5\,Whkg^{-1}$ at $450\,Wkg^{-1}$ power density.

Manganese sulfide and manganese selenides are also significant electrode materials for the fabrications of flexible supercapacitors. Since they are low cost, less toxic, and have rich electrochemical properties of Mn-based metal complexes. Pujari et al. [100] deposited MnS microfibers over flexible stainless steel substrate by the chemical bath deposition technique. Using this electrode, they designed the symmetric flexible SC, and it exhibited a substantial specific capacitance of $68\,Fg^{-1}$ with a specific energy of $18.9\,Whkg^{-1}$. Recently, Tang et al. [101] fabricated a porous nano-cellular MnSe electrode for the construction of battery-type flexible hybrid MnSe//AC supercapacitor. The device exhibited very low internal resistance ($1.34\,\Omega$), a considerable energy density ($39.6\,\mu Whcm^{-2}$), and outstanding cyclic stability over 8000 cycles. Germanium selenides (GeSe and $GeSe_2$) based chalcogenides are most commonly used in optoelectronic devices, sensors, resistive-switching memory cells, etc. But, Shen and coworkers first time used it for the fabrication of flexible SCs like stacked and planar SC configuration [102]. In their work, a 3D hierarchical nanostructure of $GeSe_2$ was prepared by thermal evaporation technique and employed for the fabrication of stacked and in-planar type symmetric SCs. Both the device displayed outstanding rate capability, excellent mechanical properties, and high capacitance retention. Interestingly, they demonstrated the practical application of the planar flexible SC in a self-powered nanosystem, i.e., to power a CdSe nanowire film-based photodetector successfully.

Oxides free materials Chapter | 6 **133**

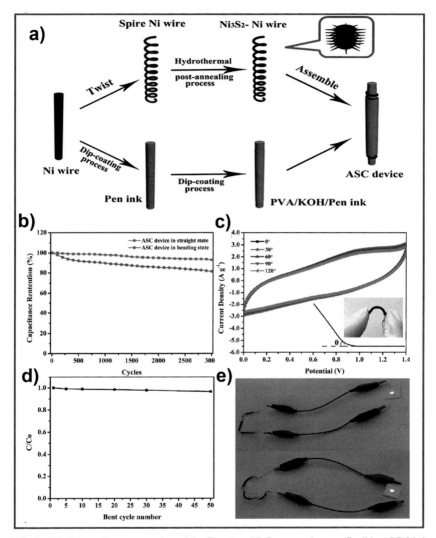

FIG. 6 (A) Schematic representation of the fiber-type Ni$_3$S$_2$ nanorods array flexible ASC fabrication steps. (B) Cycling performances for the fiber-type flexible ASC (3000 cycles) in straight and bending states, (C) CVs of ASC device in straight and different bending states at the scan rate of 60 mV s^{-1} (inset shows the photograph of a flexible ASC), (D) specific capacitance of ASC device after various bending cycle number (Co and C are the specific capacitance before and after bending). (E) Photographs three flexible fiber ASCs powering *blue LED* (white in print versions) in straight state bending state. *(Reprinted with permission from J. Wen, S. Li, K. Zhou, Z. Song, B. Li, Z. Chen, T. Chen, Y. Guo, G. Fang, Flexible coaxial-type fiber solid-state asymmetrical supercapacitor based on Ni$_3$S$_2$ nanorod array and pen ink electrodes, J. Power Sources 324 (2016) 325–333. Copyright 2016, Elsevier.)*

134 Oxide free nanomaterials for energy storage and conversion applications

Generally, single metal chalcogenides have some drawbacks like poor cycling stability and rate capability, lack of electrochemically active sites, structural instability, etc. [103]. Thus, similar to bimetallic oxides ($NiCo_2O_4$), one of the peculiar strategies to overcome these drawbacks is by forming bimetallic or mixed transition metal chalcogenides. These will exhibit the conductivity of polymetallic ions, improves the electrochemical catalytic activity through the synergistic effects of the transition metals combinations (multiple redox reactions). Thus, it makes the bimetallic or ternary transition metal chalcogenides a promising candidate for SC electrode materials [104, 105]. Recently, the ternary transition metal chalcogenides like nickel-cobalt sulfide (NiCoS) [106], nickel vanadium sulfide (NiVS) [107], nickel-cobalt selenide (NiCoSe) [108], copper-molybdenum sulfide (CuMoS) [109], cobalt nickel selenide (CoNiSe) [110], zinc cobalt sulfide (ZnCoS) [111], etc., are reported as the excellent materials for the flexible hybrid SCs electrodes with outstanding energy storage properties.

For example, Xu et al., [112] comparatively studied the flexible supercapacitor performance of the Ni-Co selenides ($Ni_{0.34}Co_{0.66}Se_2$) nanorods with other ternary metal compounds like sulfide ($NiCo_2S_4$) and oxide ($NiCo_2O_4$) nanorods based devices (Fig. 7). In this, the electrode materials are deposited on carbon fiber cloth substrate and assembled the flexible symmetric SCs using PVA/KOH electrolyte. And they achieved the areal capacitance and performances of the SCs in the order $NiCo_2O_4 < NiCo_2S_4 < Ni_{0.34}Co_{0.66}Se_2$. That is, $Ni_{0.34}Co_{0.66}Se_2$ flexible SC achieved a maximum areal capacitance of $1.16\,F\,cm^{-2}$ which is twice and trice the times higher than the $NiCo_2S_4$ and $NiCo_2O_4$ flexible SCs (Fig. 7C). This enhancement in the capacitance and electrochemical performances of the $Ni_{0.34}Co_{0.66}Se_2$ flexible device attributes is due to its higher electrical conductivity and catalytic properties than other ternary metal compounds. However, they investigated that the $Ni_{0.34}Co_{0.66}Se_2$ in the electrode suffers from a loss in Se from the structural site and transforms gradually into the $Ni_xCo_{1-x}O$ phase during repeated charge/discharge cycles. Meanwhile, the electrochemical surface area (ESA) of the electrode was found to be increased considerably. So, the increase in ESA and the lower electrode materials' internal resistance compensates for the capacitance fading due to Se loss from $Ni_{0.34}Co_{0.66}Se_2$ electrode and maintains their stability significantly.

Most recently, Pan et al., fabricated a flexible asymmetric flexible SC using the mixed metal chalcogenides of $Mn_xMoS_{2-y}Se_y$ and $MoFe_2S_{4-z}Se_z$ as the positive and negative electrodes. This flexible ASC demonstrated excellent battery-type hybrid SC performances with excellent energy density ($\sim 69\,W\,h\,kg^{-1}$) and cyclic stability ($\sim 83.5\%$) over 10,000 cycles [48]. In another work, Wang et al., [108] designed $Ni_{4.5}Co_{4.5}$-Se nanowires-NPCC/PVA-KOH/Fe_2C-CFC flexible battery-type hybrid supercapacitor. Where the positive electrode $Ni_{4.5}Co_{4.5}$-Se nanowires were grown over nickel-plated cotton cloth substrate and the negative electrode was deposited on carbon fiber cloth (CFC). The flexible device was assembled by sandwiching ultrathin silk

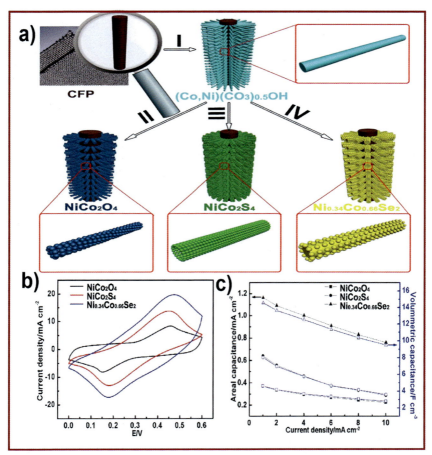

FIG. 7 (A) Schematic diagram of the formation of NiCo$_2$O$_4$, NiCo$_2$S$_4$ and Ni$_{0.34}$Co$_{0.66}$Se$_2$ from (Ni, Co) (CO$_3$)$_{0.5}$OH nanorods array, (B) CVs of NiCo$_2$O$_4$, NiCo$_2$S$_4$, and NiCo$_2$O$_4$, NiCo$_2$S$_4$, and Ni$_{0.34}$Co$_{0.66}$Se$_2$ based flexible symmetric SCs, (C) the volumetric and areal capacitances of the flexible SCs at various current density. *(Reprinted with permission from P. Xu, W. Zeng, S. Luo, C. Ling, J. Xiao, A. Zhou, Y. Sun, K. Liao, 3D Ni-Co selenide nanorod array grown on carbon fiber paper: towards high-performance flexible supercapacitor electrode with new energy storage mechanism, Electrochim. Acta 241 (2017) 41–49. Copyright 2017, Elsevier.)*

fabric (separator) with PVA/KOH electrolyte. The flexible device displayed a typical battery type supercapacitor behavior with a maximum specific capacity of ~114 C g^{-1} and ~47.4 W h kg^{-1} energy density. Moreover, the device displayed considerable cyclic stability (80%/4000 cycles) and mechanical stability over 500 bending cycles with a capacity retention of ~82.3%. Similarly, various types of flexible devices constructed using different metal chalcogenides and their composites-based electrodes are presented in Table 3, with their appropriate supercapacitor performances.

TABLE 3 The supercapacitor performances of metal chalcogenides based symmetric (SSC) and asymmetric (ASC) flexible SCs.

Electrode materials	Design	Electrolyte	Voltage (V)	Capacitance	Energy density	Cyclic stability %/cycles	Ref.
$MoSe_2NF$	SSC-stacked	PVA/KOH	1.4	$133\,Fg^{-1}$	$36.2\,Whkg^{-1}$	92/2000	[113]
VS_2	SSC-stacked	$PVA\text{-}LiClO_4$	1.6	$118\,Fg^{-1}$	$25.9\,Whkg^{-1}$	–	[114]
MoS_2	SSC-stacked	PVA-LiCl	0.8	$368\,Fg^{-1}$	$5.42\,Whkg^{-1}$	96.5/5000	[115]
ZnS/CuS	SSC-stacked	PVA/KOH	1.0	$536\,mFcm^{-2}$	$0.39\,mWhcm^{-2}$	79/2000	[116]
$G/CoS_2/Ni_3S_4$	ASC-printed planar	PVA/KOH	1.5	$840.5\ mFcm^2$	–	79.6%/ 10,000	[117]
Cu@Ni@NiCoS NFs	SSC-transparent stacked	PVA/KOH	0.8	$1.21\,mAcm^{-2}$ capacity	$0.48\,\mu Whcm^{-2}$	92/10000	[118]
$Ni_2CoS_4HS\text{-}HT\text{-}5//$ CAC	ASC-stacked	PVA/KOH	1.0	$1511\ mF\ cm^{-2}$	$13.6\,mWhcm^{-3}$	96/2000	[119]
Ni-Co-Se//AC	ASC-stacked	PVA/KOH	1.6	$108\,Fg^{-1}$	$38.5\,Whkg^{-1}$	82.3/5000	[120]
$CoNi_2S_4$	SSC-yarn	PVA/KOH	1.5	$141\,Fg^{-1}$	$40.9\,Whkg^{-1}$	80/3000	[121]
$GNR\text{-}Co_{0.85}Se//$ $GNR\text{-}Bi_2Se_3$	ASC-stretchable stacked	DMAOP-PVA/ GO-KOH	1.6	–	$30\,Whkg^{-1}$	89/5000	[122]
$NiCo_2S_4@MgS//$ FeOOH	ASC-fiber type	PVA/KOH	1.6	$134.4\,mAhcm^{-3}$	$107.5\,mWhcm^{-3}$	87.5/5000	[123]

3.4 Metal phosphides

Apart from the transition metal chalcogenides, nitride, and carbides, the transition metal phosphides are also and interesting oxide-free metal compounds for flexible SC fabrications. Since metal phosphides also have potential electrochemical properties similar to metal sulfides and selenides. Generally, the metals phosphides have been widely used in the electro-catalysts' application and battery electrodes. However, for the past 5 years, a vast concern is devoted to applying metal phosphides for the fabrications of supercapacitor electrodes [124–126]. Among various available metal phosphides, CoP, NiP, and its bimetallic combinations are normally explored as the potential electrode materials for the construction of flexible SCs. Since, these metal phosphides show the battery type properties (noncapacitive), higher electrical conductivity, reasonable stability. Moreover, the bimetallic phosphides have richer faradaic redox reactions and synergic electrochemical properties similar to their metal oxide and chalcogenides (Ni-Co-O, Ni-Co-S, or Ni-Co-Se).

For example, Ling et al. [127] fabricated the urchin-like NiCoP nanoarrays on a carbon cloth electrode for a flexible supercapacitor. In these special architectures, the 1D nanospines in structure offer an easy charge transport process and the 2D ultrathin nanoflakes on the structure render prominent active sites and shorten the diffusion paths of ions. These properties of the electrode significantly improved the performance of the device than the other designed electrode structures. Thus, using this hybrid electrode, a flexible asymmetric hybrid SC (NiCoP@C /PVA-KOH/AC) was constructed and achieved a maximum specific capacity of $\sim169C\ g^{-1}$ with a high energy density $\sim37.1\,W\,h\,kg^{-1}$ at $792.8\,W\,kg^{-1}$ power density. Further, Table 4 lists some most recent progress in metal phosphide nanostructures in the flexible SCs.

4. Conclusion and future prospects

Concerning the recent research and development in flexible energy storage systems for wearable and smart electronic appliances, this chapter provides an overview of the current trends and development in the electrode materials for flexible, wearable, and paper supercapacitors. For a better understanding of flexible and paper-based supercapacitors, a slight insight into the fundamental concepts, key components of flexible supercapacitors like flexible substrates, solid-state flexible electrolytes, and device designs were summarized. Being vast categories of energy storage materials are explored and available worldwide and considering the theme of this book, the oxide-free metal compounds-based nanomaterials were discussed in detail with some examples of device design and performances.

In the recent decade, more than thousands of research articles have been published per year in metal compounds like carbides, nitrides, sulfides, selenides, etc., for supercapacitor electrodes. These represent some great interest

TABLE 4 The supercapacitor performances of metal phosphides based flexible SCs.

Electrode materials	Design	Electrolyte	Voltage (V)	Capacitance	Energy density	Cyclic stability %/cycles	Ref.
NiVS/NiCuP//rGO/CF	ASC-fiber micro	PVA/ NaOH	1.8	$40.96\,F\,cm^{-3}$	$22.7\,mWh\,cm^{-3}$	91.5/3000	[128]
NiCoP@NF//AC	ASC-stacked	PVA/KOH	1.2	$133\,F\,g^{-1}$	$27\,Wh\,kg^{-1}$	67.2/500	[129]
CC/NiCoP@NiCo-LDH//AC	ASC-stacked	PVA/KOH	1.7	$142\,F\,g^{-1}$	$57\,Wh\,kg^{-1}$	97/10000	[130]
CoP	SSC-in planar	PVA/LiOH	0.8	$156.6\,mF\,cm^{-2}$	$0.013\,mWh\,cm^{-2}$	~95/5000	[131]
NiCoP/CNT//CNT@N-C	ASC-stacked	PVA/KOH	1.5	$23.3\,mAh\,g^{-1}$	$138.7\,Wh\,kg^{-1}$	85/5000	[132]
NiCoP@MoSe$_2$//AC	ASC-stacked	PVA/KOH	1.6	$155\,F\,g^{-1}$	$55.1\,Wh\,kg^{-1}$	95.8/8000	[133]

has committed to the development of oxide-free metal compounds. Since, these materials possessed a high electrical conductivity than metal oxides, more electrochemical activity than carbon nanostructure and conducting polymers. However, the long-term stability of flexible supercapacitors based-metal compounds (especially chalcogenides) is dubitable. So, some peculiar strategies have been adapted to improve the long-term stability of these materials. For example, fabricating composites with stable carbon, metal, or polymer nanostructures or designing a particular nanoarchitecture (core-shell) with these materials. At the same time, it is necessary to consider the materials' original electrochemical properties not been sacrifice during these modifications to attain the high-performance flexible SCs.

Recently, substantial advances have focused on the evolution of 2D metal nitride and carbide for the flexible and paper-based SCs. But, still have some challenges to overcome for a flexible supercapacitor, like cost-effectiveness, better life cycle, high energy, and power densities. Since, these materials exhibit small operating voltage (SSC less than 1 V in GPEs), volume expansion, stacking of 2D nanosheets lead to the less active site and instability during mechanical contortions. So, material optimization and detail interfacial studies are necessary to ensure the strong, and stable interaction between the electrode and solid-state flexible electrolyte. Moreover, the synthesis of metal carbide and nitride involves some complicated procedures, harsh reactions, and even some expensive techniques. Thus, a control, scalable and retainable quality of 2D nanosheets processing through optimized and feasible wet chemical approaches are of great interest.

From the overall review of flexible SCs, the inefficient energy density remains a great challenge to compete with the Li-ion-based flexible batteries in flexible and smart electronic appliances. In light of the above, the hybrid flexible SCs utilizing a battery-type positive electrode (metal chalcogenides and phosphides) and a capacitive negative electrode lift the energy density from 10 to $\sim 80 \, W \, h \, kg^{-1}$. The bimetallic chalcogenides and phosphide have great potential for the development of hybrid SCs, due to their promising synergetic electrochemical activities. However, the positive and negative electrode materials selection with a large work function difference is challenging and important to extend the working voltage of flexible SCs. Thus, a proper charge balancing between the electrodes (optimized mass ratio) and extending the potential window can further determine the high energy density and long-term stability of the flexible SCs.

In addition, an appropriate selection of solid-state flexible electrolytes with less interfacial resistance between electrode/electrolyte, good ionic conductivity, high thermal, and mechanical stabilities have a considerable impact on the device performance. Since the most used aqueous PVA-based GPEs have a limited potential range ($\sim 2 \, V$) due to oxygen and hydrogen evaluation reactions. Hence, the proper studies on ceramic or organic plasticizers or ionic liquid-based flexible electrolytes are crucial to address this issue. Moreover, there

140 Oxide free nanomaterials for energy storage and conversion applications

requires a standard method to evaluate and understand the charge storage mechanism and optimum matching between the newly developed electrode materials and solid-state flexible electrolytes. Finally, the hybrid flexible SCs with the combination of high energy (battery type) and high power (electrochemical capacitors) electrode materials with proper flexible solid-state electrolytes have a bright prospect. In the future, these types of flexible hybrid supercapacitors can be the alternative to batteries and provide much cheaper and safer power sources for flexible and wearable electronic devices.

Acknowledgments

The author H. Jung acknowledges the Basic Science Research Program through the National Research Foundation of Korea (NRF), funded by the Ministry of Education (No. NRF 2016R1D1A1B01009640) and the Korea Institute of Energy Technology Evaluation and Planning (KETEP), and the Ministry of Trade, Industry & Energy (MOTIE) of the Republic of Korea (No. 20194030202320).

The author B. C. Kim acknowledges the Creative Materials Discovery Program through the National Research Foundation of Korea (NRF) funded by the Ministry of Science, ICT, and Future (NRF-2015M3D1A1069710); and the Basic Science Research Program through the National Research Foundation of Korea (NRF) funded by the Ministry of Education (NRF-2014R1A6A1030419), Republic of Korea.

References

[1] Y. Ko, M. Kwon, W.K. Bae, B. Lee, S.W. Lee, J. Cho, Flexible supercapacitor electrodes based on real metal-like cellulose papers, Nat. Commun. 8 (2017) 536.

[2] Q. Xue, J. Sun, Y. Huang, M. Zhu, Z. Pei, H. Li, Y. Wang, N. Li, H. Zhang, C. Zhi, Recent progress on flexible and wearable supercapacitors, Small 13 (2017) 1701827.

[3] A. Yu, I. Roes, A. Davies, Z. Chen, Ultrathin, transparent, and flexible graphene films for supercapacitor application, Appl. Phys. Lett. 96 (2010) 253105.

[4] J. Balamurugan, T.T. Nguyen, V. Aravindan, N.H. Kim, J.H. Lee, Flexible solid-state asymmetric supercapacitors based on nitrogen-doped graphene encapsulated ternary metal-nitrides with ultralong cycle life, Adv. Funct. Mater. 28 (2018) 1804663.

[5] D.P. Dubal, N.R. Chodankar, D.H. Kim, P. Gomez-Romero, Towards flexible solid-state supercapacitors for smart and wearable electronics, Chem. Soc. Rev. 47 (2018) 2065–2129.

[6] B.C. Kim, J.-Y. Hong, G.G. Wallace, H.S. Park, Recent progress in flexible electrochemical capacitors: electrode materials, device configuration, and functions, Adv. Energy Mater. 5 (2015) 1500959.

[7] Y. Shao, M.F. El-Kady, J. Sun, Y. Li, Q. Zhang, M. Zhu, H. Wang, B. Dunn, R.B. Kaner, Design and mechanisms of asymmetric supercapacitors, Chem. Rev. 118 (2018) 9233–9280.

[8] L. Yu, G.Z. Chen, Redox electrode materials for supercapatteries, J. Power Sources 326 (2016) 604–612.

[9] B.C. Kim, H.T. Jeong, C.J. Raj, Y.-R. Kim, B.-B. Cho, K.H. Yu, Electrochemical performance of flexible poly(ethylene terephthalate) (PET) supercapacitor based on reduced graphene oxide (rGO)/single-wall carbon nanotubes (SWNTs), Synth. Met. 207 (2015) 116–121.

[10] C.J. Raj, R. Manikandan, W.-J. Cho, K.H. Yu, B.C. Kim, High-performance flexible and wearable planar supercapacitor of manganese dioxide nanoflowers on carbon fiber cloth, Ceram. Int. 46 (2020) 21736–21743.

[11] M.R. Benzigar, V.D.B.C. Dasireddy, X. Guan, T. Wu, G. Liu, Advances on emerging materials for flexible supercapacitors: current trends and beyond, Adv. Funct. Mater. 30 (2020) 2002993.

[12] R.K. Gupta, J. Candler, S. Palchoudhury, K. Ramasamy, B.K. Gupta, Flexible and high performance supercapacitors based on $NiCo_2O_4$ for wide temperature range applications, Sci. Rep. 5 (2015) 15265.

[13] H. Seok Jang, C. Justin Raj, W.-G. Lee, B. Chul Kim, K. Hyun Yu, Enhanced supercapacitive performances of functionalized activated carbon in novel gel polymer electrolytes with ionic liquid redox-mediated poly(vinyl alcohol)/phosphoric acid, RSC Adv. 6 (2016) 75376–75383.

[14] A. Lamberti, M. Fontana, S. Bianco, E. Tresso, Flexible solid-state CuxO-based pseudo-supercapacitor by thermal oxidation of copper foils, Int. J. Hydrogen Energy 41 (2016) 11700–11708.

[15] L.T. Le, M.H. Ervin, H. Qiu, B.E. Fuchs, W.Y. Lee, Graphene supercapacitor electrodes fabricated by inkjet printing and thermal reduction of graphene oxide, Electrochem. Commun. 13 (2011) 355–358.

[16] G. Zhu, Z. He, J. Chen, J. Zhao, X. Feng, Y. Ma, Q. Fan, L. Wang, W. Huang, Highly conductive three-dimensional MnO_2–carbon nanotube–graphene–Ni hybrid foam as a binder-free supercapacitor electrode, Nanoscale 6 (2014) 1079–1085.

[17] B. Liu, B. Liu, Q. Wang, X. Wang, Q. Xiang, D. Chen, G. Shen, New energy storage option: toward $ZnCo_2O_4$ nanorods/nickel foam architectures for high-performance supercapacitors, ACS Appl. Mater. Interfaces 5 (2013) 10011–10017.

[18] C.J. Raj, B.C. Kim, W.J. Cho, W.G. Lee, S.D. Jung, Y.H. Kim, S.Y. Park, K.H. Yu, Highly flexible and planar supercapacitors using graphite flakes/polypyrrole in polymer lapping film, ACS Appl. Mater. Interfaces 7 (2015) 13405–13414.

[19] H.T. Jeong, J.F. Du, Y.R. Kim, C.J. Raj, B.C. Kim, Electrochemical performances of highly stretchable polyurethane (PU) supercapacitors based on nanocarbon materials composites, J. Alloys Compd. 777 (2019) 67–72.

[20] H.T. Jeong, B.C. Kim, R. Gorkin, M.J. Higgins, G.G. Wallace, Capacitive behavior of latex/single-wall carbon nanotube stretchable electrodes, Electrochim. Acta 137 (2014) 372–380.

[21] Z. Bai, H. Li, M. Li, C. Li, X. Wang, C. Qu, B. Yang, Flexible carbon nanotubes-MnO_2/reduced graphene oxide-polyvinylidene fluoride films for supercapacitor electrodes, Int. J. Hydrogen Energy 40 (2015) 16306–16315.

[22] M. Rajesh, C.J. Raj, R. Manikandan, B.C. Kim, S.Y. Park, K.H. Yu, A high performance PEDOT/PEDOT symmetric supercapacitor by facile in-situ hydrothermal polymerization of PEDOT nanostructures on flexible carbon fibre cloth electrodes, Mater. Today Energy 6 (2017) 96–104.

[23] M. Liang, M. Zhao, H.-y. Wang, J. Shen, X. Song, Enhanced cycling stability of novel hierarchical $NiCo_2S_4$@Ni(OH)$_2$@PPy core-cell nanotube arrays for aqueous asymmetric supercapacitors, J. Mater. Chem. A 6 (2018) 2482–2493.

[24] L. Bao, J. Zang, X. Li, Flexible Zn_2SnO_4/MnO_2 core/shell nanocable–carbon microfiber hybrid composites for high-performance supercapacitor electrodes, Nano Lett. 11 (2011) 1215–1220.

[25] S. Tec-Yam, R. Patiño, A.I. Oliva, Chemical bath deposition of CdS films on different substrate orientations, Curr. Appl. Phys. 11 (2011) 914–920.

142 Oxide free nanomaterials for energy storage and conversion applications

[26] P.E. Lokhande, U.S. Chavan, Nanoflower-like Ni(OH)$_2$ synthesis with chemical bath deposition method for high performance electrochemical applications, Mater. Lett. 218 (2018) 225–228.

[27] D.Y. Kim, G.S. Ghodake, N.C. Maile, A.A. Kadam, D. Sung Lee, V.J. Fulari, S.K. Shinde, Chemical synthesis of hierarchical NiCo$_2$S$_4$ nanosheets like nanostructure on flexible foil for a high performance supercapacitor, Sci. Rep. 7 (2017) 9764.

[28] Y. Cheng, S. Lu, H. Zhang, C.V. Varanasi, J. Liu, Synergistic effects from graphene and carbon nanotubes enable flexible and robust electrodes for high-performance supercapacitors, Nano Lett. 12 (2012) 4206–4211.

[29] Y. He, W. Chen, X. Li, Z. Zhang, J. Fu, C. Zhao, E. Xie, Freestanding three-dimensional graphene/MnO$_2$ composite networks as ultralight and flexible supercapacitor electrodes, ACS Nano 7 (2013) 174–182.

[30] P. Yang, W. Mai, Flexible solid-state electrochemical supercapacitors, Nano Energy 8 (2014) 274–290.

[31] S. Alipoori, S. Mazinani, S.H. Aboutalebi, F. Sharif, Review of PVA-based gel polymer electrolytes in flexible solid-state supercapacitors: opportunities and challenges, J. Energy Storage 27 (2020) 101072.

[32] S. Panero, A. Clemente, E. Spila, Solid state supercapacitors using gel membranes as electrolytes, Solid State Ion. 86–88 (1996) 1285–1289.

[33] X. Liu, T. Osaka, All-solid-state electric double-layer capacitor with isotropic high-density graphite electrode and polyethylene oxide/LiClO$_4$ polymer electrolyte, J. Electrochem. Soc. 143 (1996) 3982–3986.

[34] K.-T. Lee, N.-L. Wu, Manganese oxide electrochemical capacitor with potassium poly(acrylate) hydrogel electrolyte, J. Power Sources 179 (2008) 430–434.

[35] C.-W. Huang, C.-A. Wu, S.-S. Hou, P.-L. Kuo, C.-T. Hsieh, H. Teng, Gel electrolyte derived from poly(ethylene glycol) blending poly(acrylonitrile) applicable to roll-to-roll assembly of electric double layer capacitors, Adv. Funct. Mater. 22 (2012) 4677–4685.

[36] C.J. Raj, M. Rajesh, R. Manikandan, K.H. Yu, J.R. Anusha, J.H. Ahn, D.-W. Kim, S.Y. Park, B.C. Kim, High electrochemical capacitor performance of oxygen and nitrogen enriched activated carbon derived from the pyrolysis and activation of squid gladius chitin, J. Power Sources 386 (2018) 66–76.

[37] Y.J. Kang, H. Chung, W. Kim, 1.8-V flexible supercapacitors with asymmetric configuration based on manganese oxide, carbon nanotubes, and a gel electrolyte, Synt. Met. 166 (2013) 40–44.

[38] K. Naoi, M. Morita, Advanced polymers as active materials and electrolytes for electrochemical capacitors and hybrid capacitor systems, Electrochem. Soc. Interface 17 (2008) 44–48.

[39] C. Ramasamy, J. Palma, M. Anderson, A 3-V electrochemical capacitor study based on a magnesium polymer gel electrolyte by three different carbon materials, J. Solid State Electrochem. 18 (2014) 2903–2911.

[40] X. Hu, Y.L. Chen, Z.C. Hu, Y. Li, Z.Y. Ling, All-solid-state supercapacitors based on a carbon-filled porous/dense/porous layered ceramic electrolyte, J. Electrochem. Soc. 165 (2018) A1269–A1274.

[41] Y.S. Yoon, W.I. Cho, J.H. Lim, D.J. Choi, Solid-state thin-film supercapacitor with ruthenium oxide and solid electrolyte thin films, J. Power Sources 101 (2001) 126–129.

[42] Y.G. Wang, X.G. Zhang, All solid-state supercapacitor with phosphotungstic acid as the proton-conducting electrolyte, Solid State Ion. 166 (2004) 61–67.

[43] D.R. MacFarlane, N. Tachikawa, M. Forsyth, J.M. Pringle, P.C. Howlett, G.D. Elliott, J.H. Davis, M. Watanabe, P. Simon, C.A. Angell, Energy applications of ionic liquids, Energ. Environ. Sci. 7 (2014) 232–250.

[44] K. Karuppasamy, H.-S. Kim, D. Kim, D. Vikraman, K. Prasanna, A. Kathalingam, R. Sharma, H.W. Rhee, An enhanced electrochemical and cycling properties of novel boronic ionic liquid based ternary gel polymer electrolytes for rechargeable $Li/LiCoO_2$ cells, Sci. Rep. 7 (2017) 11103.

[45] H. Yu, J. Wu, L. Fan, K. Xu, X. Zhong, Y. Lin, J. Lin, Improvement of the performance for quasi-solid-state supercapacitor by using PVA–KOH–KI polymer gel electrolyte, Electrochim. Acta 56 (2011) 6881–6886.

[46] F. Yu, M. Huang, J. Wu, Z. Qiu, L. Fan, J. Lin, Y. Lin, A redox-mediator-doped gel polymer electrolyte applied in quasi-solid-state supercapacitors, J. Appl. Polym. Sci. 131 (2014) 39784.

[47] P. Qin, X. Li, B. Gao, J. Fu, L. Xia, X. Zhang, K. Huo, W. Shen, P.K. Chu, Hierarchical TiN nanoparticles-assembled nanopillars for flexible supercapacitors with high volumetric capacitance, Nanoscale 10 (2018) 8728–8734.

[48] U.N. Pan, V. Sharma, T. Kshetri, T.I. Singh, D.R. Paudel, N.H. Kim, J.H. Lee, Freestanding $1T-Mn_xMo_{1-x}S_{2-y}Se_y$ and $MoFe_2S_{4-z}Se_z$ ultrathin nanosheet-structured electrodes for highly efficient flexible solid-state asymmetric supercapacitors, Small 16 (2020), e2001691.

[49] C. Zhang, S. Wang, S. Tang, S. Wang, Y. Li, Y. Du, Ni_3S_2 nanorods and three-dimensional reduced graphene oxide electrodes-based high-performance all-solid-state flexible asymmetric supercapacitors, Appl. Surf. Sci. 458 (2018) 656–664.

[50] D. Kim, G. Lee, D. Kim, J.S. Ha, Air-stable, high-performance, flexible microsupercapacitor with patterned ionogel electrolyte, ACS Appl. Mater. Interfaces 7 (2015) 4608–4615.

[51] C. Justin Raj, M. Rajesh, R. Manikandan, W.-g. Lee, K.H. Yu, B.C. Kim, Direct fabrication of two-dimensional copper sulfide nanoplates on transparent conducting glass for planar supercapacitor, J. Alloys Compd. 735 (2018) 2378–2383.

[52] Y. Fu, X. Cai, H. Wu, Z. Lv, S. Hou, M. Peng, X. Yu, D. Zou, Fiber supercapacitors utilizing pen ink for flexible/wearable energy storage, Adv. Mater. 24 (2012) 5713–5718.

[53] C. Yang, Y. Tang, Y. Tian, Y. Luo, X. Yin, W. Que, Methanol and diethanolamine assisted synthesis of flexible nitrogen-doped Ti_3C_2 (MXene) film for ultrahigh volumetric performance supercapacitor electrodes, ACS Appl. Energy Mater. 3 (2019) 586–596.

[54] D. Van Lam, H.C. Shim, J.-H. Kim, H.-J. Lee, S.-M. Lee, Carbon textile decorated with pseudocapacitive VC/V_xO_y for high-performance flexible supercapacitors, Small 13 (2017) 1702702.

[55] M. Shi, L. Zhao, X. Song, J. Liu, P. Zhang, L. Gao, Highly conductive Mo_2C nanofibers encapsulated in ultrathin MnO_2 nanosheets as a self-supported electrode for high-performance capacitive energy storage, ACS Appl. Mater. Interfaces 8 (2016) 32460–32467.

[56] Y. Chang, X. Sun, M. Ma, C. Mu, P. Li, L. Li, M. Li, A. Nie, J. Xiang, Z. Zhao, J. He, F. Wen, Z. Liu, Y. Tian, Application of hard ceramic materials B_4C in energy storage: design $B_4C@C$ core-shell nanoparticles as electrodes for flexible all-solid-state micro-supercapacitors with ultrahigh cyclability, Nano Energy 75 (2020) 104947.

[57] K. Krishnamoorthy, P. Pazhamalai, S. Sahoo, S.-J. Kim, Titanium carbide sheet based high performance wire type solid state supercapacitors, J. Mater. Chem. A 5 (2017) 5726–5736.

[58] T.-H. Gu, N.H. Kwon, K.-G. Lee, X. Jin, S.-J. Hwang, 2D inorganic nanosheets as versatile building blocks for hybrid electrode materials for supercapacitor, Coord. Chem. Rev. 421 (2020) 213439.

144 Oxide free nanomaterials for energy storage and conversion applications

[59] C.J. Zhang, B. Anasori, A. Seral-Ascaso, S.H. Park, N. McEvoy, A. Shmeliov, G.S. Duesberg, J.N. Coleman, Y. Gogotsi, V. Nicolosi, Transparent, flexible, and conductive 2D titanium carbide (MXene) films with high volumetric capacitance, Adv. Mater. 29 (2017) 1702678.

[60] S. Zheng, C. Zhang, F. Zhou, Y. Dong, X. Shi, V. Nicolosi, Z.-S. Wu, X. Bao, Ionic liquid pre-intercalated MXene films for ionogel-based flexible micro-supercapacitors with high volumetric energy density, J. Mater. Chem. A 7 (2019) 9478–9485.

[61] R. Li, S. Wang, S. Wang, W. Wang, M. Cao, Ultrafine Mo_2C nanoparticles encapsulated in N-doped carbon nanofibers with enhanced lithium storage performance, Phys. Chem. Chem. Phys. 17 (2015) 24803–24809.

[62] L. Qin, Q. Tao, A. El Ghazaly, J. Fernandez-Rodriguez, P.O.Å. Persson, J. Rosen, F. Zhang, High-performance ultrathin flexible solid-state supercapacitors based on solution processable $Mo_{1.33}C$ MXene and PEDOT:PSS, Adv. Fun. Mater. 28 (2018), 1703808.

[63] C. Couly, M. Alhabeb, K.L. Van Aken, N. Kurra, L. Gomes, A.M. Navarro-Suárez, B. Anasori, H.N. Alshareef, Y. Gogotsi, Asymmetric flexible MXene-reduced graphene oxide micro-supercapacitor, Adv. Electron. Mater. 4 (2018) 1700339.

[64] K. Wang, B. Zheng, M. Mackinder, N. Baule, H. Qiao, H. Jin, T. Schuelke, Q.H. Fan, Graphene wrapped MXene via plasma exfoliation for all-solid-state flexible supercapacitors, Energy Storage Mater. 20 (2019) 299–306.

[65] J. Yan, Y. Ma, C. Zhang, X. Li, W. Liu, X. Yao, S. Yao, S. Luo, Polypyrrole–MXene coated textile-based flexible energy storage device, RSC Adv. 8 (2018) 39742–39748.

[66] C.J. Zhang, M.P. Kremer, A. Seral-Ascaso, S.-H. Park, N. McEvoy, B. Anasori, Y. Gogotsi, V. Nicolosi, Stamping of flexible, coplanar micro-supercapacitors using MXene inks, Adv. Funct. Mater. 28 (2018) 1705506.

[67] H. Li, Y. Hou, F. Wang, M.R. Lohe, X. Zhuang, L. Niu, X. Feng, Flexible all-solid-state supercapacitors with high volumetric capacitances boosted by solution processable MXene and electrochemically exfoliated graphene, Adv. Energy Mater. 7 (2017) 1601847.

[68] M. Zhu, Y. Huang, Q. Deng, J. Zhou, Z. Pei, Q. Xue, Y. Huang, Z. Wang, H. Li, Q. Huang, C. Zhi, Highly flexible, freestanding supercapacitor electrode with enhanced performance obtained by hybridizing polypyrrole chains with MXene, Adv. Energy Mater. 6 (2016) 1600969.

[69] B.-S. Shen, H. Wang, L.-J. Wu, R.-S. Guo, Q. Huang, X.-B. Yan, All-solid-state flexible microsupercapacitor based on two-dimensional titanium carbide, Chin. Chem. Lett. 27 (2016) 1586–1591.

[70] P. Yang, D. Chao, C. Zhu, X. Xia, Y. Zhang, X. Wang, P. Sun, B.K. Tay, Z.X. Shen, W. Mai, H.J. Fan, Ultrafast-charging supercapacitors based on corn-like titanium nitride nanostructures, Adv. Sci. 3 (2016) 1500299.

[71] S. Ghosh, S.M. Jeong, S.R. Polaki, A review on metal nitrides/oxynitrides as an emerging supercapacitor electrode beyond oxide, Korean J. Chem. Eng. 35 (2018) 1389–1408.

[72] W. Bi, Z. Hu, X. Li, C. Wu, J. Wu, Y. Wu, Y. Xie, Metallic mesocrystal nanosheets of vanadium nitride for high-performance all-solid-state pseudocapacitors, Nano Res. 8 (2014) 193–200.

[73] H. Gao, Y.-J. Ting, N.P. Kherani, K. Lian, Ultra-high-rate all-solid pseudocapacitive electrochemical capacitors, J. Power Sources 222 (2013) 301–304.

[74] X. Lu, G. Wang, T. Zhai, M. Yu, S. Xie, Y. Ling, C. Liang, Y. Tong, Y. Li, Stabilized TiN nanowire arrays for high-performance and flexible supercapacitors, Nano Lett. 12 (2012) 5376–5381.

[75] F. Grote, H. Zhao, Y. Lei, Self-supported carbon coated TiN nanotube arrays: innovative carbon coating leads to an improved cycling ability for supercapacitor applications, J. Mater. Chem. A 3 (2015) 3465–3470.

[76] P. Sun, R. Lin, Z. Wang, M. Qiu, Z. Chai, B. Zhang, H. Meng, S. Tan, C. Zhao, W. Mai, Rational design of carbon shell endows TiN@C nanotube based fiber supercapacitors with significantly enhanced mechanical stability and electrochemical performance, Nano Energy 31 (2017) 432–440.

[77] Q. Li, Y. Chen, J. Zhang, W. Tian, L. Wang, Z. Ren, X. Ren, X. Li, B. Gao, X. Peng, P.K. Chu, K. Huo, Spatially confined synthesis of vanadium nitride nanodots intercalated carbon nanosheets with ultrahigh volumetric capacitance and long life for flexible supercapacitors, Nano Energy 51 (2018) 128–136.

[78] C. Zhu, P. Yang, D. Chao, X. Wang, X. Zhang, S. Chen, B.K. Tay, H. Huang, H. Zhang, W. Mai, H.J. Fan, All metal nitrides solid-state asymmetric supercapacitors, Adv. Mater. 27 (2015) 4566–4571.

[79] X. Wu, B. Huang, Q. Wang, Y. Wang, Evaluation of the role of nitrogen atoms in cobalt oxynitride electrodes for flexible asymmetric supercapacitors, Electrochim. Acta 353 (2020) 136603.

[80] Y. Wang, L. Sun, P. Song, C. Zhao, S. Kuang, H. Liu, D. Xiao, F. Hu, L. Tu, A low-temperature-operated direct fabrication method for all-solid-state flexible micro-supercapacitors, J. Power Sources 448 (2020) 227415.

[81] X. Hou, Q. Li, L. Zhang, T. Yang, J. Chen, L. Su, Tunable preparation of chrysanthemum-like titanium nitride as flexible electrode materials for ultrafast-charging/discharging and excellent stable supercapacitors, J. Power Sources 396 (2018) 319–326.

[82] Y. Liu, R. Xiao, Y. Qiu, Y. Fang, P. Zhang, Flexible advanced asymmetric supercapacitors based on titanium nitride-based nanowire electrodes, Electrochim. Acta 213 (2016) 393–399.

[83] C. Zhu, Y. Sun, D. Chao, X. Wang, P. Yang, X. Zhang, H. Huang, H. Zhang, H.J. Fan, A 2.0V capacitive device derived from shape-preserved metal nitride nanorods, Nano Energy 26 (2016) 1–6.

[84] D. Ruan, R. Lin, K. Jiang, X. Yu, Y. Zhu, Y. Fu, Z. Wang, H. Yan, W. Mai, High-performance porous molybdenum oxynitride based fiber supercapacitors, ACS Appl. Mater. Interfaces 9 (2017) 29699–29706.

[85] M. Yu, Y. Han, X. Cheng, L. Hu, Y. Zeng, M. Chen, F. Cheng, X. Lu, Y. Tong, Holey tungsten oxynitride nanowires: novel anodes efficiently integrate microbial chemical energy conversion and electrochemical energy storage, Adv. Mater. 27 (2015) 3085–3091.

[86] X. Xiao, X. Peng, H. Jin, T. Li, C. Zhang, B. Gao, B. Hu, K. Huo, J. Zhou, Freestanding mesoporous VN/CNT hybrid electrodes for flexible all-solid-state supercapacitors, Adv. Mater. 25 (2013) 5091–5097.

[87] J. Theerthagiri, K. Karuppasamy, G. Durai, A. Rana, P. Arunachalam, K. Sangeetha, P. Kuppusami, H.S. Kim, Recent advances in metal chalcogenides (MX; X = S, Se) nanostructures for electrochemical supercapacitor applications: a brief review, Nanomaterials 8 (2018) 256.

[88] C. Justin Raj, B.C. Kim, W.-J. Cho, W.-G. Lee, Y. Seo, K.-H. Yu, Electrochemical capacitor behavior of copper sulfide (CuS) nanoplatelets, J. Alloys Compd. 586 (2014) 191–196.

[89] R. Barik, P.P. Ingole, Challenges and prospects of metal sulfide materials for supercapacitors, Curr. Opin. Electrochem. 21 (2020) 327–334.

[90] M. Acerce, D. Voiry, M. Chhowalla, Metallic 1T phase MoS_2 nanosheets as supercapacitor electrode materials, Nat. Nanotechnol. 10 (2015) 313–318.

[91] Y. Han, Y. Ge, Y. Chao, C. Wang, G.G. Wallace, Recent progress in 2D materials for flexible supercapacitors, J. Energy Chem. 27 (2018) 57–72.

[92] J. Feng, X. Sun, C. Wu, L. Peng, C. Lin, S. Hu, J. Yang, Y. Xie, Metallic few-layered VS_2 ultrathin nanosheets: high two-dimensional conductivity for in-plane supercapacitors, J. Am. Chem. Soc. 133 (2011) 17832–17838.

146 Oxide free nanomaterials for energy storage and conversion applications

[93] X. Li, X. Li, J. Cheng, D. Yuan, W. Ni, Q. Guan, L. Gao, B. Wang, Fiber-shaped solid-state supercapacitors based on molybdenum disulfide nanosheets for a self-powered photodetecting system, Nano Energy 21 (2016) 228–237.

[94] A. Liu, H. Lv, H. Liu, Q. Li, H. Zhao, Two dimensional MoS_2/CNT hybrid ink for paper-based capacitive energy storage, J. Mater. Sci. Mater. Electron. 28 (2017) 8452–8459.

[95] C. Zhang, H. Yin, M. Han, Z. Dai, H. Pang, Y. Zheng, Y.-Q. Lan, J. Bao, J. Zhu, Two-dimensional tin selenide nanostructures for flexible all-solid-state supercapacitors, ACS Nano 8 (2014) 3761–3770.

[96] P. Yu, W. Fu, Q. Zeng, J. Lin, C. Yan, Z. Lai, B. Tang, K. Suenaga, H. Zhang, Z. Liu, Controllable synthesis of atomically thin type-II Weyl semimetal WTe_2 nanosheets: an advanced electrode material for all-solid-state flexible supercapacitors, Adv. Mater. 29 (2017) 1701909.

[97] J. Wen, S. Li, K. Zhou, Z. Song, B. Li, Z. Chen, T. Chen, Y. Guo, G. Fang, Flexible coaxial-type fiber solid-state asymmetrical supercapacitor based on Ni_3S_2 nanorod array and pen ink electrodes, J. Power Sources 324 (2016) 325–333.

[98] Y. Gu, W. Du, Y. Darrat, M. Saleh, Y. Huang, Z. Zhang, S. Wei, In situ growth of novel nickel diselenide nanoarrays with high specific capacity as the electrode material of flexible hybrid supercapacitors, Appl. Nanosci. 10 (2020) 1591–1601.

[99] Z. Tian, H. Dou, B. Zhang, W. Fan, X. Wang, Three-dimensional graphene combined with hierarchical CuS for the design of flexible solid-state supercapacitors, Electrochim. Acta 237 (2017) 109–118.

[100] R.B. Pujari, A.C. Lokhande, A.A. Yadav, J.H. Kim, C.D. Lokhande, Synthesis of MnS microfibers for high performance flexible supercapacitors, Mater. Des. 108 (2016) 510–517.

[101] H. Tang, Y. Yuan, L. Meng, W. Wang, J. Lu, Y. Zeng, T. Huang, C. Gao, Low-resistance porous nanocellular MnSe electrodes for high-performance all-solid-state battery-supercapacitor hybrid devices, Adv. Mater. Technol. 3 (2018) 1800074.

[102] X. Wang, B. Liu, Q. Wang, W. Song, X. Hou, D. Chen, Y.B. Cheng, G. Shen, Three-dimensional hierarchical $GeSe_2$ nanostructures for high performance flexible all-solid-state supercapacitors, Adv. Mater. 25 (2013) 1479–1486.

[103] H. Jiang, Z. Wang, Q. Yang, L. Tan, L. Dong, M. Dong, Ultrathin $Ti_3C_2T_x$ (MXene) nanosheet-wrapped $NiSe_2$ octahedral crystal for enhanced supercapacitor performance and synergetic electrocatalytic water splitting, Nano-Micro Lett. 11 (2019) 31.

[104] R. Manikandan, C.J. Raj, G. Nagaraju, M. Pyo, B.C. Kim, Selective design of binder-free hierarchical nickel molybdenum sulfide as a novel battery-type material for hybrid supercapacitors, J. Mater. Chem. A 7 (2019) 25467–25480.

[105] R. Manikandan, C.J. Raj, K.H. Yu, B.C. Kim, Self-coupled nickel sulfide @ nickel vanadium sulfide nanostructure as a novel high capacity electrode material for supercapattery, Appl. Surf. Sci. 497 (2019) 143778.

[106] S. Hussain, T. Liu, M.S. Javed, N. Aslam, N. Shaheen, S. Zhao, W. Zeng, J. Wang, Amaryllis-like $NiCo_2S_4$ nanoflowers for high-performance flexible carbon-fiber-based solid-state supercapacitor, Ceram. Int. 42 (2016) 11851–11857.

[107] Y. Li, X. Chen, Y. Cao, W. Zhou, H. Chai, The ultralong cycle life of solid flexible asymmetric supercapacitors based on nickel vanadium sulfide nanospheres, CrstEngComm 22 (2020) 5226–5236.

[108] C. Wang, Z. Song, H. Wan, X. Chen, Q. Tan, Y. Gan, P. Liang, J. Zhang, H. Wang, Y. Wang, X. Peng, P.A. van Aken, H. Wang, Ni-Co selenide nanowires supported on conductive wearable textile as cathode for flexible battery-supercapacitor hybrid devices, Chem. Eng. J. 400 (2020) 125955.

Oxides free materials Chapter | 6 **147**

[109] S. Sahoo, K. Krishnamoorthy, P. Pazhamalai, S.J. Kim, Copper molybdenum sulfide anchored nickel foam: a high performance, binder-free, negative electrode for supercapacitors, Nanoscale 10 (2018) 13883–13888.

[110] Q. Wang, X. Tian, D. Zhang, Flexible asymmetric supercapacitors and electrocatalytic water splitting based on $CoNiSe_2/CoNiSe_2$ nanoflowers, Mater. Lett. 276 (2020) 128245.

[111] Y. Zhang, N. Cao, S. Szunerits, A. Addad, P. Roussel, R. Boukherroub, Fabrication of ZnCoS nanomaterial for high energy flexible asymmetric supercapacitors, Chem. Eng. J. 374 (2019) 347–358.

[112] P. Xu, W. Zeng, S. Luo, C. Ling, J. Xiao, A. Zhou, Y. Sun, K. Liao, 3D Ni-Co selenide nanorod array grown on carbon fiber paper: towards high-performance flexible supercapacitor electrode with new energy storage mechanism, Electrochim. Acta 241 (2017) 41–49.

[113] Y. Qiu, X. Li, M. Bai, H. Wang, D. Xue, W. Wang, J. Cheng, Flexible full-solid-state supercapacitors based on self-assembly of mesoporous $MoSe_2$ nanomaterials, Inorg. Chem. Front. 4 (2017) 675–682.

[114] B. Pandit, L.K. Bommineedi, B.R. Sankapal, Electrochemical engineering approach of high performance solid-state flexible supercapacitor device based on chemically synthesized VS2 nanoregime structure, J. Energy Chem. 31 (2019) 79–88.

[115] M.S. Javed, S. Dai, M. Wang, D. Guo, L. Chen, X. Wang, C. Hu, Y. Xi, High performance solid state flexible supercapacitor based on molybdenum sulfide hierarchical nanospheres, J. Power Sources 285 (2015) 63–69.

[116] S. Zhai, Z. Fan, K. Jin, M. Zhou, H. Zhao, Y. Zhao, F. Ge, X. Li, Z. Cai, Synthesis of zinc sulfide/copper sulfide/porous carbonized cotton nanocomposites for flexible supercapacitor and recyclable photocatalysis with high performance, J. Colloid Interface Sci. 575 (2020) 306–316.

[117] D. Jiang, H. Liang, W. Yang, Y. Liu, X. Cao, J. Zhang, C. Li, J. Liu, J.J. Gooding, Screen-printable films of graphene/CoS_2/Ni_3S_4 composites for the fabrication of flexible and arbitrary-shaped all-solid-state hybrid supercapacitors, Carbon 146 (2019) 557–567.

[118] B.S. Soram, I.S. Thangjam, J.Y. Dai, T. Kshetri, N.H. Kim, J.H. Lee, Flexible transparent supercapacitor with core-shell Cu@Ni@NiCoS nanofibers network electrode, Chem. Eng. J. 395 (2020) 125019.

[119] L. Liu, A. Liu, A. Xu, F. Yang, J. Wang, Q. Deng, Z. Zeng, S. Deng, Fabrication of dual-hollow heterostructure of Ni_2CoS_4 sphere and nanotubes as advanced electrode for high-performance flexible all-solid-state supercapacitors, J. Colloid Interface Sci. 564 (2020) 313–321.

[120] G. Qu, X. Zhang, G. Xiang, Y. Wei, J. Yin, Z. Wang, X. Zhang, X. Xu, ZIF-67 derived hollow Ni-Co-Se nano-polyhedrons for flexible hybrid supercapacitors with remarkable electrochemical performances, Chin. Chem. Lett. 31 (2020) 2007–2012.

[121] H.T. Wang, Y.N. Liu, X.H. Kang, Y.F. Wang, S.Y. Yang, S.W. Bian, Q. Zhu, Flexible hybrid yarn-shaped supercapacitors based on porous nickel cobalt sulfide nanosheet array layers on gold metalized cotton yarns, J. Colloid Interface Sci. 532 (2018) 527–535.

[122] Z. Chen, Y. Yang, Z. Ma, T. Zhu, L. Liu, J. Zheng, X. Gong, All-solid-state asymmetric supercapacitors with metal selenides electrodes and ionic conductive composites electrolytes, Adv. Funct. Mater. 29 (2019) 1904182.

[123] X. Zhang, X. Chen, T. Bai, J. Chai, X. Zhao, M. Ye, Z. Lin, X. Liu, A simple route to fiber-shaped heterojunctioned nanocomposites for knittable high-performance supercapacitors, J. Mater. Chem. A 8 (2020) 11589–11597.

148 Oxide free nanomaterials for energy storage and conversion applications

[124] H. Liang, C. Xia, Q. Jiang, A.N. Gandi, U. Schwingenschlögl, H.N. Alshareef, Low temperature synthesis of ternary metal phosphides using plasma for asymmetric supercapacitors, Nano Energy 35 (2017) 331–340.

[125] S. Xie, J. Gou, Facile synthesis of $Ni_2P/Ni_{12}P_5$ composite as long-life electrode material for hybrid supercapacitor, J. Alloys Compd. 713 (2017) 10–17.

[126] S. Surendran, S. Shanmugapriya, A. Sivanantham, S. Shanmugam, R. Kalai Selvan, Electrospun carbon nanofibers encapsulated with NiCoP: a multifunctional electrode for supercapattery and oxygen reduction, oxygen evolution, and hydrogen evolution reactions, Adv. Energy Mater. 8 (2018) 1800555.

[127] J. Ling, H. Zou, W. Yang, S. Chen, Urchin-like NiCoP coated with a carbon layer as a high-performance electrode for all-solid-state asymmetric supercapacitors, Mater. Adv. 1 (2020) 481–494.

[128] L. Naderi, S. Shahrokhian, Nickel vanadium sulfide grown on nickel copper phosphide dendrites/Cu fibers for fabrication of all-solid-state wire-type micro-supercapacitors, Chem. Eng. J. 392 (2020) 124880.

[129] Y. Lan, H. Zhao, Y. Zong, X. Li, Y. Sun, J. Feng, Y. Wang, X. Zheng, Y. Du, Phosphorization boosts the capacitance of mixed metal nanosheet arrays for high performance supercapacitor electrodes, Nanoscale 10 (2018) 11775–11781.

[130] X. Gao, Y. Zhao, K. Dai, J. Wang, B. Zhang, X. Shen, NiCoP nanowire@NiCo-layered double hydroxides nanosheet heterostructure for flexible asymmetric supercapacitors, Chem. Eng. J. 384 (2020) 123373.

[131] J. Wen, B. Xu, J. Zhou, Toward flexible and wearable embroidered supercapacitors from cobalt phosphides-decorated conductive fibers, Nano-Micro Lett. 11 (2019) 89.

[132] G. Zhao, Y. Tang, G. Wan, X. Xu, X. Zhou, M. Zhou, C. Hao, S. Deng, G. Wang, High-performance and flexible all-solid-state hybrid supercapacitor constructed by NiCoP/CNT and N-doped carbon coated CNT nanoarrays, J. Colloid Interface Sci. 572 (2020) 151–159.

[133] X. Gao, L. Yin, L. Zhang, Y. Zhao, B. Zhang, Decoration of NiCoP nanowires with interlayer-expanded few-layer $MoSe_2$ nanosheets: a novel electrode material for asymmetric supercapacitors, Chem. Eng. J. 395 (2020) 125058.

Chapter 7

Metal oxides-free anodes for lithium-ion batteries

A. Nichelson[a], Bradha Madhavan[b], Ganesh Kumar Veerasubramani[c], Waqas Hassan Tanveer[d,e], Jayaraman Theerthagiri[f,g], A.G. Ramu[h], Dhanasekaran Vikraman[i], K. Karuppasamy[i], and Sung-Chul Yi[c,j]

[a]Department of Physics, National Engineering College, Kovilpatti, Thoothukudi, Tamil Nadu, India, [b]Department of Science and Humanities, Rathinam Technical Campus, Coimbatore, Tamil Nadu, India, [c]Department of Chemical Engineering, Hanyang University, Seoul, Republic of Korea, [d]Energy Safety Research Institute, College of Engineering, Swansea University, Bay Campus, Fabian Way, Swansea, United Kingdom, [e]Department of Mechanical and Manufacturing Engineering, National University of Sciences and Technology (NUST), Islamabad, Pakistan, [f]Centre of Excellence for Energy Research, Centre for Nanoscience and Nanotechnology, Sathyabama Institute of Science and Technology (Deemed to be University), Chennai, Tamil Nadu, India, [g]Department of Chemistry and Research Institute of Natural Sciences, Gyeongsang National University, Jinju, South Korea, [h]Department of Materials Science and Engineering, Hongik University, Sejong-City, Republic of Korea, [i]Division of Electronics and Electrical Engineering, Dongguk University-Seoul, Seoul, Republic of Korea, [j]Department of Hydrogen and Fuel Cell Technology, Hanyang University, Seoul, Republic of Korea

Chapter outline

1. Introduction	149		3.2 Selenide-based anodes	154
2. Components and operating			3.3 Sulfide-based anodes	159
mechanism of LIBs	152		3.4 Phosphide-based anodes	162
3. Anode materials for LIB	152		4. Summary, outlook, and future	
3.1 Graphene and its composites			scopes	167
as anode materials	153		References	167

1. Introduction

At present energy-storage devices such as lithium-ion batteries (LIBs), super-capacitors and sodium-ion batteries (SIB) are outperforming the overwhelming need for energy-based devices to handle our day-to-day life [1–11]. Among them, LIB has been commercialized and widely consumed by mankind for the last two decades to operate most of the day-to-day electronic accessories such as cell phones, laptops, notebooks, and electric vehicles, etc. [12]. Till

Oxide Free Nanomaterials for Energy Storage and Conversion Applications.
https://doi.org/10.1016/B978-0-12-823936-0.00008-5
Copyright © 2022 Elsevier Inc. All rights reserved.

149

150 Oxide free nanomaterials for energy storage and conversion applications

date, a huge number of studies have been performed and numerous materials have been investigated to improve the efficiency of LIBs. Based on the storage principles, the batteries are of two types: one is primary or nonrechargeable batteries and the other is secondary or rechargeable batteries in which LIB comes under the latter category. These LIBs have unique features such as high energy density, power rate, life cycle, low self-discharge rate, large operating temperature range making them superior to other secondary batteries such as lead-acid, NiCd, and NiMH [13, 14]. Apart from that, it is capable of fast charging with better load capacity, extended cycle and extended self-life; fewer repairs, high energy efficiency, large coulombic-efficiency; low self-discharge than that of commercially available NiCd and NiMH batteries [15]. The unique properties such as high operating voltage and low self-discharge of LIB lead to an increase in demand for LIB technology. As represented in Fig. 1A, the performances of energy devices in terms of energy density vs power density, the fuel cells have large energy density with truncated power density whereas traditional capacitors possess high power density. It is observed that the LIBs are outstanding in the contest exhibiting both high energy density and a very good power density. From the voltaic cell to today's LIBs, the LIBs are best suited for compact size and high performance. As a result, researchers have focused on the route to utilize LIBs in hybrid electric vehicles and successfully launched them in many countries.

Among the different components of LIBs, the negative electrode called anode plays one of the most dominant roles to decide the electrochemical behavior and performance of the LIBs. Further, the performance of LIBs can be controlled by the nature of the metals employed, chemical and physical properties of the compounds. In the past decades, lithium metal and graphite were widely employed as an anode material for LIBs because of their extraordinary specific capacities and large energy density. Among these two materials, graphite possesses an outstanding lithium storage capacity of $4200 \, mAh \, g^{-1}$ at ambient temperature and is hence considered to be a potential high-performance anode candidate in LIBs for a long time [1, 16–19]. However, the shape deformation happens when the electrons and Li^+ ions are accommodated in the graphite structure that resulted in poor columbic efficiency as well as affects the charge storage mechanism of the battery. To overcome such deficiency, several other potential electrodes which include silicon anodes, transition metal oxides, their alloys, and composites were utilized in the past 10 years. Though the silicon and metal-oxide based anodes are 10 times better than carbon-oriented anodes from a capacity point of view, it's volumetric change issue and low coulombic efficiency hinders the commercialization of LIBs [20].

In this regard, plenty of active electrochemical negative electrode materials other than carbon-based anodes, TMOs, and silicon anodes have been investigated to enhance the performance of LIBs. Transition metal chalcogenides (TMCs) which include sulfides and selenides, phosphides are of great interest because of their better electrical conductivities, superior thermal-chemical

Metal oxides-free anodes for lithium-ion batteries **Chapter | 7** **151**

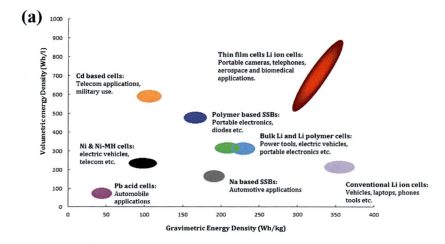

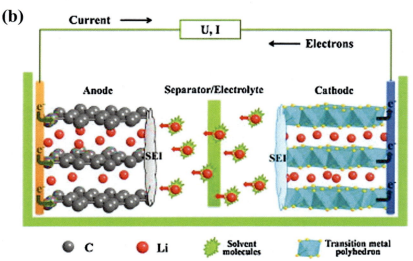

FIG. 1 (A) Theoretical energy densities of various types of batteries, (B) functioning mechanism of LIB. *((A) From J.G. Kim, B. Son, S. Mukherjee, N. Schuppert, A. Bates, O. Kwon, M.J. Choi, H.Y. Chung, S. Park, A review of lithium and non-lithium based solid state batteries, J. Power Sources 282 (2015) 299–322. Copyright, Elsevier, 2015; (B) B. Xu, D. Qian, Z. Wang, Y.S. Meng, Recent progress in cathode materials research for advanced lithium ion batteries, Mater. Sci. Eng. R: Rep. 73 (2012) 51–65. Copyright, Elsevier, 2012.)*

stabilities, and excellent lithium storage capacity in the range between 800 and 1100 mAh g^{-1}, which are almost two folds greater than that of graphite-based anode (372 mAh g^{-1}) [21–23] which made them act as the good anodes for LIBs. Recently, TMCs include CoS [24], CoS$_2$ [25], MoS$_2$ [26], SnS$_2$ [27], MoSe$_2$ [28], SnSe [29], WS$_2$ [30], NiS$_2$ [31], and Sb$_2$Se$_3$ [32] have been

152 Oxide free nanomaterials for energy storage and conversion applications

studied as the potential anodes for LIBs. Among the TMC-based anodes, the majority of studies have been focused on metal sulfides because the selenide and phosphide-based anodes' lithiation mechanism are not fully understood for the conversion products, henceforth the reports on selenides and phosphides are very scarce. Herein, we explain the current progress and developments of metal chalcogenides and metal phosphides anodes for the high performance of LIBs toward next-generation energy harvesting research community. The content of the chapter includes various wide categories as follows: (a) components and principles of LIBs; (b) anodes for LIBs; (c) transition metal selenide based anodes for LIBs and (d) transition metal sulfide and phosphides anodes for LIBs applications.

2. Components and operating mechanism of LIBs

The main components of LIBs include a cathode, a separator, and an electrolyte and anode. Each component has its role to stimulate the battery to perform. The cathode materials [33] are of different structures such as olivine phase [34], layered phase [35], and spinel phase [36]. The structural significance is the main feature for high-performance cathode material. The performance of cathode materials with different structures has been explored and tested by several researchers [37, 38]. These cathodes have an undeviating influence on the capital cost of the battery. Another component is the anode materials which are of various types like with the compounds include carbon [38, 39], silicon [40], titanium [41], etc., Alloy based anode materials are another category [42]. The electrode materials have enhancing performance when they possess nano size with specific synthetic routes [43, 44]. In the LIB, the cathode and the anode are the two electrodes that can store lithium during the charging period, better known as the intercalation/deintercalation process. Fig. 1B schematically representing the principle of working of a LIB.

The electrolyte is the component that acts as a charge carrier and provides the feasible movement of Li^+ to migrate freely from anode to cathode and vice versa. The electrolyte is of various kinds such as liquid electrolyte, solid polymer electrolyte [45], solid ceramic electrolyte [46], gel polymer electrolyte [47], and recently ionic liquids [48, 49]. And the last component is the separator which prevents the direct contact of cathode and anode available for liquid electrolyte as well as solid electrolyte [50, 51]. As the power supply is given to the battery the oxidation takes place at the anode where it loses an electron and the reduction happen at the cathode region where it gains electron which indicates the reversible reaction mechanism of a LIB system at cathode and anode compartments.

3. Anode materials for LIB

Another significant component that decides the performance of batteries is the anode or negative electrode. Already numerous reports were reported on anode-based electrode materials, however, the metal-free oxides as anode for LIBs are very scarce. In this present chapter, we specifically discuss the anode electrode materials in detail particularly about the metal-free oxides as the potential candidates for LIBs. It is well known that graphite-based anode materials have been investigated for the past three decades which shows a theoretical capacity of $375\,mAh\,g^{-1}$ and hence ruled out for high energy applications. For instance, the graphitic carbon coated on a copper foil shows a massive performance for LIBs and is considered as a potential anode for a long period [39]. In the dominant cases, the anode material is coated over the current collectors using conventional binders like PVDF, phenolic resin, and cellulose-based materials [52, 53]. The carbon black has also added along with the binders to boost the overall conductivity of the electrodes. Since the age of graphite carbon as an anode for LIBs, various materials such as metal oxides, sulfides, selenides, nitrides, and alloys have been known as the potential anodes for the same reason. In the following section, a brief note on graphene, graphene oxide, and their composite-based anodes for LIBs are discussed. Following by the metal-free oxides as potential anode and their high electrochemical and storage performance for LIBs are discussed in detail later.

3.1 Graphene and its composites as anode materials

Composites based on graphene have been prepared and studied and employed for LIB application by many researchers around the world. The composite made up of graphene and nanosized silicon was used as an anode for LIBs by Goriparti et al. [54]. These two separate samples of graphene materials were synthesized using a wide range of methods. The homogeneous mixture of graphene oxide and nano-sized silicone particles in an aqueous solution was first dried and annealed @ 500°C. In another process, graphene samples were fabricated by the rapid thermal cure of expandable graphite at 1050°C. Following the mixture of graphene and nanosized silicone was then obtained by mechanical blending. Among these two samples, the latter one is outperformed well with its cycling stability and specific capacity of $2753\,mAh\,g^{-1}$ since there were fewer defects in the graphene sheets [55]. Li et al. have prepared boron nitride (BN)/reduced graphene oxide (rGO) composite thin films using vacuum filtration followed by a thermal treatment process. These films are binder-free anode for LIBs which displayed an excellent reversible specific capacity of $278\,mAh\,g^{-1}$ @ $100\,mA\,g^{-1}$ of current density and had good capacity retention for the first 200 cycles. The film structure and coordination between layered BN and graphene lead to boost its electrochemical performance. Layer BN and graphene support the feasible movement of lithium-ion into it during the charging and discharging process [56].

Zhu et al. prepared graphene foam (GF) with the incorporation of Prussian blue (PB) nano-cubes in it using a facile immersion method which acts as the binder-free electrodes for LIBs. The framework of GF/PB composite showed an

154 Oxide free nanomaterials for energy storage and conversion applications

excellent gravimetric capacity of $514\,mAh\,g^{-1}$ ($0.47\,mAh\,cm^{-2}$) at $100\,mA\,g^{-1}$ above 150 cycles and displayed a better rate capability of $\sim150\,mAh\,g^{-1}$ at $1\,A\,g^{-1}$. These binder-free composite materials displayed excellent performance as a negative electrode in LIB [57]. With the support of silicon, another collection of anodes for LIBs has been created. The silicone film deposited on graphene using chemical vapor deposition has improved the versatility of the samples which did not need any binder or additive content. These samples have the areal capacity of $4\,mAh\,cm^{-2}$ @ $0.22\,mA\,cm^{-2}$. Further, the hollow structure of the graphene foam is reasonable for the enhanced electrochemical performance which in turn supports the deposition of Si film that accommodates the change in volume and contraction strain-made cracks of Si for the duration of the cycling process [58].

Su and his co-workers have adopted a spray drying process afterward annealing to prepare Si/graphite@graphene which showed an initial maximum charge capacity of 820.7 and $766.2\,mAh\,g^{-1}$ @ 50 and $500\,mA\,g^{-1}$ correspondingly. The initial coulombic efficiency is found to be 77.98% [59]. Using simple evaporation and leavening method, Tang et al. have made Si/porous rGO composite film. The porosity in it has effectively improved the conductivity and decreased the charge transfer resistance of the Li^+. Further, it possesses a good specific potential and cycling stability by showing $1261\,mAh\,g^{-1}$ @ $50\,mA\,g^{-1}$ up to 70 cycles with very good rate capability [60]. Si-based material has been prepared and characterized by Tao et al. They synthesized Si/carbon fibers paper (Si@rGO/CFP) using pyrolysis technique in which the rGO covered Si-nanoparticles were uniformly disseminated in the CFP. This binder free material exhibited a high-initial capacity of $2055\,mAh\,g^{-1}$ @ $100\,mA\,g^{-1}$ with outstanding rate behavior which may be attributed to the following fact that the Si@rGO interconnected with carbon fibers substrate afford a pathway for electronic conductivity and the voids in between maintain the integrity of the materials [61]. Anode comprised of Si/graphene nanocomposite through high-energy ball milling cum heat treatment process has synthesized by Chen et al. using xanthan gum as the binder. The Si/graphene showed a high performance compared to the Si anode as prepared. Improved reversible power, good cycling efficiency, and speed capability were observed in Si/graphene. The sample with xanthan gum as a binder showed higher performances than the cellulose binder [78]. The various electrochemical properties of graphene and its composite-based anodes are provided in Table 1.

3.2 Selenide-based anodes

This section confers the selenide-based anode materials for LIBs. In general, the electrode materials morphology plays a great impact on the performance of the battery. In this regard, various nanostructured selenide-based anodes have been studied which exhibits better electrochemical as well as storage performance as compared to graphene electrodes. In brief, a stable nanostructured Bi_2Se_3 was

TABLE 1 Material synthesis and cycling properties of graphene and its composites-based anode materials for LIBs.

No.	Synthesis method	Material	Performance $mAhg^{-1}$/C-rate (mAg^{-1})/ number of cycles	References
1	Ultrasound assisted process	NiO hollow microspheres	380/200/30	[62]
2	Carbonization of Cu-based metal-organic nanofibers	Cu based metal organic nanofiber	853.1/500/800	[63]
3	Electrodeless plating and thermal oxidation	Hierarchical shell/CuO nanowire/CFP composites	598.2/0.1 C/50	[64]
4	Thermal decomposition— in situ process	3D A-Fe$_2$O$_3$@CBC composite	1390/200/400	[65]
5	Electro deposition cum hydrothermal	3D Ni-Co$_3$O$_4$ NWs	714/0.5 C/100	[66]
6	Hydrothermal	NiO nano-cone arrays	969/0.5 C/120	[67]
7	Chemical bath deposition	3D porous NiO/ N-C (NiO/N-doped carbon)	712.8/0.5 C/200	[68]
8	Hydrothermal	CuO nano hexagons	575/215/100	[69]
9	Pyrolysis	CuO/C composite microcubes	510.5/100/200	[70]
10	Solution immersion	3DGN/CuO composites	409/100/50	[71]
11	Oxidation of Ni-foam	Ni/NiO	701/156/65	[72]
12	In situ filtration method	Graphene-Si composites	708/304/100	[73]
13	Temperature programmed reaction	MWCNTs/ZnO	419.8/200/100	[74]

Continued

156 Oxide free nanomaterials for energy storage and conversion applications

TABLE 1 Material synthesis and cycling properties of graphene and its composites-based anode materials for LIBs—cont'd

No.	Synthesis method	Material	Performance mAh g^{-1}/C-rate (mA g^{-1})/ number of cycles	References
14	Pyrolysis/oxidation	TMO@BNG	1554/96/480	[75]
15	Layered self-assembly route	Fe$_3$O$_4$-graphene	1427.5/1000/100	[76]
16	Solvothermal	Nanostructured MoO$_2$/graphite oxide (GO)	720/100/30	[77]

synthesized using a hydrothermal method and characterized for its anodized performance in LIB. The samples collected were in the form of nanoparticles and nanosheets with dimensions of 20–150 nm and 5–30 nm respectively. The different morphology of the material was observed in the SEM micrographs. On comparing the performance of commercially available bulk Bi$_2$Se$_3$, it has been revealed that the bulk material unveiled the initial charge and discharge capacities of 621 and 499 mAh g^{-1} correspondingly which is considerably larger than nanostructured Bi$_2$Se$_3$ whose initial charge and discharge capacities were found to be 594 and 468 mAh g^{-1}. In the contrast, these anode materials have shown excellent cycling stability over 50 cycles [79].

The hybrid tin selenide and multiwalled carbon nanotubes (MWCNT) composite anode was fabricated using a facile grinding cum solvent mixing process. The MWCNT provided enough room for the movement of Li$^+$ ions during the charge/discharge cycle that revealed the superior reversible capacity of 882–651 mAh g^{-1} after 50 cycles when related to the performance of pristine tin selenide (602–58 mAh g^{-1}) and MWCNT (339–171 mAh g^{-1}) electrodes [80]. Alternative anode-based selenide material was carbon-coated bimetal selenide composites. This material has been synthesized using a basic bimetal-organic-framework approach that yielded a very high reversible specific capacity of 949 mAh g^{-1} over 500 cycles with an effective rate-capability [81].

Ma and his co-workers prepared a very thin hybrid composition of MoSe$_2$/graphene by liquid-assisted hydrothermal approach. The micrograph images (Fig. 2A, C and E) reveal as prepared MoSe$_2$ possess a morphology of agglomerated granules. Whereas the composite MoSe$_2$/graphene as exhibits in Fig. 2B

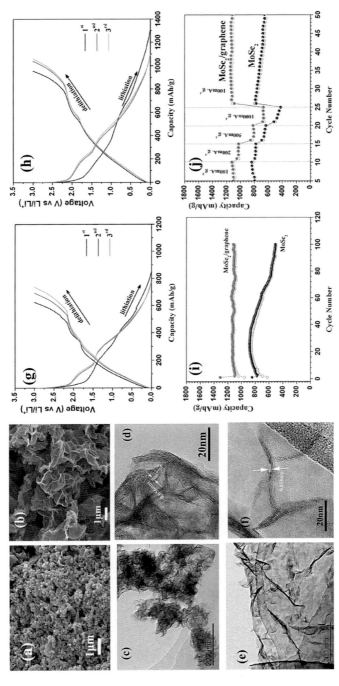

FIG. 2 (A, C, E) FE-SEM, TEM and HR-TEM micrographs of pristine MoSe$_2$, (B, D, F) FE-SEM, TEM and HR-TEM micrographs of hybrid MoSe$_2$/graphene, (G, H) galvanostatic charge-discharge (GCD) curves of pristine MoSe$_2$ and composite MoSe$_2$/graphene at 1000 mA g^{-1}, (I, J) cycling behavior and rate performance curve of pristine MoSe$_2$ and MoSe$_2$/graphene composite anodes. *(From L. Ma, X. Zhou, L. Xu, X. Xu, L. Zhang, W. Chen, Ultrathin few-layered molybdenum selenide/graphene hybrid with superior electrochemical Li-storage performance, J. Power Sources 285 (2015) 274–280. Copyright, Elsevier, 2015.)*

158 Oxide free nanomaterials for energy storage and conversion applications

showed a wrinkled paper-like morphology with quasi-3D architecture which gives a well-defined conduit for transport of electrons. Further observing the TEM and HR-TEM images as exposed in Fig. 2D and F comprises of curly nanosheets and layered sheets of $MoSe_2$ having the interlayer-spacing of 0.656 nm. Owing to these morphological evolutions, the prepared anodes show an outstanding reversible capacity with excellent capacity retention as shown in Fig. 2G–L. It is evident from Fig. 2G, the starting discharge capacity value of pristine $MoSe_2$ electrode as 846 mAh g^{-1} and a reversible capacity of 624 mAh g^{-1} exhibited a coulombic efficiency of 73.8%. In the contrast, its composite $MoSe_2$/graphene hybrid electrode has shown higher performance in the first cycle by displaying 1309 and 955 mAh g^{-1} for discharge and reversible capacities individually. The maximum coulombic efficiency was found to be 72.9%, as shown in Fig. 2H. The capacity loss can be attributed to the deformation of surface electrolyte interphase (SEI) and by the trapping of lithium ions in defects and disorders [28, 82]. The coulombic efficiency values varied greater than 94% in the upcoming cycles other than the initial cycle. From Fig. 2I and J, it has confirmed that the $MoSe_2$ showed a deprived cyclability particularly over 30 cycles. The charge capacity increases at first 624 mAh g^{-1} to the maximum of 859 mAh g^{-1} and after a series degradation was observed [83].

Manganese selenide is one of the metallic selenide materials that may act as an anodized material for LIB. The pulse laser deposited MnSe thin films provided the discharge capacity values for the first 120 cycles have ranged from 361 to 472 mAh g^{-1}. It was observed that reversible formation of β-MnSe and an irreversible decomposition of α-MnSe during discharging and charging [84]. Carbon-coated vanadium selenide (VSe_2) composites were prepared using a simple ball milling technique in which the carbon formed a layer over the VSe_2 particles. This material exhibited a capacity value of 453 mAh g^{-1} over 50 GCD cycles [85].

Tin selenides from SnSe and $SnSe_2$ exhibited a layered crystal structure have synthesized using a solid-state process. It was further coated with carbon by combining with the ball milling technique. The $SnSe_2$/C material has a power of 726 mAh g^{-1} with the highest coulombic efficiency of 82% [86]. Xiang Huang et al. have prepared crystalline tin selenides in the form of quantum dots rGO. The material was synthesized by the solvothermal method with a subsequent freeze-drying process and annealing. The samples collected were in the matrix shape of a 3D network structure. The electrochemical output of the samples was studied and has an ability of 778.5 mAh g^{-1} at a @ 50 mA g^{-1} which is larger than its theoretical capacity [87]. On the other hand, germanium and tin selenide have also been investigated as the significant anodes for LIBs by Im et al. They have confirmed the Ge and Sn conversion phases during the GCD process by XRD sequence. The material exhibits a supremacy capacity of 400–800 mAh g^{-1} over 70 cycles [88]. The electrochemical capacitive performance of some important selenide-based electrodes is listed in Table 2.

TABLE 2 Morphological feature and cycling performances of selenide-based anodes.

No.	Material	Morphology	Performance $mAhg^{-1}$/C-rate (mAg^{-1})/number of cycles	References
1	$Cu_9Sn_2Se_9$	Nanoparticle	979.8/100/100	[89]
2	Sb_2Se_3 and rGO	Nanorods	868.30/200/100	[90]
3	MnSe	Hexagonal prism	864.7/500/750	[91]
4	$MoSe_2$/graphene	Nanosheets	1100/1000/100	[83]
5	Sb_2Se_3	Nanofilms	660.7/5 $\mu A cm^{-2}$/100	[92]
6	Ga Se	Layered crystal	760/5000/50	[93]
7	$CoSe_2$	Nanolayer	638.3/50/100	[94]
8	(ZnSe/CoSe) ZCS@NC/CNTs	Nano-porous	873/5000/500	[95]
9	$Co_{0.85}Se$	Nanosheets	675/100/50	[96]
10	SnSe/C	Nanocomposite	748.5/5000/50	[97]

3.3 Sulfide-based anodes

Generally, transition metal sulfides (TMS) are considered to be conversion reaction-based electrodes and the electrochemical reactions generally happen at above 1 V. However, still unclear and distressing points are its cut-off potentials which can't allow maintaining the capacity though it is fixed. Hence, most of the reported literature on TMS as anode has chosen low cut-off voltage instead of theoretical cut-off potential and achieved high capacity. But unfortunately, its initial coulombic efficiency was mostly less than 70% which might be attributed to the physicochemical and electrochemical properties of the TMS materials. Owing to this behavior, the reports on TMS-based anodes are limited and very few of the TMS provided better electrochemical capacitive behavior without compromising the efficiency which is discussed below.

Feng et al. have developed copper sulfide nanowires/rGO (CuSNWs/rGO) composites using a single-pot template free solution synthesis that offered the reversible capacity of $620\,mAh\,g^{-1}$ @ 0.5 C over 100 cycles and $320\,mAh\,g^{-1}$ @ 4 C even after 430 cycles. The effective performance is attributed to the coordination of rGO nanosheets and CuS nanowires. The large surface-to-volume changes leads to a 2D network the trapped the polysulfide generation during the conversion reaction of CuS [98]. Xiong et al. have investigated the efficiency of molybdenum sulfide blended with RGO on carbon fiber cloth. The entire structure in the 3D pattern used improved electronic conductivity and ease of release of strain in electrochemical reactions. The prepared anode has revealed the highest specific capacity of $1225\,mAh\,g^{-1}$ with a high cycling efficiency of $680\,mAh\,g^{-1}$ after 250 cycles [99].

A different MoS_2 based anode materials have been prepared and characterized by Chao et al. In this work, 2D MoS_2-rGO film (MG) was synthesized in a 3D porous structure using a simple automatic self-assembly process followed by freeze-drying. Here, the freezing process was carried out by mixing MoS_2 and liquid crystalline graphene oxide at 70°C with one atmospheric pressure. The MG film with 75% of MoS_2 offered the reversible capacity of $800\,mAh\,g^{-1}$ @ $100\,mA\,g^{-1}$ which confirms their better rate capability and cycling stability (stable over 500 cycles @ $400\,mA\,g^{-1}$) [100].

One of the classic sulfides of cobalt's electrochemical behavior have been described in detail for a better understanding. Fig. 3A–D showed the electrochemical behaviors of CNTs@C@Co_{1-x}S(CCCS). Fig. 3A displays the cyclic voltammogram with a potential range of 0.01–$3.0\,V$ @ $0.2\,mV\,s^{-1}$. The presence of a peak at $0.96\,V$ connected to the reduction of Co_{1-x}S to metallic Co and Li_2S and Li^+ insertion process [101]. The wider minor peak observed at $0.5\,V$ was attributed to SEI formation. A shift in peaks at higher voltage was observed in the consequent cycles which were the cause of electrode polarization. The same behavior was observed in cobalt sulfide electrodes. Peaks at 2.08 and $2.35\,V$ were ascribed to the change of Co^0 to Co^{2+} and Li extraction process [102]. Fig. 3B displays the first three cycles of charging and discharging of CCCS @ $0.1\,A\,g^{-1}$. The CCCS electrode exhibited an initial high specific capacity of $1249\,mAh\,g^{-1}$ with a reasonable coulombic-efficiency of 79.4%. Their corresponding rate performance plot is displayed in Fig. 3C and D which indicates the reversible capacities ranging from 897 to $570\,mAh\,g^{-1}$ in the C-rate between 0.1 and $5\,A\,g^{-1}$. The performance rate of CCCS far better than previously reported materials [101–104]. The CV graph of CNTs@C@NiS (CCNS) is showed in Fig. 3E @ $0.2\,mV\,s^{-1}$. In the first cathodic scan, two peaks that middle at $1.31\,V$ and $1.15\,V$ is attributed to the conversion of Ni^{2+} to Ni^{3+} and then to Ni^0, respectively. The broad peak at $0.49\,V$ faded away in the next cycles, indicated the formation of an additional SEI layer while the initial scans [105, 106]. Further, the overlapping of the CV plot after the first cycle was attributed to the outstanding CCNS electrode's cycling stability. Fig. 3F represents the GCD plot of CCNS @ $0.1\,A\,g^{-1}$ in the voltage window between 0.01 and $3.0\,V$. The first GCD capacities of CCNs are 860 and $770\,mAh\,g^{-1}$ with a coulombic efficiency of 89.5%. The charge-discharge performance cycle of

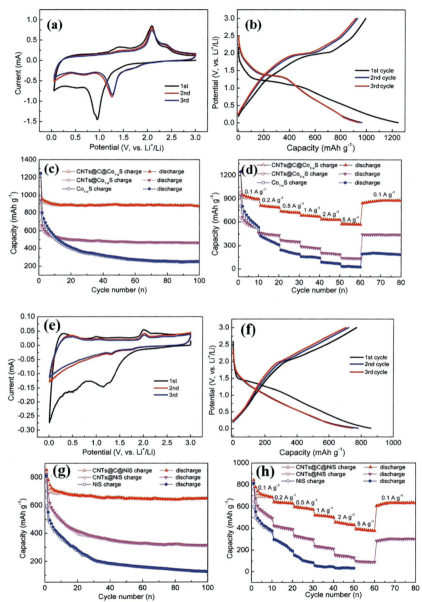

FIG. 3 (A) Cyclic voltammogram (CV) of CNTs@C@Co$_{1-x}$S for 1–3 cycles, (B) GCD plot of CNTs@C@Co$_{1-x}$S @ 0.1 A g^{-1}, (C) cycling behavior curve @ 0.1 A g^{-1}, (D) rate-capability plot of different NiS based electrodes, (E) CV plot CNTs@C@NiS (1–3 cycles), (F) GCD curves of CNTs@C@NiS @ 0.1 A g^{-1}, (G) cycling behavior curve @1 A g^{-1}, (H) corresponding rate capability plot. *(From R. Jin, Y. Jiang, G. Li, Y. Meng, Amorphous carbon coated multiwalled carbon nanotubes@transition metal sulfides composites as high performance anode materials for lithium ion batteries, Electrochim. Acta 257 (2017) 20–30. Copyright, Elsevier, 2017.)*

162 Oxide free nanomaterials for energy storage and conversion applications

CCNS at $0.1\,A\,g^{-1}$ was displayed in Fig. 3G. Of the two electrode materials CNTs@NiS and pristine NiS, the CCNS electrode demonstrated good cycling cyclability. Fig. 3H, illustrates the reversible capacities of CCNS as 687, 626, 571, 500, and $429\,mAh\,g^{-1}$ at the current densities 0.1, 0.2, 0.5, 1, and $2\,A\,g^{-1}$, respectively.

A simple solvothermal process has been employed to make layered MoS_2/graphene composites using L-cysteine as a sulfur donor. The performance of anode material has analyzed using two-electrode assembly versus lithium metal. Of the various molar ratios of MoC, the sample with 1:2 displays the excellent capacity of $1100\,mAh\,g^{-1}$ @ $100\,mA\,g^{-1}$ [107]. In addition, a novel composite of MoS_2 based material coated with a 3D graphene network anode shows good electrochemical performance with a capacity of 877 and $665\,mAh\,g^{-1}$ in the initial cycle @100 and $500\,mA\,g^{-1}$. This in turn revealed a good cycling performance [108].

Tungsten sulfide is another significant anode material for LIBs. The tungsten disulfide nanotubes (WS_2-NTs)/graphene together built as network architecture. The electrochemical performances of this material showed them as an excellent candidate. It displayed a capacity of $318.6\,mA\,g^{-1}$ after 500 cycles @ $1\,A\,g^{-1}$ [109]. The performance of different synthesis routed tin sulfides and their structural variation was discussed in detail by Wei et al. [110].

The manganese sulfide (5-007-MnS) submicro crystals have been synthesized using a simple hydrothermal process. It possessed the first discharge capacity of $1327\,mAh\,g^{-1}$ @ $0.7\,V$ versus Li/Li^+. The findings observed are corresponded to the crystallinity aspect of the material. The precursors used were hydrated manganese chloride ($MnCl_4 \cdot 4H_2O$), thiourea ($(NH_2)_2CS$), and aqueous and alkaline hydrazine ($N_2H_4 \cdot H_2O$) [111]. The solvothermal prepared MnS/graphene composite displays a uniform microsphere in the diameter of 700 nm which enables a feasible pathway for the migration of Li^+ during the GCD process. MnS exhibits the reversible capacity of $1231\,mAh\,g^{-1}$ @ $0.2\,A\,g^{-1}$ and has good retention characteristics [112]. Table 3 provides the complete electrochemical behavior information on different sulfide-based anode material for rechargeable LIBs.

From Table 3, it is observed that vanadium sulfide possessed a high-capacity value than other reported materials. The materials derived from metal-organic frame-work also reveals excellent performance for LIBs. Cobalt sulfide and manganese sulfide have also exhibited decent discharge capacity values. The use of carbon nanotubes leads to a remarkable change in the discharge capacity and has good capacity retention even over 100 cycles.

3.4 Phosphide-based anodes

Anodes based on phosphide materials have a huge contribution to the enhanced performance of LIBs. The different structures of the materials, the transition metals used and the synthesis methods are supplements for high-performance LIBs. In this section, let us address and have a comparative study of the various

TABLE 3 Morphological feature and cycling performances of sulfide-based anodes.

No.	Material	Morphology	Performance mAh g^{-1}/C-rate (mA g^{-1})/ number of cycles	References
1	Vanadium sulfide	Nano-porous	1144/100/50	[113]
2	Cobalt sulfide@N, S-co-doped carbon composites	Hexagon	754/1000/100	[114]
3	MOF derived iron sulfide	Nanorod	936.3/1000/300	[115]
4	MOF derived cobalt sulfide	Orostachys-like superstructures	791/2000/100	[116]
5	CoZnS$_x$@N-doped carbon	Olive nanosheet	527/5000/200	[117]
6	ZnCo$_2$S$_4$/NiCo$_2$S$_4$	Nanoplates	2.4 mAh cm^{-2}/ 0.36 mA cm^{-2}/25	[118]
7	CNTs@C@Co$_{1-x}$S CNTs@C@NiS	Nanowire	875/1000/100	[119]
8	α-MnS	Nanorod	772/5000/100	[120]
9	CoS$_2$	Worm-shaped structures	883/100/100	[103]
10	Zinc blend-MnS	Nano-cube	287.9/1000/600	[121]

anode materials of transition metal phosphide. Wang et al. have studied metal phosphides in detail for LIBs and sodium-ion batteries. They concluded that nano-sized materials outperform the LIBs anode competition [122].

Of the various metal phosphides, Nam and his coworkers studied germanium phosphide (GeP$_3$) in a standardized sheet using a solid-state process. As this material was used as an anode in LIBs during the lithiation and (de) lithiation process, GeP$_3$ has been shown to exhibit sequential three-step and two-step reactions. The first reversible capacity of 1526 mAh g^{-1} was delivered with superior coulombic-efficiency due to their layered structure which in turn suggests that these electrodes have a great impact in enhancing the performance of LIBs [123]. Likewise, GeP$_3$ was synthesized by the combination of a rucked-layer designed orthorhombic black P with a cubic crystal designed Ge, as shown in Fig. 4A. The XRD showed the rhombohedral GeP$_3$ phase shown in Fig. 4B. The HR-TEM images displayed the GeP$_3$ nano-crystallites with the range of size 20–30 nm. Fig. 4C shows the comparative EXAFS spectra of Ge and GeP$_3$ in which the peak at 1.87 Å corresponds to the Ge—P bond and the peak at 2.15 Å depicted to Ge—Ge bond of Ge. The electrochemical

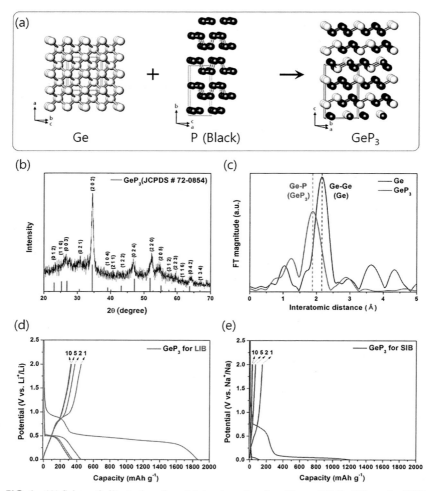

FIG. 4 (A) Schematic illustration of crystal structure of puckered layered GeP$_3$, (B) powder XRD of GeP$_3$ (JCPDS No: 72-0854). (C) EXAFS (extended X-ray absorption fine structure) spectrum of metallic Ge and their corresponding phosphide GeP$_3$, (D, E) GCD curve of GeP$_3$ electrode for the different storage devices include LIB (@ 100 mA g^{-1}) and SIB (@ 50 mA g^{-1}). *(From K.-H. Nam, K.-J. Jeon, C.-M. Park, Layered germanium phosphide-based anodes for high-performance lithium- and sodium-ion batteries, Energy Storage Mater. 17 (2019) 78–87. Copyright, Elsevier, 2019.)*

performance of various GeP$_3$ and controlled sample electrodes have tested galvanostatically @ 100 and 50 mA g^{-1} for LIBs and SIBs (Fig. 4D and E). It delivered the first reversible capacity of 1910 mAh g^{-1} @ low C-rate for LIB [123].

A facile hydrothermal preparation technique has been used to prepare cobalt phosphide nanowires and rGO (CoP/RGO). Due to the greater surface area, it facilitates better conductivity led to smooth charge transfer during GCD analysis. It revealed a reversible capacity of 960 mAh g^{-1} after 200 cycles @ 0.2 A g^{-1} with outstanding rate capability behavior [124]. Cobalt phosphide Co$_x$P was synthesized by a thermal decomposing method in the nano dimension under a controlled condition with appropriate size, phase, and shape (e.g., Co$_2$P spheres and rods, CoP solid and hollow particles) using carbonization of different organic surfactant reactants. After observing the electrochemical performance, Co$_x$P has displayed the capacity of 630 mAh g^{-1} @ 0.2 C over 100 cycles, and the reversible capacity of 256 mAh g^{-1} could be achieved at a high C-rate, i.e., @ 5 C [125].

The rich phosphorous usage in the metal phosphides was another category of materials. CuP$_2$ nanowires were prepared using the supercritical fluid-liquid-solid (SFLS) growth at 410°C and 10.2 MPa. The obtained sample of CuP$_2$ showed a single crystal structure of monoclinic CuP$_2$. The pure phased CuP$_2$ has demonstrated the capacity of 945 mAh g^{-1} over 100 cycles with much better retention properties [126]. Similarly, the solid-state reaction method at 400°C was used to prepare copper phosphide (Cu$_3$P) in the form of a thick film on copper foils. The output of this thick film is compared to the Cu$_3$P composite electrode. The thick film and the composite film have an initial capacity of 360 and 274 mAh g^{-1}, respectively [127]. Similarly, Chandrasekar et al. have synthesized Cu$_3$P in the samples via low-temperature solid-state reaction. Red phosphorous with appropriate stoichiometry was sprayed on the electrodeposited copper maintained at a temperature of 250°C in an inert atmosphere. The Cu$_3$P nucleated and agglomerated and displayed a nonuniform morphology. The Cu$_3$P has a thickness of (0.2, 0.4 m) prepared over the copper-plated of 10 mm diameter. The as-prepared carbon-free Cu$_3$P electrodes enhanced the electrochemical performance in terms of capacity retention and rate capability after 40 GCD cycles when tested against lithium at constant 20 A/cm^2 [128].

Bichat et al. have prepared Cu$_3$P with different particles sizes and crystallite sizes using various synthetic methods like solvothermal, ball milling, and spray methods. In particular, the morphology changes had a significant impact on its electrochemical performances which results in high initial discharge capacities [129]. Various methodologies have been pursued to enhance the efficiency of anode-dependent phosphide materials. Composites are made up of carbon and phosphorous materials consisting of different metal phosphides and metal phosphide composite parts. This, in turn, improves power, cycle ability, and speed efficiency. It deals with weak electronic and ion transport and volume expansion [130]. The morphological feature and electrochemical behaviors of various phosphide-based anodes are listed in Table 4.

TABLE 4 Synthesis, morphological feature and cycling performances of phosphide-based anodes.

No.	Materials	Synthesis method	Morphology	Performance mAh g^{-1}/C-rate (mA g^{-1})/ number of cycles	References
1	Fe$_2$P	Electrodeposition	Nano-hexagon	563/10 μA cm^{-2}	[131]
2	Fe$_2$P/Fe$_2$O$_3$	Scalable two step strategy	Nanosphere	227/0.2 C/50	[132]
3	CoP	Facile phosphiding technique	Nano-cubes	522.6/200/750	[133]
4	Co$_2$P/graphene	One pot solution method	Nanosphere	900/100/70	[134]
5	Co$_x$P-NC	MOF assisted calcination process	Polyhedra	1224/1000/100	[135]
6	CoP/graphene aerogel	Thermally induced phosphidation process	Nano-cubes	805.3/200/200	[136]
7	CoP/RGO	Hydrothermal	Nanowire	960/2000/200	[124]
8	Cu$_3$P	Chemical-vapor deposition	Nanofiber	320/0.2 C/20	[137]
9	InP	Commercial product	NA	475/0.2–1.5 V/20	[138]
10	FeP$_2$@CNTs	Chemical-vapor deposition	Ball-cactus like structure	736.2/0.1365/100	[139]

4. Summary, outlook, and future scopes

The large capacity values of the transition metal sulfide-based anodes as compared to conventional graphite anode are received huge consideration. However, it could not be found in the subsequent load discharge cycles in the LIBs which may be due to large volume expansion for the duration of the GCD process. In the transition metal selenides, the fabrication design plays a key role to influence the battery performance. However, the high-temperature synthesis procedure and poor coulombic efficiency restrict their utilization in conventional LIBs. Different preparation techniques lead to the enhancement of battery performance in LIBs when phosphide-based anode materials were used in them. The composites comprise carbon materials and phosphorous formed various metal phosphides and metal phosphide composite components. This in turn improves the capacity, cyclability, and rate performance. It deals with poor electronic and ionic transportation and volume expansion.

Notwithstanding all these anode materials for LIBs, nanostructured and nanocomposites have displayed much-appreciated results, however, the low cycling stability is one of the major concerns while using these electrodes. To overcome these factors, two components' matrices will be helpful in which one matrix acts as a buffer and another matrix reacts with lithium during electrochemical reactions. Further, to avoid the cracking of nanostructures and external oxidation during the cycling process, surface coating and modifications of the anode materials are highly necessary. An effective method of doping hetero atom (N, S, or O) or halogen atoms in the electrode matrix is a highly effective way to improve the electric conductivity in the electrode materials. Furthermore, current collectors or active materials made up of porous materials are highly required to accommodate volume contraction or expansion during the galvanostatic cycling process.

However, the transition metal sulfides, selenides, and phosphides commercialization are restricted by their structural variability, elevated temperature synthesis, and high cost. Henceforth, cheap, inexpensive preparation processes need to be established to minimize the capital cost of the reactants and final products. Also, suitable aprotic electrolytes that can ensure the structural stability of these anodes to employ them in LIBs needs to be identified. We trust that the present chapter will be useful to provide a complete view of the recent progress in transition metal sulfide, selenide, and phosphide-based anode materials for LIBs. Also, it helps to offer an insight into auspicious perspectives in the energy conversion and storage field.

References

[1] K. Karuppasamy, H.-S. Kim, D. Kim, D. Vikraman, K. Prasanna, A. Kathalingam, R. Sharma, H.W. Rhee, An enhanced electrochemical and cycling properties of novel boronic ionic liquid based ternary gel polymer electrolytes for rechargeable Li/LiCoO$_2$ cells, Sci. Rep. 7 (2017) 11103.

168 Oxide free nanomaterials for energy storage and conversion applications

[2] K. Karuppasamy, P.A. Reddy, G. Srinivas, A. Tewari, R. Sharma, X.S. Shajan, D. Gupta, Electrochemical and cycling performances of novel nonafluorobutanesulfonate (nonaflate) ionic liquid based ternary gel polymer electrolyte membranes for rechargeable lithium ion batteries, J. Membr. Sci. 514 (2016) 350–357.

[3] K. Karuppasamy, S. Thanikaikarasan, R. Antony, S. Balakumar, X.S. Shajan, Effect of nano-chitosan on electrochemical, interfacial and thermal properties of composite solid polymer electrolytes, Ionics 18 (2012) 737–745.

[4] J.B. Goodenough, Y. Kim, Challenges for rechargeable Li batteries, Chem. Mater. 22 (2010) 587–603.

[5] K. Karuppasamy, D. Vikraman, J.-H. Jeon, S. Ramesh, H.M. Yadav, V. Rajendiran Jothi, R. Bose, H.S. Kim, A. Alfantazi, H.-S. Kim, Highly porous, hierarchical microglobules of Co_3O_4 embedded N-doped carbon matrix for high performance asymmetric supercapacitors, Appl. Surf. Sci. 529 (2020) 147147.

[6] V.R. Jothi, K. Karuppasamy, T. Maiyalagan, H. Rajan, C.-Y. Jung, S.C. Yi, Corrosion and alloy engineering in rational design of high current density electrodes for efficient water splitting, Adv. Energy Mater. 10 (2020) 1904020.

[7] D. Vikraman, S. Hussain, K. Karuppasamy, A. Feroze, A. Kathalingam, A. Sanmugam, S.-H. Chun, J. Jung, H.-S. Kim, Engineering the novel $MoSe_2$-Mo_2C hybrid nanoarray electrodes for energy storage and water splitting applications, Appl. Catal. B Environ. 264 (2020) 118531.

[8] R. Bose, V.R. Jothi, K. Karuppasamy, A. Alfantazi, S.C. Yi, High performance multicomponent bifunctional catalysts for overall water splitting, J. Mater. Chem. A 8 (2020) 13795–13805.

[9] K. Karuppasamy, V.R. Jothi, A. Nichelson, D. Vikraman, W.H. Tanveer, H.-S. Kim, S.-C. Yi, Nanostructured transition metal sulfide/selenide anodes for high-performance sodium-ion batteries, in: A. Pandikumar, P. Rameshkumar (Eds.), Nanostructured, Functional, and Flexible Materials for Energy Conversion and Storage Systems, Elsevier, 2020, pp. 437–464 (Chapter 14).

[10] K. Karuppasamy, K. Prasanna, P.R. Ilango, D. Vikraman, R. Bose, A. Alfantazi, H.-S. Kim, Biopolymer phytagel-derived porous nanocarbon as efficient electrode material for high-performance symmetric solid-state supercapacitors, J. Ind. Eng. Chem. 80 (2019) 258–264.

[11] S. Ramesh, A. Kathalingam, K. Karuppasamy, H.-S. Kim, H.S. Kim, Nanostructured CuO/Co_2O_4@ nitrogen doped MWCNT hybrid composite electrode for high-performance supercapacitors, Compos. Part B 166 (2019) 74–85.

[12] J.B. Goodenough, Evolution of strategies for modern rechargeable batteries, Acc. Chem. Res. 46 (2013) 1053–1061.

[13] M. Armand, J.M. Tarascon, Building better batteries, Nature 451 (2008) 652–657.

[14] A. Mauger, C.M. Julien, A. Paolella, M. Armand, K. Zaghib, Building better batteries in the solid state: a review, Materials 12 (2019).

[15] N. Mohamed, N.K. Allam, Recent advances in the design of cathode materials for Li-ion batteries, RSC Adv. 10 (2020) 21662–21685.

[16] S. Komaba, Y. Matsuura, T. Ishikawa, N. Yabuuchi, W. Murata, S. Kuze, Redox reaction of Sn-polyacrylate electrodes in aprotic Na cell, Electrochem. Commun. 21 (2012) 65–68.

[17] H. Zhang, P.V. Braun, Three-dimensional metal scaffold supported bicontinuous silicon battery anodes, Nano Lett. 12 (2012) 2778–2783.

[18] K. Karuppasamy, D. Kim, Y.H. Kang, K. Prasanna, H.W. Rhee, Improved electrochemical, mechanical and transport properties of novel lithium bisnonafluoro-1-butanesulfonimidate (LiBNFSI) based solid polymer electrolytes for rechargeable lithium ion batteries, J. Ind. Eng. Chem. 52 (2017) 224–234.

[19] K. Karuppasamy, K. Prasanna, D. Kim, Y.H. Kang, H.W. Rhee, Headway in rhodanide anion based ternary gel polymer electrolytes (TILGPEs) for applications in rechargeable lithium ion batteries: an efficient route to achieve high electrochemical and cycling performances, RSC Adv. 7 (2017) 19211–19222.

[20] W. Qi, J.G. Shapter, Q. Wu, T. Yin, G. Gao, D. Cui, Nanostructured anode materials for lithium-ion batteries: principle, recent progress and future perspectives, J. Mater. Chem. A 5 (2017) 19521–19540.

[21] D. Kong, H. He, Q. Song, B. Wang, W. Lv, Q.-H. Yang, L. Zhi, Rational design of MoS_2@-graphene nanocables: towards high performance electrode materials for lithium ion batteries, Energy Environ. Sci. 7 (2014) 3320–3325.

[22] J. Xiao, D. Choi, L. Cosimbescu, P. Koech, J. Liu, J.P. Lemmon, Exfoliated MoS_2 nanocomposite as an anode material for lithium ion batteries, Chem. Mater. 22 (2010) 4522–4524.

[23] G. Guo, J. Hong, C. Cong, X. Zhou, K. Zhang, Molybdenum disulfide synthesized by hydro-thermal method as anode for lithium rechargeable batteries, J. Mater. Sci. 40 (2005) 2557–2559.

[24] Y. Gu, Y. Xu, Y. Wang, Graphene-wrapped CoS nanoparticles for high-capacity lithium-ion storage, ACS Appl. Mater. Interfaces 5 (2013) 801–806.

[25] Y. Wang, J. Wu, Y. Tang, X. Lü, C. Yang, M. Qin, F. Huang, X. Li, X. Zhang, Phase-controlled synthesis of cobalt sulfides for lithium ion batteries, ACS Appl. Mater. Interfaces 4 (2012) 4246–4250.

[26] J. Zhou, J. Qin, X. Zhang, C. Shi, E. Liu, J. Li, N. Zhao, C. He, 2D space-confined synthesis of few-layer MoS_2 anchored on carbon nanosheet for lithium-ion battery anode, ACS Nano 9 (2015) 3837–3848.

[27] J.-w. Seo, J.-t. Jang, S.-w. Park, C. Kim, B. Park, J. Cheon, Two-dimensional SnS_2 nanoplates with extraordinary high discharge capacity for lithium ion batteries, Adv. Mater. 20 (2008) 4269–4273.

[28] Y. Shi, C. Hua, B. Li, X. Fang, C. Yao, Y. Zhang, Y.-S. Hu, Z. Wang, L. Chen, D. Zhao, G.D. Stucky, Highly ordered mesoporous crystalline $MoSe_2$ material with efficient visible-light-driven photocatalytic activity and enhanced lithium storage performance, Adv. Funct. Mater. 23 (2013) 1832–1838.

[29] S.-Z. Kang, L. Jia, X. Li, Y. Yin, L. Li, Y.-G. Guo, J. Mu, Amine-free preparation of SnSe nanosheets with high crystallinity and their lithium storage properties, Colloids Surf. A Physicochem. Eng. Asp. 406 (2012) 1–5.

[30] R. Bhandavat, L. David, G. Singh, Synthesis of surface-functionalized WS_2 nanosheets and performance as Li-ion battery anodes, J. Phys. Chem. Lett. 3 (2012) 1523–1530.

[31] T. Takeuchi, H. Sakaebe, H. Kageyama, T. Sakai, K. Tatsumi, Preparation of NiS_2 using spark-plasma-sintering process and its electrochemical properties, J. Electrochem. Soc. 155 (2008).

[32] J. Ma, Y. Wang, Y. Wang, Q. Chen, J. Lian, W. Zheng, Controlled synthesis of one-dimensional Sb_2Se_3 nanostructures and their electrochemical properties, J. Phys. Chem. C 113 (2009) 13588–13592.

[33] J.B. Goodenough, A. Manthiram, B. Wnetrzewski, Electrodes for lithium batteries, J. Power Sources 43 (1993) 269–275.

[34] S. Karthikprabhu, K. Karuppasamy, D. Vikraman, K. Prasanna, T. Maiyalagan, A. Nichelson, A. Kathalingam, H.-S. Kim, Electrochemical performances of $LiNi_{1-x}Mn_xPO_4$ (x = 0.05–0.2) olivine cathode materials for high voltage rechargeable lithium ion batteries, Appl. Surf. Sci. 449 (2018) 435–444.

[35] A. Nichelson, K. Karuppasamy, S. Thanikaikarasan, P. Anil Reddy, P. Kollu, S. Karthickprabhu, X. Sahaya Shajan, Electrical, electrochemical, and cycling studies of high-power layered $Li(Li_{0.05}Ni_{0.7-x}Mn_{0.25}Co_x)O_2$ ($x = 0$, 0.1, 0.3, 0.5, and 0.7) cathode materials for rechargeable lithium ion batteries, Ionics 24 (2018) 1007–1017.

[36] A. Nichelson, S. Karthickprabhu, K. Karuppasamy, G. Hirankumar, X. Sahaya Shajan, A brief review on integrated (layered and spinel) and olivine nanostructured cathode materials for lithium ion battery applications, Mater. Focus 5 (2016) 324–334.

[37] M.S. Whittingham, Lithium batteries and cathode materials, Chem. Rev. 104 (2004) 4271–4302.

[38] B. Xu, D. Qian, Z. Wang, Y.S. Meng, Recent progress in cathode materials research for advanced lithium ion batteries, Mater. Sci. Eng. R: Rep. 73 (2012) 51–65.

[39] Y.P. Wu, E. Rahm, R. Holze, Carbon anode materials for lithium ion batteries, J. Power Sources 114 (2003) 228–236.

[40] B. Liang, Y. Liu, Y. Xu, Silicon-based materials as high capacity anodes for next generation lithium ion batteries, J. Power Sources 267 (2014) 469–490.

[41] C. Lin, C. Yang, S. Lin, J. Li, Titanium-containing complex oxides as anode materials for lithium-ion batteries: a review, Mater. Technol. 30 (2015) A192–A202.

[42] W.-J. Zhang, A review of the electrochemical performance of alloy anodes for lithium-ion batteries, J. Power Sources 196 (2011) 13–24.

[43] N. Mahmood, T. Tang, Y. Hou, Nanostructured anode materials for lithium ion batteries: progress, challenge and perspective, Adv. Energy Mater. 6 (2016) 1600374.

[44] Y. Wang, G. Cao, Developments in nanostructured cathode materials for high-performance lithium-ion batteries, Adv. Mater. 20 (2008) 2251–2269.

[45] L. Long, S. Wang, M. Xiao, Y. Meng, Polymer electrolytes for lithium polymer batteries, J. Mater. Chem. A 4 (2016) 10038–10069.

[46] J.W. Fergus, Ceramic and polymeric solid electrolytes for lithium-ion batteries, J. Power Sources 195 (2010) 4554–4569.

[47] J.Y. Song, Y.Y. Wang, C.C. Wan, Review of gel-type polymer electrolytes for lithium-ion batteries, J. Power Sources 77 (1999) 183–197.

[48] K. Karuppasamy, J. Theerthagiri, D. Vikraman, C.-J. Yim, S. Hussain, R. Sharma, T. Maiyalagan, J. Qin, H.-S. Kim, Ionic liquid-based electrolytes for energy storage devices: a brief review on their limits and applications, Polymers 12 (2020).

[49] K. Karuppasamy, D. Vikraman, I.-T. Hwang, H.-J. Kim, A. Nichelson, R. Bose, H.-S. Kim, Nonaqueous liquid electrolytes based on novel 1-ethyl-3-methylimidazolium bis (nonafluorobutane-1-sulfonyl imidate) ionic liquid for energy storage devices, J. Mater. Res. Technol. 9 (2020) 1251–1260.

[50] S.S. Zhang, A review on the separators of liquid electrolyte Li-ion batteries, J. Power Sources 164 (2007) 351–364.

[51] H. Lee, M. Yanilmaz, O. Toprakci, K. Fu, X. Zhang, A review of recent developments in membrane separators for rechargeable lithium-ion batteries, Energy Environ. Sci. 7 (2014) 3857–3886.

[52] H. Maleki, G. Deng, I. Kerzhner-Haller, A. Anani, J.N. Howard, Thermal stability studies of binder materials in anodes for lithium-ion batteries, J. Electrochem. Soc. 147 (2000) 4470.

[53] J. Drofenik, M. Gaberscek, R. Dominko, F.W. Poulsen, M. Mogensen, S. Pejovnik, J. Jamnik, Cellulose as a binding material in graphitic anodes for Li ion batteries: a performance and degradation study, Electrochim. Acta 48 (2003) 883–889.

[54] S. Goriparti, E. Miele, F. De Angelis, E. Di Fabrizio, R. Proietti Zaccaria, C. Capiglia, Review on recent progress of nanostructured anode materials for Li-ion batteries, J. Power Sources 257 (2014) 421–443.

[55] H. Xiang, K. Zhang, G. Ji, J.Y. Lee, C. Zou, X. Chen, J. Wu, Graphene/nanosized silicon composites for lithium battery anodes with improved cycling stability, Carbon 49 (2011) 1787–1796.

[56] H. Li, R.Y. Tay, S.H. Tsang, W. Liu, E.H.T. Teo, Reduced graphene oxide/boron nitride composite film as a novel binder-free anode for lithium ion batteries with enhanced performances, Electrochim. Acta 166 (2015) 197–205.

[57] M. Zhu, H. Zhou, J. Shao, J. Feng, A. Yuan, Prussian blue nanocubes supported on graphene foam as superior binder-free anode of lithium-ion batteries, J. Alloys Compd. 749 (2018) 811–817.

[58] F. Li, H. Yue, Z. Yang, X. Li, Y. Qin, D. He, Flexible free-standing graphene foam supported silicon films as high capacity anodes for lithium ion batteries, Mater. Lett. 128 (2014) 132–135.

[59] M. Su, Z. Wang, H. Guo, X. Li, S. Huang, W. Xiao, L. Gan, Enhancement of the cyclability of a Si/graphite@graphene composite as anode for lithium-ion batteries, Electrochim. Acta 116 (2014) 230–236.

[60] H. Tang, Y.J. Zhang, Q.Q. Xiong, J.D. Cheng, Q. Zhang, X.L. Wang, C.D. Gu, J.P. Tu, Self-assembly silicon/porous reduced graphene oxide composite film as a binder-free and flexible anode for lithium-ion batteries, Electrochim. Acta 156 (2015) 86–93.

[61] H. Tao, L. Xiong, S. Zhu, X. Yang, L. Zhang, Flexible binder-free reduced graphene oxide wrapped Si/carbon fibers paper anode for high-performance lithium ion batteries, Int. J. Hydrog. Energy 41 (2016) 21268–21277.

[62] D. Xie, W. Yuan, Z. Dong, Q. Su, J. Zhang, G. Du, Facile synthesis of porous NiO hollow microspheres and its electrochemical lithium-storage performance, Electrochim. Acta 92 (2013) 87–92.

[63] C. Shen, C. Zhao, F. Xin, C. Cao, W.-Q. Han, Nitrogen-modified carbon nanostructures derived from metal-organic frameworks as high performance anodes for Li-ion batteries, Electrochim. Acta 180 (2015) 852–857.

[64] W. Yuan, J. Luo, B. Pan, Z. Qiu, S. Huang, Y. Tang, Hierarchical shell/core CuO nanowire/carbon fiber composites as binder-free anodes for lithium-ion batteries, Electrochim. Acta 241 (2017) 261–271.

[65] Y. Huang, Z. Lin, M. Zheng, T. Wang, J. Yang, F. Yuan, X. Lu, L. Liu, D. Sun, Amorphous Fe_2O_3 nanoshells coated on carbonized bacterial cellulose nanofibers as a flexible anode for high-performance lithium ion batteries, J. Power Sources 307 (2016) 649–656.

[66] Q.Q. Xiong, H.Y. Qin, H.Z. Chi, Z.G. Ji, Synthesis of porous nickel networks supported metal oxide nanowire arrays as binder-free anode for lithium-ion batteries, J. Alloys Compd. 685 (2016) 15–21.

[67] L. Gu, W. Xie, S. Bai, B. Liu, S. Xue, Q. Li, D. He, Facile fabrication of binder-free NiO electrodes with high rate capacity for lithium-ion batteries, Appl. Surf. Sci. 368 (2016) 298–302.

[68] Q. Xiong, H. Chi, J. Zhang, J. Tu, Nitrogen-doped carbon shell on metal oxides core arrays as enhanced anode for lithium ion batteries, J. Alloys Compd. 688 (2016) 729–735.

[69] P. Subalakshmi, A. Sivashanmugam, CuO nano hexagons, an efficient energy storage material for Li-ion battery application, J. Alloys Compd. 690 (2017) 523–531.

172 Oxide free nanomaterials for energy storage and conversion applications

[70] H. Yin, X.-X. Yu, Q.-W. Li, M.-L. Cao, W. Zhang, H. Zhao, M.-Q. Zhu, Hollow porous CuO/C composite microcubes derived from metal-organic framework templates for highly reversible lithium-ion batteries, J. Alloys Compd. 706 (2017) 97–102.

[71] D. Ji, H. Zhou, Y. Tong, J. Wang, M. Zhu, T. Chen, A. Yuan, Facile fabrication of MOF-derived octahedral CuO wrapped 3D graphene network as binder-free anode for high performance lithium-ion batteries, Chem. Eng. J. 313 (2017) 1623–1632.

[72] X. Li, A. Dhanabalan, K. Bechtold, C. Wang, Binder-free porous core–shell structured Ni/NiO configuration for application of high performance lithium ion batteries, Electrochem. Commun. 12 (2010) 1222–1225.

[73] J.-Z. Wang, C. Zhong, S.-L. Chou, H.-K. Liu, Flexible free-standing graphene-silicon composite film for lithium-ion batteries, Electrochem. Commun. 12 (2010) 1467–1470.

[74] Y. Zou, Z. Qi, Z. Ma, W. Jiang, R. Hu, J. Duan, MOF-derived porous ZnO/MWCNTs nanocomposite as anode materials for lithium-ion batteries, J. Electroanal. Chem. 788 (2017) 184–191.

[75] H. Tabassum, R. Zou, A. Mahmood, Z. Liang, Q. Wang, H. Zhang, S. Gao, C. Qu, W. Guo, S. Guo, A universal strategy for hollow metal oxide nanoparticles encapsulated into B/N co-doped graphitic nanotubes as high-performance lithium-ion battery anodes, Adv. Mater. 30 (2018) 1705441.

[76] X. Huang, B. Sun, S. Chen, G. Wang, Self-assembling synthesis of free-standing nanoporous graphene–transition-metal oxide flexible electrodes for high-performance lithium-ion batteries and supercapacitors, Chem. Asian J. 9 (2014) 206–211.

[77] Y. Xu, R. Yi, B. Yuan, X. Wu, M. Dunwell, Q. Lin, L. Fei, S. Deng, P. Andersen, D. Wang, H. Luo, High capacity MoO_2/graphite oxide composite anode for lithium-ion batteries, J. Phys. Chem. Lett. 3 (2012) 309–314.

[78] D. Chen, R. Yi, S. Chen, T. Xu, M.L. Gordin, D. Wang, Facile synthesis of graphene–silicon nanocomposites with an advanced binder for high-performance lithium-ion battery anodes, Solid State Ionics 254 (2014) 65–71.

[79] P. Kumari, R. Singh, K. Awasthi, T. Ichikawa, M. Kumar, A. Jain, Highly stable nanostructured Bi_2Se_3 anode material for all solid-state lithium-ion batteries, J. Alloys Compd. 838 (2020) 155403.

[80] A. Gurung, R. Naderi, B. Vaagensmith, G. Varnekar, Z. Zhou, H. Elbohy, Q. Qiao, Tin selenide – multi-walled carbon nanotubes hybrid anodes for high performance lithium-ion batteries, Electrochim. Acta 211 (2016) 720–725.

[81] W. Sun, C. Cai, X. Tang, L.-P. Lv, Y. Wang, Carbon coated mixed-metal selenide microrod: bimetal-organic-framework derivation approach and applications for lithium-ion batteries, Chem. Eng. J. 351 (2018) 169–176.

[82] M. Wang, G. Li, H. Xu, Y. Qian, J. Yang, Enhanced lithium storage performances of hierarchical hollow MoS_2 nanoparticles assembled from nanosheets, ACS Appl. Mater. Interfaces 5 (2013) 1003–1008.

[83] L. Ma, X. Zhou, L. Xu, X. Xu, L. Zhang, W. Chen, Ultrathin few-layered molybdenum selenide/graphene hybrid with superior electrochemical Li-storage performance, J. Power Sources 285 (2015) 274–280.

[84] M.-Z. Xue, Z.-W. Fu, Manganese selenide thin films as anode material for lithium-ion batteries, Solid State Ionics 178 (2007) 273–279.

[85] X. Yang, Z. Zhang, Carbon-coated vanadium selenide as anode for lithium-ion batteries and sodium-ion batteries with enhanced electrochemical performance, Mater. Lett. 189 (2017) 152–155.

[86] D.-H. Lee, C.-M. Park, Tin selenides with layered crystal structures for Li-ion batteries: interesting phase change mechanisms and outstanding electrochemical behaviors, ACS Appl. Mater. Interfaces 9 (2017) 15439–15448.

Metal oxides-free anodes for lithium-ion batteries **Chapter | 7** **173**

[87] Z. Xiang Huang, B. Liu, D. Kong, Y. Wang, H. Ying Yang, $SnSe_2$ quantum dot/rGO composite as high performing lithium anode, Energy Storage Mater. 10 (2018) 92–101.

[88] H.S. Im, Y.R. Lim, Y.J. Cho, J. Park, E.H. Cha, H.S. Kang, Germanium and tin selenide nanocrystals for high-capacity lithium ion batteries: comparative phase conversion of germanium and tin, J. Phys. Chem. C 118 (2014) 21884–21888.

[89] Y. Lou, M. Zhang, C. Li, C. Chen, C. Liang, Z. Shi, D. Zhang, G. Chen, X.-B. Chen, S. Feng, Mercaptopropionic acid-capped Wurtzite $Cu_9Sn_2Se_9$ nanocrystals as high-performance anode materials for lithium-ion batteries, ACS Appl. Mater. Interfaces 10 (2018) 1810–1818.

[90] X. Wang, H. Wang, Q. Li, H. Li, J. Xu, G. Zhao, H. Li, P. Guo, S. Li, Y.-k. Sun, Antimony selenide nanorods decorated on reduced graphene oxide with excellent electrochemical properties for Li-ion batteries, J. Electrochem. Soc. 164 (2017) A2922–A2929.

[91] J. Feng, Q. Li, H. Wang, M. Zhang, X. Yang, R. Yuan, Y. Chai, Hexagonal prism structured MnSe stabilized by nitrogen-doped carbon for high performance lithium ion batteries, J. Alloys Compd. 789 (2019) 451–459.

[92] M.-Z. Xue, Z.-W. Fu, Pulsed laser deposited Sb_2Se_3 anode for lithium-ion batteries, J. Alloys Compd. 458 (2008) 351–356.

[93] J.-H. Jeong, D.-W. Jung, E.-S. Oh, Lithium storage characteristics of a new promising gallium selenide anodic material, J. Alloys Compd. 613 (2014) 42–45.

[94] N. Yu, L. Zou, C. Li, K. Guo, In-situ growth of binder-free hierarchical carbon coated $CoSe_2$ as a high performance lithium ion battery anode, Appl. Surf. Sci. 483 (2019) 85–90.

[95] J. Jin, Y. Zheng, L.B. Kong, N. Srikanth, Q. Yan, K. Zhou, Tuning ZnSe/CoSe in MOF-derived N-doped porous carbon/CNTs for high-performance lithium storage, J. Mater. Chem. A 6 (2018) 15710–15717.

[96] J. Zhou, Y. Wang, J. Zhang, T. Chen, H. Song, H.Y. Yang, Two dimensional layered $Co_{0.85}Se$ nanosheets as a high-capacity anode for lithium-ion batteries, Nanoscale 8 (2016) 14992–15000.

[97] Z. Zhang, X. Zhao, J. Li, SnSe/carbon nanocomposite synthesized by high energy ball milling as an anode material for sodium-ion and lithium-ion batteries, Electrochim. Acta 176 (2015) 1296–1301.

[98] C. Feng, L. Zhang, M. Yang, X. Song, H. Zhao, Z. Jia, K. Sun, G. Liu, One-pot synthesis of copper sulfide nanowires/reduced graphene oxide nanocomposites with excellent lithium-storage properties as anode materials for lithium-ion batteries, ACS Appl. Mater. Interfaces 7 (2015) 15726–15734.

[99] F. Xiong, Z. Cai, L. Qu, P. Zhang, Z. Yuan, O.K. Asare, W. Xu, C. Lin, L. Mai, Three-dimensional crumpled reduced graphene oxide/MoS_2 nanoflowers: a stable anode for lithium-ion batteries, ACS Appl. Mater. Interfaces 7 (2015) 12625–12630.

[100] Y. Chao, R. Jalili, Y. Ge, C. Wang, T. Zheng, K. Shu, G.G. Wallace, Self-assembly of flexible free-standing 3D porous MoS_2-reduced graphene oxide structure for high-performance lithium-ion batteries, Adv. Funct. Mater. 27 (2017) 1700234.

[101] Y. Zhou, D. Yan, H. Xu, J. Feng, X. Jiang, J. Yue, J. Yang, Y. Qian, Hollow nanospheres of mesoporous Co_9S_8 as a high-capacity and long-life anode for advanced lithium ion batteries, Nano Energy 12 (2015) 528–537.

[102] Y. Zhou, D. Yan, H. Xu, S. Liu, J. Yang, Y. Qian, Multiwalled carbon nanotube@a-C@Co_9S_8 nanocomposites: a high-capacity and long-life anode material for advanced lithium ion batteries, Nanoscale 7 (2015) 3520–3525.

[103] R. Jin, L. Yang, G. Li, G. Chen, Hierarchical worm-like CoS_2 composed of ultrathin nanosheets as an anode material for lithium-ion batteries, J. Mater. Chem. A 3 (2015) 10677–10680.

[104] A. Aijaz, J. Masa, C. Rösler, W. Xia, P. Weide, A.J.R. Botz, R.A. Fischer, W. Schuhmann, M. Muhler, Co@Co_3O_4 encapsulated in carbon nanotube-grafted nitrogen-doped carbon

174 Oxide free nanomaterials for energy storage and conversion applications

polyhedra as an advanced bifunctional oxygen electrode, Angew. Chem. Int. Ed. 55 (2016) 4087–4091.

[105] N. Mahmood, C. Zhang, Y. Hou, Nickel sulfide/nitrogen-doped graphene composites: phase-controlled synthesis and high performance anode materials for lithium ion batteries, Small 9 (2013) 1321–1328.

[106] S. Ji, L. Zhang, L. Yu, X. Xu, J. Liu, In situ carbon-coating and Ostwald ripening-based route for hollow Ni_3S_4@C spheres with superior Li-ion storage performances, RSC Adv. 6 (2016) 101752–101759.

[107] Z. Wang, L. Ma, W. Chen, G. Huang, D. Chen, L. Wang, J.Y. Lee, Facile synthesis of MoS_2/graphene composites: effects of different cationic surfactants on microstructures and electrochemical properties of reversible lithium storage, RSC Adv. 3 (2013) 21675–21684.

[108] X. Cao, Y. Shi, W. Shi, X. Rui, Q. Yan, J. Kong, H. Zhang, Preparation of MoS_2-coated three-dimensional graphene networks for high-performance anode material in lithium-ion batteries, Small 9 (2013) 3433–3438.

[109] R. Chen, T. Zhao, W. Wu, F. Wu, L. Li, J. Qian, R. Xu, H. Wu, H.M. Albishri, A.S. Al-Bogami, D.A. El-Hady, J. Lu, K. Amine, Free-standing hierarchically sandwich-type tungsten disulfide nanotubes/graphene anode for lithium-ion batteries, Nano Lett. 14 (2014) 5899–5904.

[110] Z. Wei, L. Wang, M. Zhuo, W. Ni, H. Wang, J. Ma, Layered tin sulfide and selenide anode materials for Li- and Na-ion batteries, J. Mater. Chem. A 6 (2018) 12185–12214.

[111] N. Zhang, R. Yi, Z. Wang, R. Shi, H. Wang, G. Qiu, X. Liu, Hydrothermal synthesis and electrochemical properties of alpha-manganese sulfide submicrocrystals as an attractive electrode material for lithium-ion batteries, Mater. Chem. Phys. 111 (2008) 13–16.

[112] D. Chen, H. Quan, Z. Huang, L. Guo, Mesoporous manganese sulfide spheres anchored on graphene sheets as high-capacity and long-life anode materials for lithium-ion batteries, ChemElectroChem 2 (2015) 1314–1320.

[113] N. Zhou, W. Qin, C. Wu, C. Jia, Graphene-attached vanadium sulfide composite prepared via microwave-assisted hydrothermal method for high performance lithium ion batteries, J. Alloys Compd. 834 (2020) 155073.

[114] L. Li, J. Zhou, C. Zhang, T. Liu, Confined sulfidation strategy toward cobalt sulfide@nitrogen, sulfur co-doped carbon core-shell nanocomposites for lithium-ion battery anodes, Compos. Commun. 15 (2019) 162–167.

[115] H. Wang, X. Qian, H. Wu, R. Zhang, R. Wu, MOF-derived rod-like composites consisting of iron sulfides embedded in nitrogen-rich carbon as high-performance lithium-ion battery anodes, Appl. Surf. Sci. 481 (2019) 33–39.

[116] T. Yang, D. Yang, Y. Liu, J. Liu, Y. Chen, L. Bao, X. Lu, Q. Xiong, H. Qin, Z. Ji, C.D. Ling, R. Zheng, MOF-derived carbon-encapsulated cobalt sulfides orostachys-like micro/nano-structures as advanced anode material for lithium ion batteries, Electrochim. Acta 290 (2018) 193–202.

[117] S. Guo, Y. Feng, J. Qiu, X. Li, J. Yao, Leaf-shaped bimetallic sulfides@N-doped porous carbon as advanced lithium-ion battery anode, J. Alloys Compd. 792 (2019) 8–15.

[118] H. Zhang, J. Liu, X. Lin, T. Han, M. Cheng, J. Long, J. Li, A novel binary metal sulfide hybrid Li-ion battery anode: three-dimensional $ZnCo_2S_4$/$NiCo_2S_4$ derived from metal-organic foams enables an improved electron transfer and ion diffusion performance, J. Alloys Compd. 817 (2020) 153293.

[119] R. Jin, Y. Jiang, G. Li, Y. Meng, Amorphous carbon coated multiwalled carbon nanotubes@-transition metal sulfides composites as high performance anode materials for lithium ion batteries, Electrochim. Acta 257 (2017) 20–30.

[120] X.-h. Zhou, K.-m. Su, W.-m. Kang, B.-w. Cheng, Z.-h. Li, Nanosized α-MnS homogenously embedded in axial multichannel carbon nanofibers as freestanding electrodes for lithium-ion batteries, J. Mater. Sci. 55 (2020) 7403–7416.

[121] Y. Hao, C. Chen, X. Yang, G. Xiao, B. Zou, J. Yang, C. Wang, Studies on intrinsic phase-dependent electrochemical properties of MnS nanocrystals as anodes for lithium-ion batteries, J. Power Sources 338 (2017) 9–16.

[122] X. Wang, H.-M. Kim, Y. Xiao, Y.-K. Sun, Nanostructured metal phosphide-based materials for electrochemical energy storage, J. Mater. Chem. A 4 (2016) 14915–14931.

[123] K.-H. Nam, K.-J. Jeon, C.-M. Park, Layered germanium phosphide-based anodes for high-performance lithium- and sodium-ion batteries, Energy Storage Mater. 17 (2019) 78–87.

[124] J. Yang, Y. Zhang, C. Sun, H. Liu, L. Li, W. Si, W. Huang, Q. Yan, X. Dong, Graphene and cobalt phosphide nanowire composite as an anode material for high performance lithium-ion batteries, Nano Res. 9 (2016) 612–621.

[125] D. Yang, J. Zhu, X. Rui, H. Tan, R. Cai, H.E. Hoster, D.Y.W. Yu, H.H. Hng, Q. Yan, Synthesis of cobalt phosphides and their application as anodes for lithium ion batteries, ACS Appl. Mater. Interfaces 5 (2013) 1093–1099.

[126] G.-A. Li, C.-Y. Wang, W.-C. Chang, H.-Y. Tuan, Phosphorus-rich copper phosphide nanowires for field-effect transistors and lithium-ion batteries, ACS Nano 10 (2016) 8632–8644.

[127] H. Pfeiffer, F. Tancret, T. Brousse, Synthesis, characterization and electrochemical properties of copper phosphide (Cu_3P) thick films prepared by solid-state reaction at low temperature: a probable anode for lithium ion batteries, Electrochim. Acta 50 (2005) 4763–4770.

[128] M.S. Chandrasekar, S. Mitra, Thin copper phosphide films as conversion anode for lithium-ion battery applications, Electrochim. Acta 92 (2013) 47–54.

[129] M.-P. Bichat, T. Politova, H. Pfeiffer, F. Tancret, L. Monconduit, J.-L. Pascal, T. Brousse, F. Favier, Cu_3P as anode material for lithium ion battery: powder morphology and electrochemical performances, J. Power Sources 136 (2004) 80–87.

[130] W. Liu, H. Zhi, X. Yu, Recent progress in phosphorus based anode materials for lithium/sodium ion batteries, Energy Storage Mater. 16 (2019) 290–322.

[131] M.S. Chandrasekar, S. Mitra, Electrodeposition of iron phosphide on copper substrate as conversion negative electrode for lithium-ion battery application, Ionics 20 (2014) 137–140.

[132] P.S. Veluri, S. Mitra, Iron phosphide (FeP) synthesis, and full cell lithium-ion battery study with a $[Li(NiMnCo)O_2]$ cathode, RSC Adv. 6 (2016) 87675–87679.

[133] K. Zhu, J. Liu, S. Li, L. Liu, L. Yang, S. Liu, H. Wang, T. Xie, Ultrafine cobalt phosphide nanoparticles embedded in nitrogen-doped carbon matrix as a superior anode material for lithium ion batteries, Adv. Mater. Interfaces 4 (2017) 1700377.

[134] A. Lu, X. Zhang, Y. Chen, Q. Xie, Q. Qi, Y. Ma, D.-L. Peng, Synthesis of Co_2P/graphene nanocomposites and their enhanced properties as anode materials for lithium ion batteries, J. Power Sources 295 (2015) 329–335.

[135] G. Xia, J. Su, M. Li, P. Jiang, Y. Yang, Q. Chen, A MOF-derived self-template strategy toward cobalt phosphide electrodes with ultralong cycle life and high capacity, J. Mater. Chem. A 5 (2017) 10321–10327.

[136] H. Gao, F. Yang, Y. Zheng, Q. Zhang, J. Hao, S. Zhang, H. Zheng, J. Chen, H. Liu, Z. Guo, Three-dimensional porous cobalt phosphide nanocubes encapsulated in a graphene aerogel as an advanced anode with high coulombic efficiency for high-energy lithium-ion batteries, ACS Appl. Mater. Interfaces 11 (2019) 5373–5379.

[137] V. Yarmiayev, Y. Miroshnikov, G. Gershinsky, V. Shokhen, D. Zitoun, Fast kinetics in freestanding porous Cu_3P anode for Li-ion batteries, Electrochim. Acta 292 (2018) 846–854.

[138] M.V.V.M. Satya Kishore, U.V. Varadaraju, Phosphides with zinc blende structure as anodes for lithium-ion batteries, J. Power Sources 156 (2006) 594–597.
[139] X. Chen, J. Qiu, Y. Wang, F. Huang, J. Peng, J. Li, M. Zhai, Cactus-like iron diphosphide@carbon nanotubes composites as advanced anode materials for lithium-ion batteries, Electrochim. Acta 259 (2018) 321–328.

Chapter 8

Oxides free materials as anodes for sodium-ion batteries

Chelladurai Karuppiah[a], Dhayanantha Prabu Jaihindh[b], Balamurugan Thirumalraj[c], Ahmed S. Haidyrah[d], and Chun-Chen Yang[a,e]

[a]Battery Research Center of Green Energy, Ming Chi University of Technology, New Taipei City, Taiwan, ROC, [b]Department of Chemical Engineering, National Taiwan University, Taipei City, Taiwan, ROC, [c]Department of Energy & Mineral Resources Engineering, Sejong University, Seoul, South Korea, [d]Nuclear and Radiological Control Unit, King Abdulaziz City for Science and Technology (KACST), Riyadh, Saudi Arabia, [e]Department of Chemical Engineering, Ming Chi University of Technology, New Taipei City, Taiwan, ROC

Chapter outline

1. Introduction	177	4.1 Carbon based transition metal sulfide composites 186
2. Carbon nanostructure based anode materials for SIBs	180	5. Carbon based transition metal selenide composites 187
3. Carbon nanotube (CNTs)	182	6. Carbon based transition metal telluride composites 189
3.1 Carbon nanofibers (CNFs)	182	7. Transition metal phosphides and their carbon composites 191
3.2 Graphene-based materials	185	8. Summary 193
4. Carbon materials with transition metal chalcogenides for SIBs anode	186	References 193

1. Introduction

The energy demands are the critical factors in the present situation, and the industrial power sectors are using natural resources to satisfy energy requirements. Day by day, the population and industries rise gradually, but natural resources are limited. The depletion of fossil fuels and the energy demands has attracted great significance to electrical energy storage technologies. Lithium-ion batteries (LIBs) are the most promising energy storage technology among rechargeable batteries due to their high energy density and power density. The concept of sodium-ion batteries (SIBs) and LIBs was proposed in the 1970–1980s and the LIBs have already been commercialized by Sony in 1991. Thereafter, the LIBs have been investigated with tremendous research efforts [1–3]. However, the study regarding SIBs was hardly conducted that suspended

Oxide Free Nanomaterials for Energy Storage and Conversion Applications.
https://doi.org/10.1016/B978-0-12-823936-0.00016-4
Copyright © 2022 Elsevier Inc. All rights reserved.

178 Oxide free nanomaterials for energy storage and conversion applications

the researches on SIBs for three decades. In another way, LIBs occupied the all-electronic market and electric vehicles development which leads to high demand for LIBs [4].

In the face of greatly increased demands on existing global lithium resources because of the eventual increase in these markets, concerns over potentially increasing costs and accessibility have risen. As a result, research on SIBs was again boosted due to the high abundance of sodium earth crusts ($\sim$23,600 ppm) [5]. As known, sodium (Na) and lithium (Li) belongs to the alkali metal species of group 1 element, both can readily form +1 oxidation states due to the loosely bound electron in valence shell s-orbital [6]. Hence, there is no much difference in their standard electrochemical potential of Na is -2.71 V vs. the standard hydrogen electrode (SHE) and Li is -3.04 V vs. SHE. Na compounds' structure and electrochemical properties are equally explored similar to the development of Li intercalation compounds. Even though the LIBs performed very well, the high abundance and price of Na as compared to Li were anticipated to shift fortunes in favor of Na compounds development [7].

The research suspended as long fundamental factors have not proceeded with a significant investigation about SIBs development due to the absence of suitable anode materials. In LIBs case, carbonaceous materials, i.e., graphite, have been explored as promising anode candidates that offer a high theoretical specific capacity of $372 \, \text{mAh g}^{-1}$ and exhibit a very stable potential plateau of 0.1–0.2 V vs. Li/Li$^+$ [8]. Unfortunately, the graphite anode has not been worked for SIBs due to the larger ionic radius of Na$^+$ limiting the intercalation properties into the graphite layers, resulting in poor energy storage performances. Therefore, several other anode materials have been used for Na storage performance, but they exhibit much lower energy densities than graphite anode used in LIBs. In 2000, hard carbon (HC) has been invented by Stevens and Dhan and used as an anode for SIBs in room temperature applications [9, 10]. The HC anode exhibits love voltage and a greater specific capacity of $300 \, \text{mAh g}^{-1}$ in SIBs, which is almost close to that of graphite in LIBs. This discovery was the turning point to boost the research interest in SIBs, however, it has not been commercialized immediately due to the lack of motivation about SIBs to replace the LIBs at that time. Meanwhile, the research has focused on the progress of cathode materials such as layered metal-oxides, polyanion compounds, and Prussian blue analogs (PBA) developments for SIBs. Numerous cathode materials have been reported between 2010 and 2013, which is near equals to that of existed before [11, 12]. As for the anode in SIBs, HC is of special interest due to its cost-effectiveness and abundant sources. The history and technical evolutions of LIBs and SIBs are clearly shown in Fig. 1. Furthermore, Fig. 2 illustrates the number of publications in SIBs irrespective of materials and electrodes from 2010 to 2021, based on the web of science search by using the keyword "sodium-ion-batteries."

Since 2010, the research on SIBs takes rapid growth of development. A large number of materials have been investigated to be utilized of SIBs as anode materials, which includes metal/metal alloys [14–17], metal oxide (SnO$_2$,

Oxides free materials as anodes for sodium-ion batteries **Chapter | 8** **179**

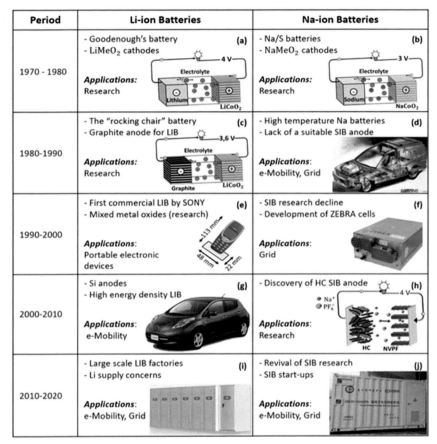

FIG. 1 The technological evolution of LIBs and SIBs; (A) Li and LiCoO$_2$ based cell, (B) Na and NaCoO$_2$ based cell, (C) graphite and LiCoO$_2$ battery, (D) Ford Ecostar designed by NIBs, (E) first portable device Nokia 3310 powered by LIBs, (F) ZEBRA sodium-ion battery for grid energy storage applications, (G) Si anode-based LIBs used in 2010 Nissan Leaf model, (H) SIBs based on HC anode and Na$_3$V$_2$(PO$_4$)$_2$F$_3$ cathode, (I) Illustration of a grid-scale LIB installation, (J) HiNa SIB installation in Liyang city in China [13].

CuO, TiO$_2$, Co$_3$O$_4$, etc.) [18–21] metal sulfide (CoS, CoS$_2$, NiS$_2$, etc.) [22–24], metal phosphide (CuP$_2$, Cu$_3$P, NiP$_3$, etc.) [25–28] and carbonaceous materials [16]. The development of anode materials with special nano/microstructures and composition greatly enhanced electrochemical performances equal to LIBs. Among the materials mentioned above, carbon-related materials played a vital role in developing a LIBs system due to their good electronic conductivity, cost-effectiveness, and high abundance. The carbon-based materials, especially graphite, graphene nanosheets, carbon nanotubes (CNTs), carbon nanofibers (CNFs), and other carbon nanostructures, have been extensively investigated

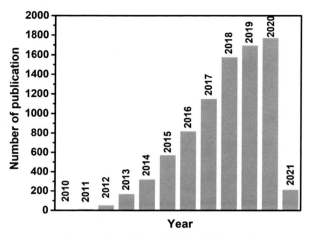

FIG. 2 The number of publications from 2010 to Feb'28, 2021 on sodium-ion batteries from the web of science database.

and reported [16, 29, 30]. The observed specific capacities range of these carbon-based materials about to 200–500 mAh g^{-1}. In practical application, carbon materials can meet the following properties: low-cost, non-toxicity, abundance, stability, and durability. This chapter briefly focuses on the recent investigations on carbon and its hybrid composite anode materials and their electrochemical performance towards SIBs, as clearly depicted in Fig. 3.

2. Carbon nanostructure based anode materials for SIBs

The materials' structure, shape, and size are effective strategies to enhance mass transport and storage, significantly improving sodium's electrochemical performance. Developing and design of nanomaterials with morphology and size-controlled electrode materials is one of the essential factors [31–33]. Fabrication of different nanostructures has been an attractive area of significant consideration in LIBs. The distinctive structure, shape, and size may enhance mass transport by delivering a large surface area and a short diffusion distance [34–36]. Such nanosized structure engineering has been prolonged to SIBs and great accomplishments have been made in various nanostructured materials such as hollow nanostructure, [37, 38] nanofibers, [37] nanotubes, [39, 40] nanosheets, [41, 42] mesoporous carbon, [43, 44] graphene and modified graphene materials [45]. Hence, gradual progress on nanosized carbon materials in SIBs applications has been increased since 2011. Fig. 4 depicted the number of research publications on carbon-based SIBs from 2011 to 2021, based on the web of science search with the keyword "carbon materials in sodium-ion-batteries".

Oxides free materials as anodes for sodium-ion batteries **Chapter | 8** **181**

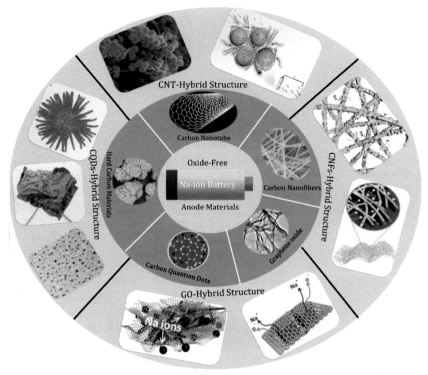

FIG. 3 Schematic representation of oxide-free anode materials used in SIBs applications.

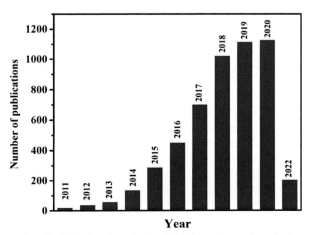

FIG. 4 The number of publications from 2011 to Feb'28, 2021 on sodium-ion batteries from the web of science database.

3. Carbon nanotube (CNTs)

Commonly using carbon nanoparticles is unsuitable for secondary battery development due to its unstable surface morphology and rapid aggregation of particles, limiting their nanostructure advantages with high surface-to-volume ratios, resulting in poor capacity and cyclability. In recent years, one-dimensional (1D) carbon nanomaterials, such as CNTs and CNFs have paid enormous consideration as anode materials in SIBs due to their extraordinary physical and chemical properties [46]. Since its discovery in 1991, CNT has successfully developed nanomaterials by attracting a great deal of interest in academia and industry [47]. The high specific surface area, high electrical conductivity, and hollow structure provide enough space for Na^+ storage [48]. Like single-walled CNTs (SWCNTs), the multi-walled CNTs (MWCNTs) have also provided more space for Na^+ ion intercalation through the inner and outside walls open ends of the multilayers [30]. Sharma and co-workers carried out a comparative electrochemical and mechanistic study of Na^+ ion insertion into MWCNTs and SWCNTs [49]. The SWCNTs deliver the specific capacity of $70 \, mAh \, g^{-1}$ after 100 cycles when discharged to 0.1 V vs. Na/Na^+ and $126 \, mAh \, g^{-1}$ when discharged further to 0.01 V vs. Na/Na^+. The MWCNTs show the relatively poor performance of $30 \, mAh \, g^{-1}$ after 100 cycles when discharged to 0.1 V vs. Na/Na^+ [49]. The corresponding experimental output illustrates in Fig. 5.

The surface-modified SWCNTs can provide more active sites for Na^+ coordination. Meanwhile, the sodiation and desodiation process can be easier while expanding the interlayer spacing of MWCNTs. To increase the effective contact at the nanoscale, the CNTs can directly grow on the active materials, whether it is cathode or anode, thus increasing the performance.

3.1 Carbon nanofibers (CNFs)

CNFs are also considered as high Na^+ storage anode materials due to their 1D nanostructure that provides high surface area and shortens the Na^+ ion transport distance. The Na^+ ions can easily be stored into the distorted graphite interlayers through the intercalation phenomenon during the sodiation process [50]. Fabricating the microporous CNFs with heteroatoms doping can offer more surface area and active sites to permeate the electrolyte and Na^+ ion transport. This modified CNFs nanostructure can extensively provide high electronic and ionic conductivity properties for Na^+ ion storage, therefore, the CNFs can be known as a potential anode for SIBs. Owing to its greater mechanical strength and flexibility, CNFs can also be used as supporting matrixes for hybrid materials preparation [51, 52]. In general, the electrospinning technique is the very efficient method for CNFs preparation owing to its simplicity and low cost and various polymer precursors, such as poly(acrylonitrile) (PAN) and poly(vinyl chloride) (PVC), are mainly used to form the 1D porous nanostructures. Wen and co-workers employed an electrospinning method to synthesize telluride

Oxides free materials as anodes for sodium-ion batteries **Chapter | 8 183**

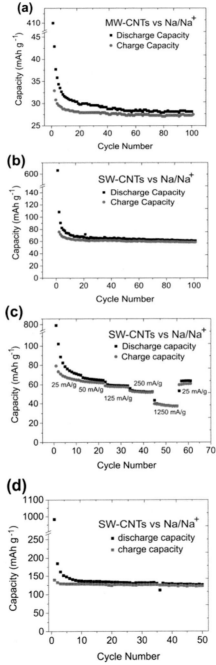

FIG. 5 (A) Cycle stability of MWCNTs electrodes. Cycling performance of SWCNTs electrodes when discharged to (B) 0.1 V vs. Na/Na$^+$ and (D) 0.01 V vs. Na/Na$^+$. (C) High-rate performance of SWCNTs with a voltage cut-off of 0.1 V vs. Na/Na$^+$ [49].

184 Oxide free nanomaterials for energy storage and conversion applications

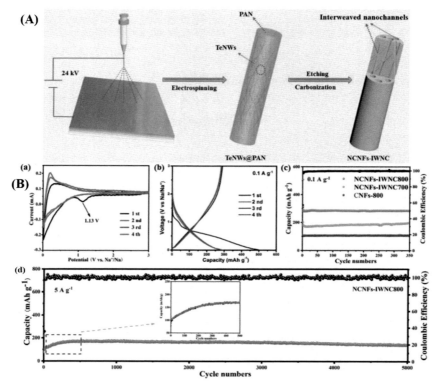

FIG. 6 (A) Graphical representation of the fabrication process of N-doped CNFs, and (B) CV curves (a), charge-discharge plots of the NCNFs-IWNC800 electrode at $0.1\,A\,g^{-1}$ (b), cycling stability curves at $0.1\,A\,g^{-1}$ for 350 cycles (c), and cycle stability of NCNFs-IWNC800 electrode at $5\,A\,g^{-1}$ for 5000 cycles (the inset is the initial 500 cycles at $5\,A\,g^{-1}$ (d) [53].

nanowires integrated PAN composites (TeNWs@PAN) hallow porous fibers by applying the voltage of 24 kV followed TeNWs etching. The fabricated nanofibers were calcined at a different temperature to remove TeNWs and exhibit a well-ordered 1D porous structure with a large surface area and good electronic conductivity for the outstanding Na$^+$ storage performance of both half and full cell SIBs. Fig. 6 illustrates the process of preparation of CNFs and the sodium storage properties.

Systematic electrochemical investigation indicates the NCNFs-IWNC shows a better rate capability, exhibiting a high capacity of $148\,mAh\,g^{-1}$ at $10\,A\,g^{-1}$, and outstanding cyclability at $5\,A\,g^{-1}$ for 5000 cycles. The NCNFs-IWNC anode coupled with PBA-based FeFe(CN)$_6$ cathode for full cell SIBs delivers excellent cycle stability and maintains the capacity of $97\,mAh\,g^{-1}$ over 100 cycles [53]. Amine et al. reported direct pyrolysis of pristine PVC particles and electrospun PVC nanofibers anode materials at different temperatures that produce a high discharge capacity of 389 mAh g^{-1} and initial coulombic efficiency of 69.9%. The electrospun PVC nanofibers anode materials retain excellent capacity retention of 211 mAh g^{-1} after 120 cycles. Developing and introducing a new methodology in electrospinning and controlling the preparation temperature, melting of polymers, cross-linking structure, and graphitization degree will help obtain high-performance carbon nanofibers in SIBs applications.

3.2 Graphene-based materials

Two-dimensional energy storage materials such as graphene, graphene oxide (GO) have much attention due to the large surface area, chemical stability, and good electrical conductivity features favorable in use as anode materials electrochemical energy storage devices [54]. Layered structured graphene can adsorb sodium ions on both sides of the sheets, possible to produce greater capacities of 300–550 mAh g^{-1}. Kang and his co-worker developed thionyl chloride treated randomly oriented GO as an electrode for SIBs [55]. The power density of nearly 20,000 W kg^{-1} was produced by randomly folded and crumbled graphene anode, which is greater than the Li storage capability of conventional graphene paper electrode and also has more excellent stability of 500 charge/discharge cycles and 1000 cycles of repeated pending as depicted in Fig. 7.

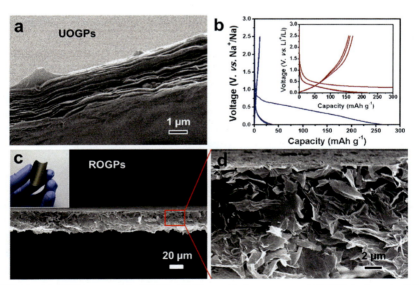

FIG. 7 (A) SEM image of the UOGP. (B) Charge-discharge curves of UOGPs in the potential range of 0.01–2.5 V vs. Na$^+$/Na at 100 mA g^{-1}. The inset shows charge-discharge profiles vs. Li+/Li. (C) and (D) cross-sectional SEM photographs of ROGP in different magnifications. The inset of (C) displays the corresponding optical images [55].

186 Oxide free nanomaterials for energy storage and conversion applications

Zhao et al. developed a few-layered metal-reduced GO as a negative electrode material for SIBs and provided a discharge capacity of $272\,mAh\,g^{-1}$ at $50\,mA\,g^{-1}$ good cycling durability over 300 cycles and a greater rate capability [54]. The different kinds of alkali metal ions functionalized with the GO can alter the surface properties and electronic conductivity. Among the alkali metal ions, the Na-ion functionalized GO exhibited outstanding electrochemical performance attributed to the large mesoporous surface with significant disorder and improved electronic conductivity.

Still, the carbon materials alone are insufficient and should carry over plenty of modifications to enhance the charge capacity, stability, charge/discharge cycle, and long-term cycle performances. Modify the carbon nanostructures morphology, doped with heteroatoms such as N, P, S, etc. Preparation Oxide-free anode materials with unique heterostructure/composites are an efficient and easy way to improve the performance of the materials in SIBs applications.

4. Carbon materials with transition metal chalcogenides for SIBs anode

4.1 Carbon based transition metal sulfide composites

Among the availability of anode materials in SIBs, the transition metal sulfide (TMS)-based anode materials are offered numerous advantages in terms of capacity, cycling performance, and durability due to their greater theoretical capacity, low cost, and amended mechanical constancy and electrical conductivity [56, 57]. Such TMS are paid more attention recently because their remarkable redox flexibilities and structural peculiarities lead to a better candidate for SIBs anode than their metal oxide counterparts [58, 59]. Recently, few metal sulfides were extensively studied as a promising anode for SIBs such as MoS_2 [60], SnS_2 [61], MnS [62], CoS_2 [63], VS_2 [64], and WS_2 [65] due to their appropriable layer spacing to accommodate the sodium ion. Unfortunately, the TMS suffers from reduction caused by huge volume expansion and sluggish electrode kinetics during sodium charge/discharge processes, which lead to inadequate cycling performance [66, 67]. Therefore, it is mandatory to find attractive anode materials with good cycling stability, high-rate performance, and high capacity with reasonable cycle life. The outstanding remarkable properties of carbon-based materials are a subject of extended studies for the energy storage system, which is proven to be viable and efficient for improving electrical conductivity and structural features to suffer the volume changes. Numerous reports were reported that the carbon-supported TMS significantly improve the sodium ion cycling features in terms of specific capacity, cycle life, and C-rate.

For instance, MoS_2, as 2D TMS, proved as promising potential anode materials for SIBs owing to its cost-effectiveness and high theoretical specific capacity ($670\,mAh\,g^{-1}$) in SIBs. Unfortunately, bulk MoS_2 shows large volume changes and low intrinsic electronic conductivity during reversible sodiation/desodiation cycling, which leads to poor electrochemical performance; therefore, its potential practical applications are restricted. Veerasubramani et al. [68] reported heterostructure anode consists of $MoS_2@C$ nanosheets over MoS_2 nanorods ($MoS_2@C@MoS_2$) for SIBs application (Fig. 8). The $MoS_2@C@MoS_2$ nanocomposite electrode was delivered a greater specific capacity, excellent cycling stability, and good rate performance compared with pristine MoS_2. After 200 cycles, $MoS_2@C@MoS_2$ nanocomposite offered $352\,mAh\,g^{-1}$ at $1000\,mA\,g^{-1}$. Likewise, a nanocomposite of three-dimensional porous MoS_2/nitrogen-doped graphene aerogels (MoS_2/NGA) exhibited very high cyclability, long service life, and high sodium intercalation rate. MoS_2/NGA anode material provides a high specific capacity of $673\,mAh\,g^{-1}$ at a current density of $100\,mA\,g^{-1}$ with long cycle stability [69]. Besides, Yong Cheng and Co-workers proved that SnS nanosheets with sulfur-doped amorphous carbon layer showed a Na-storage capacity of $349\,mAh\,g^{-1}$ at $1.0\,A\,g^{-1}$ even after 300 cycles [70]. Ultrasmall SnS quantum dots (QDs) supported nitrogen-incorporated carbon (NC) nanospheres were also reported by Veerasubramani et al. [71] for SIBs anode material and thus achieved a discharge capacity of $281\,mAh\,g^{-1}$ at $100\,mA\,g^{-1}$ and shown exceptional cycling stability with a capacity retention of 75% after 500 cycles at a high current density of $1000\,mA\,g^{-1}$. Lastly, a double-layer WS_2/hollow carbon composite designed as an anode for SIBs by Liu et al. The hollow carbon spheres containing WS_2 has significantly enhanced the specific surface area and electrical conductivity, and thus attain a high specific capacity of $575\,mAh\,g^{-1}$ at $100\,mA\,g^{-1}$ and $379\,mAh\,g^{-1}$ at $1000\,mA\,g^{-1}$ for SIBs [72].

5. Carbon based transition metal selenide composites

As the excellent electrochemical behavior of metal sulfides, the transition metal selenides were also a decent candidate for anode material in SIBs application. The 2D layered selenides-based anode materials have been extensively investigated for the past few years due to their high theoretical capacity and electronic conductivity than metal oxides-based anode materials. These materials have a large interlayer spacing and weak van der Waals interaction between the layers, thus promoting the enhanced Na^+ ion transport properties [23, 73, 74]. The remarkable factor of transition metal selenide is that the density of Se is about 2.5 times that of S, which indicates that the volumetric theoretical capacity of metal selenide is equivalent to that of metal sulfides [75]. Though, these metal selenides even hurt from electrode reductions and lower Na^+ diffusion kinetics during the sodium plating-stripping process, which is improved by modification of carbon-based microstructure nanomaterials. Based on these findings, nanostructured surface engineering and carbon decorating have been

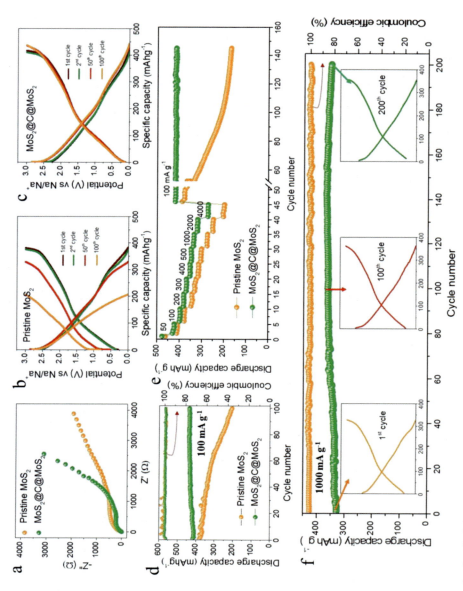

FIG. 8 (A) Nyquist plot of pristine MoS_2 and $MoS_2@C@MoS_2$. Charge-discharge curves of (B) pristine MoS_2 and (C) $MoS_2@C@MoS_2$. (D) Cycling at 100 mA g^{-1} and (E) high-rate performances of pristine MoS_2 and $MoS_2@C@MoS_2$. (F) Cycle stability of the $MoS_2@C@MoS_2$ electrode at 1 A g^{-1}, the corresponding inset shows the charge-discharge curves at 1st, 100th and 200th cycles [68].

considered efficient methodologies for improving the issues described above on the transition metal selenides [76–78].

Several carbon-based metal selenides were reported as electrode materials for SIBs in recent years. For example, the CoSe/C@C composite was prepared by Zhang et al. using the freeze-drying and in-situ selenization process [79]. The CoSe/C@C anode material exhibited excellent cycling stability with a capacity of $332.3\,mAh\,g^{-1}$ as well as the capacity retention of 63.1% at $200\,mA\,g^{-1}$ after 500 cycles. Besides, the CoSe/C@C also displays a capacity of $403.4\,mAh\,g^{-1}$ at $2\,A\,g^{-1}$ with an outstanding rate capability. Hierarchically nanostructured $FeSe_2$ nanoparticles encapsulated bifunctional carbon cuboids ($FeSe_2$/NC@G) were prepared by Fe-based PBA adapted GO as precursor materials. Here, the $FeSe_2$ nanoparticles are completely incorporated into the NC and graphene nanosheets surface [80]. The $FeSe_2$/NC@G anode material reveals tremendous storage performance towards sodium ion cycling with a reversible capacity of $331\,mAh\,g^{-1}$ at $5\,A\,g^{-1}$ and the highest cycling stability of $323\,mAh\,g^{-1}$ at $2\,A\,g^{-1}$ after 1000 cycles (82% capacity retention). Like, a hybrid material of ultrasmall Fe_7Se_8 nanoparticles/N-doped carbon nanofibers (Fe_7Se_8/N-CNFs) was reported by Zhang et al. [81] The Fe_7Se_8/N-CNF hybrid anode material was outperformed that of other metal selenides and exhibited superior cycling rate performance with a high specific capacity of $286.3\,mAh\,g^{-1}$ at $20\,A\,g^{-1}$. Also, Wang et al. [82] designed the nitrogen-doped graphene decorated with amorphous SnSe quantum dots (a-SnSe/rGO) for SIBs (Fig. 9). The a-SnSe/rGO electrode offers superior cycling reversibility of $397\,mAh\,g^{-1}$ at $1\,A\,g^{-1}$ after 1400 cycles with the lowest cycling capacity fading of 0.014% per cycle. Similarly, tin selenide/N-doped carbon composite anode was developed by Shaji et al. for SIBs [83]. The tin selenide/N-doped carbon exhibited the initial discharge capacity of $460\,mAh\,g^{-1}$ and retained the specific discharge capacity of $348\,mAh\,g^{-1}$ at $200\,mA\,g^{-1}$ for 100 cycles. Moreover, it delivers a discharge capacity of $234\,mAh\,g^{-1}$ at a high current density of $1600\,mA\,g^{-1}$. Finally, the carbon composite of WSe_2/C was also investigated by Zhang et al. as anode materials for SIBs [84]. The WSe_2/C developed a specific capacity of 270 and $208\,mAh\,g^{-1}$ at 200 and $1000\,mA\,g^{-1}$, respectively.

6. Carbon based transition metal telluride composites

Among all the chalcogens, Te exhibits numerous advantages in terms of its conductivity, and the volume change during sodiation is lesser due to its larger atomic size. The smaller volume change inhibits electrode reduction upon repeated charge-discharge cycles [85, 86]. Due to the fast alloying behavior of Te, it can rapidly produce Na_2Te species during sodiation and delivers high theoretical gravimetric ($420\,mAh\,g^{-1}$) and volumetric ($2621\,mAh\,cm^{-3}$) capacity in SIBs. Numerous Te-based anode materials have been investigated, such as SnTe, GeTe, ZnTe, and $CoTe_2$ for LIBs and SIBs [87–91]. Nevertheless, considering the volume change of Te, the electrochemical features of the

190 Oxide free nanomaterials for energy storage and conversion applications

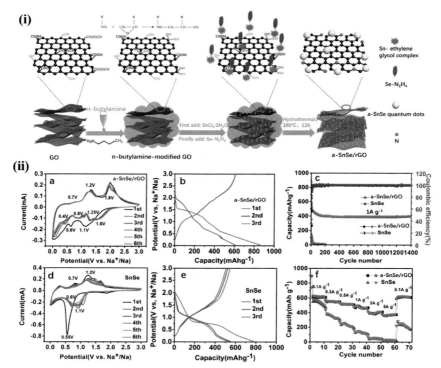

FIG. 9 (i) The schematic illustration for the fabrication of a-SnSe/rGO. (ii)-(a) CV curves of a-SnSe/rGO and (d) SnSe electrode at a scan rate of $0.1\,\mathrm{mV\,s^{-1}}$. (ii)-(b) Charge-discharge profiles of a-SnSe/rGO and (e) SnSe electrode at $0.1\,\mathrm{A\,g^{-1}}$. (c) Cycle stability curves of the SnSe and a-SnSe/rGO electrode at $1\,\mathrm{A\,g^{-1}}$. (f) High-rate capability of the a-SnSe/rGO and SnSe electrode [82].

Te-based composite could be further improved by using carbon-supported nanomaterials and thus relaxing the volume strain of the composite within the carbon matrix. For example, Sb_2Te_3 with functionalized CNTs composite anode was reported by Ihsan-Ul-Haq et al. [92] The Sb_2Te_3/CNT composite offers a uniform and thin solid electrolyte interface (SEI) layer of $\sim$19.1 nm, while the pristine Sb_2Te_3 produces an abnormal SEI with a higher thickness of $\sim$67.3 nm. Further, the Sb_2Te_3/CNT composite electrode provides exceptional reversible gravimetric and volumetric capacities of $422\,\mathrm{mAh\,g^{-1}}$ and $1232\,\mathrm{mAh\,cm^{-3}}$, respectively, at $100\,\mathrm{mA\,g^{-1}}$ with the capacity retention of $\sim$97.5% after 300 cycles. Likewise, the homogeneous agglomeration of $CoTe_2$ nanocrystallites on polyhedral carbon matrix ($CoTe_2$-C) can avoid the recombination of $CoTe_2$ during the electrochemical reaction and control the volume change towards the cycling performance. In addition, the $CoTe_2$-C anode exhibits superior Li- and Na-ion storage performances concerning high reversible capacities and C-rate capability with excellent cyclic stability at a high rate of 1C [91].

7. Transition metal phosphides and their carbon composites

Phosphorous-based materials have also been considered as oxide-free and high-capacity anodes in both LIBs and SIBs applications. Among the three phosphorous (P) allotropes, the red P (amorphous) and black P (orthorhombic), since, the white P is highly reactive and toxic, have been extensively attracted by many researchers due to their advantages of low price, excellent chemical stability and high theoretical capacity (Na_3P: 2595 mAh g^{-1}) [93]. However, these P-based materials mainly suffer from a huge volume change (>400%) during sodiation/desodiation process, it is also accompanied by pulverization and aggregation of particles, leading to poor electrochemical cycling stability [93, 94]. Recently, owing to the high specific capacity and low intercalation potentials, TMPs, such as V, Fe, Ni, Co, Cu, Zn, Mo, Sn, and so on, have been used as anode in SIBs and paid more attention to alleviate the volume changes and to improve the electrochemical conductivity [93–106]. However, the low electron transfer capability has strongly limited their further development that can be significantly overcome by incorporating the conductive carbon support, thus enhances the electrochemical performances of SIBs.

Carbon materials such as graphene, CNTs, CNFs, and other hetero-atom-doped carbon have been utilized as effective conducting and supporting matrixes for improving the electronic and chemical stability properties of TMPs materials. The rate capability and cycle stability of the carbon with TMPs have been extensively improved as compared to pristine TMPs anode. For example, Yin et al. evaluated the sodium storage properties of the $Cu_3P@P/N$-C nanosheets anode at 5 A g^{-1} that exhibits an outstanding cycling performance up to 2000 cycles yet maintained the stable specific capacity of over 118 mAh g^{-1} [103]. Meanwhile, a very rapid capacity loss that drops to 30 mAh g^{-1} after 500 cycles, indicates the structural stability of the materials. Sn_4P_3 has also emerged as a promising SIBs anode in recent years due to its high specific capacity; however, poor electronic and ionic conductivity and large volume change during sodiation limit its practicality [107, 108]. On the other hand, the nanostructure of Sn_4P_3 embedded with a large surface area of best conductive and mechanical support carbon materials has been examined to suppress the volume change and improve cycle stability high current densities [109, 110]. Zhao et al. reported an electrospun synthesis of $Sn_4P_3@C$ nanospheres anchored carbon nanofiber ($Sn_4P_3@CNF$) electrodes for SIBs [111]. In this structural design, the nano-chains are uniformly constructed with hollow porous Sn_4P_3 nanospheres particles, which produce a highly conductive network with good mechanical support. Hence, the $Sn_4P_3@CNF$ electrodes displayed cycle stability over 4700 cycles at 2 A g^{-1} and 14,000 cycles at 5 A g^{-1}. The $Sn_4P_3@CNF$ electrode performed better rate capability than the pure Sn_4P_3 electrode at different current densities range of 0.1–10 A g^{-1}, indicating a superior electron and Na$^+$ ion transport properties of $Sn_4P_3@CNF$ electrode. It clearly indicates that the synergistic role between Sn_4P_3 nanospheres and nano-chain networks boosts the electrochemical performances of SIBs (Fig. 10). The proposed structures also can afford a significant volume change

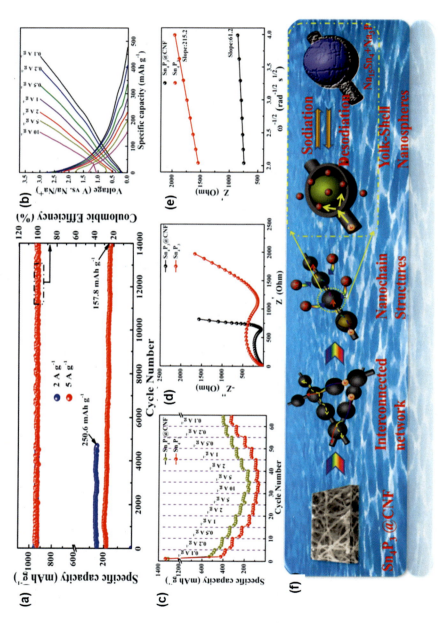

FIG. 10 (A) Cycle stability curve of Sn$_4$P$_3$@CNF electrode at 2 and 5 A g^{-1}. (B) Charge-discharge curves of Sn$_4$P$_3$@CNF electrode at different rates. (C) Rate performances of Sn$_4$P$_3$@CNF and pure Sn$_4$P$_3$ electrode. (D) Nyquist plots of Sn$_4$P$_3$@CNF and pure Sn$_4$P$_3$ electrode and corresponding linear plot of Z' vs. $\omega^{-1/2}$ (E). (F) Schematic illustration of Na$^+$ ion and electron transport at Sn$_4$P$_3$@CNF electrode [111].

during the sodiation/desodiation process. Due to this superior cycling stability and rate performances, the carbon-based TMPs composite anode especially low-cost and high-capacity materials, are a great choice for practical application in SIBs system.

8. Summary

Significant features and progress of oxide-free anode materials in SIBs application have been reviewed and discussed in this chapter. In oxide-free anode materials, carbon-based materials meet significant attention to commercializing the secondary batteries, either LIBs or SIBs. Due to their availability, cost, and appreciated storage capacity, these carbon-based materials brought SIBs development to some extent. To increase the practical capacity of SIBs, the focusing of carbon hybrid materials, i.e., carbon-TMDs or carbon-TMPs, showed noteworthy improvement in electrochemical performance. Because the TMDs and TMPs materials exhibited a very high specific capacity than the only carbon-based anode materials. The chapter clearly reveals that the strategy of carbon coverage and large surface area conductive carbon support provides excellent mechanical support and helps to reduce the volume change properties of TMDs and TMPs materials. The nanostructured carbon materials with different morphology and size effects can afford the volume change of TMDs and TMPs materials during charge-discharge cycling due to the formation of a stable electrode/electrolyte interface layer on the hybrid materials. However, there is a need for fundamental investigation on the interface properties of carbon-based hybrid materials during the cycling process. Such detailed investigation can help us construct potential oxide-free hybrid materials with appropriate interface properties to enhance the Na^+ ion storage in SIBs application. By concerning these significant requirements or advancements in finding effective oxide-free anode materials, there is a still need for more time and effort for SIBs to use practical application.

References

[1] N. Yabuuchi, K. Kubota, M. Dahbi, S. Komaba, Research development on sodium-ion batteries, Chem. Rev. 114 (2014) 11636–11682.

[2] W. Luo, F. Shen, C. Bommier, H. Zhu, X. Ji, L. Hu, Na-ion battery anodes: materials and electrochemistry, Acc. Chem. Res. 49 (2016) 231–240.

[3] M.D. Slater, D. Kim, E. Lee, C.S. Johnson, Sodium-Ion Batteries, Adv. Funct. Mater. 23 (2013) 947–958.

[4] D. Kundu, E. Talaie, V. Duffort, L.F. Nazar, The emerging chemistry of sodium ion batteries for electrochemical energy storage, Angew. Chem. Int. Ed. Engl. 54 (2015) 3431–3448.

[5] C. Nithya, S. Gopukumar, Sodium ion batteries: a newer electrochemical storage, WIREs Energy Environ. 4 (2015) 253–278.

[6] Y. Marcus, Ion Properties, Marcel Dekker, New York, 1997.

[7] C. Delmas, J.-J. Braconnier, C. Fouassier, P. Hagenmuller, Electrochemical intercalation of sodium in NaxCoO2 bronzes, Solid State Ionics 3-4 (1981) 165–169.

[8] H. Pan, Y.-S. Hu, L. Chen, Room-temperature stationary sodium-ion batteries for large-scale electric energy storage, Energy Environ. Sci. 6 (2013) 2338–2360.

[9] D.A. Stevens, J.R. Dahn, High capacity anode materials for rechargeable sodium-ion batteries, J. Electrochem. Soc. 147 (2000) 1271.

[10] D.A. Stevens, J.R. Dahn, The mechanisms of lithium and sodium insertion in carbon materials, J. Electrochem. Soc. 148 (2001) A803.

[11] D. Larcher, J.M. Tarascon, Towards greener and more sustainable batteries for electrical energy storage, Nat. Chem. 7 (2015) 19–29.

[12] K. Chayambuka, G. Mulder, D.L. Danilov, P.H.L. Notten, Sodium-ion battery materials and electrochemical properties reviewed, Adv. Energy Mater. 8 (2018) 1800079.

[13] K. Chayambuka, G. Mulder, D.L. Danilov, P.H.L. Notten, From Li-ion batteries toward Na-ion chemistries: challenges and opportunities, Adv. Energy Mater. 10 (2020) 2001310.

[14] X. Li, A. Dhanabalan, L. Gu, C. Wang, Three-dimensional porous Core-Shell Sn@carbon composite anodes for high-performance lithium-ion battery applications, Adv. Energy Mater. 2 (2012) 238–244.

[15] S.Y. Sayed, W.P. Kalisvaart, E.J. Luber, B.C. Olsen, J.M. Buriak, Stabilizing tin anodes in sodium-ion batteries by alloying with silicon, ACS Appl. Energy Mater. 3 (2020) 9950–9962.

[16] M.-S. Balogun, Y. Luo, W. Qiu, P. Liu, Y. Tong, A review of carbon materials and their composites with alloy metals for sodium ion battery anodes, Carbon 98 (2016) 162–178.

[17] D. Lan, W. Wang, Q. Li, Cu4SnP10 as a promising anode material for sodium ion batteries, Nano Energy 39 (2017) 506–512.

[18] L. Wang, Z. Wei, M. Mao, H. Wang, Y. Li, J. Ma, Metal oxide/graphene composite anode materials for sodium-ion batteries, Energy Storage Mater. 16 (2019) 434–454.

[19] Y. Jiang, M. Hu, D. Zhang, T. Yuan, W. Sun, B. Xu, M. Yan, Transition metal oxides for high performance sodium ion battery anodes, Nano Energy 5 (2014) 60–66.

[20] S. Yuan, X.-l. Huang, D.-l. Ma, H.-g. Wang, F.-z. Meng, X.-b. Zhang, Electrodes: engraving copper foil to give large-scale binder-free porous CuO arrays for a high-performance sodium-ion battery anode (Adv. Mater. 14/2014), Adv. Mater. 26 (2014) 2284.

[21] Y.M. Chang, H.W. Lin, L.J. Li, H.Y. Chen, Two-dimensional materials as anodes for sodium-ion batteries, Mater. Today Adv. 6 (2020) 100054.

[22] Y. Liu, C. Yang, Q. Zhang, M. Liu, Recent progress in the design of metal sulfides as anode materials for sodium ion batteries, Energy Storage Mater. 22 (2019) 66–95.

[23] Y. Xiao, S.H. Lee, Y.-K. Sun, The application of metal sulfides in sodium ion batteries, Adv. Energy Mater. 7 (2017) 1601329.

[24] M. Hu, Z. Ju, Z. Bai, K. Yu, Z. Fang, R. Lv, G. Yu, Revealing the critical factor in metal sulfide anode performance in sodium-ion batteries: an investigation of polysulfide shuttling issues, Small Methods 4 (2020) 1900673.

[25] G. Chang, Y. Zhao, L. Dong, D.P. Wilkinson, L. Zhang, Q. Shao, W. Yan, X. Sun, J. Zhang, A review of phosphorus and phosphides as anode materials for advanced sodium-ion batteries, J. Mater. Chem. A 8 (2020) 4996–5048.

[26] D. Sun, X. Zhu, X. Luo, B. Luo, Y. Zhang, Y. Tang, H. Wang, L. Wang, New binder-free metal phosphide–carbon felt composite anodes for sodium-ion battery, Adv. Energy Mater. 8 (2018) 1801197.

[27] W. Liu, H. Zhi, X. Yu, Recent progress in phosphorus based anode materials for lithium/sodium ion batteries, Energy Storage Mater. 16 (2019) 290–322.

[28] H.-S. Shin, K.-N. Jung, Y.N. Jo, M.-S. Park, H. Kim, J.-W. Lee, Tin phosphide-based anodes for sodium-ion batteries: synthesis via solvothermal transformation of Sn metal and phase-dependent Na storage performance, Sci. Rep. 6 (2016) 26195.

[29] C. Bommier, X. Ji, Recent development on anodes for Na-ion batteries, Isr. J. Chem. 55 (2015) 486–507.

[30] H. Zhang, Y. Huang, H. Ming, G. Cao, W. Zhang, J. Ming, R. Chen, Recent advances in nanostructured carbon for sodium-ion batteries, J. Mater. Chem. A 8 (2020) 1604–1630.

[31] A.S. Aricò, P. Bruce, B. Scrosati, J.-M. Tarascon, W. van Schalkwijk, Nanostructured materials for advanced energy conversion and storage devices, Nat. Mater. 4 (2005) 366–377.

[32] P.G. Bruce, B. Scrosati, J.-M. Tarascon, Nanomaterials for rechargeable lithium batteries, Angew. Chem. Int. Ed. Engl. 47 (2008) 2930–2946.

[33] Y.-G. Guo, J.-S. Hu, L.-J. Wan, Nanostructured materials for electrochemical energy conversion and storage devices, Adv. Mater. 20 (2008) 2878–2887.

[34] K.-l. Hong, L. Qie, R. Zeng, Z.-q. Yi, W. Zhang, D. Wang, W. Yin, C. Wu, Q.-j. Fan, W.-x. Zhang, Y.-h. Huang, Biomass derived hard carbon used as a high performance anode material for sodium ion batteries, J. Mater. Chem. A 2 (2014) 12733–12738.

[35] Z. Wang, L. Zhou, X.W. Lou, Metal oxide hollow nanostructures for lithium-ion batteries, Adv. Mater. 24 (2012) 1903–1911.

[36] X.W. Lou, L.A. Archer, Z. Yang, Hollow Micro-/nanostructures: synthesis and applications, Adv. Mater. 20 (2008) 3987–4019.

[37] H. Han, X. Chen, J. Qian, F. Zhong, X. Feng, W. Chen, X. Ai, H. Yang, Y. Cao, Hollow carbon nanofibers as high-performance anode materials for sodium-ion batteries, Nanoscale 11 (2019) 21999–22005.

[38] D.-S. Bin, Y. Li, Y.-G. Sun, S.-Y. Duan, Y. Lu, J. Ma, A.-M. Cao, Y.-S. Hu, L.-J. Wan, Structural engineering of multishelled hollow carbon nanostructures for high-performance Na-ion battery anode, Adv. Energy Mater. 8 (2018) 1800855.

[39] X. Yan, H. Ye, X.-L. Wu, Y.-P. Zheng, F. Wan, M. Liu, X.-H. Zhang, J.-P. Zhang, Y.-G. Guo, Three-dimensional carbon nanotube networks enhanced sodium trimesic: a new anode material for sodium ion batteries and Na-storage mechanism revealed by ex situ studies, J. Mater. Chem. A 5 (2017) 16622–16629.

[40] A.P.V.K. Saroja, M. Muruganathan, K. Muthusamy, H. Mizuta, R. Sundara, Enhanced sodium ion storage in interlayer expanded multiwall carbon nanotubes, Nano Lett. 18 (2018) 5688–5696.

[41] L. Gao, J. Ma, S. Li, D. Liu, D. Xu, J. Cai, L. Chen, J. Xie, L. Zhang, 2D ultrathin carbon nanosheets with rich N/O content constructed by stripping bulk chitin for high-performance sodium ion batteries, Nanoscale 11 (2019) 12626–12636.

[42] G. Zhao, D. Yu, H. Zhang, F. Sun, J. Li, L. Zhu, L. Sun, M. Yu, F. Besenbacher, Y. Sun, Sulphur-doped carbon nanosheets derived from biomass as high-performance anode materials for sodium-ion batteries, Nano Energy 67 (2020) 104219.

[43] X. Li, X. Hu, L. Zhou, R. Wen, X. Xu, S. Chou, L. Chen, A.-M. Cao, S. Dou, A S/N-doped high-capacity mesoporous carbon anode for Na-ion batteries, J. Mater. Chem. A 7 (2019) 11976–11984.

[44] A. Raj K, M.R. Panda, D.P. Dutta, S. Mitra, Bio-derived mesoporous disordered carbon: an excellent anode in sodium-ion battery and full-cell lab prototype, Carbon 143 (2019) 402–412.

[45] K.C. Wasalathilake, H. Li, L. Xu, C. Yan, Recent advances in graphene based materials as anode materials in sodium-ion batteries, J. Energy Chem. 42 (2020) 91–107.

[46] Z. Yang, J. Ren, Z. Zhang, X. Chen, G. Guan, L. Qiu, Y. Zhang, H. Peng, Recent advancement of nanostructured carbon for energy applications, Chem. Rev. 115 (2015) 5159–5223.

196 Oxide free nanomaterials for energy storage and conversion applications

[47] S. Iijima, Helical microtubules of graphitic carbon, Nature 354 (1991) 56–58.

[48] M. Moniruzzaman, K.I. Winey, Polymer nanocomposites containing carbon nanotubes, Macromolecules 39 (2006) 5194–5205.

[49] D. Goonetilleke, J.C. Pramudita, M. Choucair, A. Rawal, N. Sharma, Sodium insertion/extraction from single-walled and multi-walled carbon nanotubes: the differences and similarities, J. Power Sources 314 (2016) 102–108.

[50] X. Dou, I. Hasa, D. Saurel, C. Vaalma, L. Wu, D. Buchholz, D. Bresser, S. Komaba, S. Passerini, Hard carbons for sodium-ion batteries: structure, analysis, sustainability, and electrochemistry, Mater. Today 23 (2019) 87–104.

[51] Y. Wang, N. Xiao, Z. Wang, Y. Tang, H. Li, M. Yu, C. Liu, Y. Zhou, J. Qiu, Ultrastable and high-capacity carbon nanofiber anodes derived from pitch/polyacrylonitrile for flexible sodium-ion batteries, Carbon 135 (2018) 187–194.

[52] S. Qiu, L. Xiao, M.L. Sushko, K.S. Han, Y. Shao, M. Yan, X. Liang, L. Mai, J. Feng, Y. Cao, X. Ai, H. Yang, J. Liu, Manipulating adsorption–insertion mechanisms in nanostructured carbon materials for high-efficiency sodium ion storage, Adv. Energy Mater. 7 (2017) 1700403.

[53] W. Zhao, X. Hu, S. Ci, J. Chen, G. Wang, Q. Xu, Z. Wen, N-doped carbon nanofibers with interweaved nanochannels for high-performance sodium-ion storage, Small 15 (2019) 1904054.

[54] N.A. Kumar, R.R. Gaddam, S.R. Varanasi, D. Yang, S.K. Bhatia, X.S. Zhao, Sodium ion storage in reduced graphene oxide, Electrochim. Acta 214 (2016) 319–325.

[55] Y.S. Yun, Y.-U. Park, S.-J. Chang, B.H. Kim, J. Choi, J. Wang, D. Zhang, P.V. Braun, H.-J. Jin, K. Kang, Crumpled graphene paper for high power sodium battery anode, Carbon 99 (2016) 658–664.

[56] X. Wei, X. Wang, X. Tan, Q. An, L. Mai, Nanostructured conversion-type negative electrode materials for low-cost and high-performance sodium-ion batteries, Adv. Funct. Mater. 28 (2018) 1804458.

[57] Y. Liu, H. Kang, L. Jiao, C. Chen, K. Cao, Y. Wang, H. Yuan, Exfoliated-SnS_2 restacked on graphene as a high-capacity, high-rate, and long-cycle life anode for sodium ion batteries, Nanoscale 7 (2015) 1325–1332.

[58] C. Zhu, P. Kopold, W. Li, P.A. van Aken, J. Maier, Y. Yu, A general strategy to fabricate carbon-coated 3D porous interconnected metal sulfides: case study of SnS/C nanocomposite for high-performance lithium and sodium ion batteries, Adv. Sci. 2 (2015) 1500200.

[59] H. Yuan, L. Kong, T. Li, Q. Zhang, A review of transition metal chalcogenide/graphene nanocomposites for energy storage and conversion, Chin. Chem. Lett. 28 (2017) 2180–2194.

[60] W. Ren, H. Zhang, C. Guan, C. Cheng, Ultrathin MoS_2 nanosheets@metal organic framework-derived N-doped carbon nanowall arrays as sodium ion battery anode with superior cycling life and rate capability, Adv. Funct. Mater. 27 (2017) 1702116.

[61] Z. Liu, A. Daali, G.-L. Xu, M. Zhuang, X. Zuo, C.-J. Sun, Y. Liu, Y. Cai, M.D. Hossain, H. Liu, K. Amine, Z. Luo, Highly reversible sodiation/desodiation from a carbon-sandwiched SnS_2 nanosheet anode for sodium ion batteries, Nano Lett. 20 (2020) 3844–3851.

[62] N. Zhang, X. Li, T. Hou, J. Guo, A. Fan, S. Jin, X. Sun, S. Cai, C. Zheng, MnS hollow microspheres combined with carbon nanotubes forenhanced performance sodium-ion battery anode, Chin. Chem. Lett. 31 (2020) 1221–1225.

[63] W. Zhang, Z. Yue, Q. Wang, X. Zeng, C. Fu, Q. Li, X. Li, L. Fang, L. Li, Carbon-encapsulated CoS2nanoparticles anchored on N-doped carbonnanofibers derived from ZIF-8/ZIF-67 as anode for sodium-ion batteries, Chem. Eng. J. 380 (2020) 1225482.

[64] R. Sun, Q. Wei, J. Sheng, C. Shi, Q. An, S. Liu, L. Mai, Novel layer-by-layer stacked VS_2 nanosheets with intercalation pseudocapacitance for high-rate sodium ion charge storage, Nano Energy 35 (2017) 396–404.

[65] S. Xu, X. Gao, Y. Hua, A. Neville, Y. Wang, K. Zhang, Rapid deposition of WS_2 platelet thin films as additive-free anode for sodium ion batteries with superior volumetric capacity, Energy Storage Mater. 26 (2020) 534–542.

[66] H. Kim, M.K. Sadan, C. Kim, S.-H. Choe, K.-K. Cho, K.-W. Kim, J.-H. Ahn, H.-J. Ahn, Simple and scalable synthesis of CuS as an ultrafast and long-cycling anode for sodium ion batteries, J. Mater. Chem. A 7 (2019) 16239–16248.

[67] Q. Guo, Y. Ma, T. Chen, Q. Xia, M. Yang, H. Xia, Y. Yu, Cobalt sulfide quantum dot embedded N/S-doped carbon nanosheets with superior reversibility and rate capability for sodium-ion batteries, ACS Nano 11 (2017) 12658–12667.

[68] G.K. Veerasubramani, M.-S. Park, G. Nagaraju, D.-W. Kim, Unraveling the Na-ion storage performance of a vertically aligned interlayer-expanded two-dimensional $MoS_2@C@MoS_2$ heterostructure, J. Mater. Chem. A 7 (2019) 24557–24568.

[69] X. Dong, Z. Xing, G. Zheng, X. Gao, H. Hong, Z. Ju, Q. Zhuang, MoS_2/N-doped graphene aerogles composite anode for highperformance sodium/potassium ion batteries, Electrochim. Acta 339 (2020) 1359322.

[70] Y. Cheng, Z. Wang, L. Chang, S. Wang, Q. Sun, Z. Yi, L. Wang, Sulfur-mediated interface engineering enables fast SnS nanosheet anodes for advanced lithium/sodium-ion batteries, ACS Appl. Mater. Interfaces 12 (2020) 25786–25797.

[71] G.K. Veerasubramani, M.-S. Park, J.-Y. Choi, D.-W. Kim, Ultrasmall SnS quantum dots anchored onto nitrogen-enriched carbon nanospheres as an advanced anode material for sodium-ion batteries, ACS Appl. Mater. Interfaces 6 (2020) 7114–7124.

[72] W. Liu, M. Wei, L. Ji, Y. Zhang, Y. Song, J. Liao, L. Zhang, Hollow carbon sphere based WS_2 anode for high performance lithium andsodium ion batteries, Chem. Phys. Lett. 741 (2020) 1370612.

[73] J. Li, D. Yan, T. Lu, Y. Yao, L. Pan, An advanced CoSe embedded within porous carbon polyhedra hybrid for high performance lithium-ion and sodium-ion batteries, Chem. Eng. J. 325 (2017) 14–24.

[74] H. Kong, C. Lv, Y. Wu, C. Yan, G. Chen, Integration of cobalt selenide nanocrystals with interlayer expanded 3D Se/N Co-doped carbon networks for superior sodium-ion storage, J. Energy Chem. 55 (2021) 169–175.

[75] M. Luo, H. Yu, F. Hu, T. Liu, X. Cheng, R. Zheng, Y. Bai, M. Shui, J. Shu, Metal selenides for high performance sodium ion batteries, Chem. Eng. J. 380 (2020) 122557.

[76] H. Zhang, I. Hasa, S. Passerini, Beyond insertion for Na-ion batteries: nanostructured alloying and conversion anode materials, Adv. Energy Mater. 8 (2018) 1702582.

[77] B. Li, Y. Liu, X. Jin, S. Jiao, G. Wang, B. Peng, S. Zeng, L. Shi, J. Li, G. Zhang, Designed formation of hybrid nanobox composed of carbon sheathed $CoSe_2$ anchored on nitrogen-doped carbon skeleton as ultrastable anode for sodium-ion batteries, Small 15 (2019) 1902881.

[78] Y. Yun, J. Shao, X. Shang, W. Wang, W. Huang, Q. Qu, H. Zheng, Simultaneously formed and embedding-type ternary $MoSe_2/MoO_2$/nitrogen-doped carbon for fast and stable Na-ion storage, Nanoscale Adv. 2 (2020) 1878–1885.

[79] X. Zhang, P. He, B. Dong, N. Mu, Y. Liu, T. Yang, R. Mi, Synthesis and characterization of metal–organic framework/biomass-derived CoSe/C@C hierarchical structures with excellent sodium storage performance, Nanoscale 13 (2021) 4167–4176.

[80] S. Jiang, M. Xiang, J. Zhang, S. Chu, A. Marcelli, W. Chu, D. Wu, B. Qian, S. Tao, L. Song, Rational design of hierarchical $FeSe_2$ encapsulated with bifunctional carbon cuboids as an advanced anode for sodium-ion batteries, Nanoscale 12 (2020) 22210–22216.

198 Oxide free nanomaterials for energy storage and conversion applications

[81] D. Mei, Z. Jian, H. Chun, C. Yang, Q. Jiang, Fe_7Se_8 nanoparticles anchored on N-doped carbon nanofibers as high-rate anode for sodium-ion batteries, Energy Storage Mater. 24 (2020) 439–449.

[82] M. Wang, A. Peng, H. Xu, Z. Yang, L. Zhang, J. Zhang, H. Yang, J. Chen, Y. Huang, X. Li, Amorphous SnSe quantum dots anchoring on graphene as high performance anodes for battery/capacitor sodium ion storage, J. Power Sources 496 (2020) 228414.

[83] N. Shaji, P. Santhoshkumar, H.S. Kang, M. Nanthagopal, J.W. Park, S. Praveen, G.S. Sim, C. Senthil, C.W. Lee, Tin selenide/N-doped carbon composite as a conversion and alloying type anode for sodium-ion batteries, J. Alloys Compd. 834 (2020) 154304.

[84] Z. Zhang, X. Yang, Y. Fu, Nanostructured WSe_2/C composites as anode materials for sodium-ion batteries, RSC Adv. 6 (2016) 12726–12729.

[85] D.A. Grishanov, A.A. Mikhaylov, A.G. Medvedev, J. Gun, P.V. Prikhodchenko, Z.J. Xu, A. Nagasubramanian, M. Srinivasan, O. Lev, Graphene oxide-supported β-tin telluride composite for sodium- and lithium-ion battery anodes, Energy Technol. 6 (2018) 127–133.

[86] J. Zhang, Y.-X. Yin, Y.-G. Guo, High-capacity Te anode confined in microporous carbon for long-life Na-ion batteries, ACS Appl. Mater. Interfaces 7 (2015) 27838–27844.

[87] J.S. Cho, S.Y. Lee, J.K. Lee, Y.C. Kang, Iron telluride-decorated reduced graphene oxide hybrid microspheres as anode materials with improved Na-ion storage properties, ACS Appl. Mater. Interfaces 8 (2016) 21343–21349.

[88] W. Zhang, Q. Zhang, Q. Shi, S. Xin, J. Wu, C.L. Zhang, L. Qiu, C. Zhang, Facile synthesis of carbon-coated porous Sb_2Te_3 nanoplates with high alkali metal ion storage, ACS Appl. Mater. Interfaces 11 (2019) 29934–29940.

[89] A.-R. Park, C.-M. Park, Cubic crystal-structured SnTe for superior Li- and Na-ion battery anodes, ACS Nano 11 (2017) 6074–6084.

[90] K.-H. Nam, G.-K. Sung, J.-H. Choi, J.-S. Youn, K.-J. Jeon, C.-M. Park, New high-energy-density GeTe-based anodes for Li-ion batteries, J. Mater. Chem. A 7 (2019) 3278–3288.

[91] V. Ganesan, K.-H. Nam, C.-M. Park, Robust polyhedral $CoTe_2$–C nanocomposites as high-performance Li- and Na-ion battery anodes, ACS Appl. Energy Mater. 3 (2020) 4877–4887.

[92] M. Ihsan-Ul-Haq, H. Huang, J. Wu, J. Cui, S. Yao, W.G. Chong, B. Huang, J.-K. Kim, Thin solid electrolyte interface on chemically bonded Sb_2Te_3/CNT composite anodes for high performance sodium ion full cells, Nano Energy 71 (2020) 104613.

[93] K.-H. Nam, Y. Hwa, C.-M. Park, Zinc phosphides as outstanding sodium-ion battery anodes, ACS Appl. Mater. Interfaces 12 (2020) 15053–15062.

[94] X. Ren, Y. Zhao, Q. Li, F. Cheng, W. Wen, L. Zhang, Y. Huang, X. Xia, X. Li, D. Zhu, K. Huo, R. Tai, A novel multielement nanocomposite with ultrahigh rate capacity and durable performance for sodium-ion battery anodes, J. Mater. Chem. A 8 (2020) 11598–11606.

[95] K.-H. Kim, J. Choi, S.-H. Hong, Superior electrochemical sodium storage of V_4P_7 nanoparticles as an anode for rechargeable sodium-ion batteries, Chem. Commun. 55 (2019) 3207–3210.

[96] Y.V. Lim, S. Huang, Y. Zhang, D. Kong, Y. Wang, L. Guo, J. Zhang, Y. Shi, T.P. Chen, L.K. Ang, H.Y. Yang, Bifunctional porous iron phosphide/carbon nanostructure enabled high-performance sodium-ion battery and hydrogen evolution reaction, Energy Storage Mater. 15 (2018) 98–107.

[97] Q. Li, J. Yuan, Q. Tan, G. Wang, S. Feng, Q. Liu, Q. Wang, Mesoporous FeP/RGO nanocomposites as anodes for sodium ion batteries with enhanced specific capacity and long cycling life, New J. Chem. 44 (2020) 5396–5403.

Oxides free materials as anodes for sodium-ion batteries **Chapter | 8 199**

[98] C. Wu, P. Kopold, P.A. van Aken, J. Maier, Y. Yu, High performance graphene/Ni_2P hybrid anodes for lithium and sodium storage through 3D yolk-shell-like nanostructural design, Adv. Mater. 29 (2017) 1604015.

[99] Z. Zhao, H. Li, Z. Yang, S. Hao, X. Wang, Y. Wu, Hierarchical Ni_2P nanosheets anchored on three-dimensional graphene as self-supported anode materials towards long-life sodiumion batteries, J. Alloys Compd. 817 (2020) 152751.

[100] W. Zhao, X. Ma, G. Wang, X. Long, Y. Li, W. Zhang, P. Zhang, Carbon-coated CoP_3 nanocomposites as anode materials for high-performance sodium-ion batteries, Appl. Surf. Sci. 445 (2018) 167–174.

[101] Q. Chang, Y. Jin, M. Jia, Q. Yuan, C. Zhao, M. Jia, Sulfur-doped CoP@ nitrogen-doped porous carbon hollow tube as an advanced anode with excellent cycling stability for sodium-ion batteries, J. Colloid Interface Sci. 575 (2020) 61–68.

[102] S. Chen, F. Wu, L. Shen, Y. Huang, S.K. Sinha, V. Srot, P.A. van Aken, J. Maier, Y. Yu, Cross-linking hollow carbon sheet encapsulated CuP_2 nanocomposites for high energy density sodium-ion batteries, ACS Nano 12 (2018) 7018–7027.

[103] Y. Yin, Y. Zhang, N. Liu, B. Sun, N. Zhang, Biomass-derived P/N-co-doped carbon nanosheets encapsulate Cu_3P nanoparticles as high-performance anode materials for sodium–ion batteries, Front. Chem. 8 (2020) 316.

[104] X. Li, W. Li, J. Yu, H. Zhang, Z. Shi, Z. Guo, Self-supported Zn_3P_2 nanowires-assembly bundles grafted on Ti foil as an advanced integrated electrodes for lithium/sodium ion batteries with high performances, J. Alloys Compd. 724 (2017) 932–939.

[105] Z. Huang, H. Hou, C. Wang, S. Li, Y. Zhang, X. Ji, Molybdenum phosphide: a conversion-type anode for ultralong-life sodium-ion batteries, Chem. Mater. 29 (2017) 7313–7322.

[106] C. Fu, H. Yang, G. Feng, L. Wang, T. Liu, In-situ reducing synthesis of MoP@nitrogen-doped carbon nanofibers as an anode material for lithium/sodium-ion batteries, Electrochim. Acta 358 (2020) 136921.

[107] J. Liu, S. Wang, K. Kravchyk, M. Ibáñez, F. Krumeich, R. Widmer, D. Nasiou, M. Meyns, J. Llorca, J. Arbiol, M.V. Kovalenko, A. Cabot, SnP nanocrystals as anode materials for Na-ion batteries, J. Mater. Chem. A 6 (2018) 10958–10966.

[108] J. Saddique, X. Zhang, T. Wu, H. Su, S. Liu, D. Zhang, Y. Zhang, H. Yu, Sn_4P_3-induced crystalline/amorphous composite structures for enhanced sodium-ion battery anodes, J. Mater. Sci. Technol. 55 (2020) 73–80.

[109] L. Ran, B. Luo, I.R. Gentle, T. Lin, Q. Sun, M. Li, M.M. Rana, L. Wang, R. Knibbe, Biomimetic Sn_4P_3 anchored on carbon nanotubes as an anode for high-performance sodium-ion batteries, ACS Nano 14 (2020) 8826–8837.

[110] L. Ran, I. Gentle, T. Lin, B. Luo, N. Mo, M. Rana, M. Li, L. Wang, R. Knibbe, Sn_4P_3@porous carbon nanofiber as a self-supported anode for sodium-ion batteries, J. Power Sources 461 (2020) 228116.

[111] W. Zhao, X. Ma, L. Gao, Y. Li, G. Wang, Q. Sun, Engineering carbon-nanochain concatenated hollow Sn_4P_3 nanospheres architectures as ultrastable and high-rate anode materials for sodium ion batteries, Carbon 167 (2020) 736–745.

Chapter 9

Oxides free materials as anodes for zinc-bromine batteries

Prabhakarn Arunachalam[a], Mabrook S. Amer[a],
Govindhasamy Murugadoss[b], and Abdullah M. Al-Mayouf[a]
[a]*Electrochemical Science Research Chair, Chemistry Department, College of Science, King Saud University, Riyadh, Saudi Arabia,* [b]*Centre for Nanoscience and Nanotechnology, Sathyabama Institute of Science and Technology (Deemed to be University), Chennai, Tamil Nadu, India*

Chapter outline

1. Introduction	201	3.1 Carbon electrode materials in ZBBs	205
2. Principle and structure of Zn-Br$_2$ RFBs	203	4. Future prospects for ZnBr$_2$ RFBs	211
3. Electrodes in ZBBs	205	5. Conclusions	212
		References	212

1. Introduction

Energy storage systems, energy can store in the method of chemical, thermal, electric, or kinetic is absorbed and kept for a while before releasing it to provide energy or power services [1–4]. The main utilization of the energy generated from sustainable and renewable energy resources has a great significance as the probable way out for recent global energy and eco-friendly problems. Regrettably, intrinsic intermittency and trouble in guessing natural renewable resource sources, with sunlight, ocean tides, and wind, can, in the lack of effective energy storage systems [5–7]. EES is a key empowering agent of the smarter grid, and it relies upon coordinating a substantial quantity of renewable energy production, transmission, and distribution. Also, these EES can transform transportation systems where EES devices might supplant the power train systems of current transportation advancements from a chemical fuel-based force train to an electricity-based power train. These EES can connect global and geographical gaps between energy supply and demand when combined with the other energy infrastructure constituents. In this regard, battery-based effective and rapid charge/discharge response are considered major EES, which provides electric energy storage from renewable resources. Its on-demand release must be energy-efficient, safe, consistent, and economical. By the way, redox

Oxide Free Nanomaterials for Energy Storage and Conversion Applications.
https://doi.org/10.1016/B978-0-12-823936-0.00012-7
Copyright © 2022 Elsevier Inc. All rights reserved.

flow batteries (RFBs) are recognized as an encouraging choice for larger-scale. RFBs are identified to be the most compatible with grid ESS [8,9]. Further, RFBs have received huge attention in EESs, particularly for larger-scale electricity storage, owing to greater energy efficiency (EE), cost-effectiveness, and lengthier life [10–13]. In general, RFBs ideally have no activity loss since they mostly depend on the electrolytes' electrochemical reactions and not between the electrode and electrolyte. Also, one of the foremost benefits of this battery is its 100% depth of charge-discharge, and it can be left in an entirely discharged state deprived of loss, specifying that its rated capacity is its real capacity [14,15]. Other benefits are decoupled energy and power, as with fuel cells. RFBs are rechargeable batteries that employ two different liquid electrolyte mediums—one with a positive charge and the other with a negative charge—as energy carriers. More importantly, the electrolyte mediums are separated through an ion-selective membrane, permitting selected ions to pass and complete chemical reactions [16–19]. The distinctive of this method is the total decoupling between power as well as energy ratings. Notably, the power rate is dogged by the membrane's active surface and management of the hydraulic pump. Also, the energy capability is mostly governed by the number of electrolytes employed and the tanks' capacity. The employed electrolyte mediums are separately kept in individual tanks and are propelled into the battery when mandatory [20–22]. The storage capacity of RFBs can be advanced by purely operating larger storage tanks for the electrolytes. Numerous groupings of chemical constituents are probable for the RFBs. Since both the electrolytes, safety issues from these two active materials can be significantly diminished [23,24]. Dependent on the engaged redox couple, RFB makes numerous forms. Particularly, these RFBs have been recognized for ESSs usages, including all-vanadium and other V-based systems [25], hydrogen/bromine (Br_2) [26], zinc/Br_2 [27], all-chromium [28], all-cobalt, zinc/cerium [29,30], sodium/iodine [31], all-copper [32], iron/air [33], zinc/polyiodide [34], iron/Br_2, polythiophene, aqueous-based lithium/Br_2 [35], and even an iron/vanadium system functioning on various redox couples. On the other hand, the current high industrial, working, and maintenance costs of certain RFBs decrease their desirability for commercial usages. To overcome these issues, zinc/Br_2 RFBs battery (ZBB) [27] have many advantages that make them appropriate for a larger scale and provide a comparatively priced battery. The ZBBs can also be the potential route to provide greater energy density ($70\,Wh\,kg^{-1}$). Regardless, ZBBs have revealed favorable prospects for stationary EES applications. The ZBBs is considered as one of the potential technologies for larger-scale EES credited to its high energy density and lower operational costs. Though, it hurts from lower power density, mostly owing to larger internal resistances affected by the lower conductive nature of electrolyte and great polarization in the positive electrode. Further, some common issues still occur that all the ZBBs have met because of the similar plating-stripping process of the zinc couple in the negative half-cell, assisting through charging-discharging of the battery. These disputes contain zinc dendrite and their accumulation [36–40], a limited areal capacity [41], and a comparatively lower working current density.

In recent years, several significant research efforts have been carried out in EES to reduce the operations costs (per unit energy stored). These research works have primarily been dedicated to emerging cost-efficient catalytic materials while still employing typical cell systems that need classical "balance-of plant materials" [1,4,42,43,44a] On the other hand, the unpredicted and undesired effect of merging low-cost catalytic materials with classical passive constituents is that the cost outweighs the active material's cost by a wide-ranging margin. Further, a lower rate of energy ($ per kWh), an extended lifetime, and the least maintenance and working costs are utmost for the achievement of EESs. Vanadium RFBs and ZBBs have been revealed to perform with <20% capacity fade for 10,000 cycles [16], though no distinct battery kind has met the cost-effectiveness, lifespan, and performance targets mandatory for effective EES operation. In terms of holistic techno-economic understanding when seeing EES system strategy, examining: (a) are luxurious passive electrodes compulsory to defend the economical electrochemically active electrode materials? (b) is it usually required to inhibit both the electrode materials from reacting mutually? All the global researchers employ this theoretical outline to re-design the ZBBs and benefit from the physical features of electrochemically active electrode materials.

Herein, we examine and discuss the improvement in energy storage sodium-Br_2 batteries for energy storage systems. This chapter focused on a brief discussion of ZBBs, followed by a comprehensive debate about the problems related to ZBBs. Also, the innovative materials for numerous varieties of ZBBs will be discussed. This book chapter discusses the recent advances in developing carbon-based cathode materials of exceptional features for ZBBs, enhancing the power density, decreasing the stack size of the ZBBs, and increasing its potential for marketable usage. Notably, the influence of these electrode materials on ZBBs features will be elucidated in detail. This chapter's main plan is to provide the researchers with appropriate scenarios of this rapidly emerging field.

2. Principle and structure of Zn-Br$_2$ RFBs

In 1885, the ZBBs were first patented [45] and then demonstrated as hybrid RFBs by Exxon, Gould, and NASA in the 1970s [46]. Theoretically, these ZBBs have a great theoretic energy density ($440\,W\,h\,kg^{-1}$) and a greater cell voltage (ca. 1.8 V). Afterward, several companies have commercialized ZBBs and installed them [47]. These commercial systems' specific energy density is still restricted to $60–80\,Wh\,kg^{-1}$ [48], which is nearly only 20% of the theoretic value.

The device configuration of the ZBBs battery is clearly demonstrated in Fig. 1. The electrodes, electrolyte, and ion-selective membrane are the major

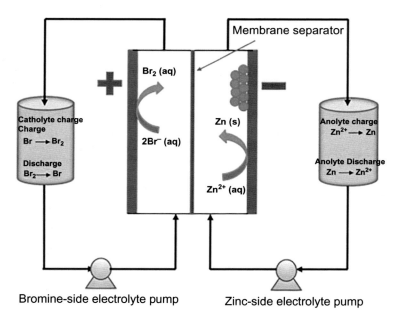

FIG. 1 Charging principal of Redox flow battery.

constituents of ZBBs, and their physical features can significantly influence Zn-dendrite formation. Moreover, the electrolyte medium is stowed externally and circulated over every half-cell. Membranes are usually applied to inhibit Br- and Br_2 crossover between the anode and cathode half-cells. Membrane-free structures are also obtainable, e.g., if the active species concentration is low and a solid-phase electrode reaction is involved [42]. The battery involves $ZnBr_2$ as the active electrode candidate for both the electrode half-cells, avoiding contamination concerns from either side.

At the positive eletrode:

$$2Br^+ \rightarrow Br_2 + 2e^- \ (E° = 1.08V \text{ versus SHE}) \quad (1)$$

At the negative eletrode:

$$Zn^{2+} + 2e^- \rightarrow Zn \ (E° = -0.76 \text{ V versus SHE}) \quad (2)$$

Overall reaction:

$$Zn^{2+} + 2\,Br^- \rightarrow Br_2 + Zn \ (E° = 1.84 \text{ V}) \quad (3)$$

Br_2 molecules produced from the oxidation of bromide anions (Eq. 1) grow to polybromide anions (Br_3^-, Br_5^-, Br_7^-, etc.) (Eq. 2) at the positive electrode through the EC mechanism. At the negative electrode, the Zn^{2+}/Zn redox reactions take place (Eq. 3). The standard rate constants for the Zn^{2+}/Zn and Br_2/Br^- redox couples are 7.5×10^{-5} and $4 \times 10 \text{ s}^{-1}$, correspondingly [49]. The observed difference in the rate constants clearly revealed that the Br_2/Br^- redox couple's kinetics is much more sluggish than that of the Zn^{2+}/Zn, subsequent in incompatible reaction kinetics in ZBBs [50]. For commercial systems, the working pH of ZBBs ranges between 1 and 3.5 and needs to be regulated throughout the operation [51]. At lesser pH (<3), corrosion and hydrogen evolution reactions become more governing in the cathode and become even more exhaustive when pH < 1, while a higher pH might result in a lesser current efficiency [52,53].

The effect of supporting electrolytes on the electrodes [54,55], solubilities, and ionic conductivities [52,56], have been assessed for a wide variety of salts (Na^+, NH_4^+, Cl^-, Br^-, SO_4^{2-}, $H_2PO_4^-$ and NO_3^-). Notably, chloride supporting electrolytes (up to 0.5 M) tend to implement poorer than other electrolytes (Cl^-, Br^-, SO_4^{2-}, $H_2PO_4^-$) counterparts, while NO_3^- electrolytes are not appropriate. Conversely, zinc complexes' ligands might influence the half-cell electrode reactions' electrochemical features [49].

In a classical ZBB system, a porous membrane is placed amongst the positive and negative parts of ZBB acted as a hurdle for Br_2 crossover while permitting the ionic conductivity of Zn^{2+} and Br^- [57]. Some hundred microndense hydrophilic-processed porous polyethylene membranes, namely SF-600 (Asahi Kasei) and Daramic membranes, are usually introduced, seeing the balance amongst ionic conduction and Br_2 crossover. Notably, Nafion no-porous membranes, which are typically employed in all vanadium RFBs [58], can be applied for ZBBs owing to its greater Br_2 blocking capability, as reported by Lai et al. [59] Further, they demonstrated Nafion 115 (127 μm) and Daramic membrane for ZBBs and revealed that the employed Nafion membrane has a greater coulombic efficiency (CE) of 15% however a lower voltage efficiency (VE) of 12% is credited to their higher membrane resistance. Concerning the large membrane resistance, Nafion membranes' production cost can inhibit their usage in ZBBs.

3. Electrodes in ZBBs

The ZBFB has a great theoretic energy density of up to $440\,Wh\,kg^{-1}$ with a higher cell voltage of 1.84 V, due to the great solubility of $ZnBr_2$ and the enormous potential gap between Zn and Br_2, making it favorable materials for industrial EES. Nevertheless, owing to the larger polarization of the positive electrode, the lower power density, and great stack cost [60–63]. In general, metallic electrodes are mostly applied in ZBBs owing to their lower charge transfer resistance. Still, they hurt it from production costs and acute degradation, resulting in the longstanding features and working function of the RFBs [64].

3.1. Carbon electrode materials in ZBBs

Carbonaceous electrode materials with huge specific surface area (SSA), remarkable electrochemical durability are consequently employed as alternate electrode materials [65,66], but as a problem, the exchange current density of the electrochemically active species on the carbonaceous materials is typically one to two orders inferior to that of metallic electrodes [67]. This works offers some basic details about the Br_2/Br^- redox couple reactions for successive studies. Vitreous carbon, graphitic black, carbon felt materials have substantial charge transfer [68]; e.g., ZBBs with Rayon-based carbon felt displayed a double CE of 92.26% than without a felt [69]. Also, they reveal the remarkable potential to perform as the cathode material in ZBBs. The larger polarization of the Br_2 electrode is comparatively high, resulting in a lower power density as well as working current density ($20\,mA\,cm^{-2}$). Numerous research works have been dedicated to improving the electrochemical features of Br_2/Br^- redox couple (the electrochemical activity of $Br_2/Br^- \ll Zn^{2+}/Zn$) to balance the mismatch reaction rates to enhance the power density. Carbonaceous materials such as SWCNT [70], MWCNT [71], and mesostructured carbons [72] are widely applied on the Br_2 electrode reactions to improve the features for Br_2/Br^-. More importantly, the activity of oxygen-functionalized single-wall carbon nanotubes (FSWCNTs) toward Br_2/Br^- reactions was examined [70,71,73]. Bipolar plate type of medium made of polypropylene electrode filling with graphitic black carbon and CNTs have also been reported [74]. Antisymmetric ZBBs with the carbon-surface electrode as zinc electrode and carbon-volume electrode as Br_2 electrode have verified enriched EE and stability [75]. Similarly, significant research efforts have been demonstrated for an asymmetrical cell with a thin and electrochemically active carbon-paper (CP) Br_2 electrode as an alternative for classical thick felt [76]. Lastly, general necessities for the ZBBs materials might be lower charge transfer resistance, lower polarization and cost-effectiveness, good physical durability, and huge SSA, and higher reaction rate.

Mainly, many carbonaceous-based electrode materials have been produced to enhance the redox reaction kinetics of Br_2 reactions. For instance, Wang et al. examined the four different commercially available carbon materials (i.e., expanded graphite, acetylene black, CNTs, and BP2000) for the RFBs. Due to its high SSA, favorable pore size distribution, and degree of graphitization, BP2000 has been found to exhibit the best behavior for Br_2/Br amongst the carbonaceous-based materials. Munaiah et al. have produced electrodes based on CNTs for ZBBs, with which greater rate capacity and efficiency are obtained [71,77]. It should be noted that ZBBs performance is still very poor even with single-wall CNTs compared to VFB, with an EE of just 56.3% at $50\,mA\,cm^{-2}$. Bimodal highly ordered mesostructured carbons for Br_2 reactions have recently been developed and manufactured by Wang et al. Using these mesoporous carbons, the Br_2/Br^- redox reaction kinetic is shown to be substantially improved, and the ZBBs with the obtained carbon-based materials has a greater EE of 80.1% at $80\,mA\,cm^{-2}$ [72].

In recent years, Wu and co-workers examined the features of ZBBs organized with nitrogen-rich carbonized tubular polypyrrole and they achieved the EE of 76.0% at $80\,mA\,cm^{-2}$ [78]. Similarly, the same group reported the nitrogen-doped carbon spheres for the Br_2 reactions, and they revealed the EE of 83.0% at 80 mA cm^{-2} [79]. In contrast, low-dimensional carbon reveals remarkable features when it is introduced to a porous electrode. For instance, Wan et al. demonstrated that 2D mesostructured carbon can efficiently improved the catalytic interface, thereby revealing more exposed electrocatalytic reactive sites and favoring reactants' mass transfer [80].

Wu et al. demonstrated the highly active ZBBs by improving the design of electrolytes and electrodes [54]. Particularly, they reported the EE enhancement of ZBBs from 60.4% to 74.3% at 40 mA cm^{-2} with the introduction of NH_4Cl as a supporting electrolyte. Also, with the combination of thermally treated graphite-felt (GF) electrodes, EE reached up to 81.8% (Fig. 2). These combined approaches over ZBBs demonstrated that it can even operate at a high current density till 80 mA cm^{-2}, thereby delivering the EE of 70%, showing the superior performances of ZBBs.

Recently, Biswas et al. demonstrated minimal-architecture (MA) ZBBs that exclude the luxurious components in traditional methods [17]. Fig. 3 represents the membrane-free MA-ZBBs cell by employing the CF electrodes. They also

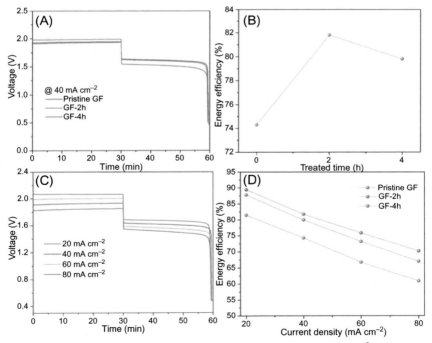

FIG. 2 (A) Charge-discharge plots and (B) energy efficiency (EE) at 40 mA cm^{-2}; (C) charge-discharge plots and (D) EE at numerous current densities. *(Reproduced from M.C. Wu, T.S, Zhao, H.R. Jiang, Y.K. Zeng, Y.X. Ren, High-performance zinc bromine flow battery via improved design of electrolyte and electrode, J. Power Sources 355 (2017) 62–68. Copyright (2021), Elsevier.)*

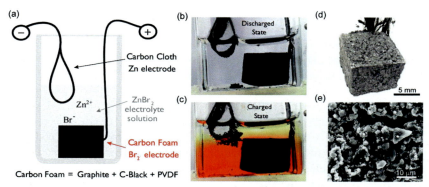

FIG. 3 Minimal architecture (MA)-ZBBs cell (A) schematic representations. Photographic representation of (B) discharged state and (C) charged states of the ZnBr$_2$(aq) electrolyte (clear), dissolved Br$_2$(aq) *(yellow)*, and Br$_2$(l) *(red)*. As the MA-ZBBs cell is membrane-free, Zn-dendrites can grow toward the positive electrode. (D) Photographic representation and (E) SEM photographs of the carbon foam electrode. *(Reproduced from S. Biswas, A. Senju, R. Mohr, T. Hodson, N. Karthikeyan, K.W. Knehr, D.A. Steingart, Minimal architecture zinc–bromine battery for low cost electrochemical energy storage, Energ. Environ. Sci. 10(1) (2017) 114–120, with permission from the Royal Society of Chemistry.)*

reported the membrane-free and low-cost single chamber MA-ZBBs cell that functions steadily with 490% CE and 460% EE for over 1000 cycles. These MA-ZBBs cells can realize closely 9 WhL^{-1} with a production cost of <\$100 kWh^{-1} at scale. The assembled MA-ZBBs cells will be an exciting choice for grid-scale EES systems in terms of techno-economical point of view (Fig. 3).

Table 1 summarizes the recent development of carbon-based electrode materials employed in ZBBs and their energy efficiencies. Consequently, an organized overview of the factors influencing carbonaceous materials' features is generally required to create a theoretical approach to the global researchers for fabricating particular electrode materials with greater electrocatalytically active cathode materials for Br$_2$ reactions. This is particularly noteworthy for reducing battery polarization, enriching EE and CE, and further dropping production cost for developing stack size and efficiently hurrying commercial development.

Quite Recently, Jin and co-workers reported the low-dimensional N$_2$-doped carbon (NOMC) for Br$_2$ reaction in ZBBs [84]. Fig. 4 represents the charge-discharge plots for two-dimensional NOMC materials and their equivalent estimated efficiencies. Also, they reported that fabricated NOMC-2D materials display outstanding durability for the functioning of ZBBs. Further, the assembled ZBBs with NOMC materials obtained a remarkably higher EE of 84.3% at 80 mA cm^{-2}. Also, Yuan et al. reported the aqueous ZBBs EES and compared the performance with Li-ion batteries [91]. They obtained the ZBBs system with

TABLE 1 Review on characteristics comparison of the ZBBs employing carbon-based electrode materials.

Electrode nature	Columbic efficiency CE (%)	Voltage efficiency VE (%)	Energy efficiency EE (%)	Area of active electrode (cm^2)	Current density (mA cm^{-2})	Electrolyte composition	Ref.
AC-graphite felt	99	83	82	5 × 5	30	3M ZnBr$_2$, 3M KCl MEP: MEM	[78]
Graphite felt	98.0	77.5	78.0	2 × 2.5	40	2M ZnBr$_2$	[78]
Graphite felt	97.4	74.9	73.0	2 × 2	40	2M ZnBr$_2$, 2M NaCl	[76]
Graphite felt	97.0	65.0	64.0	–	40	2M ZnBr$_2$	[81]
Pt@graphite felt	99.96	88.12	88.02	5	50	3M ZnBr$_2$	[82]
Graphite felt	–	–	77%	6 × 5	20	3M ZnBr$_2$, 1M KCl	[83]
N doped-activated carbon	–	–	84.3	5 × 5	80	2.0M ZnBr$_2$ + 4.0M NH$_4$Cl	[84]
Mesoporous carbon	–	–	78.5	0.5	120	0.1M ZnBr$_2$ + 1M HCl	[85]
rGO/carbon felt	95	–	80	3 × 3	80	3M ZnBr$_2$ + 1M ZnCl$_2$ + 1:1M MEP	[86]
N-doped graphene nanoplatelets	–	–	84.2	2 × 2.5	120	2M ZnBr$_2$, 4M NH$_4$Cl	[87]

Continued

TABLE 1 Review on characteristics comparison of the ZBBs employing carbon-based electrode materials—cont'd

Electrode nature	Columbic efficiency CE (%)	Voltage efficiency VE (%)	Energy efficiency EE (%)	Area of active electrode (cm^2)	Current density (mA cm^{-2})	Electrolyte composition	Ref.
A porous carbon foam electrode	95	–	75	1×1	5–10	2M ZnBr$_2$	[17]
Graphite felts	97	–	77	5	10	3M ZnBr$_2$, 1M KCl	[83]
Functionalized carbon paper	95	86	78	3×3	20	3M ZnBr$_2$+1M ZnCl$_2$+1:1M MEP: MEM	[88]
Graphite electrode	95.3	84	65	5×7	20	2.25M ZnBr$_2$, 0.5M ZnCl$_2$, 0 (MEP-Br)	[89]
Nafion/PP composite membrane	95.8	77.7	82.10	2×3	20	2.25M ZnBr$_2$, 0.5M ZnCl$_2$, 0.8M MEP, and 5 mL L^{-1} Br$_2$	[90]

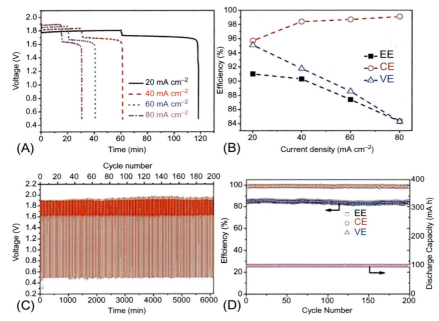

FIG. 4 (A) Charging-discharging plots of ZBBs with 2D NOMCD, (B) Equivalent VE, CE, and EE. (C) Cycling performance of ZBBs with 2D NOMC at 80 mA cm^{-2}, (D) equivalent estimated efficiencies and discharge capacity. *(Reproduced from C.X. Jin, H.Y. Lei, M.Y. Liu, A.D. Tan, J. H. Piao, Z.Y. Fu, ..., H.H. Wang, Low-dimensional nitrogen-doped carbon for Br$_2$/Br$^-$ redox reaction in zinc-bromine flow battery, Chem. Eng. J. 380 (2020) 122606. Copyright, Elsevier.)*

a higher average discharge voltage of 2.15 V and achieved an energy density of 276.7 W kg^{-1}. This overview of the materials mainly offers an effective route to promote the output voltage as well as energy density and remarkably inspires further studies on RFBs (Fig. 4).

4. Future prospects for ZnBr$_2$ RFBs

The ZBBs remain a sustainable substitute for electrical EESs in the marketplace for 10 kW to 10 MW in a techno-economical perspective, quick response to electricity demands, dependability, and resilience. Due to its higher reversibility, specific energy, cell voltage, and EE, the system remains popular in research institutes and the industrial sector, particularly in Asia and Australia. New types of advancements in electrode materials, namely composites, CNTs [92], and electrolyte materials [93], have been anticipated to upturn efficiency. However, there are numerous setbacks, comprising electrode fabrication, reversibility of the electrode (specifically at the Zn electrode owing to phase changes), and Br$_2$ preservation to overcome leakages and sustain active health

212 Oxide free nanomaterials for energy storage and conversion applications

and safer working situations. Despite its challenges, the ZBBs have reached an industrialized scale and been commercialized by numerous companies.

5. Conclusions

The research works on the ZBBs still stay at the earlier stage. In both acidic and alkaline mediums, significant advancement in both electrodes' chemistry and electrochemistry is necessary before becoming commercially applied. These ZBBs are advanced for more than 4 decades and are effectively commercialized in the range of MW scales. This book chapter is titled to give an overview of recent advancements in ZBBs components beyond the realm of science fiction and their role in the improvement of energy storage systems. However, in-depth knowledge of the electrode materials and their energy-level mechanisms and the latest nanotechnology-based applications in energy-storage systems will assists the scientific community understands the selection of carbon-based materials for ZBBs and design new kinds of innovative materials. At the positive electrode, it is essential to agree with sizable losses due to overpotentials. The major setback is to assemble a gas diffusion electrode that can offer an adequate current density at moderate overpotentials and preserve the features through lengthy cycling. However, zinc negative electrodes' storage capacities are still restricted to $<1000\,mAh\,cm^{-2}$ (mostly $<500\,mAh\,cm^{-2}$). This value desires to be further elevated by recognizing appropriate electrolyte compositions and charging approaches.

References

[1] Z. Yang, J. Zhang, M.C. Kintner-Meyer, X. Lu, D. Choi, J.P. Lemmon, J. Liu, Electrochemical energy storage for green grid, Chem. Rev. 111 (5) (2011) 3577–3613.

[2] M. Armand, J.M. Tarascon, Building better batteries, Nature 451 (7179) (2008) 652–657.

[3] J.R. Miller, P. Simon, Electrochemical capacitors for energy management, Sci. Mag. 321 (5889) (2008) 651–652.

[4] B. Dunn, H. Kamath, J.M. Tarascon, Electrical energy storage for the grid: a battery of choices, Science 334 (6058) (2011) 928–935.

[5] G.L. Soloveichik, Flow batteries: current status and trends, Chem. Rev. 115 (20) (2015) 11533–11558.

[6] T. Janoschka, N. Martin, U. Martin, C. Friebe, S. Morgenstern, H. Hiller, U.S. Schubert, An aqueous, polymer-based redox-flow battery using non-corrosive, safe, and low-cost materials, Nature 527 (7576) (2015) 78–81.

[7] D. Larcher, J.M. Tarascon, Towards greener and more sustainable batteries for electrical energy storage, Nat. Chem. 7 (1) (2015) 19–29.

[8] B. Hu, C. DeBruler, Z. Rhodes, T.L. Liu, Long-cycling aqueous organic redox flow battery (AORFB) toward sustainable and safe energy storage, J. Am. Chem. Soc. 139 (3) (2017) 1207–1214.

[9] A. Mauger, M. Armand, C.M. Julien, K. Zaghib, Challenges and issues facing lithium metal for solid-state rechargeable batteries, J. Power Sources 353 (2017) 333–342.

Oxides free materials as anodes for zinc-bromine batteries **Chapter | 9** **213**

[10] D. Chen, M.A. Hickner, E. Agar, E.C. Kumbur, Optimized anion exchange membranes for vanadium redox flow batteries, ACS Appl. Mater. Interfaces 5 (15) (2013) 7559–7566.

[11] C. Ding, H. Zhang, X. Li, T. Liu, F. Xing, Vanadium flow battery for energy storage: prospects and challenges, J. Phys. Chem. Lett. 4 (8) (2013) 1281–1294.

[12] S.H. Shin, S.H. Yun, S.H. Moon, A review of current developments in non-aqueous redox flow batteries: characterization of their membranes for design perspective, RSC Adv. 3 (24) (2013) 9095–9116.

[13] W. Wang, Q. Luo, B. Li, X. Wei, L. Li, Z. Yang, Recent progress in redox flow battery research and development, Adv. Funct. Mater. 23 (8) (2013) 970–986.

[14] A. Patil, V. Patil, D.W. Shin, J.W. Choi, D.S. Paik, S.J. Yoon, Issue and challenges facing rechargeable thin film lithium batteries, Mater. Res. Bull. 43 (8–9) (2008) 1913–1942.

[15] J.M. Tarascon, M. Armand, Issues and challenges facing rechargeable lithium batteries, in: Materials for Sustainable Energy: A Collection of Peer-Reviewed Research and Review Articles From Nature Publishing Group, Nature Publishing Group, 2011, pp. 171–179.

[16] M. Skyllas-Kazacos, M.H. Chakrabarti, S.A. Hajimolana, F.S. Mjalli, M. Saleem, Progress in flow battery research and development, J. Electrochem. Soc. 158 (8) (2011) R55.

[17] S. Biswas, A. Senju, R. Mohr, T. Hodson, N. Karthikeyan, K.W. Knehr, D.A. Steingart, Minimal architecture zinc–bromine battery for low cost electrochemical energy storage, Energ. Environ. Sci. 10 (1) (2017) 114–120.

[18] A.Z. Weber, M.M. Mench, J.P. Meyers, P.N. Ross, J.T. Gostick, Q. Liu, Redox flow batteries: a review, J. Appl. Electrochem. 41 (10) (2011) 1137.

[19] W. Li, Z. Liang, Z. Lu, X. Tao, K. Liu, H. Yao, Y. Cui, Magnetic field-controlled lithium polysulfide semiliquid battery with ferrofluidic properties, Nano Lett. 15 (11) (2015) 7394–7399.

[20] B. Zakeri, S. Syri, Electrical energy storage systems: a comparative life cycle cost analysis, Renew. Sustain. Energy Rev. 42 (2015) 569–596.

[21] G.P. Rajarathnam, M.E. Easton, M. Schneider, A.F. Masters, T. Maschmeyer, A.M. Vassallo, The influence of ionic liquid additives on zinc half-cell electrochemical performance in zinc/bromine flow batteries, RSC Adv. 6 (33) (2016) 27788–27797.

[22] X. Wu, S. Liu, N. Wang, S. Peng, Z. He, Influence of organic additives on electrochemical properties of the positive electrolyte for all-vanadium redox flow battery, Electrochim. Acta 78 (2012) 475–482.

[23] X. Chen, B.J. Hopkins, A. Helal, F.Y. Fan, K.C. Smith, Z. Li, Y.M. Chiang, A low-dissipation, pumpless, gravity-induced flow battery, Energ. Environ. Sci. 9 (5) (2016) 1760–1770.

[24] W.A. Braff, M.Z. Bazant, C.R. Buie, Membrane-less hydrogen bromine flow battery, Nat. Commun. 4 (1) (2013) 1–6.

[25] A. Parasuraman, T.M. Lim, C. Menictas, M. Skyllas-Kazacos, Review of material research and development for vanadium redox flow battery applications, Electrochim. Acta 101 (2013) 27–40.

[26] Y.A. Hugo, W. Kout, G. Dalessi, A. Forner-Cuenca, Z. Borneman, K. Nijmeijer, Techno-economic analysis of a kilo-watt scale hydrogen-bromine flow battery system for sustainable energy storage, Processes 8 (11) (2020) 1492.

[27] ZBB Energy Corp, ZBB Energy: Zn-Br Flow Battery Technology, 2014, Available at http://www.zbbenergy.com.

[28] C. Bae, E.P.L. Roberts, M.H. Chakrabarti, M. Saleem, All-chromium redox flow battery for renewable energy storage, Int. J. Green Energy 8 (2) (2011) 248–264.

[29] X. Xing, D. Zhang, Y. Li, A non-aqueous all-cobalt redox flow battery using 1, 10-phenanthrolinecobalt (II) hexafluorophosphate as active species, J. Power Sources 279 (2015) 205–209.

214 Oxide free nanomaterials for energy storage and conversion applications

[30] F.C. Walsh, C. Ponce de Leon, L. Berlouis, G. Nikiforidis, L.F. Arenas-Martínez, D. Hodgson, D. Hall, The development of Zn-Ce hybrid redox flow batteries for energy storage and their continuing challenges, ChemPlusChem 80 (2) (2015) 288–311.

[31] S. Bhavaraju, Development of sodium-iodine battery for large-scale energy storage, in: Proceedings of the 2013 Electrical Energy Storage Applications & Technologies (EESAT) Biennial International Conference, 2013.

[32] D. Lloyd, E. Magdalena, L. Sanz, L. Murtomäki, K. Kontturi, Preparation of a cost-effective, scalable and energy efficient all-copper redox flow battery, J. Power Sources 292 (2015) 87–94.

[33] R.D. McKerracher, C. Ponce de Leon, R.G.A. Wills, A.A. Shah, F.C. Walsh, A review of the iron–air secondary battery for energy storage, ChemPlusChem 80 (2) (2015) 323–335.

[34] B. Li, Z. Nie, M. Vijayakumar, G. Li, J. Liu, V. Sprenkle, W. Wang, Ambipolar zinc-polyiodide electrolyte for a high-energy density aqueous redox flow battery, Nat. Commun. 6 (1) (2015) 1–8.

[35] Z. Chang, X. Wang, Y. Yang, J. Gao, M. Li, L. Liu, Y. Wu, Rechargeable Li//Br battery: a promising platform for post lithium ion batteries, J. Mater. Chem. A 2 (45) (2014) 19444–19450.

[36] J. Fu, Z.P. Cano, M.G. Park, A. Yu, M. Fowler, Z. Chen, Electrically rechargeable zinc–air batteries: progress, challenges, and perspectives, Adv. Mater. 29 (7) (2017) 1604685.

[37] Z. Liu, T. Cui, G. Pulletikurthi, A. Lahiri, T. Carstens, M. Olschewski, F. Endres, Dendrite-free nanocrystalline zinc electrodeposition from an ionic liquid containing nickel triflate for rechargeable Zn-based batteries, Angew. Chem. Int. Ed. 55 (8) (2016) 2889–2893.

[38] J.F. Parker, C.N. Chervin, E.S. Nelson, D.R. Rolison, J.W. Long, Wiring zinc in three dimensions re-writes battery performance—dendrite-free cycling, Energ. Environ. Sci. 7 (3) (2014) 1117–1124.

[39] S.J. Banik, R. Akolkar, Suppressing dendrite growth during zinc electrodeposition by PEG-200 additive, J. Electrochem. Soc. 160 (11) (2013) D519.

[40] J.F. Parker, C.N. Chervin, I.R. Pala, M. Machler, M.F. Burz, J.W. Long, D.R. Rolison, Rechargeable nickel–3D zinc batteries: an energy-dense, safer alternative to lithium-ion, Science 356 (6336) (2017) 415–418.

[41] K.W. Knehr, S. Biswas, D.A. Steingart, Quantification of the voltage losses in the minimal architecture zinc-bromine battery using GITT and EIS, J. Electrochem. Soc. 164 (13) (2017) A3101.

[42] B. Huskinson, M.P. Marshak, C. Suh, S. Er, M.R. Gerhardt, C.J. Galvin, M.J. Aziz, A metal-free organic–inorganic aqueous flow battery, Nature 505 (7482) (2014) 195–198.

[43] Y. Ito, M. Nyce, R. Plivelich, M. Klein, D. Steingart, S. Banerjee, Zinc morphology in zinc–nickel flow assisted batteries and impact on performance, J. Power Sources 196 (4) (2011) 2340–2345.

[44] S.W. Kim, D.H. Seo, X. Ma, G. Ceder, K. Kang, Electrode materials for rechargeable sodium-ion batteries: potential alternatives to current lithium-ion batteries, Adv. Energy Mater. 2012 (2) (2013) 710. b MD Slater, D. Kim, E. Lee, CS Johnson. Sodium-ion batteries, Adv. Funct. Mater., 23, 947.

[45] C.S. Beadley, Secondary Battery, US Pat. 312802, 1885.

[46] P.C. Butler, D.W. Miller, A.E. Verardo, Flowing-Electrolyte-Battery Testing and Evaluation (No. SAND-82-0381C; CONF-820814-6), Sandia National Labs, Albuquerque, NM, 1982.

[47] A. Khor, P. Leung, M.R. Mohamed, C. Flox, Q. Xu, L. An, A.A. Shah, Review of zinc-based hybrid flow batteries: from fundamentals to applications, Mater. Today Energy 8 (2018) 80–108.

Oxides free materials as anodes for zinc-bromine batteries Chapter | 9 **215**

[48] G.P. Rajarathnam, The Zinc/Bromine Flow Battery: Fundamentals and Novel Materials for Technology Advancement, University of Sydney, 2016.

[49] X. Li, Modeling and simulation study of a metal free organic–inorganic aqueous flow battery with flow through electrode, Electrochim. Acta 170 (2015) 98–109.

[50] S. Suresh, T. Kesavan, Y. Munaiah, I. Arulraj, S. Dheenadayalan, P. Ragupathy, Zinc–bromine hybrid flow battery: effect of zinc utilization and performance characteristics, RSC Adv. 4 (71) (2014) 37947–37953.

[51] G.P. Rajarathnam, A.M. Vassallo, The Zn-Br Flow Battery-Materials Challenges and Practical Solutions for Technology Advancement, University of Sydney, 2015.

[52] H.S. Lim, A.M. Lackner, R.C. Knechtli, Zinc-bromine secondary battery, J. Electrochem. Soc. 124 (8) (1977) 1154.

[53] C. Cachet, R. Wiart, Zinc deposition and passivated hydrogen evolution in highly acidic sulphate electrolytes: depassivation by nickel impurities, J. Appl. Electrochem. 20 (6) (1990) 1009–1014.

[54] M.C. Wu, T.S. Zhao, H.R. Jiang, Y.K. Zeng, Y.X. Ren, High-performance zinc bromine flow battery via improved design of electrolyte and electrode, J. Power Sources 355 (2017) 62–68.

[55] G.P. Rajarathnam, M. Schneider, X. Sun, A.M. Vassallo, The influence of supporting electrolytes on zinc half-cell performance in zinc/bromine flow batteries, J. Electrochem. Soc. 163 (1) (2015) A5112.

[56] R. Zito Jr., U.S. Patent No. 4,482,614, U.S. Patent and Trademark Office, Washington, DC, 1984.

[57] J. Noack, N. Roznyatovskaya, T. Herr, P. Fischer, The chemistry of redox-flow batteries, Angew. Chem. Int. Ed. 54 (34) (2015) 9776–9809.

[58] B. Jiang, L. Wu, L. Yu, X. Qiu, J. Xi, A comparative study of Nafion series membranes for vanadium redox flow batteries, J. Membr. Sci. 510 (2016) 18–26.

[59] Q. Lai, H. Zhang, X. Li, L. Zhang, Y. Cheng, A novel single flow zinc–bromine battery with improved energy density, J. Power Sources 235 (2013) 1–4.

[60] C. Wang, Q. Lai, P. Xu, D. Zheng, X. Li, H. Zhang, Cage-like porous carbon with superhigh activity and Br2-complex-entrapping capability for bromine-based flow batteries, Adv. Mater. 29 (22) (2017) 1605815.

[61] K. Cedzynska, Properties of modified electrolyte for zinc-bromine cells, Electrochim. Acta 40 (8) (1995) 971–976.

[62] G. Bauer, J. Drobits, C. Fabjan, H. Mikosch, P. Schuster, Raman spectroscopic study of the bromine storing complex phase in a zinc-flow battery, J. Electroanal. Chem. 427 (1–2) (1997) 123–128.

[63] L. Zhang, H. Zhang, Q. Lai, X. Li, Y. Cheng, Development of carbon coated membrane for zinc/bromine flow battery with high power density, J. Power Sources 227 (2013) 41–47.

[64] R.A. Putt, Assessment of Technical and Economic Feasibility of Zinc/Bromine Batteries for Utility Load Leveling. Final Report Gould, 1979.

[65] H.R. Jiang, M.C. Wu, Y.X. Ren, W. Shyy, T.S. Zhao, Towards a uniform distribution of zinc in the negative electrode for zinc bromine flow batteries, Appl. Energy 213 (2018) 366–374.

[66] H.R. Jiang, W. Shyy, M.C. Wu, L. Wei, T.S. Zhao, Highly active, bi-functional and metal-free B4C-nanoparticle-modified graphite felt electrodes for vanadium redox flow batteries, J. Power Sources 365 (2017) 34–42.

[67] Y. Shao, M. Engelhard, Y. Lin, Electrochemical investigation of polyhalide ion oxidation–reduction on carbon nanotube electrodes for redox flow batteries, Electrochem. Commun. 11 (10) (2009) 2064–2067.

216 Oxide free nanomaterials for energy storage and conversion applications

[68] A.G. Pandolfo, A.F. Hollenkamp, Carbon properties and their role in supercapacitors, J. Power Sources 157 (1) (2006) 11–27.

[69] S. Suresh, M. Ulaganathan, N. Venkatesan, P. Periasamy, P. Ragupathy, High performance zinc-bromine redox flow batteries: role of various carbon felts and cell configurations, J. Energy Storage 20 (2018) 134–139.

[70] Y. Munaiah, S. Dheenadayalan, P. Ragupathy, V.K. Pillai, High performance carbon nanotube based electrodes for zinc bromine redox flow batteries, ECS J. Solid State Sci. Technol. 2 (10) (2013) M3182.

[71] Y. Munaiah, S. Suresh, S. Dheenadayalan, V.K. Pillai, P. Ragupathy, Comparative Electrocatalytic performance of single-walled and multiwalled carbon nanotubes for zinc bromine redox flow batteries, J. Phys. Chem. C 118 (27) (2014) 14795–14804.

[72] C. Wang, X. Li, X. Xi, W. Zhou, Q. Lai, H. Zhang, Bimodal highly ordered mesostructure carbon with high activity for Br2/Br − redox couple in bromine based batteries, Nano Energy 21 (2016) 217–227.

[73] X. Rui, A. Parasuraman, W. Liu, D.H. Sim, H.H. Hng, Q. Yan, M. Skyllas-Kazacos, Functionalized single-walled carbon nanotubes with enhanced electrocatalytic activity for Br-/Br3- redox reactions in vanadium bromide redox flow batteries, Carbon 64 (2013) 464–471.

[74] W.I. Jang, J.W. Lee, Y.M. Baek, O.O. Park, Development of a PP/carbon/CNT composite electrode for the zinc/bromine redox flow battery, Macromol. Res. 24 (3) (2016) 276–281.

[75] Y. Kim, J. Jeon, An antisymmetric cell structure for high-performance zinc bromine flow battery, J. Phys. Conf. Ser. 939 (1) (2017) 012021. IOP Publishing.

[76] M. Wu, T. Zhao, R. Zhang, H. Jiang, L. Wei, A zinc–bromine flow battery with improved design of cell structure and electrodes, Energ. Technol. 6 (2) (2018) 333–339.

[77] Y. Munaiah, P. Ragupathy, V.K. Pillai, Single-step synthesis of halogenated graphene through electrochemical exfoliation and its utilization as electrodes for zinc bromine redox flow battery, J. Electrochem. Soc. 163 (14) (2016) A2899.

[78] M.C. Wu, T.S. Zhao, R.H. Zhang, L. Wei, H.R. Jiang, Carbonized tubular polypyrrole with a high activity for the Br2/Br − redox reaction in zinc-bromine flow batteries, Electrochim. Acta 284 (2018) 569–576.

[79] H.X. Xiang, A.D. Tan, J.H. Piao, Z.Y. Fu, Z.X. Liang, Efficient nitrogen-doped carbon for zinc–bromine flow battery, Small 15 (24) (2019) 1901848.

[80] K. Wan, A.D. Tan, Z.P. Yu, Z.X. Liang, J.H. Piao, P. Tsiakaras, 2D nitrogen-doped hierarchically porous carbon: key role of low dimensional structure in favoring electrocatalysis and mass transfer for oxygen reduction reaction, Appl. Catal. Environ. 209 (2017) 447–454.

[81] M.C. Wu, T.S. Zhao, L. Wei, H.R. Jiang, R.H. Zhang, Improved electrolyte for zinc-bromine flow batteries, J. Power Sources 384 (2018) 232–239.

[82] K. Mariyappan, R. Velmurugan, B. Subramanian, P. Ragupathy, M. Ulaganathan, Low loading of Pt@ graphite felt for enhancing multifunctional activity towards achieving high energy efficiency of Zn–Br2 redox flow battery, J. Power Sources 482 (2021) 228912.

[83] K.S. Archana, R. Pandiyan Naresh, H. Enale, V. Rajendran, A.V. Mohan, A. Bhaskar, D. Dixon, Effect of positive electrode modification on the performance of zinc-bromine redox flow batteries, J. Energy Storage 29 (2020) 101462.

[84] C.X. Jin, H.Y. Lei, M.Y. Liu, A.D. Tan, J.H. Piao, Z.Y. Fu, H.H. Wang, Low-dimensional nitrogen-doped carbon for Br$_2$/Br − redox reaction in zinc-bromine flow battery, Chem. Eng. J. 380 (2020) 122606.

[85] M.C. Wu, R.H. Zhang, K. Liu, J. Sun, K.Y. Chan, T.S. Zhao, Mesoporous carbon derived from pomelo peel as a high-performance electrode material for zinc-bromine flow batteries, J. Power Sources 442 (2019) 227255.

[86] S. Suresh, M. Ulaganathan, R. Pitchai, Realizing highly efficient energy retention of Zn–Br_2 redox flow battery using rGO supported 3D carbon network as a superior electrode, J. Power Sources 438 (2019) 226998.

[87] M.C. Wu, H.R. Jiang, R.H. Zhang, L. Wei, K.Y. Chan, T.S. Zhao, N-doped graphene nanoplatelets as a highly active catalyst for Br2/Br − redox reactions in zinc-bromine flow batteries, Electrochim. Acta 318 (2019) 69–75.

[88] S. Suresh, M. Ulaganathan, R. Aswathy, P. Ragupathy, Enhancement of bromine reversibility using chemically modified electrodes and their applications in zinc bromine hybrid redox flow batteries, ChemElectroChem 5 (22) (2018) 3411–3418.

[89] S. Bae, J. Lee, D.S. Kim, The effect of Cr^{3+}-functionalized additive in zinc-bromine flow battery, J. Power Sources 413 (2019) 167–173.

[90] R. Kim, H.G. Kim, G. Doo, C. Choi, S. Kim, J.H. Lee, H.T. Kim, Ultrathin Nafion-filled porous membrane for zinc/bromine redox flow batteries, Sci. Rep. 7 (1) (2017) 1–8.

[91] X. Yuan, J. Mo, J. Huang, J. Liu, C. Liu, X. Zeng, Y. Wu, An aqueous hybrid zinc-bromine battery with high voltage and energy density, ChemElectroChem 7 (2020) 1531–1536.

[92] J.H. Yang, H.S. Yang, H.W. Ra, J. Shim, J.D. Jeon, Effect of a surface active agent on performance of zinc/bromine redox flow batteries: improvement in current efficiency and system stability, J. Power Sources 275 (2015) 294–297.

[93] D. Kim, J. Jeon, Study on durability and stability of an aqueous electrolyte solution for zinc bromide hybrid flow batteries, J. Phys. Conf. Ser. 574 (1) (2015) 012074. IOP Publishing.

Chapter 10

Metal nitrides and carbides as advanced counter electrodes for dye-sensitized solar cells

Meenakshamma Ambapuram, Gurulakshmi Maddala, and Raghavender Mitty
Department of Physics, Yogi Vemana University, Kadapa, Andhra Pradesh, India

Chapter outline

1. Introduction 219
 1.1 Characterization of counter electrodes 220
 1.2 Photovoltaic measurements 227
2. Advancement of metal nitrides and carbides as a counter
electrode for dye-sensitized solar cells 228
 2.1 Metal nitrides 229
 2.2 Metal carbides 235
3. Summary 249
References 255

1. Introduction

Photovoltaics (PV) are widely studied for the last five decades, commercialized for the usage of domestic and electrical energy production processes. Though, the cost of solar power obtained from existing well-established silicon technology is still suggestively higher than the electrical grid, consequently prohibiting its large-scale usage. Photovoltaics are directly transmuted solar energy to electrical energy, epitomize promising renewable substitutes to fossil fuels. Third-generation solar cell technologies like organic photovoltaics (OPV), dye-sensitized solar cells (DSSCs), quantum-dot solar cells (Q-DSC), and perovskite solar cells (PSC) emerged as cost-effective substitutes to replace silicon-based solar cells. In which, DSSCs with attributes like lower production costs, simple and easier fabrication protocols, improved performance under diffused light drive the DSSCs to supersede in PV field.

To realize the wide-scale applicability of DSSCs technology in various applications, fabrication of prototype device, its subsequent modules have to be developed. A breakthrough has been made and reported in Nature paper

Oxide Free Nanomaterials for Energy Storage and Conversion Applications.
https://doi.org/10.1016/B978-0-12-823936-0.00013-9
Copyright © 2022 Elsevier Inc. All rights reserved.

by O'Regan and Grätzel [1] in 1991, and recently the efficiency levels 14.3% [2] at lab scale and 10% in its modules [3,4] under 1 sun irradiation (100mW/cm^2, AM 1.5G) condition have been obtained by using the well-researched Ru(II) polypyridyl complexes. DSSCs fit in environment-friendly devices, consists of (i) the photoelectrode (working electrode) made by use of a fluorine doped tin oxide (FTO) substrate with a thin and mesoporous semiconductor TiO$_2$ film, which is anchored with dye sensitizers; (ii) the counter electrode, a thin catalyzer film (often Platinum) over FTO substrate; and (iii) a redox couple (generally I_3^-/I^-) based electrolyte solution filled between the photoelectrode and counter electrode. Related to the counter electrode catalysts are concerned, (i) superior catalytic action, (ii) larger specific surface area, (iii) higher electrical conductivity, (iv) more chemical stability, (v) higher corrosion resistance, (vi) healthier matching factor, etc., are prerequired towards efficient DSSCs performance, are essential for supplant of a platinum-free catalyzer in order to achieve best counter electrode functionality.

Counter electrodes development strategies for DSSCs application are reported, presented in Figs. 1–3.

1.1 Characterization of counter electrodes

To realize the better reaction processed mechanism, as the charge generation, its transfer, recombination progression [5–8], towards in design of novel catalytic materials and improve photovoltaic concert of dye-sensitized solar cells, an assortment of characterization approaches have been advanced [9–11]. Electrochemical/photoelectrochemical characterization has been explored as powerful

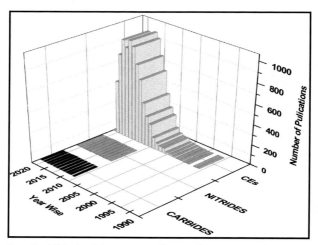

FIG. 1 Number of published articles per year. *(Source: Scopus, search keywords: Counter electrodes for dye-sensitized solar cells (as on 04th November 2020).)*

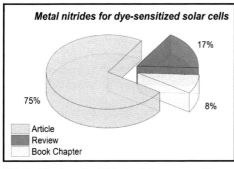 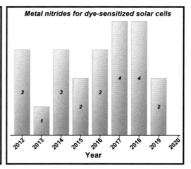

FIG. 2 Number of published articles per year. *(Source: Scopus, search keywords: Nitrides counter electrodes for dye-sensitized solar cells (as on 04th November 2020).)*

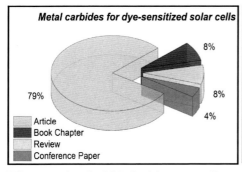 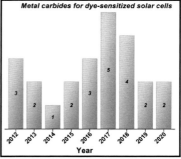

FIG. 3 Number of published articles per year. *(Source: Scopus, search keywords: Carbides counter electrodes for dye-sensitized solar cells (as on 04th November 2020).)*

technique in examining the process mechanism, evaluating constituent interactions, and estimating the concerts of DSSCs [7,12,13]. Includes cyclic-voltammetry *(C-V)*, electrochemical impedance spectroscopy *(EIS)*, intensity-modulated photovoltage spectroscopy *(IMVS)*, intensity-modulated photocurrent spectroscopy *(IMPS)*, and current density-voltage *(J-V)* characterization, etc.

1.1.1 Cyclic-voltammetry

Electrochemical (EC) studies are well used to characterize all the components of the DSSCs. EC analysis provides significant evidence of data on the energy levels of the components, the reversibility of electrochemical reactions, and the kinetics of electrochemical processes. Cyclic-voltammetry is an imperative tool for examining counter electrodes [14–16]. Three electrode arrangement is extensively used to carry out *C-V* contains a working electrode, reference electrode, and auxiliary (counter) electrode. A constant rate of potential is swept, inverted at a certain point, though the current is supervised incessantly. The potential is noted among the working electrode, reference electrode, whereas the current is noted between working and counter electrodes; and obtained resulted information is plotted as current (i) versus scanned potential.

A typical C-V is revealed in Fig. 4, the wave shape of even reversible redox couples is complex due to collective effects of diffusion, polarization, and rate of electron transfer. The potential difference among the two peaks as $\Delta E_P = |E_{PC} - E_{PA}|$ is vital data. As per the theory proposed by Nicholson in the year 1965, the ΔE_P is negatively associated with the redox reaction rate, and the electron transfer rate could be evaluated using ΔE_P, and this concept suggests for a lesser ΔE_P in a quicker redox reaction, signifies an improved electrocatalytic performance [15]. In an idyllic reversible couple for an "n" electron course system the $\Delta E_P = E_{PA} - E_{PC} = 56.51$ mV/n [14] and through an experiment results are higher. Nevertheless, in one electron processed system, ΔE_P comes near ≈ 75 mV. The difference among theory, practical resulted value is considered as 'activation barrier' or the "overpotential" for electron transfer. Higher overpotential is a disadvantage for redox reaction.

Although for current, the reversible pairs are considered by $i_{PA}/i_{PC} = 1$, which indicates the large revocable redox couple is, the analogous the oxidation peak could be in shape to reduction peak. While in reversible peak is evidenced, the thermodynamic potential is a half-cell potential which is equal to $E^0_{1/2}$ could be evaluated. Well researched studies reveal redox processes detected through C-V characterization are quasi-reversible, in such cases the potential $E^0_{1/2}$ is evaluated through simulation. The irreversibility is

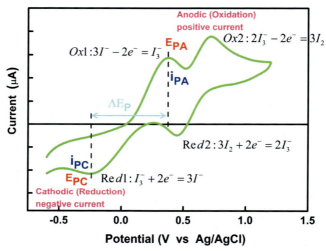

FIG. 4 Typical cyclic-voltammogram of the counter electrode of dye-sensitized solar cell with iodide/tri-iodide electrolyte.

designated as $i_{PA}/i_{PC} \neq 1$, and irreversible deviation is attributed to a successive chemical reaction is impelled by the transfer of electrons. These electrochemical progressions are complex, involved like dissociation, deposition, isomerization, etc., and the scan rate regulates the peak current. As the transfer of electrons at the surface of electrodes are faster, and current is minimal through the diffusion of measured species to electrodes surfaces, and the peak current is proportional to the square root of the scan rate. The relation of peak diffusion coefficient (D_n), current density (J_{red}), scan rate (n) could be articulated by an Eq. (1) [17,18]

$$J_{red} = KCAn^{1.5}D_n^{0.5}\gamma^{0.5} \tag{1}$$

Here K is constant (2.69×10^5), n is the number of electrodes contributive to the transfer of charges, C is the concentration of redox species, A is the active area of the electrode. In DSSCs, the iodide/tri-iodide couple is generally used as a redox medium in liquid electrolytes. In general, the C-V curves relate to iodide/tri-iodide systems consists of two pairs of redox peaks. The left pair represents a reaction denoted as Eq. (2) and the right pair refers to the reaction signified Eq. (3). The reaction of Eq. (2) had a substantial impact on DSSCs performance, and this could be provoked more, and in the anode reaction [19,20].

$$I_3^- + 2e^- \rightarrow 3I^- \tag{2}$$

$$3I_2 + 2e^- \rightarrow 2I_3^- \tag{3}$$

1.1.2 Tafel polarization

Tafel-polarization measurement is a potential characterization is to be carried for a symmetrical dummy cell analogous to a cell used in *EIS* studies for obtains counter electrode performance, a resulted Tafel curve shown in Fig. 5 of the dummy cell. It expresses the logarithmic current density (Log J) versus applied voltage (V). At a lower sweep rate, the limit in diffusion current density (J_{lim}) depends on the diffusion coefficient of tri-iodide species in an electrolyte [21,22].

The J_{lim} values of a CEs are the equal magnitude conferring to Eq. (4). This enlightens that there is an insignificant difference of diffusion coefficients among counter electrodes.

$$D = \left(\frac{l}{2nFC}\right)J_{lim} \tag{4}$$

Here D is represents the diffusion coefficient of tri-iodide, and l is the thickness of the spacer which placed between the counter electrodes, n represents the number of electrons involved in the reduction of tri-iodide at the electrode side, C is the tri-iodide concentration, F relates the Faraday's constant. The anodic

224 Oxide free nanomaterials for energy storage and conversion applications

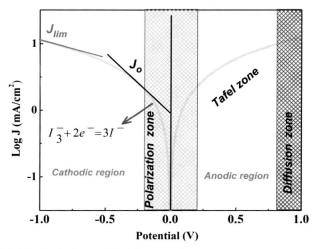

FIG. 5 Tafel plot of typical symmetrical dummy cell.

and cathodic twigs of the Tafel plot demonstrate a large slope for potential counter electrodes, indicating a higher exchange current density (J_o) on the CE surface. This illustration the better electrodes could catalyze the reduction of I_3^- to I^-. The J_o is inverse proportion to R_{CT} associated on Eq. (5). The collective *EIS* results, change propensity of R_{CT} is in agreement with the Tafel studies. Momentarily, the *EIS* outcomes and Tafel results well explain the DSSCs performance.

$$J_o = RT/nFR_{CT} \qquad (5)$$

Here, R represents the gas constant.

1.1.3 Electrochemical impedance spectroscopy

Electrochemical impedance spectroscopy (*EIS*) is an influential method to study the kinetics of hole-electron recombination, charge transport for evaluating the performance of electrocatalytic behavior of CEs of DSSCs [23–26]. A small potential is applied to the device, is distorted by a small sine wave variation, and the results in sinusoidal current response (amplitude, phase shift) is noted as a function of modulation frequency. The applied perturbation and noted results response, the impedance is evaluated. Generally, impedance is defining the frequency domain ratio of the voltage to the current. By choosing an equivalent/suitable circuit, measured data are fitted using various software, the electrochemical parameters of charge transfer resistance (R_{CT}), series resistance (R_S), constant phase element (*CPE*), and diffusion resistance (Z_W) are attained [26]. Results of *EIS* could be represented in the Nyquist plot (Fig. 6) or a Bode-Phase plot (Fig. 7). In a Nyquist plot belongs to DSSC, usually, three semicircles are witnessed. The contact of the first semicircle intercept on the

Metal nitrides and carbides as advanced counter electrodes **Chapter | 10** 225

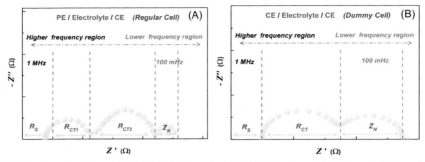

FIG. 6 Nyquist plots of (A) DSSCs fabricated with PE-CE (regular cell); (B) DSSC fabricated with CE-CE (dummy cell).

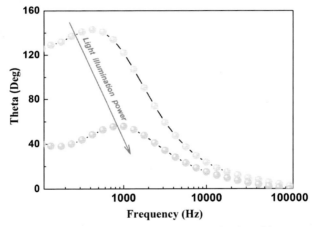

FIG. 7 Bode-phase curves of DSSC under different illumination intensities.

X-axis at the higher frequency side represents series resistance triggered by the conductive substrate, wires, etc.); the same first semicircle intercept on X-axis at the lower frequency side denotes charge transfer resistance (R_{CT1}) belongs to CE/electrolyte interface. The second semicircle intercept on X-axis (middle frequency) relates to charge transfer resistance (R_{CT2}) of the photoelectrode/dye/electrolyte interface. Whereas the third semicircle (at low frequency) intercept represents redox species diffusion resistance (Z_W) [18,26]. Due to the insignificant distance among two electrodes, and lower viscosity (of the electrolyte), the third semicircle is every so frequently not detected. Bode-phase plots are also used for evaluating lifetime of charge carriers (electrons) in DSSCs. By changing intensities the lifetime could be very. As the decrease in light intensity, lower frequency peak shifts to low-frequency region of Bode plot, inferring a fashion of longer time constant.

226 Oxide free nanomaterials for energy storage and conversion applications

1.1.4 Intensity-modulated photocurrent and photovoltage spectroscopy

Intensity-modulated photovoltage spectroscopy (*IMVS*) and intensity-modulated photocurrent spectroscopy (*IMPS*) are photoelectrochemical methods, equitably analogous to *EIS* studies. These are belonging to the control modulation of incident light intensity. An input signal is made of stable background light signals combined with a beam of light disturbed by a small sine wave modulation, these are overlaid on the test sample. The resulted output signal is relating a steady-state photocurrent, and modulated photocurrent (short-circuit mode) or steady-state photovoltage and modulated photovoltage (open-circuit mode). By connecting the frequency response of input, output signal amplitude, phase, and equivalent fitted data obtained from results of *IMVS*, *IMPS*, the electron lifetime (τ_n), electron diffusion coefficient (D_n), absorption coefficient (α) could be gained, are gave useful tackles to study the electron transporting nature and charge transfer kinetics in dye-sensitized solar cells [27–34]. In a short-circuit mode, the photocurrent response is restrained at different frequencies in *IMPS*, electron transport time (τ_d) could be defined as Eq. (6)

$$\tau_d \frac{1}{2\pi f_{min\,IMPS}} \tag{6}$$

Here $f_{min\,IMPS}$ is the frequency of the lower point of the imaginary part of the *IMPS* spectrum. In an open-circuit situation, the photovoltage response is restrained at diverse frequencies in *IMVS* studies, the electron lifetime (τ_n) express as Eq. (7)

$$\tau_n \frac{1}{2\pi f_{min\,IMVS}} \tag{7}$$

$f_{min\,IMVS}$ is the frequency of the lower point of the imaginary part of the *IMVS* spectrum. Charge collection efficiency (η_{cc}) is presented as Eq. (8)

$$\eta_{cc} = \frac{1}{1 + (\tau_d + \tau_n)} \tag{8}$$

The lower the τ_d, and higher τ_n, and the greater is the η_{cc}, which is conducive to enhance the DSSCs performance.

EIS and *IMVS* or *IMPS* belong to frequency domain practice. In the *IMVS/IMPS* trials the voltage/current response to a modulated light intensity is superimposed on steady light intensity, while *EIS* deals the current retort to a modulated applied bias which superimposed on a constant applied voltage.

The results of *IMPS/IMVS* are erected built on models, the equations; in case of *EIS* is evaluated using capacitance, resistance elements as its equivalent circuit [32,34–37]. The EIS and *IMVS/IMPS* contains forte, faintness in the study of transport, recombination. *EIS* is a technique that relates the research on the electron transporting nature in TiO$_2$ films and may combine with other procedures [38–40]. Nevertheless, the *IMPS* is appropriate to study electron transport in TiO_2 films in presence of light intensities [41,42]. Whereas, the electron transport across TiO_2/dye/electrolyte boundaries and electrolyte/Pt-TCO (transparent conductive oxide) interfaces, be sensed through *EIS* at any bias under illumination conditions or dark conditions. The *IMVS* primarily comprises the electron transfer progression across the TiO_2/dye/electrolyte interface at open-circuit under illumination. The *IMVS* obtained from the characteristic time for recombination of an electron, this method is not precious by TCO assisted glass substrates or the CE [26]. Dai et al and his team members reported [43] a relation of series resistance and dynamic process of Dye-sensitized solar cells by merging *IMPS*, *IMVS*, and *EIS*. The results exhibited at short-circuit mode, electron transport nature is conquered by the series resistance in TCO (Rs) and at electrolyte/Pt-TCO interface (R_{CT}). As Rs or R_{CT} increased the electron transit time (delay time, τ_d) converted higher and the electron lifetime (τ_n) remains constant. To examine about the impact of Nernst diffusion resistance (R), they speckled the electrolyte layer thicknesses and attained R exaggerated the characteristics of electron transfer and charge transport [43]. R directs limited the electron extraction from TiO_2 film to the external circuit, predisposed the electron transfer from TiO_2 to tri-iodides available in an electrolyte.

1.2 Photovoltaic measurements

The current-voltage (*I-V*) characteristic is a concern with output current, the voltage of a solar cell in irradiation of full spectrum of AM 1.5 condition, is mandated in evaluating the solar photovoltaic performance [5,27,44]. The *I-V* characteristic plot depicted as Fig. 8, a short-circuit current (I_{SC})/short-circuit current density (J_{SC}), open-circuit voltage (V_{OC}) are the intercepts of

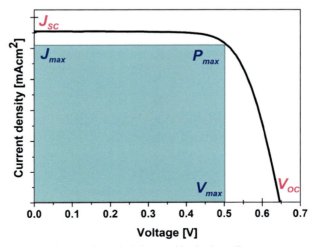

FIG. 8 *J-V* characteristic plot of a typical dye-sensitized solar cell.

228 Oxide free nanomaterials for energy storage and conversion applications

the *I-V* or *J-V* plots of lateral and vertical axes, respectively. The J_{SC} belongs to the current density of the test device which is in short-circuit condition, that is $V=0$, specifies maximum photocurrent output test device; V_{OC} relates the voltage in open-circuit mode as $I/J=0$, which represents maximum output photovoltage. At the inflection point of plot respective to photovoltage (V_{max}), photocurrent (I_{max}) gave maximum power output (P_{max}), as shown in the Fig. 8 by rectangular area of the inflection point is real maximum output power.

The ratio of P_{max} to the product of J_{SC}, V_{OC} results in fill factor (*FF*), it contains 0–1 numeric. A higher fill factor represents an added desirable rectangular shape in *J-V* plots and gave higher power conversion efficiency to the test device. Overall light to electric power conversion efficiency (PCE) or η of the developed DSSCs is represented by an Eq. (9)

$$\eta = \frac{I_{max}V_{max}}{P_{in}A} = \frac{V_{OC}J_{SC}FF}{P_{in}} \tag{9}$$

Here, P_{in} is incident light power density; A is the total active area of the test device.

The external quantum efficiency (*EQE*) called incident photon-to-current conversion efficiency (*IPCE*) is the important parameter of photovoltaic, resembles the number of electrons (n_e) restrained in response of photocurrent in external circuit divided by a number of monochromatic photons (n_p) which strike on top of test devices of solar cell in a unit of time, could be express as Eq. 10.

$$IPCE(\lambda) = \frac{n_e}{n_p} = \frac{1240J_{SC}(\lambda)}{\lambda P_{in}} \tag{10}$$

$J_{SC}(\lambda)$ is short-circuit photocurrent density (mA/cm^2) in monochromatic light illumination wavelength (λ, nm); and P_{in} represents incident light power (mW/cm^2). The incident photon to current conversion efficiency could be measure in a short-circuit mode, and there are two methods of DC or AC modes used in determining IPCE. Pioneer research group belongs to Prof. Han promoted AC mode having a frequency lower than 5 Hz and the DC mode with a large photon flux (1015 cm^2/s) of monochromatic light in the measurement of IPCE [45].

2. Advancement of metal nitrides and carbides as a counter electrode for dye-sensitized solar cells

Alternatives to the noble metal for counter electrodes development, (TMCs) transition metal compounds (TMCs) contain carbides and nitrides, have attracted due to its electronic structures comparable to Pt noble metal, with interstitial phase like structures/compounds, and witnessed Pt-like performance [46–50]. TMCs have been applied in dye-sensitized solar cells as counter electrodes catalysts substitute in the graceful Pt. In addition, metal nitrides and carbides have exceptional physical and chemical properties, like in high chemical stability and wear resistance, electronic conductivity, and platinum-like electrocatalytic activity, etc., and these compounds are typically synthesized using oxides, or metal halides are converts into nitrides through high-temperature treatment in $NH_3(g)$ or N_2 atmosphere (Fig. 9). Nitrides and Carbides have a wide range of applications, like supercapacitors, hydrogenation, and also perform excellent photovoltaic performance represented by Fig. 10 [35,51,52].

2.1 Metal nitrides

The usage of nitrides for the energy field is initiated as its technological advancement. Nitrides have significant ionic, covalent, and metallic characteristics relative to the oxides and carbides. Synthesis of Nitrides offers remarkable challenges as the incorporation of nitrogen into the interstitial sites of metals. Transition metal nitrides NiN, MoN, TiN, WN, W_2N, Fe_2N, NbN, Mo_2N, CrN, Ta_4N_5, VN are witnessed as CE catalysts for DSSCs applications [20,53–64]. A study on TiN nanotube arrays for CE preparation, its application for DSSCs has been initiated by Jiang et al. in the year 2009, results revealed in lower R_S and R_{CT} at counter electrode/E/electrolyte interface and evidenced PCE of 7.73%, compared with test cells made with platinum (7.45%) [53]. Li et al. team members tested platinum-like catalytic activity with WN, MoN, Fe_2N for the application of DSSCs, in which MoN performed higher electrocatalytic activity and PCE. This report initiated an advancement in cost-effective CEs for DSSCs [57]. A progress, which have been made by Zhang et al., contains TiN sphere, TiN particulates, and flat TiN based counter electrodes. DSSCs made by TiN sphere counter electrode evidenced higher power conversion efficiency of 7.83%, which is 30% more than DSSCs made with Platinum counter electrode CE (6.04%) [55]. Surface-nitrided Ni foil has developed by Gao et al., used it as counter electrode in dye-sensitized solar cells fabrication, performed a power conversion efficiency (5.68%), is lesser than the platinum assisted counter electrode dye-sensitized solar cell (8.41%) [56], and is ascribed to the low surface area of the compact NiN films. In the process of its performance optimization and enhanced surface area of NiN, mesoporous structured NiN film is developed using NiN particles and fabricated DSSC contributed a higher power conversion efficiency of 8.31%, and evidenced that a large surface area is a significant factor for high performance. In addition, Gao et al. group synthesized Fe_2N, WN, and MoN through nitridation of its oxides (Fe_2O_3, WO_3, MoO_2) in NH_3 atmospheric condition. The nitride-assisted test devices performed 2.65% (Fe_2N), 3.67% (WN), and 5.57% (MoN) of power conversion efficiency [57]. titanium nitride (TiN) nanoplates over a carbon fiber (CF) based counter electrode developed by Chen et al. team. They effectively produced porous single-crystalline titanium nitride (TiN) nanoplates over a carbon fiber (CF) and tested its electrocatalytic activity for DSSCs. A dye-sensitized

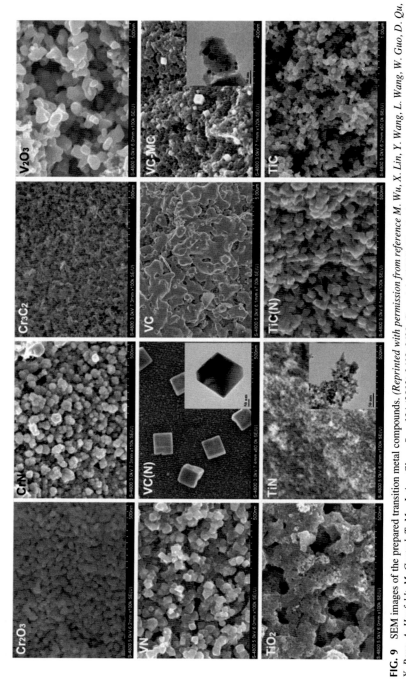

FIG. 9 SEM images of the prepared transition metal compounds. (Reprinted with permission from reference M. Wu, X. Lin, Y. Wang, L. Wang, W. Guo, D. Qu, X. Peng, A. Hagfeldt, M. Gratzel, T. Ma, J. Am. Chem. Soc. 134 (2012) 3419–3428. Copyright (2018) Elsevier.)

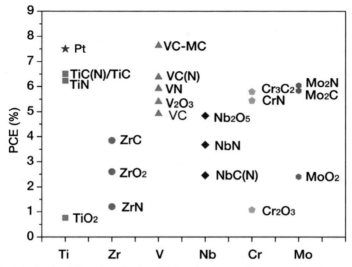

FIG. 10 Distribution PCE values for the DSSCs using the metal compounds as CEs. *(Reprinted with permission from reference M. Wu, X. Lin, Y. Wang, L. Wang, W. Guo, D. Qu, X. Peng, A. Hagfeldt, M. Gratzel, T. Ma, J. Am. Chem. Soc. 134 (2012) 3419–3428. Copyright (2018) Elsevier.)*

solar cell test devices contain fiber-shaped TiN-CF witnessed 7.20% of efficiency, which is higher than the performance of DSSC made with Pt wire (6.23%) as CE [58]. Efficient nanohybrids contains a metal-nitride nanoparticle altered nitrogen-doped graphene (NG) is fabricated by Wen et al., the prepared titanium nitride assisted NG (TiN-NG) nanohybrids are tested its electrocatalytic behavior, the results exhibited good electrocatalytic performance for tri-iodide (I_3^-) ions reduction to Pt which is desired characteristic of a counter electrode for DSSC. The test devices with TiN/NG counter electrode revealed $V_{oc} = 0.728$ (V), $FF = 0.64$, and performed 20% higher J_{SC} than the Pt-based DSSCs, in results of higher efficiency of 5.78% and is higher than DSSCs of Pt (5.03%). This study demonstrated the TiN/NG has a potential CE catalyst for DSSCs for substituting the Pt counter electrode [62].

MoN, Mo$_2$N, WN, W$_2$N, TiN, NiN, and Fe$_2$N are incorporated into DSSCs as catalysts of counter electrodes [56,57,63]. Through the sputtering method W$_2$N and Mo$_2$N thin films are grown over Ti sheets, the resulted electrodes are used for fabricating the DSSCs. The lower power conversion efficiency of 5.81% of W$_2$N and 6.38% of Mo$_2$N are obtained compared to platinum-based devices (7.01%). The low performance is accredited to the higher diffusion resistance of redox couple (I_3^-/I^-) in an electrolyte, higher charge transfer resistance (R_{CT}), which are also reflected in lower V_{OC}. Due to the light generated photon electrons of W$_2$N based DSSC need to overcome fairly higher resistance to cross the TiO$_2$ films, creating higher electron density on the conduction band

232 Oxide free nanomaterials for energy storage and conversion applications

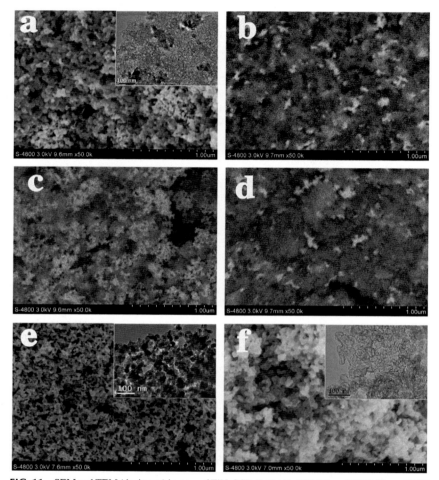

FIG. 11 SEM and TEM (the insets) images of TiN-CCB (1:1) (A), TiN (E), and CCB (F); and BSE images of TiN-CCB (1:1) (B), TiN-CCB (3:1) (C), and TiN-CCB (1:3) (D). *(Reprinted with permission from reference G.R. Li, F. Wang, J. Song, F.Y. Xiong, X.P. Gao, Electrochim. Acta 65 (2012) 216–220. Copyright (2012) Elsevier.)*

of TiO_2, which results in the rise of Fermi level [65,66]. A study contains highly dispersed Titanium Nitride (TiN) nanoparticles onto CCB/Ti (conductive carbon black/titanium) electrodes are developed, its SEM (Scanning Electron Microscope), TEM (Transmission Electron Microscope) results are presented as Fig. 11, demonstrated superior electrochemical performance compared to its individual electrodes of CCB/Ti and TiN/Ti electrodes. DSSC using TiN-CCB (1:1, mass ratio)/Ti electrode gave higher power conversion efficiency of 7.92%, compared to platinum-based devices (6.59%). *C-V* and *EIS* results show improved electrochemical performance of TiN-CCB/Ti based counter electrodes. These results are ascribed to thoroughly dispersed Titanium Nitride nanoparticles in the conductive carbon black matrix delivered in enhanced diffusion nature and electrocatalytic activity and of tri-iodide ions. This work promoted the metal nitrides - carbon composites as CEs for dye-sensitized solar cells [67].

A facile and cost-effective reduced graphene oxide assisted nanohybrid materials belongs to transition-metal nitride (TiN, MoN, VN) fashioned conducting network proved electrocatalytic performance for tri-iodide reduction, this study has initiated by Zhang et al. The developed nanostructured hybrids counter electrodes for DSSCs relieved an enhanced photovoltaic performance, witnessed power conversion efficiency (PCE) of 7.49% for TiN-NG, 6.27% for VN-NG, and 7.913% for MoN-NG, and the PCE of these electrodes is comparable to Pt-based CE devices (7.858%). [68] Song et al. evidenced improvement of diffusion kinetics for electrochemical process with MoN CE. A porous Molybdenum Nitride (MoN) nanorods thin film on a metallic Ti foil substrate is fabricated through nitridation of $Mo_3O_{10}(C_6H_8N)_2 \cdot 2H_2O$, is prepared hydrothermally by reacting $(NH_4)_6Mo_7O_{24}$ (ammonium heptamolybdate) and aniline. DSSCs fabricated using MoN are revealed 7.29% of power conversion efficiency, is higher of 6.48% of devices made with MoN electrode with sphere-like particles. The PCE is comparable with Pt-based devices 7.42%. The electrochemical impedance spectra confirmed diffusion resistance of porous MoN nanorod electrode is much lower than MoN nanoparticle electrode and the results achieved is related to interconnected channels with larger porosity of porous nano rod-structured electrode [69].

A newly designed counter electrode composed of a hybridized structure of iron nitride (FeN) core-shell nanoparticles grown on nitrogen-doped graphene (NG), applied for DSSCs. The FeN core-shell nanoparticles on NG performed an exceptional electrocatalytic activity and lower charge transfer resistance towards the CE I_3^-/I^- redox reaction in DSSC. The hybridized electrode exhibited notable cycling stability even after 500 cycles. The power conversion efficiency for core-shell FeN/NG CE revealed 10.86%, which is higher than a Pt-assisted DSSC (9.93%) [70]. TiN nanotube arrays for counter electrodes develop and its application for photovoltaics have been studied by Jiang et al. and reported that the performance is comparable to platinum counter electrode DSSC, suggested the TiN could be a better option to replace any electro-catalytically active materials [53]. In another study made by Zhang et al., developed nitride-based catalysts, tested for tri-iodide reduction in DSSCs applications, proved nitride materials are potential [68] Wu et al. used a simple urea precursor and synthesized vanadium nitride (VN) peas and applied them for a counter electrode. The VN peas assisted counter electrode exhibited a better PCE of 5.57% than Platinum-based CE (3.69%) [61].

Further, a study based on the effect of particle size, morphology on the performance of an electrocatalytic activity, different particle sizes of VN peas and cubes is developed by adopting a suitable molar ratio of the $VOCl_3$ and urea [61]. The VN peas assisted counter electrode demonstrated higher catalytic activity, followed by small-sized VN cubes and larger VN cubes in iodide redox couple electrolyte solution. The respective DSSCs produced a higher PCE of 7.29% (VN peas), small-sized VN cubes (6.53%), and larger VN cubes (5.59%). The achieved results are attributed to its different specific surface areas, the diverse electron structures depending on the synthesized particle size and shape. Nitrides combined nanostructured materials produced composite counter electrodes become potential for DSSCs application. TiN thin films are prepared on CNTs via thermal hydrolysis of $TiOSO_4$ and then nitridation in ammonia is made at 800 °C [64]. On the surface of the CNTs, 10 nm TiN nanoparticles are decorated. In this study, aggregation of TiN is prohibited by the usage of low concentration Ti precursor, and slow hydrolysis reaction. The TiN-CNT composite counter electrode initiated an efficiency of 5.41% is higher compared to either individual materials or Pt. A work initiated by Divya et al. on Graphitic carbon nitride-Graphene hybrid assisted Pt nanoparticles (Pt-GCN-G) for CE in DSSCs. This group has adopted a facile method for preparation of Pt-GCN-G hybrid, achieved incorporation of Graphene layers in Graphitic carbon nitride, and resulted in powder materials are characterized well. The DSSC fabricated using Pt-GCN-G-based CEs proved PCE of 5.2%, is comparable performance of Platinum CE (∼6%). Enhanced electrocatalytic activity in the synthesized material has been confirmed by results of C-V, EIS studies, attributed the improved performance of DSSCs to the synergic effects of Graphene, GCN, and Pt nanoparticles [71]. Superior photovoltaic performance in DSSCs has witnessed a study made by Zhang et al., and the synthesis method is represented as Fig. 12, results are accredited to the catalytic activity and morphology of the CE [72].

The electrocatalytic performance of NbN compounds and its photovoltaic performance of DSSCs is examined, achieved efficiency of 5.19% is compared to Pt (7.19%). Carbon nitride (CNx) thin films (90–100 nm) are grown by radio frequency magnetron sputtering method on cleaned FTO (fluorine-doped tin

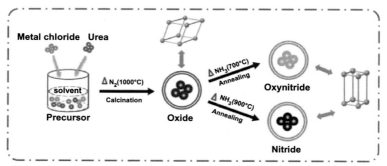

FIG. 12 Schematic illustrations of synthesis of Nb-based compounds. *(Reprinted with permission from reference V.-D. Dao, J. Power Sources 337 (2016) 125-129. Copyright (2017) Elsevier.)*

oxide) glass substrate, the element components in the CNx films are restricted as $x = 0.15$–0.25. As-prepared CNx CEs showed a good light transmittance in the visible light region. The CNx films performed electrocatalytic activity and were used for DSSCs application. XPS (X-ray photoelectron spectroscopy) study revealed a higher proportion of sp^2 C=C and sp^3 C—N hybridized bonds in CNx-500 (treated at 500°C) than the CNx-RT (without heat treatment). This study proposed the sp^{3}C-N and sp^2 C=C hybridization bonds available in the CNx films are supporting improved electrocatalytic activity. A Raman studies also proven CNx-500 has a relatively higher graphitization level which is helping in the increase of its electrical conductivity. This study witnessed the samples after heat treatment result in higher electrochemical performance [73]. Unique mesoporous counter electrodes with Titanium Nitride microspheres on a Titanium substrate [74] has been developed by Wang team members. The prepared mesoporous TiN electrode used for DSSCs results in a power conversion efficiency of 6.8%, which is higher than the flat TiN based DSSCs (2.4%). It is attributed to mesoporous TiN electrode showed lower R_{CT} compared to a flat TiN electrode. Sung et al. developed surface-nitride Ni foam film for DSSCs as a cost-effective counter electrode [60]. For the Nitridation of Ni foam in an ammonia atmosphere forms nitride Ni foam. The prepared mesoporous Ni foam had potentiality as interrelated metallic nature is desirable easy electron transport-ability in 3D structures. The nitrided Ni foam counter electrode performed higher conductivity, higher catalytic activity for I_3^-/I^- redox couple regeneration in Dye-sensitized solar cells. The test cells witnessed a power conversion efficiency of 3.88%. Further, the catalytic activity of nitrided Ni foam electrode is optimized by treating in HNO_3 and H_2SO_4. As a result, improved specific surface area to the I_3^-/I^- redox couple available in an electrolyte solution. DSSCs with Ni foam CE treated with HNO_3 and H_2SO_4 improved from 3.88% to 4.73% and 4.45%, respectively. A study based on urea-assisted metal path prepared ZrN, TiN, Mo$_2$N, VN, NbN, and CrN nanoparticles contains different morphologies [20]. Dye-sensitized solar cells made using synthesized nitrides revealed power conversion efficiency of 3.68% (ZrN), 6.23% (TiN), 6.04% (Mo$_2$N), 5.92% (VN), 1.20% (NbN), and 5.44% (CrN).

The performance of DSSCs made with nitrides as counter electrodes are presented in Table 1.

2.2 Metal carbides

The electrocatalytic activity of carbides-based research explored after 1960 [66]. Transition-metal carbides (TMCs) prepared via a facile acetamide-metal route and usage for CEs in dye-sensitized solar cells are progressed initially. In this study, Acetamide is used as a source of carbon due to it being cost-effective, acts as a good chelating agent, and is an environment-friendly material. Levy and Boudart in the year 1973 stated the tungsten carbide (WC) behaves as a

TABLE 1 Photovoltaic parameters related to Nitrides based counter electrode DSSCs.

CE materials	Jsc (mA/cm^2)	V_{oc} (V)	FF	PCE (%)	PCE (%) (Pt)	Reference
TiN nanotube	15.78	0.76	0.64	7.73	7.45	[53]
TiN-sphere	16.57	0.759	0.62	7.83	6.04	[55]
Mesoporous NiN	15.76	0.766	0.69	8.31	7.93	[56]
TiN-carbon fiber	19.35	0.64	0.58	7.20	6.23	[58]
FeN/N-doped graphene	18.83	0.74	0.78	10.86	9.93	[70]
FeN	14.02	0.69	0.70	6.79	9.93	
MoN	11.55	0.735	0.66	5.57	6.56	[57]
WN	9.75	0.7	0.54	3.67	6.56	
Fe$_2$N	12.20	0.535	0.41	2.65	6.56	
TiN	16.57	0.759	0.622	7.83	6.04	[55]
Ni foil/Pt	15.79	0.77	0.69	8.41	7.93	[56]
Surface nitride Ni foil	15.74	0.766	0.47	5.68	7.93	
NiN	15.76	0.766	0.69	8.31	7.93	
TiN/NG	12.24	0.72	0.64	5.78	5.03	[62]
Flexible Mo$_2$N	14.09	0.743	0.61	6.38	7.01	[63]
Flexible W$_2$N	12.96	0.786	0.57	5.81	7.01	
TiN-CCB (conductive carbon black)	14.29	0.791	0.70	7.92	6.59	[67]
VN-nanoparticle N-doped reduced graphene oxide	12.58	0.78	0.63	6.27	7.85	[68]
TiN-NG	14.16	0.79	0.66	7.49	7.85	
MoN-NG	14.13	0.788	0.71	7.91	7.85	
MoN porous nanorod	15.26	0.74	0.65	7.29	7.42	[69]
MoN-nano particle	14.76	0.705	0.62	6.48	7.42	

Metal nitrides and carbides as advanced counter electrodes Chapter | 10 **237**

TABLE 1 Photovoltaic parameters related to Nitrides based counter electrode DSSCs—cont'd

CE materials	Jsc (mA/ cm^2)	V_{oc} (V)	FF	PCE (%)	PCE (%) (Pt)	Reference
VN(T$^-$/T$_2^-$)	12.97	0.632	0.68	5.57	3.69	[61]
TiN-CNT	12.74	0.750	0.57	5.41	5.68	[64]
Pt-GCN-G	13.25	0.70	0.56	5.2	6.01	[71]
NbN	13.21	0.71	0.55	5.19	7.19	[72]
TiN flat film	10.6	0.65	0.34	2.4	7.2	[74]
Mesoporous TiN	15.3	0.69	0.65	6.8	7.2	
Nitride Ni	9.62	0.67	0.60	3.88	5.57	[60]
TiN	12.83	0.796	0.61	6.23	7.50	[20]
VN	11.74	0.788	0.64	5.92	7.50	
CrN	10.39	0.818	0.64	5.44	7.5	
NbN	9.81	0.798	0.47	3.68	7.5	
Mo$_2$N	11.68	0.784	0.66	6.04	7.5	
ZrN	8.20	0.733	0.2	1.2	7.5	
S-VN cubes	14.43	0.718	0.63	6.53	7.68	[61]
L-VN cubes	13.38	0.721	0.58	5.59	7.68	
VN peas	15.06	0.733	0.66	7.29	7.68	
S-VN cubes (T$^-$/T$_2^-$)	12.32	0.628	0.64	4.95	3.69	
L-VN cubes (T$^-$/T$_2^-$)	10.84	0.619	0.59	3.96	3.69	

platinum-like characteristic for catalytic performance due to its distinct electronic structure [46]. In the progress of Carbide based research, *Lee et al.* fabricated a polymer-derived WC (PD-WC) and microwave-assisted WC (MW-WC), tested in dye-sensitized solar cells as CE catalysts [75]. Test cells made using CEs proved power conversion efficiency of 6.61% (PC-WC) and 7.01% (MW-WC). These results are lesser in values compared to the DSSC fabricated using platinum CE (8.23%).

238 Oxide free nanomaterials for energy storage and conversion applications

The crystallinity has impacted the performance of the catalytic activity, in view of this, an investigation made by Park et al., synthesized tungsten carbide through sinter the precursor of $WO_3 \cdot H_2O$ in a CH_4/H_2 atmosphere at different temperature conditions. The study reveals the sintering temperature could control the crystallinity and consequent impression on its catalytic performance [76]. Temperature at 900°C results in pure WC is realized, the DSSC showed 3.0% of power conversion efficiency is lower than platinum (5.22%) assisted DSSC. Further, in its performance optimization, tungsten and molybdenum carbides are applied for DSSCs [77]. WC commercial material contain DSSC witnessed an efficiency of 5.35% and 5.70% for Mo_2C, this performance is lower and the results are ascribed to the huge size of both carbide materials. For its further optimization, synthesize catalysts materials as trivial in particle size, and uniformly disperse catalyst particles over their carrier. For this, a trial has been made and synthesized nanometer-sized WC and W_2C by use of the metal-urea path [78]. For uniform distribution, the synthesized WC and MoC are embedded in order to mesoporous carbon (WC-OMC, MoC-OMC). Fabricated dye-sensitized solar cells are made using two nanocomposites as counter electrode catalysts material, results yielded higher power conversion efficiency of 8.18% for WC-OMC and 8.34% for MoC-OMC. A report consists of CrN, Cr_3C_2, VN, VC(N), TiN, TiC(N), TiC and V_2O_3 materials assisted counter electrodes are performed a superior catalytic performance for reduction of I_3^- to I^- in DSSC system. VC decorated over mesoporous carbon (VC-MC) is developed via in situ synthesis route. I_3^-/I^- redox couple-based electrolyte-filled DSSCs made by using VC-MC attained higher efficiency of 7.63%, which is analogous to the performance of Pt-based DSSCs (7.50%). Further its optimization using new-fangled organic redox couple T_2/T^-, TiC, and VC-MC CEs witnessed efficiency 4.96% and 5.15% respectively, which is higher than Pt CE (3.66%). This study demonstrated cost-effective, potential CE catalysts could be useful for DSSCs applications [20]. Synthesis of Ta_3N_5, Ta_4C_3, and TaOx counter electrode catalysts through of urea-metal protocol, and its characterization has made by Yun et al. group. DSSCs have been fabricated and the photovoltaic performance for use of TaOx and Ta_4C_3 CEs revealed PCE of 6.79% and 7.39% respectively and is compared with Pt-based devices (7.57%). The achieved results are compared to the early work, in which TaO/MC and TaC/MC hybrid CEs in DSSCs performed excellent catalytic activity and also photovoltaic performance than TaOx and Ta_4C_3 and MC CEs [79], presented as Fig. 13.

Cost-effective platinum-free counter electrode catalysts have been developed by Ma et al. group and studied the effect of tungsten carbide (WC) and molybdenum carbide (Mo_2C) for counter electrodes catalysts for DSSCs development. The test cells based on WC and Mo_2C assisted CEs performed efficiency is 5.35% (WC) and 5.70% (Mo_2C) and are lower than the Pt-based DSSC (7.89%). The relatively low efficiency is attributed to large size two carbides particles (Mo_2C, 300 nm; WC, 190 nm) [77]. For optimization of its performance, nano-size WC and W_2C are synthesized through the metal-urea

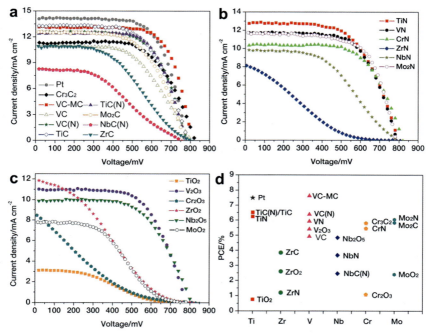

FIG. 13 *J-V* curves of DSSCs with I_3^-/I^- based electrolyte using (A) Pt or carbide, (B) nitride, or (C) Oxide CE catalysts, (D) PCEs distribution plot. *(Reprinted with permission from reference S. Yun, H. Zhang, H. Pu, J. Chen, A. Hagfeldt, T. Ma, Adv. Energy Mater. 3(11) (2013) 1407-1413. Copyright (2012) American Chemical Society.)*

preparation route. DSSCs are fabricated using nano-sized carbides material as CEs catalysts, the results of 6.68% (W_2C) and 6.23% (WC) power conversion efficiency are achieved. This study revealed compared to large WC particles, nano-sized WC performed excellent and enhanced catalytic activity [80]. This group also conducted another work, in which methodical study on transition metal carbides (VC, ZrC, TiC, Cr_3C_2, NbC) [20,81]. All of the synthesized carbides demonstrated higher catalytic activity except NbC and ZrC. Cost-effective carbon, carbides-based counter electrodes (CEs) having good adhesion by addition of conductive carbon (CC) paste as a binder for development of DSSCs has been reported by Gao et al. A superior electrochemical catalytic activity is achieved for CC and carbon dye (Cd)/CC for reduction of triiodide to iodide, is confirmed via conducting cyclic-voltammetry and electrochemical impedance spectroscopy. This group also synthesized numerous transition metal carbides, which includes tungsten carbide (WC), molybdenum carbide (MoC), ordered mesoporous carbon (OMC) MoC embedded in OMC (MoC-OMC) and WC embedded in OMC (WC-OMC) as catalytic materials of CEs for development of dye-sensitized solar cells. Efficiencies are achieved

240 Oxide free nanomaterials for energy storage and conversion applications

in between 4.50% and 6.81%. Obtained outcome specifying CC is performed as a better binder for Pt-free catalysts in DSSCs [82]. Carbides-based platinum-free TiC electrodes are successfully deposited through the hydrothermal deposited method over the conductive FTO glass substrates and developed dye-sensitized solar cells. The power conversion efficiency of 3.07% is attained for hydrothermal TiC based counter electrode DSSC prepared with a disulfide/thiolate electrolyte solution. Optimized performance of 3.59% is reached by combining TiC with carbon, the results are comparable to a Pt DSSC (3.84%). Development of graphite on nanoparticles of TiC played a vital role in improving reduction current to $10.12\,mA/cm^2$ and reducing the impedance of film to $237.63\,\Omega$, which gave a higher power efficiency [83].

Theerthagiri et al. performed a study based on nanowire-shaped α-MoO_3 and different metal-assisted NB, W, Ta, and α-MoO_3, which are synthesized through the hydrothermal method. Oxides' materials are converted into respective carbides via the carburization route using urea as carbon source. Dye-sensitized solar cells are fabricated using synthesized W-α-Mo_2C counter electrode and achieved a better power conversion efficiency of 2.53% than pure α-Mo_2C, Ta-doped α-Mo_2C, and Nb. This study revealed the W-doped α-Mo_2C is potential electrocatalyst for development of low-cost effective DSSCs [84]. A facile, feasible synthetic method is adopted by Guo et al. and successfully synthesized carbon assisted transition metal carbide composites of Mo_2C-C, Cr_3C_2-C, VC-C, WC-C, TiC-C, TaC-C, NbC-C, etc., by use of metal chlorides as sources of metals, phenolic resin acted as source of carbon, the resulted compounds microstructures are presented in Fig. 14. Results of Cyclic-Voltammetry gave carbon-supported carbide composites performed higher peak current densities, lower peak-to-peak separation, shown in Fig. 15. The electrochemical impedance spectroscopy studies performed and result in inferior charge transfer resistance compared to pristine carbides. And compared carbides, carbon assisted carbide-based composites CEs performed higher catalytic activities towards cobalt redox couple regeneration in the test cells of dye-sensitized solar cells. Devices made with WC-C, VC-C, and TiC-C composite CE demonstrated PCE of 9.75%, 9.42%, and 8.85% respectively, higher than counter electrodes of VC, WC and TiC [85], and the respective plot is demonstrated as Fig. 16.

A study composed of Co-TiC NPs (cobalt-titanium carbide nanoparticles) embedded carbon nanofibers (composite) are successfully tailored through electrospinning of solution contains polyvinylpyrrolidone (PVP), titanium(IV) isopropoxide (TIIP), cobalt acetate tetrahydrate (CoAc) in ethanol, and acetic acid, followed by carbonation process at temperature (850°C). The resulted composite, efficient counter electrode used as cost-electrodes for the development of dye-sensitized solar cells. The cyclic-voltammetry, chronoamperatory (CA) studies are conducted to measure the performance of composite electrodes. Results of composite CE gave highly electrochemical stability as low onset potential (189 mV) and high current density ($\sim$90 mA/cm^2 vs. Ag/AgCl). The achieved results provide stable

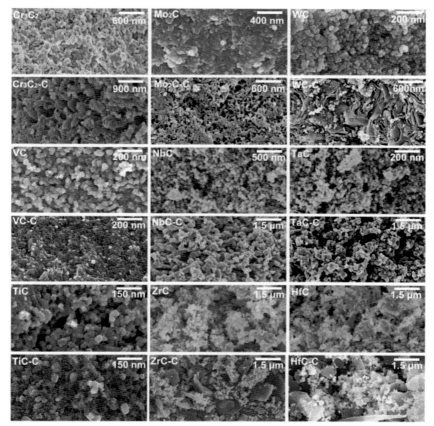

FIG. 14 SEM images of the various carbides, carbon-supported composites. *(Reprinted with permission from reference H. Guo, Q. Han, C. Gao, H. Zheng, Y. Zhu, M. Wu, J. Power Sources 332 (2016) 399–405. Copyright (2018) Elsevier.)*

electrocatalytic activity (ECA), good conductivity which is reflected in improved catalytic activity in tri-iodide reduction. The short-circuit current density (J_{SC}), fill factor (*FF*), open-circuit voltage (V_{OC}), and power conversion efficiency are found to be 0.75 V, ~9.98 mA/cm^2, 0.50, and 3.87%, respectively [86]. Wang and his team members are designed, developed efficient, nonprecious, stable metal-based catalysts for counter electrode (CE) preparation for DSSC application and accomplished great attention. This work demonstrates the Mo$_2$C nanoparticles are evenly distributed in the nitrogen-rich carbon matrix (@NC) which is synthesized using inexpensive raw materials (dicyandiamide and polyoxometalate) through a simple, single step, solid phase synthesis route. The Mo$_2$C@NC hybrid CE revealed excellent catalytic performance, and PCE is achieved 6.49%, which is comparable to Pt (6.38%) based on results

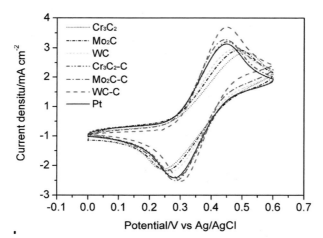

FIG. 15 Cyclic-voltammograms of various carbides assisted CEs. *(Reprinted with permission from reference H. Guo, Q.Han, C. Gao, H. Zheng, Y. Zhu, M. Wu., J. Power Sources 332 (2016) 399–405. Copyright (2018) Elsevier.)*

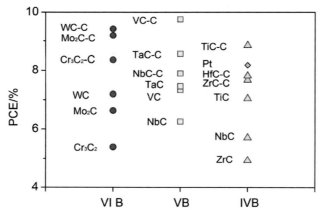

FIG. 16 PCE variation plot with different group materials. *(Reprinted with permission from reference H. Guo, Q.Han, C. Gao, H. Zheng, Y. Zhu, M. Wu, J. Power Sources 332 (2016) 399–405. Copyright (2018) Elsevier.)*

obtained from the respective DSSC [87]. Wu et al. demonstrated WC, WO_2, Cr_3C_2, Cr_2O_3, MoO_2, and Mo_2C based counter electrodes satisfactory catalytic activity for reduction of I_3^-/I^- in DSSCs. The evidenced PCE of 4.83% is attained for WC as CE, whereas the other CEs are performed 4.67% (WO_2), 4.44% (Cr_3C_2), 2.73% (Cr_2O_3), 2.53% (MoO_2), and 3.24% (Mo_2C) of power conversion efficiency [88].

Kang et al. reported based on the synthesis of nano-porous tungsten carbide and its application for dye-sensitized solar cells. This method involves tungsten

Metal nitrides and carbides as advanced counter electrodes **Chapter | 10** **243**

foil anodization followed by post heat treatment under an atmosphere of CO to achieve highly crystalline tungsten carbide film with the interconnected nanostructure. The resulted electrodes performed higher catalytic activity for the reduction of cobalt bipyridine species in DSSCs. The performance of tungsten carbide gave a substantial increase in conversion efficiency [89]. Jin et al. group members reported the transition metal carbides are proved as auspicious alternatives to noble metal-based catalysts for energy conversion applications. With morphology controlling of the TMCs the catalytic performance is enhanced in the study. A substrate-mediated route to direct grown of VC (vanadium carbide) using graphene oxide as nucleation and grown substrate. Sensibly selecting synthesis conditions the group members obtained VC nanoparticles from cuboctahedron (VC-ch) to cubic (VC-cb) structure. Attained hybrid VC-ch over Graphene sheets (VC-ch/GS) expressed outstanding catalytic activity towards the reaction of tri-iodide reduction, and lower charge transfer resistance ($R_{CT}=0.27\,\Omega\,cm^2$), performing its counterpart VC-cb/GS contained R_{CT} of $1.69\,\Omega\,cm^2$. The power conversion efficiency of DSSCs for VC-ch CE realized 7.92%, is analogous performance to Pt-based CE of 7.79% [90]. Ko et al. demonstrated the synthesis of tungsten carbide (WC) based counter electrodes through simple heating of layered tungsten oxide at different temperatures (700°C, 800°C, and 900°C). Short-circuit current densities are achieved the values of $2.51\,mA/cm^2$ (700°C) $10.54\,mA/cm^2$ (800°C) and $10.39\,mA/cm^2$ (900°C) for WC CEs and the harvested the efficiencies of 0.20%, 4.20% and 3.03% respectively. Which revealed the WC of 800°C sintered electrode gave higher efficiency compared to the other two electrodes and is lower than Pt CE (5.22%). This WC (800°C) performed dark current density as high and gave optimized electrocatalytic activity [91].

Nanoparticle Co_3O_4-WC-CN/rGO (CN=nitrogen-doped carbon; rGO = reduced Graphene oxide) are successfully prepared by Chen et al., the synthesis protocol is given as Fig. 17, and tested its performance for counter electrodes application. In the synthesis, calcination of $Na_6H_2W_{12}O_{40}\cdot H_2O$ (H_2W_{12}) at high temperature the embedded metal-organic framework in argon gas and air

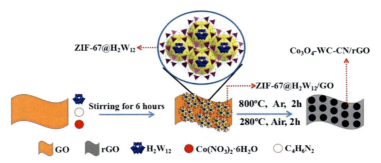

FIG. 17 Synthesis process of Co_3O_4-WC-CN/rGO compound. *(Reprinted with permission from reference L. Chen, W. Chen, E. Wang, J. Power Sources, 380 (2018) 18–25. Copyright (2018) Elsevier.)*

atmosphere is maintained. The achieved CEs gave higher catalytic activity (Fig. 18) and Tafel performance (Fig. 19), fabricated dye-sensitized solar cells. The PCE of DSSC made with Co_3O_4-WC-CN/rGO is 7.38%, superior to CE of Pt-assisted DSSCs ($\eta = 6.85\%$) [92].

Based on first-principles calculations, Majid et al. revealed work on the modification of materials that are used for counter electrode preparation towards

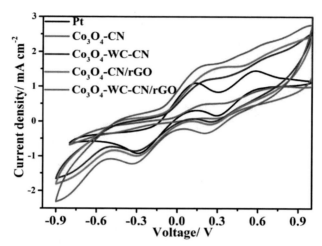

FIG. 18 Cyclic-Voltammograms in I_3^-/I^- redox coupled electrolyte solution for various CEs. *(Reprinted with permission from reference L. Chen, W. Chen, E. Wang, J. Power Sources 380 (2018) 18–25. Copyright (2018) Elsevier.)*

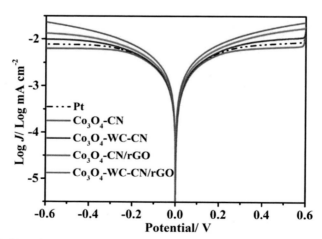

FIG. 19 Tafel polarization performance of different materials assisted CE. *(Reprinted with permission from reference L. Chen, W. Chen, E. Wang, J. Power Sources 380 (2018) 18–25. Copyright (2018) Elsevier.)*

Metal nitrides and carbides as advanced counter electrodes **Chapter | 10** **245**

dye-sensitized solar cells performance improvement. The slab replicas of pure *SiC* and its doping with Pt and Cr are examined to investigate its structural variation (Fig. 20), adsorption (Fig. 21), and catalytic activity. Iodide reduction reaction results are exhibited to study the splitting of tri-iodide into iodine and also into iodide ions which obtain negative charges. Results of Cr doped *SiC* demonstrated excellent catalytic activity, and the outcome of this study revealed the TM doped *SiC* slabs become alternates to platinum assisted counter electrodes [93].

Zhang et al. conducted work for the improvement of characteristics of a composite counter electrode, the work consists of γ-MoC/Ni@NC material is used as a counter electrode for DSSC application. Synthesis is carried out

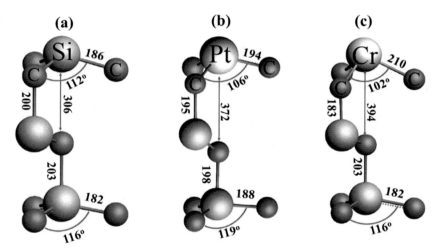

FIG. 20 Variation of structural morphology of (A) pure *SiC*, (B) Platinum, (C) Cr in two top layers of tri-layered slabs. *(Reprinted with permission from reference A. Majid, I. Ullah, K.T. Kubra, S.U.-D. Khan, S. Haider, Surf. Sci. 687 (2019) 41–47. Copyright (2019) Elsevier.)*

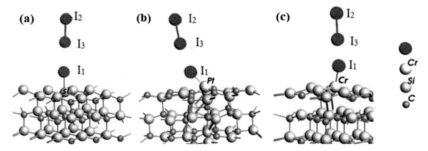

FIG. 21 Adsorption results of tri-iodide over tri-layer slabs of (A) pure *SiC*, (B) Platinum doped *SiC*, (C) Cr doped *SiC*. *(Reprinted with permission from reference A. Majid, I. Ullah, K.T. Kubra, S.U.-D. Khan, S. Haider, Surf. Sci. 687 (2019) 41–47. Copyright (2019) Elsevier.)*

246 Oxide free nanomaterials for energy storage and conversion applications

through the solid-state pyrolysis method using oxalate, molybdate, nickel, and melamine salts. DSSC made with composite catalyst achieved PCE of 5.26%, which is almost analogous performance to traditional Pt CE-based DSSCs (5.65%). Though the PCE of DSSCs based on γ-MoC/Ni@NC is somewhat lower than the Pt [94].

Ma et al. stated a two-dimensional layer structured Ti_3C_2, which is synthesized by using the method of etching under lower temperature, assisted counter electrodes are giving an outstanding performance in dye-sensitized cells, is higher activity when compared to TiC particles assisted counter electrodes. The DSSCs fabricated using a counter electrode of two-dimensional layer structured Ti_3C_2 realizes efficiency of 9.57%, is higher than the device fabricated using TiC particles counter electrode (7.37%). The outstanding activity of the Ti_3C_2 electrode is ascribed to the superior charge transfer owing to specific two-dimensional layer structured Ti_3C_2 [95]. Chen et al. developed a highly efficient electrocatalyst of a novel heterostructure of Co-embedded and N-doped carbon nanotubes assisted Mo_2C nanoparticles (Mo_2C/NCNTs@Co). By adoption of one-step metal-catalyzed carbonization-nitridation strategy, the final compounds are successfully synthesized, the developed protocol is shown in Fig. 22, and microstructural results are shown in Fig. 23. The prepared conductive NCNTs@Co substrate with Mo_2C performed high affinity for I_3^-

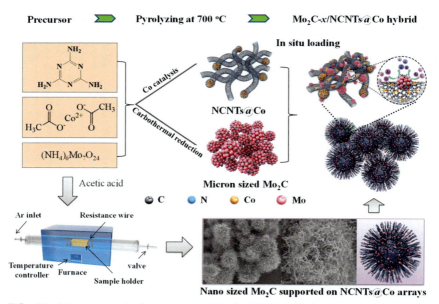

FIG. 22 Schematic synthesis process of Mo_2C-x/NCNTs@Co hybrids through one-step pyrolysis route. *(Reprinted with permission from reference Y. Zhang, S. Yun, Z. Wang, Y. Zhang, C. Wang, A. Arshad, F. Han, Y. Si, W. Fang, Ceram. Int. (2020). https://doi.org/10.1016/j.ceramint.2020.03.128. Copyright (2019) American Chemical Society.)*

Metal nitrides and carbides as advanced counter electrodes **Chapter | 10** **247**

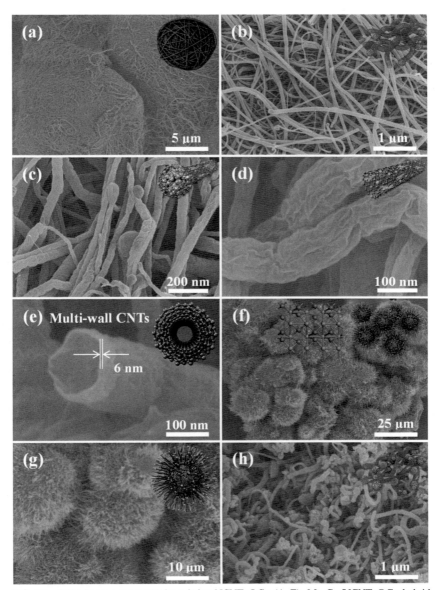

FIG. 23 SEM images represent the pristine NCNTs@Co (A–E), Mo$_2$C-$_4$/NCNTs@Co hybrid (F–H). *(Reprinted with permission from reference Y. Zhang, S. Yun, Z. Wang, Y. Zhang, C. Wang, A. Arshad, F. Han, Y. Si, W. Fang, Ceram. Int. (2020). https://doi.org/10.1016/j.ceramint.2020.03.128. Copyright (2019) American Chemical Society.)*

248 Oxide free nanomaterials for energy storage and conversion applications

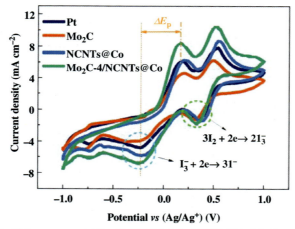

FIG. 24 Cyclic-Voltammograms of Mo$_2$C-$_4$/NCNTs@Co, NCNTs@Co, Mo$_2$C, and Pt CEs. *(Reprinted with permission from reference Y. Zhang, S. Yun, Z. Wang, Y. Zhang, C. Wang, A. Arshad, F. Han, Y. Si, W. Fang, Ceram. Int. (2020). https://doi.org/10.1016/j.ceramint.2020.03.128. Copyright (2019) American Chemical Society.)*

adsorption, higher charge transfer capability for I_3^- reduction, shown in Fig. 24. The DSSC with Mo$_2$C/NCNTs@Co CE performed a high conversion efficiency of 8.82%, unique electrochemical stability, and residual efficiency of 7.95% (after 200h of continuous illumination) [96].

Zhang et al. successfully synthesized bio-assisted porous carbon (C$_{MA}$ and C$_{TA}$) from waste cartons via two-step chemical activation (TA) and microwave-assisted activation (MA) routes, denoted in Fig. 25. The higher specific surface

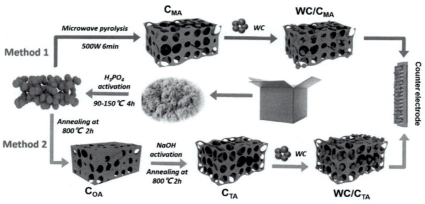

FIG. 25 Synthesis pathways of waste carton collected from bio-based carbon, hybridized with WC. *(Reprinted with permission from reference M. Chen, G.-C. Wang, W.-Q. Yang, Z.-Y. Yuan, X. Qian, J.-Q. Xu, Z.-Y. Huang, A.-X. Ding, ACS Appl. Mater. Interfaces 11 (2019) 42156–42171. Copyright (2020) Elsevier.)*

Metal nitrides and carbides as advanced counter electrodes **Chapter | 10** **249**

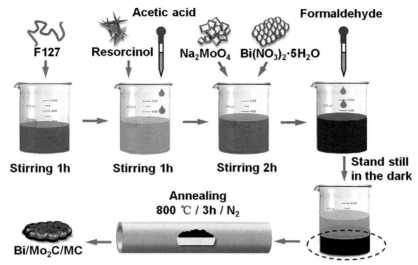

FIG. 26 Synthesis route of Bi/Mo$_2$C/MC preparation. *(Reprinted with permission from reference S. Yun, Y. Hou, C. Wang, Y. Zhang, X. Zhou, Ceram. Int. 45 (2019) 15589–15595. Copyright (2019) Elsevier.)*

area of prepared counter electrodes is achieved, which provide more catalytic active sites, delivered PCE of 6.76% (C$_{TA}$) exceed the C$_{MA}$-based DSSC (6.19%). Further, tungsten carbide (WC) introduced into C$_{TA}$ and C$_{MA}$ (WC/C$_{TA}$ and WC/C$_{MA}$) improves catalytic activities, this synergistic effect contributed superior PCE of 7.32% (WC/C$_{TA}$) and 6.85% (WC/C$_{MA}$), is closure to Pt (7.51%) [97].

Mo-based binary/ternary nanocomposites are developed through the soft template synthesis route, represented as Fig. 26, by combining the Mo$_2$C and Bi/Mo$_2$C into mesoporous carbon (MC) material. The Mo$_2$C/MC composites counter electrodes (CEs) are developed (Fig. 27) and evaluated their performance for dye-sensitized solar cell applications. The result in PCE of 7.29% (Mo$_2$C/MC), 8.06% for Bi/Mo$_2$C/MC is witnessed, is higher than Pt CE based DSSC (7.08%) [98].

The performance of DSSCs fabricated using Carbides as counter electrodes are tabulated in Table 2.

3. Summary

Dye-sensitized solar cells are cost-effective, potential to convert solar energy into electric energy. It belongs to third-generation solar cells concern of environment-friendly, simple fabrication procedure. Prof. M. Grätzel demonstrated the first dye-sensitized solar cell in the year 1991. DSSCs are made with three key components, are photoelectrodes, counter electrodes,

250 Oxide free nanomaterials for energy storage and conversion applications

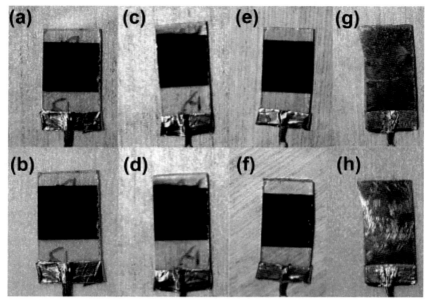

FIG. 27 Photograph images of counter electrodes developed using (A) Bi/Mo$_2$C/MC, (C) Mo$_2$C/MC, (E) MC and (G) Pt; (B) Bi/ Mo$_2$C/MC, (D) Mo$_2$C/MC, (F) MC and (H) Pt after 100 cycles of C-V scanning. *(Reprinted with permission from reference S. Yun, Y. Hou, C. Wang, Y. Zhang, X. Zhou, Ceram. Int. 45 (2019) 15589–15595. Copyright (2019) Elsevier.)*

TABLE 2 Photovoltaic parameter performance of various Carbides based dye-sensitized solar cells.

CE materials	J_{SC} (mA/cm^2)	V_{OC} (V)	FF	PCE (%)	PCE (%) Pt	Reference
Mesoporous WC	14.17	0.76	0.65	7.01	8.23	[75]
VC-mesoporous carbon	13.11	0.80	0.72	7.63	7.5	[20]
TaC/MC	14.51	0.8	0.68	7.93	7.32	[79]
Mo$_2$C@NC	14.72	0.76	0.58	6.49	6.38	[87]
Mo$_2$C	11.48	0.76	0.52	4.59	6.38	
NC	12.69	0.76	0.58	5.59	6.38	
WC-PC	14.17	0.766	0.61	6.61	8.23	[75]
WC-MW	14.17	0.763	0.65	7.01	8.23	

Metal nitrides and carbides as advanced counter electrodes Chapter | 10 **251**

TABLE 2 Photovoltaic parameter performance of various Carbides based dye-sensitized solar cells—cont'd

CE materials	J_{SC} (mA/cm^2)	V_{OC} (V)	FF	PCE (%)	PCE (%) Pt	Reference
WC (commercial)	12.66	0.807	0.52	5.35	7.89	[77]
Mo_2C	13.12	0.804	0.54	5.70	7.89	
WC-OMC	14.59	0.804	0.70	8.18	7.89	
MoC-OMC	15.50	0.787	0.68	8.34	7.89	
W_2C	12.72	0.807	0.65	6.68	7.65	[80]
WC/W_2C	12.92	0.799	0.60	6.23	7.65	
VC-MC(T_2/T^-)	12.86	0.633	0.63	5.15	3.66	[20]
Cr_3C_2 (T_2/T^-)	10.84	0.641	0.65	4.54	3.66	
VC (T_2/T^-)	11.79	0.647	0.54	4.06	3.66	
TiC(T_2/T^-)	12.44	0.629	0.63	4.96	3.66	
Ta_4C_3	12.92	0.78	0.74	7.39	7.57	[79]
MoC/CC	11.10	0.76	0.53	4.50	7.06	[82]
WC/CC	11.73	0.77	0.57	5.14	7.06	
MoC-OMC/CC	11.74	0.78	0.67	6.10	7.06	
WC-OMC/CC	11.98	0.79	0.65	6.10	7.06	
OMC/CC	11.88	0.79	0.68	6.40	7.06	
TiC	10.12	0.65	0.55	3.59	3.84	[83]
α- Mo_2C	5.8	0.60	0.28	0.97	5.14	[84]
W-α-Mo_2C	9.81	0.63	0.41	2.23	5.14	
Nb-α-Mo_2C	4.38	0.56	0.18	0.45	5.14	
Ta-α-Mo_2C	6.05	0.60	0.18	0.65	5.14	
TiC (cobalt redox couples)	13.00	0.831	0.652	7.05	8.18	[85]
TiC-C (cobalt redox couples)	14.91	0.847	0.701	8.85	8.18	
TaC (cobalt redox couples)	13.29	0.833	0.674	7.45	8.18	

Continued

252 Oxide free nanomaterials for energy storage and conversion applications

TABLE 2 Photovoltaic parameter performance of various Carbides based dye-sensitized solar cells—cont'd

CE materials	J_{SC} (mA/cm^2)	V_{OC} (V)	FF	PCE (%)	PCE (%) Pt	Reference
TaC-C (cobalt redox couples)	14.59	0.835	0.703	8.56	8.18	
ZrC (cobalt redox couples)	12.08	0.825	0.494	4.93	8.18	
ZrC-C (cobalt redox couples)	13.75	0.837	0.665	7.66	8.18	
HfC (cobalt redox couples)	12.43	0.835	0.55	5.71	8.18	
HfC-C (cobalt redox couples)	13.60	0.846	0.678	7.81	8.18	
Cr$_2$C$_3$ (cobalt redox couples)	12.1	0.834	0.535	5.40	8.18	
Cr$_2$C$_3$-C (cobalt redox couples)	14.47	0.841	0.686	8.36	8.18	
Mo$_2$C (cobalt redox couples)	12.94	0.823	0.624	6.64	8.18	
Mo$_2$C-C (cobalt redox couples)	15.38	0.838	0.713	9.19	8.18	
VC (cobalt redox couples)	13.21	0.837	0.663	7.33	8.18	
VC-C (cobalt redox couples	15.86	0.840	0.732	9.75	8.18	
WC (cobalt redox couples)	13.16	0.839	0.651	7.19	8.18	
WC-C (cobalt redox couples	15.52	0.842	0.721	9.42	8.18	
NbC (cobalt redox couples)	12.27	0.836	0.610	6.26	8.18	
NbC-C (cobalt redox couples)	13.78	0.842	0.679	7.89	8.18	
Co-TiC (nanoparticles)	9.98	0.75	0.50	3.87	–	[86]

TABLE 2 Photovoltaic parameter performance of various Carbides based dye-sensitized solar cells—cont'd

CE materials	J_{SC} (mA/cm^2)	V_{OC} (V)	FF	PCE (%)	PCE (%) Pt	Reference
WC	9.5	0.759	0.67	4.83	5.07	[88]
Mo$_2$C	8.69	0.718	0.52	3.2	5.07	
Cr$_2$C$_3$	8.94	0.7	0.63	4.44	5.07	
Mo$_2$C@NC	14.72	0.764	0.58	6.49	6.38	[87]
VC-ch	16.71	0.69	0.72	7.92	7.79	[90]
Co$_3$O$_4$-WC-CN/rGO	16.27	0.77	0.59	7.38	6.85	[92]
Υ-MoC/Ni@NC	16.13	0.702	0.46s	5.26	5.65	[94]
Ti$_3$C$_2$(2D) (cobalt redox couples)	15.09	0.859	0.74	9.57	9.25	[95]
Ti$_3$C$_2$(T$_2$/T$^-$)	13.47	0.696	0.65	6.12	3.65	
Ti$_3$C$_2$(I$_3^-$/I$^-$)	15.28	0.772	0.69	8.12	7.86	
TiC	11.66	0.842	0.75	7.37	7.47	[95]
TiC (commercial)	13.30	0.788	0.62	6.50	7.5	[20]
VC	12.56	0.782	0.65	6.38	7.50	
VC (Commercial)	10.91	0.803	0.56	4.92	7.5	
VC-MC	13.11	0.808	0.72	7.63	7.5	
Cr$_2$C$_3$	11.31	0.825	0.62	5.79	7.5	
ZrC-commercial	10.91	0.803	0.44	3.85	7.5	
NbC	8.22	0.748	0.40	2.46	7.5	
Mo$_2$C	12.42	0.769	0.61	5.83	7.5	
Mo$_2$C	14.28	0.72	0.58	5.93	7.69	[96]
NCNT@Co	15.79	0.74	0.59	6.88	7.69	
Mo$_2$C-1/NCNTs@Co	15.78	0.75	0.59	7.10	7.69	
Mo$_2$C-2/NCNTs@Co	17.13	0.77	0.62	8.17	7.69	

Continued

254 Oxide free nanomaterials for energy storage and conversion applications

TABLE 2 Photovoltaic parameter performance of various Carbides based dye-sensitized solar cells—cont'd

CE materials	J_{SC} (mA/ cm^2)	V_{OC} (V)	FF	PCE (%)	PCE (%) Pt	Reference
Mo$_2$C-4/ NCNTs@Co	17.73	0.78	0.64	8.82	7.69	
Mo$_2$C-6/ NCNTs@Co	16.52	0.77	0.61	7.78	7.69	
Mo$_2$C-8/ NCNTs@Co	16.00	0.76	0.61	7.37	7.69	
WC/C$_{MA}$	13.18	0.77	0.68	6.85	7.51	[97]
WC/C$_{TA}$	14.56	0.74	0.68	7.32	7.51	
WC	12.8	0.76	0.60	5.83	7.51	
Mo$_2$C/MC	17.10	0.67	0.64	7.29	7.08	[98]
Bi/Mo$_2$C/MC	18.11	0.71	0.63	8.06	7.08	
MC	16.24	0.70	0.57	6.52	7.08	
TiC	12.36	0.835	0.75	7.74	7.51	[92]
TiC-PEDOT	13.21	0.840	0.75	8.26	7.51	

and solution/solid electrolytes. In which, the counter electrode is a significant component, could influence on device cost and performance. As a characteristic of the counter electrode, it has to perform higher conductivity, superior catalytic activity for electrolyte regeneration, with good stability performance. Platinum is well studied electrocatalytic active material for the preparation of counter electrodes, demonstrated higher power conversion efficiency. Due to its higher cost, limited stability prohibiting its diversified application. In search of Pt-free CEs, various materials have proved for better DSSC performance. The study of TMC CE has become a potential research field of the area in recent years due to its Pt-like catalytic performance, material diversity, simple synthesis route, and its optimization steps. Though the performance of TMCs assisted DSSCs, especially Nitrides and Carbides is quite good due to their conductivity, electron transportation at the interface of TMCs and substrates. The power conversion efficiencies for Nitrides assisted DSSCs are 10.86 and 9.75 for Carbide based DSSCs. The power conversion efficiency of DSSCs could be optimize using Nitrides / Carbides in near future.

References

[1] B. O'Regan, M. Grätzel, Nature 353 (1991) 737–740.

[2] K. Kakiage, Y. Aoyama, T. Yano, K. Oya, J.-I. Fujisawa, M. Hanaya, Chem. Commun. 51 (88) (2015) 15893–15897.

[3] S. Mathew, A. Yella, P. Gao, R. Humphry-Baker, F.E. Curchod Basile, N. Ashari-Astani, I. Tavernelli, U. Rothlisberger, M.K. Nazeeruddin, M. Gratzel, Nat. Chem. 6 (2014) 242.

[4] S. Ahmad, E. Guillen, L. Kavan, M. Gratzel, M.K. Nazeeruddin, Energy Environ. Sci. 6 (2013) 3439–3466.

[5] A. Hagfeldt, G. Boschloo, L. Sun, L. Kloo, H. Pettersson, Chem. Rev. 110 (2010) 6595–6663.

[6] J. Halme, P. Vahermaa, K. Miettunen, P. Lund, Adv. Mater. 22 (2010) 210–234.

[7] A. Listorti, B. O'Regan, J. Durrant, Chem. Mater. 23 (2011) 3381–3399.

[8] P. Barnes, K. Miettunen, X. Li, A. Anderson, T. Bessho, M. Gratzel, B. O'Regan, Adv. Mater. 25 (2013) 1881–1922.

[9] S. Fantacci, F. De Angelis, Coord. Chem. Rev. 255 (2011) 2704–2726.

[10] A. Mishra, M. Fischer, P. Bauerle, Angew. Chem. 48 (2009) 2474–2499.

[11] J. Ondersma, T. Hamann, Coord. Chem. Rev. 257 (2013) 1533–1543.

[12] H. Dunn, L. Peter, J. Phys. Chem. C 113 (2009) 4726–4731.

[13] W. Liu, L. Hu, Z. Huo, S. Dai, Prog. Chem. 21 (2009) 1085–1093.

[14] R. Nicholson, S. Irving, Anal. Chem. 36 (1964) 706–723.

[15] R. Nicholson, Anal. Chem. 37 (1965) 1351–1355.

[16] J. Heinze, Angew. Chem. 23 (1984) 831–847.

[17] Y. Xiao, J. Lin, W. Wang, S. Tai, G. Yue, J. Wu, Electrochim. Acta 90 (2013) 468–474.

[18] K. Saranya, M. Rameez, A. Subramania, Eur. Polym. J. 66 (2015) 207–227.

[19] M. Wu, Y. Wang, X. Lin, N. Yu, L. Wang, A. Hagfeldt, T. Ma, Phys. Chem. Chem. Phys. 13 (2011) 19298–19301.

[20] M. Wu, X. Lin, Y. Wang, L. Wang, W. Guo, D. Qu, X. Peng, A. Hagfeldt, M. Gratzel, T. Ma, J. Am. Chem. Soc. 134 (2012) 3419–3428.

[21] L. Steve, Renewable Energy World, U.N. Secretary-General: Renewables Can End Energy Poverty, 2021. http://www.renewableenergyworld.com/articles/2011/08/u-n-secretary-generalrenewables-can-end-energy-poverty.html. (Accessed 25 August 2011).

[22] A. Polman, M. Knight, E. Garnett, B. Ehrler, W. Sinke, Science 352 (2016), aad4424.

[23] F. Fabregat-Santiago, J. Bisquert, L. Cevey, P. Chen, M. Wang, S. Zakeeruddin, M. Gratzel, J. Am. Chem. Soc. 131 (2008) 558–562.

[24] L. Han, N. Koide, Y. Chiba, A. Islam, R. Komiya, F. Nobuhiro, A. Fukui, R. Yamanaka, Appl. Phys. Lett. 86 (2005) 213501.

[25] Q. Wang, J. Moser, M. Gratzel, J. Phys. Chem. B 109 (2005) 14945–14953.

[26] Z. Huang, G. Natu, Z. Ji, P. Hasin, Y. Wu, J. Phys. Chem. C 115 (2011) 25109–25114.

[27] D. Zheng, M. Ye, X. Wen, N. Zhang, C. Lin, Sci. Bull. 60 (2015) 850–863.

[28] L. Peter, J. Phys. Chem. C 111 (2007) 6601–6612.

[29] G. Schlichthorl, N. Park, A. Frank, J. Phys. Chem. B 103 (1999) 782–791.

[30] J. van de Lagemaat, N. Park, A. Frank, J. Phys. Chem. B 104 (2000) 2044–2052.

[31] F. Cao, G. Oskam, G. Meyer, P. Searson, J. Phys. Chem. A 100 (1996) 17021–17027.

[32] L. Dloczik, O. Ileperuma, I. Lauermann, L. Peter, E. Ponomarev, G. Redmond, N. Shaw, I. Uhlendorf, J. Phys. Chem. B 101 (1997) 10281–10289.

[33] J. Kruger, R. Plass, M. Gratzel, P. Cameron, L. Peter, J. Phys. Chem. B 107 (2003) 7536–7539.

[34] G. Schlichthorl, S. Huang, J. Sprague, A. Frank, J. Phys. Chem. B 101 (1997) 8141–8155.

[35] N. Koide, A. Islam, Y. Chiba, L. Han, J. Photochem. Photobiol. A Chem. 182 (2006) 296–305.

256 Oxide free nanomaterials for energy storage and conversion applications

[36] R. Kern, R. Sastrawan, J. Ferber, R. Stangl, J. Luther, Electrochim. Acta 47 (2002) 4213–4225.

[37] L. Han, N. Koide, Y. Chiba, A. Islam, T. Mitate, C. R. Chim. 9 (2006) 645–651.

[38] J. Bisquert, J. Phys. Chem. B 106 (2002) 325–333.

[39] M. Adachi, M. Sakamoto, J. Jiu, Y. Ogata, S. Isoda, J. Phys. Chem. B 110 (2006) 13872–13880.

[40] Q. Wang, S. Ito, M. Gratzel, F. Fabregat-Santiago, I. Mora-Sero, J. Bisquert, T. Bessho, H. Imai, J. Phys. Chem. B 110 (2006) 25210–25221.

[41] L. Peter, K. Wijayantha, Electrochim. Acta 45 (2000) 4543–4551.

[42] T. Oekermann, D. Zhang, T. Yoshida, H. Minoura, J. Phys. Chem. B 108 (2004) 2227–2235.

[43] W. Liu, L. Hu, S. Dai, L. Guo, N. Jiang, D. Kou, Electrochim. Acta 55 (2010) 2338–2343.

[44] J. Wu, Z. Lan, J. Lin, M. Huang, Y. Huang, L. Fan, G. Luo, Chem. Rev. 115 (2015) 2136–2173.

[45] X. Yang, M. Yanagida, L. Han, Energy Environ. Sci. 6 (2013) 54–66.

[46] R. Levy, M. Boudart, Science 181 (1973) 547–549.

[47] E. Furimsky, Appl. Catal. A Gen. 240 (2003) 1–28.

[48] C. Giordano, M. Antonietti, Nano Today 6 (2011) 366–380.

[49] J. Hargreaves, Coord. Chem. Rev. 257 (2013) 2015–2031.

[50] R. Ningthoujam, N. Gajbhiye, Prog. Mater. Sci. 70 (2015) 50–154.

[51] B. Zhang, D. Wang, Y. Hou, S. Yang, X. Yang, J. Zhong, J. Liu, H. Wang, P. Hu, H. Zhao, H. Yang, Sci. Rep. 3 (2013) 1836–1843.

[52] D. Pysch, A. Mette, S. Glunz, Sol. Energy Mater. Sol. Cells 91 (2007) 1698–1706.

[53] Q. Jiang, G. Li, X. Gao, Chem. Commun. 44 (2009) 6720–6722.

[54] K. Soo, M. Park, J. Kim, S. Park, Y. Dong, S. Yu, J. Kim, J. Park, J. Choi, J. Lee, Sci. Rep. 5 (2015) 10450.

[55] X. Zhang, X. Chen, S. Dong, Z. Liu, X. Zhou, J. Yao, S. Pang, H. Xu, Z. Zhang, L. Lia, G. Cui, J. Mater. Chem. A 22 (2012) 6067–6071.

[56] Q. Jiang, G. Li, S. Liu, X. Gao, J. Phys. Chem. C 114 (2010) 13397–13401.

[57] G. Li, J. Song, G. Pan, X. Gao, Energy Environ. Sci. 4 (2011) 1680–1683.

[58] L. Chen, H. Dai, Y. Zhou, Y. Hu, T. Yu, J. Liu, Z. Zou, Chem. Commun. 50 (2014) 14321–14324.

[59] J. He, J. Pringle, Y. Cheng, J. Phys. Chem. C 118 (2014) 16818–16824.

[60] S. Park, Y. Cho, M. Choi, H. Choi, J. Kang, J. Um, J. Choi, H. Choe, Y. Sung, Surf. Coat. Technol. 259 (2014) 560–569.

[61] M. Wu, H. Guo, Y. Lin, K. Wu, T. Ma, A. Hagfeldt, J. Phys. Chem. C 118 (2014) 12625–12631.

[62] Z. Wen, S. Cui, H. Pu, S. Mao, K. Yu, X. Feng, J. Chen, Adv. Mater. 23 (2011) 5445–5450.

[63] M. Wu, Q. Zhang, J. Xiao, C. Ma, X. Lin, C. Miao, Y. He, Y. Gao, A. Hagfeldt, T. Ma, J. Mater. Chem. A 21 (2011) 10761–10766.

[64] G. Li, F. Wang, Q. Jiang, X. Gao, P. Shen, Angew. Chem. Int. Ed. 49 (2010) 3653–3656.

[65] M. Wang, P. Chen, R. Humphry-Baker, S.M. Zakeeruddin, M. Gratzel, ChemPhysChem 10 (2009) 290.

[66] M. Wang, C. Gratzel, S.-J. Moon, R. Humphry-Baker, N. Rossier-Iten, S.M. Zakeeruddin, M. Gratzel, Adv. Funct. Mater. 19 (2009) 2163.

[67] G.R. Li, F. Wang, J. Song, F.Y. Xiong, X.P. Gao, Electrochim. Acta 65 (2012) 216–220.

[68] X. Zhang, X. Chen, K. Zhang, S. Pang, X. Zhou, H. Xu, S. Dong, P. Han, Z. Zhang, C. Zhanga, G. Cui, J. Mater. Chem. A 1 (2013) 3340–3346.

[69] J. Song, G.R. Li, K. Xi, B. Lei, X.P. Gao, R. Vasant Kumar, J. Mater. Chem. A 2 (2014) 10041–10047.

[70] J. Balamurugan, T.D. Thanh, N.H. Kim, J.H. Lee, Adv. Mater. Interfaces 3 (2016) 1500348.

Metal nitrides and carbides as advanced counter electrodes **Chapter | 10** **257**

[71] N. Divya, M. Baro, S. Ramaprabhu, J. Nanosci. Nanotechnol. 16 (2016) 9583–9590.

[72] V.-D. Dao, J. Power Sources 337 (2016) 125–129.

[73] C.Y. Wu, G.R. Li, X.Q. Cao, B. Lei, X.P. Gao, Green Energy Environ. 2 (3) (2017) 302–309.

[74] G. Wang, S. Liu, Mater. Lett. 161 (2015) 294–296.

[75] J.S. Jang, D.J. Ham, E. Ramasamy, J. Lee, J.S. Lee, Chem. Commun. 46 (2010) 8600–8602.

[76] A.-R. Ko, J.-K. Oh, Y.-W. Lee, S.-B. Han, K.-W. Park, Mater. Lett. 65 (2011) 2220–2223.

[77] M. Wu, X. Lin, A. Hagfeldt, T. Ma, Angew. Chem. 123 (2011) 3582. Angew. Chem. Int. Ed. 50 (2011) 3520–3524.

[78] C. Giordano, C. Erpen, W. Yao, M. Antonietti, Nano Lett. 8 (2008) 4659–4663.

[79] S. Yun, H. Zhang, H. Pu, J. Chen, A. Hagfeldt, T. Ma, Adv. Energy Mater. 3 (11) (2013) 1407–1413.

[80] M. Wu, T. Ma, ChemSusChem 5 (2012) 1343–1357.

[81] S. Yun, M. Wu, Y. Wang, J. Shi, X. Lin, A. Hagfeldt, T. Ma, ChemSusChem 6 (3) (2013) 411–416.

[82] Y. Gao, L. Chu, M. Wu, L. Wang, W. Guo, T. Ma, J. Photochem. Photobiol. A Chem. 245 (2012) 66–71.

[83] M. Towannang, P. Kumlangwan, W. Maiaugree, K. Ratchaphonsaenwong, V. Harnchana, W. Jarenboon, S. Pimanpang, V. Amornkitbamrung, Electron. Mater. Lett. 11 (2015) 643–649.

[84] J. Theerthagiri, R.A. Senthil, M.H. Buraidah, J. Madhavan, A.K. Arof, Mater. Today: Proc. 3 (2016) 65–72.

[85] H. Guo, Q. Han, C. Gao, H. Zheng, Y. Zhu, M. Wu, J. Power Sources 332 (2016) 399–405.

[86] A. Yousef, R.M. Brooks, M.H. El-Newehy, S.S. Al-Deyab, H.Y. Kim, Int. J. Hydrog. Energy 42 (2017) 10407–10415.

[87] T. Wang, J. Wang, W. Chen, X. Zheng, E. Wang, Chem Eur J 23 (68) (2017) 17311–17317.

[88] K. Wu, Y. Wang, W. Cui, B. Ruan, M. Wu, Ionics 24 (3) (2017) 883–890.

[89] J.S. Kang, J. Kim, M.J. Lee, Y.J. Son, J. Jeong, D.Y. Chung, A. Lim, H. Choe, H.S. Park, Y.E. Sung, Nanoscale 9 (2017) 5413–5424.

[90] J. Jin, Z. Wei, X. Qiao, H. Fan, L. Cui, RSC Adv. 7 (2017) 26710–26716.

[91] M.Z. Iqbal, S. Khan, Sol. Energy 160 (2018) 130–152.

[92] L. Chen, W. Chen, E. Wang, J. Power Sources 380 (2018) 18–25.

[93] A. Majid, I. Ullah, K.T. Kubra, S.U.-D. Khan, S. Haider, Surf. Sci. 687 (2019) 41–47.

[94] X. Zhang, T. Wang, X. Li, D. Wu, W. Chen, J. Coord. Chem. (2020) 1–11.

[95] J.-y. Ma, M. Sun, Y.-a. Zhu, H. Zhou, K. Wu, J. Xiao, M. Wu, ChemElectroChem (2020) 2020, https://doi.org/10.1002/celc.201902159.

[96] Y. Zhang, S. Yun, Z. Wang, Y. Zhang, C. Wang, A. Arshad, F. Han, Y. Si, W. Fang, Ceram. Int. (2020), https://doi.org/10.1016/j.ceramint.2020.03.128.

[97] M. Chen, G.-C. Wang, W.-Q. Yang, Z.-Y. Yuan, X. Qian, J.-Q. Xu, Z.-Y. Huang, A.-X. Ding, ACS Appl. Mater. Interfaces 11 (2019) 42156–42171.

[98] S. Yun, Y. Hou, C. Wang, Y. Zhang, X. Zhou, Ceram. Int. 45 (2019) 15589–15595.

Chapter 11

Metal chalcogenide-based counter electrodes for dye-sensitized solar cells

Subalakshmi Kumar[a], Senthilkumar Muthu[b], Sankar Sekar[c,d], Chinna Bathula[e], Ashok Kumar Kaliamurthy[a,f], and Sejoon Lee[c,d]

[a]*Department of Nuclear Physics, University of Madras, Chennai, Tamil Nadu, India,* [b]*Crystal Growth Centre, Anna University, Chennai, India,* [c]*Division of Physics & Semiconductor Science, Dongguk University-Seoul, Seoul, Republic of Korea,* [d]*Quantum-functional Semiconductor Research Center, Dongguk University-Seoul, Seoul, Republic of Korea,* [e]*Division of Electronics and Electrical Engineering, Dongguk University-Seoul, Seoul, Republic of Korea,* [f]*Department of Energy and Materials Engineering, Dongguk University-Seoul, Seoul, Republic of Korea*

Chapter outline

1. Introduction   259
2. Sulfide based counter electrodes   261
3. Selenides and tellurides based counter electrodes   267
4. Composite based counter electrodes   272
5. Conclusion and future prospects   279
References   280

1. Introduction

The widespread global energy demand has surged the need for dependency on fossil fuels to a dramatic extent in recent decades. The impact of this dependency on fossil fuels lead to an increase in carbon emission and accelerated global warming, which is already in an alarming stage. The prevailing research on renewable energy technology is a promising solution to address the current global energy demand. The emerging third-generation solar cells are considered to be the potential candidate for the replacement of conventional c—Si solar cells as an inexpensive and promising alternative. Among the other third-generation solar cells, the dye-sensitized solar cells (DSSCs) have attracted tremendous research interest in the past few decades. Owing to their inexpensiveness, ease of fabrication, and competing power conversion efficiency (PCE), the DSSCs are believed to be the promising cost-effective solar cells [1, 2].

The device architecture of the DSSC consists of four major components. The mesoporous photoanode, light-absorbing layer (sensitizer), redox coupled

Oxide Free Nanomaterials for Energy Storage and Conversion Applications.
https://doi.org/10.1016/B978-0-12-823936-0.00015-2
Copyright © 2022 Elsevier Inc. All rights reserved.

electrolyte, and the counter electrode (CE). In general, mesoporous titanium dioxide (TiO$_2$) is the most commonly used photoanode. Different organic dyes were utilized as a sensitizer in DSSCs for effective light harvesting and conversion into electrical energy. A wide variety of dyes (ruthenium, other metal complex-based dyes, natural dyes, etc.) were explored to achieve the maximum PCE in DSSCs. Among various dyes, the molecularly engineered porphyrin-based dyes showed the maximum power conversion efficiency of 13% [3]. The iodide/triiodide (I$_3^-$/I$^-$) redox couple electrolyte is the most efficient charge transferring electrolyte in DSSCs. However, since the charge transfer depends also on the sensitizer used in DSSCs, the electrolyte should be chosen with the most required band alignments. In conventional DSSCs, the most commonly used counter electrode is the platinum (Pt) coated transparent conducting substrate [4].

The counter electrode (CE) is a crucial DSSC component, which has significant influences on the power conversion efficiency, stability, and cost of the device. The major role of the counter electrode is to collect the electrons from the external circuit and so as to subsequently catalyze the reduction of I$_3^-$ to I$^-$ ions, as shown in Fig. 1. The counter electrode material should possess some major advantages for the efficient performance of the DSSC such as high electrocatalytic activity for effective regeneration of electrolyte, high electrical conductivity for enhanced charge transportation, high surface area for better adhesion to fluorine-doped tin oxide (FTO), high chemical stability for the long-term usage, and low cost for commercialization [5, 6]. Owing to the high conductivity and the swift charge transferring characteristics, Pt CEs were widely used in DSSCs. Nevertheless, the scarcity and the cost of Pt have turned the research interest of DSSCs into the search for inexpensive and efficient Pt-free CE alternatives. An immense range of materials has been employed as an inexpensive alternative for the replacement of Pt CEs in DSSCs [4].

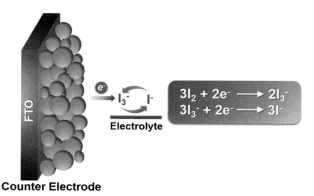

FIG. 1 Schematic representation of the redox reaction at the counter electrode interface in the DSSC.

Carbonaceous materials (e.g., graphene, reduced graphene oxide [rGO], carbon nanotubes [CNTs], activated carbon [AC], carbon black [7], etc.), inexpensive polymers [6], transition metal oxides, carbides, nitrides, chalcogenides, and their alloys [8] are the potential candidate materials that can alternate Pt CEs. Among the other materials, the semiconducting metal chalcogenides have exhibited the favorable characteristics such as tunable chemical composition, good electrocatalytic properties, and optimum bandgap energy values. Therefore, various transition metal chalcogenide CEs were employed in DSSCs, and their superior performances over conventional Pt-based CEs were studied extensively in recent years [9].

In this chapter, we review the recent developments and the research works in the field of the metal chalcogenides-based Pt-free CEs. Various chalcogenides-based CEs, including binary, ternary, multinary metal chalcogenides, and their composites with carbonaceous materials, are compiled in this review. The photovoltaic performances of the various metal chalcogenide-based CEs are tabulated to compare and address their potentials as an inexpensive CE, possessing the outstanding electrocatalytic properties.

2. Sulfide based counter electrodes

Transition metal sulfides such as CoS, MoS_2, NiS, and WS_2 are one of the promising alternative materials that can replace the Pt CEs becasue of their cost-effectiveness, superior electrocatalytic activity, and excellent electrical conductivity for the reduction of I_3^- ions in electrolyte (Table 1). Gratzel et al. demonstrated that CoS could be very effective in catalyzing the reduction of triiodide to iodide in DSSC [13]. The CoS deposited ITO/PEN films were used in the DSSC as a counter electrodes in conjunction with the Z907 sensitizer and the eutectic melt electrolyte; and the DSSC eventually yielded a 6.5% efficiency under the 1-sun intensity. Huo et al. fabricated CoS and NiS electrodes by the simple electrodeposition method, and obtained the overall PCE of about 9.23% and 9.65% for CoS and NiS CEs-based DSSCs, respectively [12]. They revealed that adding different amounts of ammonia into deposition solution can significantly influence the morphologies of the CoS and NiS films. In addition, the electrochemical impedance spectroscopy (EIS), cyclic voltammetry (CV), and Tafel studies showed that the electrodeposited CoS and NiS CEs had an excellent catalytic activity towards the I^-/I_3^- electrolyte. Li et al. fabricated the well-designed NiS nanoflower arrays and used those as a counter electrode to assemble the I^-/I_3^- mediated DSSC [14]. Fig. 2A and B shows the scanning electron microscopy (SEM) images of the oriented NiS nanoflower arrays. In Fig. 2F, the increased IPCE at 350–650 nm indicates that the form of the oriented nanoflower arrays could be a promising strategy to construct the highly efficient Pt-free CEs. Resultantly, the DSSC showed the superior electrocatalytic activity for the I_3^- reduction and the efficient charge transport properties, resulting in a higher PCE of 9.49% than that of Pt (7.88%) with a better

TABLE 1 Summary of photovoltaic parameters for various DSSCs with different types of sulfide CEs.

CE catalysts	Substrate	Redox couple	Dye	FF (%)	η (%)	η (Pt) (%)	Ref.
CoS	FTO	I^-/I_3^-	N719	56.83	8.39	7.88	[10]
CoS	FTO	I^-/I_3^-	N719	63.65	5.74	6.44	[11]
CoS	FTO	I^-/I_3^-	N719	63.1	9.23	8.12	[12]
CoS	FTO	I^-/I_3^-	Z907	73	6.5	6.5	[13]
NiS	FTO	I^-/I_3^-	N719	75	9.49	7.88	[14]
NiS_2	FTO	I^-/I_3^-	N719	70	8.46	8.04	[15]
NiS	FTO	I^-/I_3^-	N719	65.4	9.65	8.12	[12]
NiS nanotube	FTO	I^-/I_3^-	N719	73	98	8.5	[16]
Ultrathin MoS_2	FTO	I^-/I_3^-	N719	66	8.28	7.53	[17]
MoS_2	FTO	I^-/I_3^-	N719	73	7.59	7.64	[18]
MoS_2	FTO	I^-/I_3^-	N719	67	7.19	7.42	[19]
MoS_2	FTO	I^-/I_3^-	N719	61	7.14	8.73	[20]
MoS_2	FTO	I^-/I_3^-	N719	54	6.32	6.38	[21]
WS_2	FTO	I^-/I_3^-	N719	70	7.73	7.64	[18]
VS_2	FTO	I^-/I_3^-	Z907	63	6.24	6.44	[22]
Digenite Cu_9S_5	FTO	$[Co(bpy)_3]^{2+/3+}$	LEG4	53	5.7	–	[23]
Sb_2S_3	FTO	I^-/I_3^-	N719	52.8	5.37	5.36	[24]
Cu_2FeSnS_4	FTO	I^-/I_3^-	N719	57	7.36	8.15	[25]
Cu_2CdSnS_4	FTO	I^-/I_3^-	N719	56	7.12	8.15	[25]
2D leaf-likeCu_2ZnSnS_4	FTO	I^-/I_3^-	N719	51.07	8.67	6.01	[26]
$NiCo_2S_4$	FTO	I^-/I_3^-	N719	71	8.9	8.6	[27]
$NiCo_2S_4$	FTO	I^-/I_3^-	N719	68.1	7.03	5.71	[28]
$CoNi_2S_4$	FTO	I^-/I_3^-	N719	66.62	4.03	4.59	[29]
$Ni_{0.9}5Mo_{0.05}S$	FTO	I^-/I_3^-	N3	64	7.15	7.20	[30]
Cu_2SnS_3	FTO	I^-/I_3^-	C101	72	10.18	9.24	[31]
Cu2WS4	FTO	I^-/I_3^-	N719	64	8.94	8.0	[32]
$CuFeS_2$	FTO	I^-/I_3^-	N719	70.01	7.74	8.10	[33]
Cu_2CdGeS_4	FTO	I^-/I_3^-	N719	66.8	7.67	7.54	[34]
Co-Cu-WS_x	FTO	I^-/I_3^-	N719	68	9.61	8.24	[35]
Co-Ni-MoS_x	FTO	I^-/I_3^-	N719	68.1	9.76	8.24	[36]

Metal chalcogenide-based counter electrodes **Chapter | 11 263**

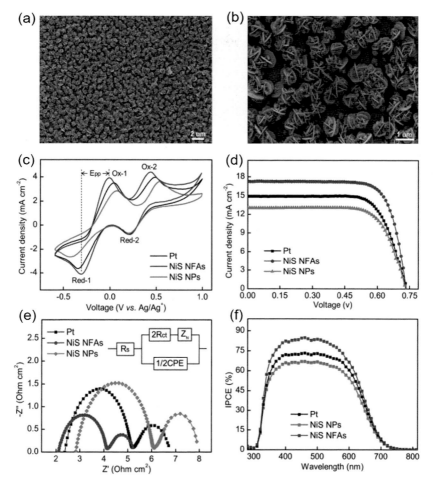

FIG. 2 (A, B) FE-SEM images of the NiS nanoflower arrays, (C) CV profiles, (D) EIS Nyquist plots of the CEs with standard Pt, NiS nanoflowers, and NiS nanoparticles, (E) *J–V* curves at the 1-sun intensity, and (F) IPCE data of the DSSCs, Copyright, Elsevier, 2019.

electrochemical stability (Fig. 2C–E). To enhance the electrocatalytic activity, different morphologies of CoS and NiS nanostructures (e.g., 3D nanospheres, 2D nanoflakes, etc.) have been prepared and investigated. The 3D hierarchical NiS$_2$ microspheres were prepared through the facile chemical etching/anion exchange reaction by Huang et al. [15]. The transmission electron microscopy (TEM) and SEM images showed that the prepared NiS$_2$ hierarchical microspheres were constructed by the 2D nanoflakes building blocks (Fig. 3A–E). Due to their favorable structural features, the NiS$_2$ hierarchical microspheres possess the large surface area, the high structural void porosity, and the accessible

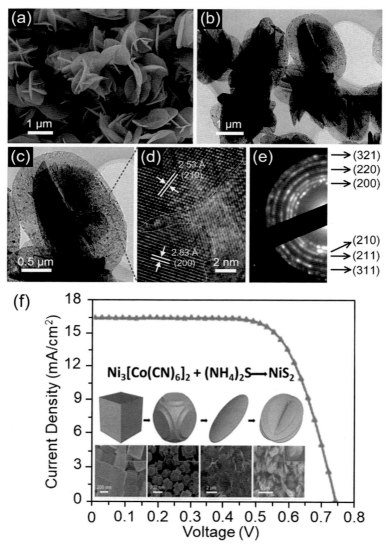

FIG. 3 (A) SEM, (B, C) TEM, (D) lattice fringe images, (E) SAED patterns of the NiS$_2$ microspheres annealed at 350°C, (F) *J–V* curve of the DSSC with the NiS$_2$ microsphere-based CE, Copyright, Royal Society of Chemistry, 2018.

inner surface, which could facilitate the mass ion diffusion and the effective charge transport between the electrolyte and the counter electrode material. As a result, the NiS$_2$ hierarchical microspheres exhibited the excellent electrocatalytic activity and the impressive PCE of 8.46%, which is higher than that of Pt electrodes (8.04%) (Fig. 3F). Besides, molybdenum sulfide (MoS$_2$) has also been considered as one of the promising materials for the CE applications because of its 2D layered structure analogous to that of graphene. The sandwiched S-Mo-S layers are stacked in the hexagonally packed MoS_2 lattices with weak van der Waals interactions. MoS_2 possesses some obvious characteristics, including the high specific surface area and the rich exposed active edges. Several research groups have studied on 2D MoS_2 as a CE material for high-performance DSSCs. Liang et al. have reported the facile one-step solution-phase process for the growth of ultrathin MoS_2, and showed the energy conversion efficiency of 8.28%, which is higher than that of the Pt CE-based DSSC (7.53%). Wu et al. found that both MoS_2 and WS_2 exhibited the good electrocatalytic activity towards the enhanced reduction of I_3^- ions, resulting in high PCE of 7.73% and 7.59%, respectively.

More recently, metal sulfides in the form of ternary and quaternary compounds (e.g., $NiCo_2S_4$, Cu_2ZnSnS_4, Cu_2SnS_3, etc.) have been attracted intensive interest due to their enormous active sites and rich redox reactions, compared with binary compounds [26, 27, 31]. The co-existence of two different cations in ternary metal sulfides increases the charge transport pathways, providing the high electrical conductivity and the excellent electrocatalytic activity for the reduction of the redox electrolyte (I^-/I_3^-). Many research groups have reported that the formation of multiple metal elements-included sulfides can efficiently enhance the electrocatalytic performance of the CEs with respect to single metal sulfides becasue of the synergetic effects from the multiple elements [31, 37]. Using the facile hydrothermal technique, the tremella-like mesoporous $NiCo_2S_4$ nanosheets were prepared by Chen et al. [27]. The tremelliform $NiCo_2S_4$ nanosheets-based CE resulted in the low charge transfer resistance and the excellent electrocatalytic activity for the I_3^- reduction. Thus, the $NiCo_2S_4$ based DSSC showed a higher conversion efficiency (8.9%), compared to the Pt-based device (8.6%). Chen et al. prepared the semitransparent 2D leaf-like Cu_2ZnSnS_4 plate arrays (PLAr) on the FTO glass substrate (Fig. 4A–E) by using in situ solvothermal synthesis without any posttreatment. The photoconversion efficiency of 7.09% was obtained when using the leaf-like CZTS PLAr as the DSSC counter electrode, which is about 18% higher than the efficiency of Pt-based device ($\eta = 6.01\%$) [26]. Furthermore, they used a reflective mirror to enhance the visible light utilization rate of the CZTS CE to achieve the high transmittance; and thus, the PCE was increased to 8.67%.

Liu et al. reported the highly efficient Cu_2SnS_3 counter electrode that showed the superior electrocatalytic performance in both I^-/I_3^- and $Co^{3+/2+}$ based DSSCs, and they demonstrated the photovoltaic efficiency up to 10.18% for iodine electrolyte-mediated DSSC, which is higher than that of expensive Pt (9.31%) [31]. Recently, Xing Quan's group synthesized a new quaternary type of the Co-Cu-WS_x double-shelled ball-in-ball hollow nanospheres with a uniform size of 600 nm by using the mild two-step solvothermal method. They found that those ball-in-ball hollow nanospheres exhibited the superior electrocatalytic properties of the DSSCs (Fig. 5A–D). Compared to

266 Oxide free nanomaterials for energy storage and conversion applications

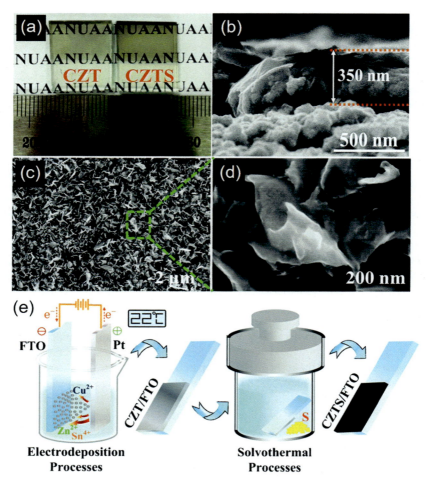

FIG. 4 (A) Photographic images of the CZT and CZTS thin-films prepared on the FTO substrates, (B) cross-sectional FE-SEM image of the CZTS thin film-based CE, (C, D) FE-SEM images of the leaf-like CZTS arrays prepared on the FTO substrate, and (E) schematic representation of the CZTS deposition process for preparing the CZT/FTO and CZTS/FTO CEs, Copyright, Royal Society of Chemistry, 2016.

ternary Co-WS$_x$ and Co-CuS$_x$ CEs, the quaternary Co-Cu-WS$_x$ nanospheres showed the excellent electrocatalytic behavior with the accelerated reduction of I$_3^-$. The counter electrode, composed of the Co-Cu-WS$_x$ nanospheres, exhibited the higher conversion efficiency of 9.61% than those of ternary Co-WS$_x$ (9.14%) and Co-CuS$_x$ (8.56%) as well as Pt (8.24%) [35]. The quaternary Co-Ni-MoS$_x$ yolk-shell nanospheres were prepared through the facile two-step solvothermal method by Qian et al. [36]. They reported the splendid photoconversion efficiency up to 9.76% from the DSSC with the Co-Ni-MoS$_x$-based

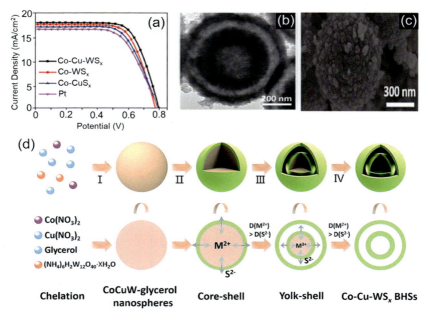

FIG. 5 (A) J–V curves of the DSSCs fabricated with the Co–Cu–WS$_x$, Co–WS$_x$, Co–CuS$_x$, and Pt CEs, (B) TEM, (C) SEM images of the Co–Cu–WS$_x$ ball-in-ball hollow nanospheres, and (D) formation mechanism of the Co–Cu–WS$_x$ ball-in-ball hollow nanospheres, Copyright, Royal Society of Chemistry, 2019.

CE, which was attributed to the increased active surface area and the decreased charge-transfer resistance.

3. Selenides and tellurides based counter electrodes

Metal selenides are also one of the competitive Pt-free CE candidates because of their distinctive electrocatalytic activity for triiodide reduction; hence, they are regarded as a highly effective CE material in DSSCs (Table 2). For instance, Hongmei Li and co-workers engineered and synthesized a tubular structured CoSe$_2$ nanomaterial by using the facile precursor transformation method. Benefiting from the advantageous structural features including functional shells and well-defined interior voids, the tubular structured CoSe$_2$ as the CE catalyst showed a high power conversion efficiency up to 9.34%, which is superior to that of the Pt CE (8.15%) [38]. Gong et al. fabricated the metal selenides such as Co$_{0.85}$Se and Ni$_{0.85}$Se on the conductive glass substrates by using the one-step in situ growth technique based on the low-temperature hydrothermal method. They directly grew the metal selenides on the conductive substrates, used the as-prepared sample as the transparent CEs for the DSSCs without any posttreatments, and found that graphene-like Co$_{0.85}$Se exhibited the higher

TABLE 2 Summary of photovoltaic parameters for various DSSCs with different types of selenides- and tellurides-based CEs.

CE catalysts	Substrate	Redox couple	Dye	FF (%)	η (%)	η (Pt) (%)	Ref.
$CoSe_2$	FTO	I^-/I_3^-	N719	70	9.34	8.15	[38]
$Co_{0.85}Se$	FTO	I^-/I_3^-	N719	75	9.40	8.64	[39]
$Co_{1.2}Se$	FTO	I^-/I_3^-	N719	64	9.28	6.84	[40]
$CoSe_2$	FTO	I^-/I_3^-	N719	66.2	8.38	7.83	[41]
$CoSe_{0.85}$	FTO	I^-/I_3^-	N719	68	6.03	6.45	[42]
NiSe	FTO	I^-/I_3^-	N719	72	7.82	7.62	[43]
$NiSe_2$	FTO	I^-/I_3^-	N719	74.3	8.69	8.04	[44]
$Ni_{0.85}Se$	FTO	I^-/I_3^-	N719	67	7.24	7.56	[45]
$Ni_{0.85}Se$	FTO	I^-/I_3^-	N719	72	8.32	8.64	[39]
$Ni_{0.85}Se$	FTO	I^-/I_3^-	N719	69.5	8.88	8.13	[46]
Ni_3Se_2	Ni foam	I^-/I_3^-	N719	67	4.62	5.34	[47]
$MoSe_2$	FTO	I^-/I_3^-	N719	73	7.28	7.40	[48]
Ru-Co-Se	FTO	I^-/I_3^-	N719	67.3	9.07	7.03	[49]
Co-Sn-Se nanocages	FTO	I^-/I_3^-	N719	65	9.25	8.19	[50]
Co-Fe-Se	FTO	I^-/I_3^-	N719	68.9	9.58	8.16	[51]
$NiCoSe_2$	FTO	I^-/I_3^-	N719	63.8	8.76	8.22	[52]
$Co_{0.42}Ni_{0.58}Se$	FTO	I^-/I_3^-	N719	63	6.15	5.53	[53]
$Ni_{0.5}Co_{0.5}Se_2$	FTO	I^-/I_3^-	N719	54	6.02	6.11	[54]
$CoTe_2$	FTO	I^-/I_3^-	N719	60	8.10	8.20	[55]
$MoTe_2$	FTO	I^-/I_3^-	N719	65.64	7.25	8.15	[56]

electrocatalytic activity for the reduction of I_3^-, compared to Pt. Therefore, the DSSC with $Co_{0.85}Se$ provided a higher short circuit current and a power conversion efficiency (9.40%) than Pt [39]. In other works, Zhang et al. also used $Ni_{0.85}Se$ as the CE for the DSSCs. They demonstrated a facile self-assembly method to fabricate mesoporous $Ni_{0.85}Se$ spheres, consisting of numerous primary particles, as shown in Fig. 6A–C. Their mesoporous structures and large internal reaction areas endowed the mesoporous $Ni_{0.85}Se$ spheres with the excellent CE performances (Fig. 6D–F). Mesoporous $Ni_{0.85}Se$ spheres

Metal chalcogenide-based counter electrodes Chapter | 11 **269**

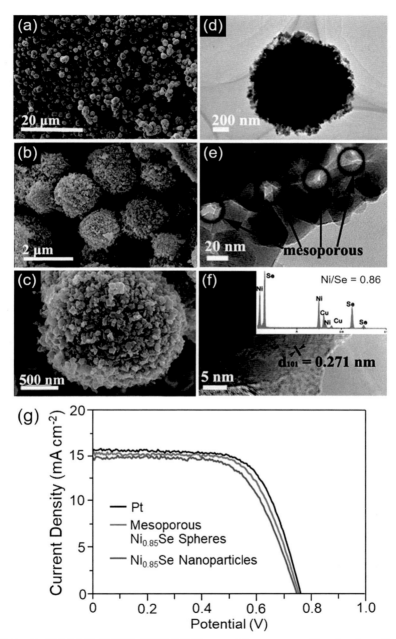

FIG. 6 (A–C) SEM, (D, E) TEM, (F) HR-TEM images of the mesoporous $Ni_{0.85}Se$ spheres (Inset: EDAX spectra), and (G) *J–V* curves of the DSSCs with various CEs based on $Ni_{0.85}Se$ spheres, $Ni_{0.85}Se$ nanoparticles, and Pt, Copyright, Royal Society of Chemistry, 2016.

exhibited the good electrocatalytic activity and the improved power conversion efficiency of 7.24% (Fig. 6G), which is comparable to the performance of Pt-based DSSC device ($=7.56$) [45]. Feng Gong's group synthesized nickel diselenide ($NiSe_2$), and used it as a CE of the DSSC. The device based on the $NiSe_2$ CE produced a higher PCE (8.69%) than that of the Pt CE-based DSSC (8.04%) [44]. Vikraman et al. prepared the $MoSe_2$ CE with the controlled uniform morphology by using a wet chemical route. The DSSCs with the $MoSe_2$ CE exhibited the remarkable PCE of 7.28%, which is comparable to that if the Pt CE ($\sim 7.40\%$) [48]. In addition, metal tellurides have also been considered as an emerging Pt-free CE material for the DSSC application. Patil et al. prepared the self-standing $CoTe_2$ nanotubes on the FTO substrate in order to use is as a Pt-free CE in DSSCs. The device with the $CoTe_2$ CE exhibited a power conversion efficiency of 8.10%, which is comparable to that of the Pt-based DSSC (8.10%) [55]. Hussain et al. presented the "sputtering-CVD-postannealing" route to prepare the $MoTe_2$ CEs for the high-performance DSSCs [56]. The $MoTe_2$ CE-based DSSC revealed the photoconversion efficiency of 7.25%, which is comparable to the efficiency observed from the DSSC with the Pt CE (8.15%).

To enhance both the electrocatalytic activity of the CE and the photovoltaic performance of the DSSC, the Pt-free ternary alloys such as Ru-Co-Se, Co-Sn-Se, and Co-Fe-Se have been developed by several research groups [49–51]. Zhao et al. have demonstrated that the electrodeposited RuCoSe ternary alloy could be an effective approach to elevate the electrocatalytic activity of the CE for the reduction of triiodides ions to iodides, and it could eventually enhance the photovoltaic performance of the DSSCs [49]. The DSSCs with the RuCoSe ternary alloy-based CEs showed an impressive power conversion efficiency of 9.07%, which is higher than 7.03% from the DSSC with the Pt-only CE. A very interesting study was conducted by Qian He's research group. They developed the CE materials by using the uniformly distributed Co-Sn-X (X=S, Se, Te) nanocages on the FTO substrates so as to increase the electrocatalytic activity. In short, they developed a simple and general anion exchange method for the synthesis of Co-Sn-X (X=S, Se, Te) nanocages with a uniform morphology (Fig. 7A–I) and a good structural stability by using the $CoSn(OH)_6$ nanocubes as the functional templates (Fig. 7J).

Benefiting from advantageous compositional features and well-designed hollow architectures, the obtained Co-Sn-X (X=S, Se, Te) nanocages displayed the enhanced electrocatalytic activity for the DSSCs (Fig. 7L). The Co-Sn-Se nanocages as the CE catalyst delivered a prominent power conversion efficiency (PCE) of 9.25% for the DSSCs (Fig. 7K), compared with the Pt CE (8.19%) [50]. Chen et al. synthesized a series of novel $NiCoSe_2$ microspheres by using a one-step hydrothermal method. The DSSCs with the $NiCoSe_2$ CEs exhibited the PCE values up to 8.76%, which is higher than that of the Pt CE-based devices (8.31%) [43]. Profiting from the coexistence of Ni and Co in $NiCoSe_2$, the ternary metal selenide offered the richer redox reaction, excellent electrochemical stability, and good charge-transfer ability (Fig. 8A–F).

Metal chalcogenide-based counter electrodes **Chapter | 11** **271**

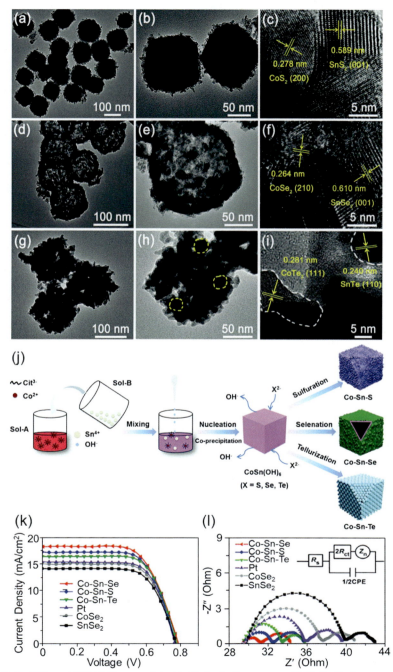

FIG. 7 TEM and HRTEM images of the (A–C) Co-Sn-S nanocages, (D–F) Co-Sn-Se nanocages, (G–I) Co-Sn-Te nanocages, (J) schematic illustration for the formation of the Co–Sn–X (X=S, Se, Te) nanocages, (K, L) *J–V* curves, and EIS curves of the DSSCs with the CEs based on the Co–Sn–X (X=S, Se, Te) nanocages, Copyright, Royal Society of Chemistry, 2018.

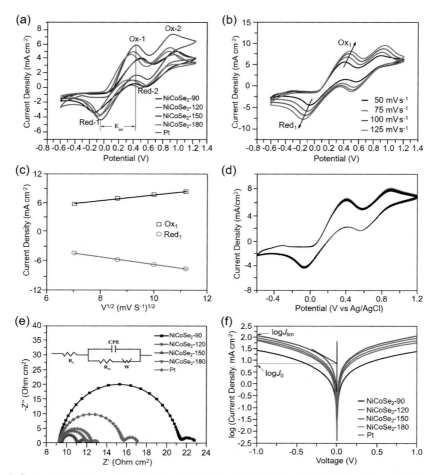

FIG. 8 (A) CV curves of the NiCoSe$_2$-90, NiCoSe$_2$-120, NiCoSe$_2$-150, NiCoSe$_2$-180, and Pt CEs at a scan rate of 50 mV/s, (B) CV curves of the optimized NiCoSe$_2$-150 CE recorded at various scan rates of 50, 75, 100, and 125 mV/s, and (C) redox current density as a function of the square root of the scan rate for the NiCoSe$_2$-150 CE, (D) 100-stacking CV curves for the NiCoSe$_2$-150 CE at a scan rate of 50 mV/s, (E) EIS Nyquist plots (Inset: equivalent circuit model), and (F) Tafel polarization curves for the symmetric dummy cells fabricated with the NiCoSe$_2$-90, NiCoSe$_2$-120, NiCoSe$_2$-150, NiCoSe$_2$-180, and Pt CEs, Copyright, Royal Society of Chemistry, 2018.

4. Composite based counter electrodes

The auspicious outcome of the metal chalcogenide-based CEs based DSSCs encouraged the extensive studies on developing the novel metal sulfides/selenides/tellurides for seeking the inexpensive Pt-free CEs (Table 3). Although the metal chalcogenide CEs have been proved as a promising alternative

TABLE 3 Summary of photovoltaic parameters of various DSSCs with different types of composites-based CEs.

CE catalysts	Substrate	Redox couple	Dye	FF (%)	η (%)	η (Pt) (%)	Ref.
NiS/CNTs	FTO	I^-/I_3^-	N719	65	10.82	8.03	[57]
Carbon/Ni$_3$S$_2$	FTO	I^-/I_3^-	N719	62	9.64	8.38	[58]
CoS/rGO	FTO	I^-/I_3^-	N719	63.3	9.39	7.34	[59]
CoS$_2$/carbon nanocages	FTO	I^-/I_3^-	N719	65.74	8.20	7.88	[60]
rGO-CoS	ITO	I^-/I_3^-	N719	63	8.34	6.27	[61]
Co$_9$S$_8$/CNFs	FTO	I^-/I_3^-	N719	51.8	8.37	8.50	[62]
MoS$_2$/N doped carbon shell-core	FTO	I^-/I_3^-	N719	61.3	6.2	7.0	[63]
MoS$_2$/graphene	FTO	I^-/I_3^-	N719	65.74	8.01	8.21	[64]
FeS$_2$/carbon hybrids	Carbon cloth	I^-/I_3^-	N719	72.2	8.15	7.61	[65]
Bi$_2$S$_3$/CNFs	FTO	I^-/I_3^-	N719	65	7.64	7.12	[66]
In$_{2.77}$S$_4$/Graphene	FTO	I^-/I_3^-	N719	66	7.32	6.48	[67]
C-SnS$_2$	FTO	I^-/I_3^-	N719	64.99	7.06	7.09	[68]
CoFeS$_2$/rGO	FTO	I^-/I_3^-	N719	71	8.82	8.40	[69]
NiCo$_2$S$_4$/graphene	FTO	I^-/I_3^-	N719	70.5	7.98	8.01	[70]
NiCo$_2$S$_4$/CNFs	FTO	I^-/I_3^-	N719	73	9.0	7.48	[71]
Cu$_2$ZnSnS$_4$-graphene	FTO	I^-/I_3^-	N719	63.0	7.34	8.12	[72]
Ni-Mo-S @ N doped rGO	FTO	I^-/I_3^-	N719	80	9.89	8.73	[73]
CESM-CuInS$_2$	FTO	I^-/I_3^-	N719	60	5.79	6.53	[74]
CoS$_2$-CoSe$_2$@NC	FTO	I^-/I_3^-	N719	70	8.45	8.07	[75]

Continued

TABLE 3 Summary of photovoltaic parameters of various DSSCs with different types of composites-based CEs—cont'd

CE catalysts	Substrate	Redox couple	Dye	FF (%)	η (%)	η (Pt) (%)	Ref.
NiSe/GN$_{0.50}$	ITO	I^-/I_3^-	N719	68.7	8.62	7.68	[76]
CoSe/GN$_{0.50}$	ITO	I^-/I_3^-	N719	68.8	9.27	7.68	[77]
CoSe$_2$/porous carbon shell	FTO	I^-/I_3^-	N719	69	7.54	7.40	[78]
TiO$_{1.1}$Se$_{0.9}$	Carbon cloth	I^-/I_3^-	N719	70	9.47	7.75	[79]
CoTe$_2$@NCNTs	FTO	I^-/I_3^-	N719	64	9.02	8.03	[80]
CoTe/rGO	FTO	I^-/I_3^-	N719	68.5	9.18	8.17	[81]
CoS$_2$/NC@Co-WS$_2$	FTO	I^-/I_3^-	N719	67	9.21	8.18	[82]
ZIF-ZnSe-NC	FTO	I^-/I_3^-	N719	69	8.69	8.26	[83]
Ni-Ni$_3$Se$_2$/graphene nanosheets	FTO	I^-/I_3^-	N719	64	7.76	7.42	[84]
NiS/PEDOT-PSS	FTO	I^-/I_3^-	N719	67	8.18	8.62	[85]
MoSe$_2$/PANI	ITO	I^-/I_3^-	N719	68.7	8.04	7.68	[86]
MoS$_2$/Ni-CoS$_x$	FTO	I^-/I_3^-	N719	67.6	9.80	8.19	[87]
MoS$_2$ coated CoS$_2$	FTO	I^-/I_3^-	N3	42	7.6	6.6	[88]
NiS$_2$/CoS$_2$	FTO	I^-/I_3^-	N719	65.4	8.22	6.61	[89]
NiCo$_2$S$_4$/NiS	FTO	I^-/I_3^-	N719	67	8.8	8.1	[90]
Ni-MoSex@CoSe$_2$ CSNs	FTO	I^-/I_3^-	N719	68.3	9.58	8.32	[91]
MoS$_2$/MoTe$_2$	FTO	I^-/I_3^-	N719	73.9	8.07	8.33	[92]
WS$_2$/MoTe$_2$	FTO	I^-/I_3^-	N719	69.69	7.99	8.50	[93]

Metal chalcogenide-based counter electrodes **Chapter | 11** **275**

choice to replace the expensive Pt CE, the comparatively low electrical conductivity of the metal chalcogenides leads to the high series resistance, which in turn hinders effective electron transportation. Therefore, it is imperative to enhance the conductivity of the metal chalcogenide CEs by tailoring their structure and composition as well as integrating the metal sulfides/selenides/tellurides with other conductive materials for making their practical applications more feasible. Recently, several works have been demonstrated that the hybridization of metal chalcogenides with other conductive carbon-based materials (e.g., graphene, AC, CNTs, etc.) could not only effectively increase the conductivity but also enhance the electrocatalytic activity, both of which in turn could improve the DSSC performances. For example, the DSSC fabricated with the sponge-like CoS/rGO hybrid film manifested a superior efficiency of 9.39% (Fig. 9A–E), yielding 21.63% and 27.93% enhancements, compared to the bare CoS (7.72%) and the Pt (7.34%) CEs, respectively. Such a result could be attributed to the synergistic effects from both electrocatalytically active CoS and conductive rGO [59]. Huo et al. prepared the cobalt sulfide/reduced graphene oxide (CoS/ rGO) hybrid film by using the electrophoretic deposition and the ion exchange deposition methods. The sponge structure of the CoS CE resulted in a large specific surface area; and the addition of rGO provided a small charge-transfer resistance at the electrode/electrolyte interface. Therefore, the CoS/rGO CE showed the improved the electrocatalytic property, compared to the

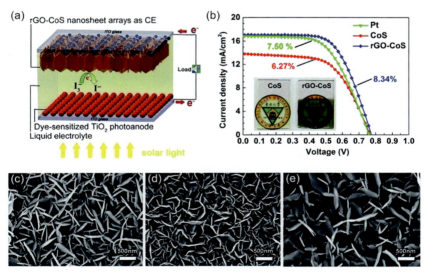

FIG. 9 (A) Schematic of the DSSC assembled with the rGO-CoS composite-based CE, (B) *J–V* curves of the DSSCs constructed with the Pt, CoS nanosheet, and rGO-CoS composite CEs under simulated solar illumination with the 1-sun intensity of 100 mW/cm^2 (AM 1.5G) (Inset: photos of CEs with pure CoS and rGO-CoS composites),and (C–E) SEM images of as-prepared rGO-CoS directly electrodeposited on GO, Copyright, Royal Society of Chemistry, 2015.

Pt CE. Wang et al. prepared the active-site-enriched NiS/multiwalled CNT composite by the hot-injection method using small size NiS nanoparticles embedded in CNTs. The DSSC fabricated with the NiS/CNT CE displayed the excellent activity of I_3^- reduction as well as the remarkable PCE of 10.82% (Fig. 10A and B) [57]. Maiaugree et al. synthesized the robust cocatalyst CE from the bilayer carbon/Ni_3S_2 nanowall composite, and used it for the high-performance DSSC that could yield a high PCE of 9.64 % [58]. Shengli Li and co-workers prepared the MoS_2/graphene composite film via the electrodeposition method; and the prepared composite exhibited a high efficiency of 8.01% [64]. Using the electrospinning and hydrothermal methods, the $NiCo_2S_4$/carbon nanofibers ($NiCo_2S_4$/CNFs) were prepared by Ling Li's group. [71]. SEM and TEM images (Fig. 11A–F) clearly reveal that the carbon nanofibers were uniformly and densely covered by the $NiCo_2S_4$ nanocrystals. The fabricated DSSC with the $NiCo_2S_4$/CNFs-based CE attained a high photoconversion efficiency of 9.00%, which is much greater than that of the Pt CE-based device (7.48%).

The FeS_2/carbon hybrids on carbon cloth (FeS_2/C@CC), where the FeS_2 nanoparticles were embedded in the carbon shells, were fabricated by Li et al. using the in situ deposition-carbonization-sulfurization approach [65]. The as-prepared FeS_2/C@CC sample showed a higher efficiency of 8.15% and a higher long-term stability than that of the Pt CE (7.61%). The enhanced efficiency could be ascribed to the intrinsic catalytic activity of the FeS_2/C nanoparticles, which possess a suitable surface potential matching well with the redox couple, a tight adhesion, a high surface area, and a rapid mass transport. Using the facile hydrothermal process, Murugadoss et al. developed the in situ growth of NiSe onto the graphene nanosheets so as to use it as the CE for high-performance DSSCs. The NiSe/graphene composite exhibited a higher efficiency of 8.62% than the that of Pt- based device ($\eta = 7.68\%$) [76], and the improved PCE mainly originated from the homogeneously immobilized catalytic sites of NiSe and the multiple interfacial electron transfer pathways in graphene, both of which could result in the increased charge transfer and the fast I_3^-

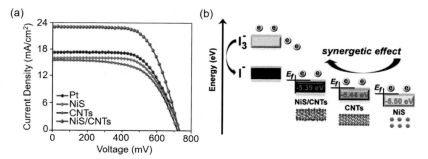

FIG. 10 (A) J–V curves of the DSSCs fabricated with various CEs under 100 mWcm^2 and (B) illustration of the lowered work function for NiS/CNTs, Copyright, Royal Society of Chemistry, 2019.

Metal chalcogenide-based counter electrodes **Chapter | 11** **277**

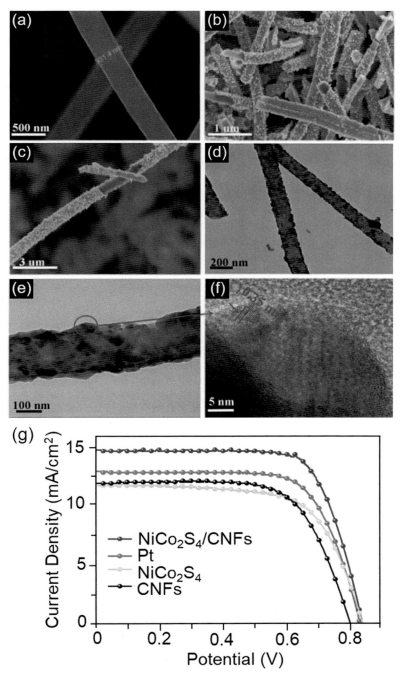

FIG. 11 SEM images of the (A) pure CNFs and the (B, C) NiCo$_2$S$_4$/CNFs composites, (D–F) TEM images of the NiCo$_2$S$_4$/CNFs, and (G) J–V curves of the DSSCs with different CEs, Copyright, Elsevier, 2020.

278 Oxide free nanomaterials for energy storage and conversion applications

reduction kinetics at the counter electrode/electrolyte interface. The $CoSe_2$ nanoparticles-embedded porous carbon shells ($CoSe_2$/CS) were explored as a highly efficient catalyst towards the I_3^- reduction reaction by Weidan Li's group [78]. The $CoSe_2$/CS composites were fabricated using the soft-template method, cinclusing the carbonization and the selenization processes. The $CoSe_2$/CS composites exhibited the high electrocatalytic performance as well as the long-term stability, which could lead to the improved efficiency of 7.54%, exceeding that of the Pt CE (7.40%). They reported that the enhanced photovoltaic efficiency could be attributed to more catalytic reactive sites from embedded $CoSe_2$ nanoparticles into porous carbon shells as well as better band matching with the potential of I_3^-/I^- redox couples. In addition, many researchers have been alternatively attempted to develop the efficient CE material by using metal chalcogenides with polymer-based composites. Sowbakkiyavathi et al. reported that the $MoSe_2$/PANI composite acted as an efficient photocatalyst due to their fast electron transport rate with the increased current flow and the good electrocatalytic activity [86]. The $MoSe_2$/PANI composite nanofibers-based DSSC showed a higher cell efficiency of 8.04% than that of the pure $MoSe_2$ (7.03%)- and Pt (7.68%)-based devices. Maiaugree et al. prepared high-quality CEs by coating the composites with NiS nanoparticles with PEDOT-PSS slurries on the FTO substrates through the doctor blading technique [85]. The DSSC with the NiS(NPs)/PEDOT-PSS CE provided the efficiency of 8.18%, which is comparable to that of the Pt CE-based DSSC (8.62%).

Recently, the heterostructured electrocatalysts with multiple active components have also demonstrated the excellent electrocatalytic activity for the reduction of tri-iodide ions compared with their single component counterparts. The main mechanism is associated with both well-exposed active sites and their matched energy band structure at the interface. For example, Li et al. demonstrated that the heterojunction of NiS_2/CoS_2 could synergistically improve the electrocatalytic property owing to their same crystal structure, matched energy band structure, and high-quality heterointerface. The DSSC based on the NiS_2/CoS_2 CE showed the impressive photovoltaic conversion efficiency of 8.22%, which is 24.36% higher than that of the DSSC with the Pt CE (6.61%) [89]. Subbiah et al. have modified the electrocatalytic properties of CoS_2 by synthesizing MoS_2-coated CoS_2 nanocomposites, and they obtained a higher conversion efficiency of 8.45% than that of the Pt CE-based DSSC (8.07%) [88]. Hussain et al. constructed the hybrid bilayer of MoS_2/$MoTe_2$ on the FTO substrate via sequential sputtering and postdeposition annealing. The DSSC with the MoS_2/$MoTe_2$ hybrid CE exhibited the PCE of 8.07%, which is comparable to that of the Pt CE-based devices (8.33%) [92]. Yudi Niu's group reported that the MoS_2/Ni-CoS_x NCs possessed numerous nanochannels and effectively active sites and thus revealed the qualified electrocatalytic property as a highly efficient electrocatalyst for the redox of I^-/I_3^- [87]. The MoS_2/Ni-CoS_x nanocubes-based DSSC provided a PCE of 9.80%, which is higher than that of Pt (8.19%). Huang et al. prepared the CoS_2-$CoSe_2$ heterostructured nanocrystals through simultaneous sulfurization and selenization of polydopamine-coated Prussian blue analogs. The CoS_2-$CoSe_2$ encapsulated by N-doped carbon hollow nanocubes (CoS_2-$CoSe_2$@NC) attained a photoconversion efficiency (PCE) of 8.45%, which is better than that of Pt (8.07%) [75]. Benefiting from both structural and compositional features (i.e., synergistic effects from the CoS_2-$CoSe_2$ heterojunction and the porous nanocube structure of the conductive N-doped carbon), the CoS_2-$CoSe_2$@NC hybrid showed the excellent electro-catalytic activity and the good cycling stability for the reduction of I_3^- ions in DSSCs.

Liu et al. reported the design and the synthesis of the hierarchical core-shell nanospheres (Ni-MoSex@$CoSe_2$ CSNs) with the enhanced performances via the facile wrapping and selenizing processes [91]. The Ni-MoSe$_x$@$CoSe_2$ delivered an excellent performance of 9.58% conversion efficiency, which is much better than that of Pt-based device (8.32%) (Fig. 12D). The FE-SEM and HR-TEM images (Fig. 12B and C) reveal the hierarchical structure and the spherical mor-phology of the Ni-MoSex@$CoSe_2$ CSNs, in which the surface of the nanospheres are covered with many fine particles. Thus, the resultant NiMoSe$_x$@$CoSe_2$ CSNs have the large number of active sites as well as the good ability for adsorbing/ transferring iodine ions (Fig. 12A).

5. Conclusion and future prospects

The surge of finding the inexpensive alternatives to the conventional Pt CEs has unveiled a wide range of potential candidate materials. In this chapter, the recent findings of the alternative Pt-free CEs were compiled to help understand more insights into the characteristics of the potential CE materials and their novel com-posites. Particularly, this chapter focused on the metal chalcogenide-based CEs as an emerging and inexpensive CE that can replace the conventional Pt-based CEs. Various metal chalcogenides such as metal sulfides, selenides, and tellurides in the structural forms of binary, ternary, quaternary compounds, and their composites with other functional materials were introduced as a good CE material for high-performance DSSCs. From enormous previous studies, the metal sulfides- and selenides-based CEs are suggested as a promising alternative, which can yield the better performance than the conventional Pt CE-based DSSCs.

In addition, the most recent researches on the metal chalcogenides-composite materials have attracted tremendous attention because of the improved electrocatalytic characteristics. The carbonaceous materials (e.g., graphene, rGO, CNTs, etc.) possesses a large surface area, which can provide the enhanced electrocatalytic properties for the nanocomposites. These charac-teristics are the important criteria for the ideal CE. Owing to the synergetic effects from metal chalcogenides and carbonaceous materials, the composites can satisfy the required ideal CE characteristics for high-performance DSSCs.

The substantial research works in this field of alternative CEs have led to the highly potential replacement of Pt-free CEs with the superior electrical and

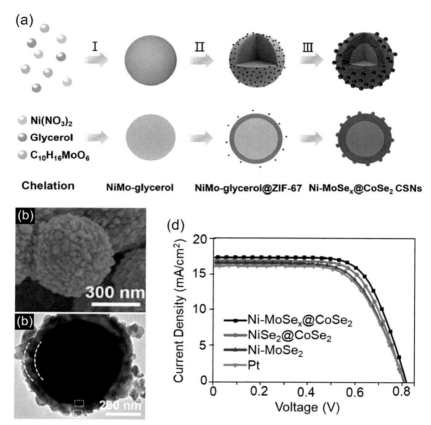

FIG. 12 (A) Formation mechanism, (B) SEM, (C) TEM images of the Ni-MoSe$_x$@CoSe$_2$ CSNs, and (D) J–V curves of the DSSCs with different CEs, Copyright, Elsevier, 2020.

electrocatalytic properties. In this chapter, the primary motivation has turned towards the alternative emerging materials and technologies from the up-to-date research findings. However, the further step in the search of the alternative CE materials should also be focused on the long-term stability and the easier fabrication of these novel inexpensive CEs. In view of major criteria for the third-generation solar cells (e.g., inexpensiveness, durability, efficiency, etc.), the further findings should shed more light on these trending nanocomposites of metal chalcogenide-carbonaceous materials.

References

[1] M. Grätzel, Photoelectrochemical cells, Nature 414 (6861) (2001) 338–344.
[2] B. O'Regan, M. Grätzel, A low-cost, high-efficiency solar cell based on dye-sensitized colloidal TiO$_2$ films, Nature 353 (6346) (1991) 737–740.

[3] S. Mathew, A. Yella, P. Gao, R. Humphry-Baker, B.F. Curchod, N. Ashari-Astani, I. Tavernelli, U. Rothlisberger, M.K. Nazeeruddin, M. Grätzel, Dye-sensitized solar cells with 13% efficiency achieved through the molecular engineering of porphyrin sensitizers, Nat. Chem. 6 (3) (2014) 242–247.

[4] G. Calogero, P. Calandra, A. Irrera, A. Sinopoli, I. Citro, G. Di Marco, A new type of transparent and low cost counter-electrode based on platinum nanoparticles for dye-sensitized solar cells, Energ. Environ. Sci. 4 (5) (2011) 1838–1844.

[5] M.Z. Iqbal, S. Khan, Progress in the performance of dye sensitized solar cells by incorporating cost effective counter electrodes, Sol. Energy 160 (2018) 130–152.

[6] S. Yun, A. Hagfeldt, T. Ma, Pt-free counter electrode for dye-sensitized solar cells with high efficiency, Adv. Mater. 26 (36) (2014) 6210–6237.

[7] M. Chen, L.-L. Shao, Review on the recent progress of carbon counter electrodes for dye-sensitized solar cells, Chem. Eng. J. 304 (2016) 629–645.

[8] E. Singh, K.S. Kim, G.Y. Yeom, H.S. Nalwa, Two-dimensional transition metal dichalcogenide-based counter electrodes for dye-sensitized solar cells, RSC Adv. 7 (45) (2017) 28234–28290.

[9] M. Wu, T. Ma, Recent Progress of counter electrode catalysts in dye-sensitized solar cells, J. Phys. Chem. C 118 (30) (2014) 16727–16742.

[10] L.T. Gularte, C.D. Fernandes, M.L. Moreira, C.W. Raubach, P.L.G. Jardim, S.S. Cava, In situ microwave-assisted deposition of CoS counter electrode for dye-sensitized solar cells, Sol. Energy 198 (2020) 658–664.

[11] K. Ashok Kumar, A. Pandurangan, S. Arumugam, M. Sathiskumar, Effect of bi-functional hierarchical flower-like CoS nanostructure on its interfacial charge transport kinetics, magnetic and electrochemical behaviors for supercapacitor and DSSC applications, Sci. Rep. 9 (1) (2019) 1228.

[12] J. Huo, J. Wu, M. Zheng, Y. Tu, Z. Lan, Effect of ammonia on electrodeposition of cobalt sulfide and nickel sulfide counter electrodes for dye-sensitized solar cells, Electrochim. Acta 180 (2015) 574–580.

[13] M. Wang, A.M. Anghel, B. Marsan, N.-L. Cevey Ha, N. Pootrakulchote, S.M. Zakeeruddin, M. Grätzel, CoS supersedes Pt as efficient electrocatalyst for triiodide reduction in dye-sensitized solar cells, J. Am. Chem. Soc. 131 (44) (2009) 15976–15977.

[14] Y. Li, H.-X. Zhang, F.-T. Liu, X.-F. Dong, X. Li, C.-W. Wang, New design of oriented NiS nanoflower arrays as platinum-free counter electrode for high-efficient dye-sensitized solar cells, Superlattices Microstruct. 125 (2019) 66–71.

[15] S. Huang, H. Wang, Y. Zhang, S. Wang, Z. Chen, Z. Hu, X. Qian, Prussian blue-derived synthesis of uniform nanoflakes-assembled NiS_2 hierarchical microspheres as highly efficient electrocatalysts in dye-sensitized solar cells, RSC Adv. 8 (11) (2018) 5992–6000.

[16] Y. Li, Y. Chang, Y. Zhao, J. Wang, C.-w. Wang, In situ synthesis of oriented NiS nanotube arrays on FTO as high-performance counter electrode for dye-sensitized solar cells, J. Alloys Compd. 679 (2016) 384–390.

[17] J. Liang, J. Li, H. Zhu, Y. Han, Y. Wang, C. Wang, Z. Jin, G. Zhang, J. Liu, One-step fabrication of large-area ultrathin MoS_2 nanofilms with high catalytic activity for photovoltaic devices, Nanoscale 8 (35) (2016) 16017–16025.

[18] M. Wu, Y. Wang, X. Lin, N. Yu, L. Wang, L. Wang, A. Hagfeldt, T. Ma, Economical and effective sulfide catalysts for dye-sensitized solar cells as counter electrodes, Phys. Chem. Chem. Phys. 13 (43) (2011) 19298–19301.

[19] H. Jeong, K. Jae-Yup, B. Koo, H. Son, D. Kim, M. Ko, Rapid sintering of MoS_2 counter electrode using near-infrared pulsed laser for use in highly efficient dye-sensitized solar cells, J. Power Sources 330 (2016) 104–110.

282 Oxide free nanomaterials for energy storage and conversion applications

[20] D. Vikraman, S.A. Patil, S. Hussain, N. Mengal, H.-S. Kim, S.H. Jeong, J. Jung, H.-S. Kim, H. J. Park, Facile and cost-effective methodology to fabricate MoS_2 counter electrode for efficient dye-sensitized solar cells, Dyes Pigments 151 (2018) 7–14.

[21] S. Ramasamy, One-step hydrothermal synthesis of marigold flower-like nanostructured MoS2 as a counter electrode for dye-sensitized solar cells, J. Solid State Electrochem. 22 (2018) 3331–3341.

[22] X. Liu, G. Yue, H. Zheng, A promising vanadium sulfide counter electrode for efficient dye-sensitized solar cells, RSC Adv. 7 (21) (2017) 12474–12478.

[23] M. Hu, Z. Yu, J. Li, X. Jiang, J. Lai, X. Yang, M. Wang, L. Sun, Low-cost solution-processed digenite Cu_9S_5 counter electrode for dye-sensitized solar cells, RSC Adv. 7 (61) (2017) 38452–38457.

[24] P. Sun, F. Yao, X. Ban, N. Huang, X. Sun, Directly hydrothermal growth of antimony sulfide on conductive substrate as efficient counter electrode for dye-sensitized solar cells, Electrochim. Acta 174 (2015) 127–132.

[25] K. Mokurala, S. Mallick, Correction: effect of annealing atmosphere on quaternary chalcogenide-based counter electrodes in dye-sensitized solar cell performance: synthesis of Cu_2FeSnS_4 and Cu_2CdSnS_4 nanoparticles by thermal decomposition process, RSC Adv. 7 (31) (2017) 18892.

[26] S. Chen, A. Xu, J. Tao, H. Tao, Y. Shen, L. Zhu, J. Jiang, T. Wang, L. Pan, In-situ synthesis of two-dimensional leaf-like Cu_2ZnSnS_4 plate arrays as Pt-free counter electrode for efficient dye-sensitized solar cells, Green Chem. 18 (2016) 2793–2801.

[27] X. Chen, J. Ding, X. Chen, X. Liu, G. Zhuang, Z. Zhang, Porous tremella-like $NiCo_2S_4$ networks electrodes for high-performance dye-sensitized solar cells and supercapacitors, Sol. Energy 176 (2018) 762–770.

[28] K.S. Anuratha, S. Mohan, S.K. Panda, Pulse reverse electrodeposited $NiCo_2S_4$ nanostructures as efficient counter electrodes for dye-sensitized solar cells, New J. Chem. 40 (2) (2016) 1785–1791.

[29] K. Subalakshmi, K.A. Kumar, O.P. Paul, S. Saraswathy, A. Pandurangan, J. Senthilselvan, Platinum-free metal sulfide counter electrodes for DSSC applications: structural, electrochemical and power conversion efficiency analyses, Sol. Energy 193 (2019) 507–518.

[30] J. Theerthagiri, R.A. Senthil, M.H. Buraidah, J. Madhavan, A.K. Arof, M. Ashokkumar, One-step electrochemical deposition of $Ni_{1-x}Mo_xS$ ternary sulfides as an efficient counter electrode for dye-sensitized solar cells, J. Mater. Chem. A 4 (41) (2016) 16119–16127.

[31] F. Liu, S. Hu, X. Ding, J. Zhu, J. Wen, X. Pan, S. Chen, M.K. Nazeeruddin, S. Dai, Ligand-free nano-grain Cu_2SnS_3 as a potential cathode alternative for both cobalt and iodine redox electrolyte dye-sensitized solar cells, J. Mater. Chem. A 4 (38) (2016) 14865–14876.

[32] M. Gulen, A. Sarilmaz, I.H. Patir, F. Ozel, S. Sonmezoglu, Ternary copper-tungsten-disulfide nanocube inks as catalyst for highly efficient dye-sensitized solar cells, Electrochim. Acta 269 (2018) 119–127.

[33] Y. Wu, B. Zhou, C. Yang, S. Liao, W.-H. Zhang, C. Li, $CuFeS_2$ colloidal nanocrystals as an efficient electrocatalyst for dye sensitized solar cells, Chem. Commun. 52 (77) (2016) 11488–11491.

[34] S. Huang, J. Zai, D. Ma, Q. He, Y. Liu, Q. Qiao, X. Qian, Colloidal synthesis of wurtz-stannite Cu_2CdGeS_4 nanocrystals with high catalytic activity toward iodine redox couples in dye-sensitized solar cells, Chem. Commun. 52 (72) (2016) 10866–10869.

[35] X. Qian, H. Liu, J. Yang, H. Wang, J. Huang, C. Xu, Co–Cu–WS_x ball-in-ball nanospheres as high-performance Pt-free bifunctional catalysts in efficient dye-sensitized solar cells and alkaline hydrogen evolution, J. Mater. Chem. A 7 (11) (2019) 6337–6347.

[36] X. Qian, H. Liu, Y. Huang, Z. Ren, Y. Yu, C. Xu, H. Linxi, Co-Ni-MoS$_x$ yolk-shell nanospheres as superior Pt-free electrode catalysts for highly efficient dye-sensitized solar cells, J. Power Sources 412 (2019) 568–574.

[37] I. Charles, S. Muthu, M. Gandhi, S. Moorthy Babu, Surface-treated Cu$_2$ZnSnS$_4$ nanoflakes as Pt-free inexpensive and effective counter electrode in DSSC, J. Mater. Sci. Mater. Electron. 27 (2020) 18164–18174.

[38] H. Li, X. Qian, C. Zhu, X. Jiang, L. Shao, L. Hou, Template synthesis of CoSe$_2$/Co$_3$Se$_4$ nanotubes: tuning of their crystal structures for photovoltaics and hydrogen evolution in alkaline medium, J. Mater. Chem. A 5 (9) (2017) 4513–4526.

[39] F. Gong, H. Wang, X. Xu, G. Zhou, Z.-S. Wang, In situ growth of Co$_{0.85}$Se and Ni$_{0.85}$Se on conductive substrates as high-performance counter electrodes for dye-sensitized solar cells, J. Am. Chem. Soc. 134 (26) (2012) 10953–10958.

[40] S. Zhou, Q. Jiang, J. Yang, W. Chu, W. Li, X. Li, Y. Hou, J. Hou, Regulation of microstructure and composition of cobalt selenide counter electrode by electrochemical atomic layer deposition for high performance dye-sensitized solar cells, Electrochim. Acta 220 (2016) 169–175.

[41] J. Dong, J. Wu, J. Jia, S. Wu, P. Zhou, Y. Tu, Z. Lan, Cobalt selenide nanorods used as a high efficient counter electrode for dye-sensitized solar cells, Electrochim. Acta 168 (2015) 69–75.

[42] Q. Jiang, G. Hu, Co$_{0.85}$Se hollow nanoparticles as Pt-free counter electrode materials for dye-sensitized solar cells, Mater. Lett. 153 (2015) 114–117.

[43] H. Wu, Y. Wang, L. Zhang, Z. Chen, C. Wang, S. Fan, Comparison of two nickel selenides materials with different morphologies as counter electrodes in dye-sensitized solar cells, J. Alloys Compd. 745 (2018) 222–227.

[44] F. Gong, X. Xu, Z. Li, G. Zhou, Z.-S. Wang, NiSe$_2$ as an efficient electrocatalyst for a Pt-free counter electrode of dye-sensitized solar cells, Chem. Commun. 49 (14) (2013) 1437–1439.

[45] X. Zhang, J. Bai, B. Yang, G. Li, L. Liu, Self-assembled mesoporous Ni$_{0.85}$Se spheres as high performance counter cells of dye-sensitized solar cells, RSC Adv. 6 (64) (2016) 58925–58932.

[46] J. Dong, J. Wu, J. Jia, J. Ge, Q. Bao, C. Wang, L. Fan, A transparent nickel selenide counter electrode for high efficient dye-sensitized solar cells, Appl. Surf. Sci. 401 (2017) 1–6.

[47] Q. Jiang, N. Xiong, K. Pan, M. Wu, Y. Zhou, Vertically aligned Ni$_3$Se$_2$ arrays with dendritic-like structure as efficient counter electrode of dye-sensitized solar cells, Mater. Sci. Semicond. Process. 66 (2017) 241–246.

[48] D. Vikraman, S.A. Patil, S. Hussain, N. Mengal, S.H. Jeong, J. Jung, H.J. Park, H.-S. Kim, H.-S. Kim, Construction of dye-sensitized solar cells using wet chemical route synthesized MoSe$_2$ counter electrode, J. Ind. Eng. Chem. 69 (2019) 379–386.

[49] Y. Zhao, J. Duan, Y. Duan, H. Yuan, Q. Tang, 9.07%-Efficiency dye-sensitized solar cell from Pt-free RuCoSe ternary alloy counter electrode, Mater. Lett. 218 (2018) 76–79.

[50] Q. He, S. Li, S. Huang, L. Xiao, L. Hou, Construction of uniform Co–Sn–X (X = S, Se, Te) nanocages with enhanced photovoltaic and oxygen evolution properties via anion exchange reaction, Nanoscale 10 (46) (2018) 22012–22024.

[51] Y. Jiang, X. Qian, Y. Niu, L. Shao, C. Zhu, L. Hou, Cobalt iron selenide/sulfide porous nanocubes as high-performance electrocatalysts for efficient dye-sensitized solar cells, J. Power Sources 369 (2017) 35–41.

[52] X. Chen, J. Ding, Y. Li, Y. Wu, G. Zhuang, C. Zhang, Z. Zhang, C. Zhu, P. Yang, Size-controllable synthesis of NiCoSe$_2$ microspheres as a counter electrode for dye-sensitized solar cells, RSC Adv. 8 (46) (2018) 26047–26055.

[53] K. Pan, C.-S. Lee, G. Hu, Y. Zhou, Cobalt-nickel based ternary selenides as high-efficiency counter electrode materials for dye-sensitized solar cells, Electrochim. Acta 235 (2017) 672–679.

284 Oxide free nanomaterials for energy storage and conversion applications

[54] J. Theerthagiri, R.A. Senthil, M.H. Buraidah, M. Raghavender, J. Madhavan, A.K. Arof, Synthesis and characterization of $(Ni_{1-x}Co_x)Se_2$ based ternary selenides as electrocatalyst for triiodide reduction in dye-sensitized solar cells, J. Solid State Chem. 238 (2016) 113–120.

[55] S.A. Patil, E.-K. Kim, N.K. Shrestha, J. Chang, J.K. Lee, S.-H. Han, Formation of semimetallic cobalt telluride nanotube film via anion exchange tellurization strategy in aqueous solution for electrocatalytic applications, ACS Appl. Mater. Interfaces 7 (46) (2015) 25914–25922.

[56] S. Hussain, S.A. Patil, D. Vikraman, N. Mengal, H. Liu, W. Song, K.-S. An, S.H. Jeong, H.-S. Kim, J. Jung, Large area growth of $MoTe_2$ films as high performance counter electrodes for dye-sensitized solar cells, Sci. Rep. 8 (1) (2018) 29.

[57] X. Wang, Y. Xie, Y. Jiao, K. Pan, B. Bateer, J. Wu, H. Fu, Carbon nanotubes in situ embedded with NiS nanocrystals outperform Pt in dye-sensitized solar cells: interface improved activity, J. Mater. Chem. A 7 (17) (2019) 10405–10411.

[58] W. Maiaugree, A. Tangtrakarn, S. Lowpa, N. Ratchapolthavisin, V. Amornkitbamrung, Facile synthesis of bilayer carbon/Ni_3S_2 nanowalls for a counter electrode of dye-sensitized solar cell, Electrochim. Acta 174 (2015) 955–962.

[59] J. Huo, J. Wu, M. Zheng, Y. Tu, Z. Lan, High performance sponge-like cobalt sulfide/reduced graphene oxide hybrid counter electrode for dye-sensitized solar cells, J. Power Sources 293 (2015) 570–576.

[60] X. Cui, Z. Xie, Y. Wang, Novel CoS_2 embedded carbon nanocages by direct sulfurizing metal–organic frameworks for dye-sensitized solar cells, Nanoscale 8 (23) (2016) 11984–11992.

[61] C. Zhu, H. Min, F. Xu, J. Chen, H. Dong, L. Tong, Y. Zhu, L. Sun, Ultrafast electrochemical preparation of graphene/CoS nanosheet counter electrodes for efficient dye-sensitized solar cells, RSC Adv. 5 (104) (2015) 85822–85830.

[62] J. Qiu, D. He, R. Zhao, B. Sun, H. Ji, N. Zhang, Y. Li, X. Lu, C. Wang, Fabrication of highly dispersed ultrafine Co_9S_8 nanoparticles on carbon nanofibers as low-cost counter electrode for dye-sensitized solar cells, J. Colloid Interface Sci. 522 (2018) 95–103.

[63] G. Zhu, H. Xu, H. Wang, W. Wang, Q. Zhang, L. Zhang, H. Sun, Microwave assisted synthesis of MoS_2/nitrogen-doped carbon shell–core microspheres for Pt-free dye-sensitized solar cells, RSC Adv. 7 (22) (2017) 13433–13437.

[64] S. Li, H. Min, F. Xu, L. Tong, J. Chen, C. Zhu, L. Sun, All electrochemical fabrication of MoS_2/graphene counter electrodes for efficient dye-sensitized solar cells, RSC Adv. 6 (41) (2016) 34546–34552.

[65] L. Li, P. Ma, S. Hussain, L. Jia, D. Lin, X. Yin, Y. Lin, Z. Cheng, L. Wang, FeS_2/carbon hybrids on carbon cloth: a highly efficient and stable counter electrode for dye-sensitized solar cells, Sustain. Energy Fuels 3 (7) (2019) 1749–1756.

[66] X. Zhao, D. Wang, S.A. Liu, Z. Li, J. Meng, Y. Ran, Y. Zhang, L. Li, Bi_2S_3 nanoparticles densely grown on electrospun-carbon-nanofibers as low-cost counter electrode for liquid-state solar cells, Mater. Res. Bull. 125 (2020) 110800.

[67] B. Zhou, X. Zhang, P. Jin, X. Li, X. Yuan, J. Wang, L. Liu, Synthesis of $In_{2.77}S_4$ nanoflakes/graphene composites and their application as counter electrode in dye-sensitized solar cells, Electrochim. Acta 281 (2018) 746–752.

[68] H. Xu, G. Zhu, Facile one-step synthesis of uniformity carbon-mixed tin sulfide hexagonal nanodisks as low-cost counter electrode material for dye-sensitized solar cells, Mater. Lett. 171 (2016) 174–177.

[69] M. Zhang, J. Zai, J. Liu, M. Chen, Z. Wang, G. Li, X. Qian, L. Qian, X. Yu, A hierarchical $CoFeS_2$/reduced graphene oxide composite for highly efficient counter electrodes in dye-sensitized solar cells, Dalton Trans. 46 (29) (2017) 9511–9516.

[70] R. Krishnapriya, S. Praneetha, A.M. Rabel, A. Vadivel Murugan, Energy efficient, one-step microwave-solvothermal synthesis of a highly electro-catalytic thiospinel $NiCo_2S_4$/graphene nanohybrid as a novel sustainable counter electrode material for Pt-free dye-sensitized solar cells, J. Mater. Chem. C 5 (12) (2017) 3146–3155.

[71] L. Li, X. Zhang, S.A. Liu, B. Liang, Y. Zhang, W. Zhang, One-step hydrothermal synthesis of $NiCo_2S_4$ loaded on electrospun carbon nanofibers as an efficient counter electrode for dye-sensitized solar cells, Sol. Energy 202 (2020) 358–364.

[72] Y. Li, H. Guo, X. Wang, N. Yuan, J. Ding, Suppression of charge recombination by application of Cu_2ZnSnS_4-graphene counter electrode to thin dye-sensitized solar cells, Sci. Bull. 61 (15) (2016) 1221–1230.

[73] J. Balamurugan, S.G. Peera, M. Guo, T.T. Nguyen, N.H. Kim, J.H. Lee, A hierarchical 2D Ni–Mo–S nanosheet@nitrogen doped graphene hybrid as a Pt-free cathode for high-performance dye sensitized solar cells and fuel cells, J. Mater. Chem. A 5 (34) (2017) 17896–17908.

[74] L. Wang, J. He, M. Zhou, S. Zhao, Q. Wang, B. Ding, Copper indium disulfide nanocrystals supported on carbonized chicken eggshell membranes as efficient counter electrodes for dye-sensitized solar cells, J. Power Sources 315 (2016) 79–85.

[75] S. Huang, H. Wang, S. Wang, Z. Hu, L. Zhou, Z. Chen, Y. Jiang, X. Qian, Encapsulating CoS_2–$CoSe_2$ heterostructured nanocrystals in N-doped carbon nanocubes as highly efficient counter electrodes for dye-sensitized solar cells, Dalton Trans. 47 (15) (2018) 5236–5244.

[76] V. Murugadoss, J. Lin, H. Liu, X. Mai, T. Ding, Z. Guo, S. Angaiah, Optimizing graphene content in a NiSe/graphene nanohybrid counter electrode to enhance the photovoltaic performance of dye-sensitized solar cells, Nanoscale 11 (38) (2019) 17579–17589.

[77] V. Murugadoss, N. Wang, S. Tadakamalla, B. Wang, Z. Guo, S. Angaiah, In situ grown cobalt selenide/graphene nanocomposite counter electrodes for enhanced dye-sensitized solar cell performance, J. Mater. Chem. A 5 (28) (2017) 14583–14594.

[78] W. Li, P. Ma, F. Chen, R. Xu, Z. Cheng, X. Yin, Y. Lin, L. Wang, $CoSe_2$/porous carbon shell composites as high-performance catalysts toward tri-iodide reduction in dye-sensitized solar cells, Inorg. Chem. Front. 6 (9) (2019) 2550–2557.

[79] C.T. Li, I.T. Chiu, R. Vittal, Y.-J. Huang, T.Y. Chen, H.W. Pang, J. Lin, K.-C. Ho, Hierarchical $TiO_{1.1}Se_{0.9}$-wrapped carbon cloth as the TCO-free and Pt-free counter electrode for iodide-based and cobalt-based dye-sensitized solar cells, J. Mater. Chem. A 5 (2017) 14079–14091.

[80] S. Huang, S. Li, Q. He, H. An, L. Xiao, L. Hou, Formation of $CoTe_2$ embedded in nitrogen-doped carbon nanotubes-grafted polyhedrons with boosted electrocatalytic properties in dye-sensitized solar cells, Appl. Surf. Sci. 476 (2019) 769–777.

[81] J. Jia, J. Wu, J. Dong, J. Lin, Cobalt telluride/reduced graphene oxide using as high performance counter electrode for dye-sensitized solar cells, Electrochim. Acta 185 (2015) 184–189.

[82] J. Huang, X. Qian, J. Yang, Y. Niu, C. Xu, L. Hou, Construction of Pt-free electrocatalysts based on hierarchical CoS_2/N-doped C@co-WS_2 yolk-shell nano-polyhedrons for dye-sensitized solar cells, Electrochim. Acta 340 (2020) 135949.

[83] S.-L. Jian, Y.-J. Huang, M.-H. Yeh, K.-C. Ho, A zeolitic imidazolate framework-derived ZnSe/N-doped carbon cube hybrid electrocatalyst as the counter electrode for dye-sensitized solar cells, J. Mater. Chem. A 6 (12) (2018) 5107–5118.

[84] X. Zhang, J. Bai, M. Zhen, L. Liu, Ultrathin Ni–Ni_3Se_2 nanosheets on graphene as a high-performance counter electrode for dye-sensitized solar cells, RSC Adv. 6 (92) (2016) 89614–89620.

[85] W. Maiaugree, P. Pimparue, W. Jarernboon, S. Pimanpang, V. Amornkitbamrung, E. Swatsitang, NiS(NPs)-PEDOT-PSS composite counter electrode for a high efficiency dye sensitized solar cell, Mater. Sci. Eng. B 220 (2017) 66–72.

286 Oxide free nanomaterials for energy storage and conversion applications

[86] E. Sowbakkiyavathi, V. Murugadoss, R. Sittaramane, S. Angaiah, Development of $MoSe_2$/PANI composite nanofibers as an alternative to Pt counter electrode to boost the photoconversion efficiency of dye sensitized solar cell, J. Solid State Electrochem. 24 (2020) 2289–2300.

[87] Y. Niu, X. Qian, J. Zhuang, H. Liu, L. Hou, MoS_2/Ni-CoS_x porous nanocubes derived from Ni–Co prussian-blue analogs as enhanced Pt-free electrode catalysts for high-efficiency dye-sensitized solar cells, J. Power Sources 440 (2019) 227121.

[88] V. Subbiah, G. Landi, J.J. Wu, S. Anandan, MoS_2 coated CoS_2 nanocomposites as counter electrodes in Pt-free dye-sensitized solar cells, Phys. Chem. Chem. Phys. 21 (45) (2019) 25474–25483.

[89] F. Li, J. Wang, L. Zheng, Y. Zhao, N. Huang, P. Sun, L. Fang, L. Wang, X. Sun, In situ preparation of NiS_2/CoS_2 composite electrocatalytic materials on conductive glass substrates with electronic modulation for high-performance counter electrodes of dye-sensitized solar cells, J. Power Sources 384 (2018) 1–9.

[90] J. Huo, J. Wu, M. Zheng, Y. Tu, Z. Lan, Flower-like nickel cobalt sulfide microspheres modified with nickel sulfide as Pt-free counter electrode for dye-sensitized solar cells, J. Power Sources 304 (2016) 266–272.

[91] H. Liu, X. Qian, Y. Niu, M. Chen, C. Xu, K.-Y. Wong, Hierarchical Ni-$MoSe_x$@$CoSe_2$ core-shell nanosphere as highly active bifunctional catalyst for efficient dye-sensitized solar cell and alkaline hydrogen evolution, Chem. Eng. J. 383 (2020) 123129.

[92] S. Hussain, S.A. Patil, D. Vikraman, I. Rabani, A.A. Arbab, S.H. Jeong, H.-S. Kim, H. Choi, J. Jung, Enhanced electrocatalytic properties in MoS_2/$MoTe_2$ hybrid heterostructures for dye-sensitized solar cells, Appl. Surf. Sci. 504 (2020) 144401.

[93] S. Hussain, S.A. Patil, A.A. Memon, D. Vikraman, H.G. Abbas, S.H. Jeong, H.-S. Kim, H.-S. Kim, J. Jung, Development of a WS_2/$MoTe_2$ heterostructure as a counter electrode for the improved performance in dye-sensitized solar cells, Inorg. Chem. Front. 5 (12) (2018) 3178–3183.

Chapter 12

Oxide free materials for perovskite solar cells

Ramya Krishna Battula[a,b,c], Easwaramoorthi Ramasamy[a], P. Bhyrappa[b], C. Sudakar[c], and Ganapathy Veerappan[a]

[a]*Centre for Solar Energy Materials, International Advanced Research Centre for Powder Metallurgy and New Materials (ARCI), Hyderabad, India,* [b]*Department of Chemistry, Indian Institute of Technology Madras, Chennai, India,* [c]*Multifunctional Materials Lab, Department of Physics, Indian Institute of Technology Madras, Chennai, India*

Chapter outline

1. Introduction	**287**	3.6	Carbazole based HTMs	295
2. Oxide free-electron transport materials	**289**	3.7	PEDOT:PSS (PEDOT = poly (3,4-ethylene dioxythiophene) and PSS = polystyrene sulfonate)	296
2.1 Fullerene and its derivatives	289			
2.2 Others	290	3.8	Poly (3-hexylthiophene) (P3HT) based HTMs	297
3. Oxide free hole transporting materials (HTMs)	**291**	3.9	Lead sulfide (PbS)	298
3.1 Spiro-OMeTAD	291	**4. HTM free carbon-based PSCs**		**299**
3.2 Poly [bis(4-phenyl) (2,5,6-trimethylphenyl)amine] (PTAA)	292	**5. Oxide free TCO for PSCs**		**301**
3.3 Copper thiocyanate (CuSCN)	292	**6. Conclusion**		**302**
3.4 Copper iodide (CuI)	293	**Acknowledgements**		**302**
3.5 Copper chalcogenides	294	**References**		**302**

1. Introduction

Perovskite solar cells grabbed the attention of the solar community as their efficiencies shot from 3.9% [1] in 2009 to 25.5% in 2020, in a matter of only 10 years [2,3] This soaring performance can be credited to the high optical absorption of the material in films as thin as 500 nm, long carrier recombination lifetimes, a good ambipolar charge-transporting properties along long diffusion lengths, which is reflected in its impressive solar cell efficiency (PCE) [4,5] Apart from efficiency, the flexibility of choosing materials, substrates, good electrical and optical properties have made PSCs a great competitor to the

Oxide Free Nanomaterials for Energy Storage and Conversion Applications.
https://doi.org/10.1016/B978-0-12-823936-0.00001-2
Copyright © 2022 Elsevier Inc. All rights reserved.

existing technologies. PSCs can be solution-processed, are earth-abundant, inexpensive, have direct bandgap, and require no extensive processing. Thus, perovskite-based solar cells are highly efficient and cost-effective. Traditionally methylammonium lead iodide ($CH_3NH_3PbI_3$ or $MAPbI_3$) is used as absorber material in photovoltaic applications. The most striking aspect of the $MAPbI_3$ PSC is its high voltage due to its optimum bandgap of 1.55 eV. Apart from photovoltaics, perovskites also find applications in light-emitting diodes, lasers, photo-detectors, X-ray detectors, and ambipolar phototransistors, etc., due to their good optoelectronic properties [6,7].

A typical PSC device configuration and charge transfer process are shown in Fig. 1. The fabrication process consists of depositing a thin layer of TiO_2 (compact layer) on a fluorine-doped tin oxide (FTO) substrate, which is the front contact. The thin layer of TiO_2 averts the recombination of carriers between the absorber and FTO. The n-type material, which is the electron transport material (ETM) consists of mesoporous TiO_2 is deposited on top of the thin layer TiO_2 (compact layer). The perovskite is the light-absorbing material in a PSC. The p-type material is the hole transporting material (HTM), which is usually Spiro-OMeTAD. The device architecture is completed by depositing gold (Au) as the back contact. When light is shone on a PSC, charges (electrons and holes) are generated in the perovskite absorber layer. The carriers are extracted by the mesoporous TiO_2 and the HTM. The front and back contacts then collect the carriers and transport them to the external load.

Initial results have showcased the importance of a mesoscopic oxide ETM/scaffold such as TiO_2, ZrO_2 or Al_2O_3, for the superior functioning of PSCs [8–11]. However, for the usage of oxide ETMs, the required sintering temperatures are usually above 450°C. This limits their applications in flexible solar cells, which can potentially be used in lightweight wearable electronics,

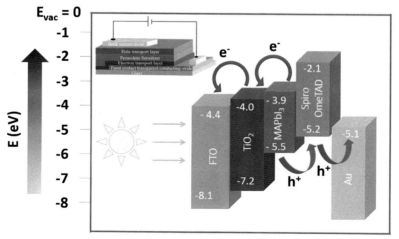

FIG. 1 Schematic representation of a conventional cell configuration and charge transfer process.

Oxide free materials for perovskite solar cells **Chapter | 12 289**

roll-to-roll processing, etc. Also, choosing an apt ETM for PSCs is very challenging. Since the electron diffusion length and carrier conductivity values are lesser than that of holes, causes a major bottleneck in improving the performance of PSCs [12,13]. Hence, a variety of ETMs has been explored by researchers which have been discussed below.

2. Oxide free-electron transport materials

ETMs play a crucial role in PSCs as they have to extract the carriers generated from the perovskite absorber material and transfer them to the electrode by minimizing the potential barrier for electron transport, as well to block the holes to reduce carrier recombination at the interfaces. Hence, ETMs should have suitable energy band matching with the absorber material and electrode, high electron mobility, and good stability. Nonoxide ETMs offer the advantage of low temperature processing of PSCs; hence choice of substrates and ease of fabrication [14]. The following section highlights the advancements in oxide free ETMs.

2.1 Fullerene and its derivatives

Fullerenes have proven to be useful as electron charge transporters owing to their electron accepting ability, energy levels matching with perovskite and good stability. Fullerenes have the ability to passivate traps at the interfaces (surface and grain boundaries). C_{60} is the most used charge extractor of all fullerenes due to good charge mobility of $1.3\,cm^2\,V^{-1}\,s^{-1}$ [15]. Wang et al., have used double layers of fullerene, which have drastically reduced the dark current leakage and passivated the traps thereby improving the fill factor and ensuing 12.2% PCE [16]. Yoon et al., have developed hysteresis-free MAPbI$_3$ devices employing ambient temperature vacuum processed C_{60} as ETM and obtained an efficiency of 19.1% on glass substrate and 16% on flexible PEN substrate [17]. Meng et al., have established a novel fullerene C5-NCMA as ETM in p-i-n architecture devices with an efficiency of 17.6% with enhanced moisture endurance and negligible hysteresis [18]. However, its poor solubility has limited its use in solution-processable PSCs. To address the solubility issue, the fullerenes have been functionalized resulting in adduct forms, which are soluble in various polar solvents. However, the multiadducts affect the packing of the cage resulting in reduced charge mobility [19]. In this context, Xing et al. [20], have developed hexakis [di(ethoxycarbonyl)methano]-C_{60} (HEMC), a multiadduct of fullerene with increased charge mobility suppressing charge accumulation and recombination, efficiency, and stability. They have reported a PCE of ~20% in MAPbI$_3$, MAPbCl$_x$I$_{(3-x)}$, and CsFAMAPbIBr-based PSCs in mesoporous device architecture shown in Fig. 2. They have reported 3840 h of storage stability, 300 h of thermal, and 240 h of light soaking stability.

Phenyl-C_{61}-butyric acid methyl ester (PCBM) is an n-type material, extensively employed in inverted architectures. Jeng et al., have reported the

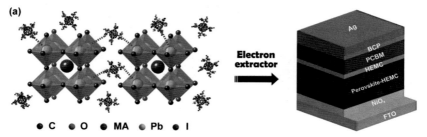

FIG. 2 Schematic of the hexakis [di(ethoxycarbonyl)methano]-C$_{60}$ (HEMC) based PSCs. *(Reproduced from Z. Xing, S.-H. Li, Y. Hui, B.-S. Wu, Z.-C. Chen, D.-Q. Yun, L.-L. Deng, M.-L. Zhang, B.-W. Mao, S.-Y. Xie, R.-B. Huang, L.-S. Zheng, Star-like hexakis[di(ethoxycarbonyl)methano]-C$_{60}$ with higher electron mobility: an unexpected electron extractor interfaced in photovoltaic perovskites, Nano Energy 74 (2020) 104859, Copyright (2020), with permission from Elsevier.)*

feasibility of using PCBM as ETM in inverted architectures with an efficiency of 3.9% [21]. Tian et al., have developed a new dimeric C$_{60}$ derivative with 2 covalently linked PCBM units as ETM and compared the performance with one unit of PCBM. A device using D$_{60}$ as ETM resulted in 16.6% PCE due to better surface defect passivation and electron extraction compared to the 14.7% of PCBM [22]. Nevertheless, PCBM has not been used in n-i-p architectures owing to the challenge in the formation of a compact perovskite absorber film on PCBM. This has been overcome by Ryu et al., who have for the first time reported its usage in n-i-p architecture for the preparation of low temperature processed MAPbI$_3$ PSCs by appropriate solvent engineering to result in a dense perovskite layer on the PCBM film as shown in Fig. 3 [23]. FTO/polyethyleneimine(PEI)/PCBM/MAPbI$_3$/polytriarylamine(PTAA)/Au device configuration has been employed and 15.3% PCE was obtained.

In one study by Malinkiewicz et al., PCBM$_{60}$ and tris(2,4,6-trimethyl-3-(pyridin-3-yl)phenyl)borane (3TPYMB) were compared as ETMs [24]. ETMs were deposited on a thermally evaporated MAPbI$_3$ layer using a meniscus coating of 10 nm. Initially, the cells were fabricated without PEDOT:PSS HTM to check for the better performing ETM. PCBM$_{60}$ as ETM yielded 10% PCE while 3TPYBM yielded only 5.5% owing to the favorable band position of the conduction band of perovskite with the LUMO of PCBM$_{60}$ which promotes the flow of electrons. Finally, devices were fabricated using ITO/PEDOT:PSS/poly(4-butylphenyl-diphenyl-amine)(polyTPD)/MAPbI$_3$/PCBM$_{60}$/Au configuration. A PCE of 14.8% for labscale-devices was yielded with a perovskite thickness of 285 nm.

2.2 Others

Naphthalene-based ETM has been demonstrated as a nonfullerene ETM by Jung et al., in inverted PSCs with CH(NH$_2$)$_2$PbI$_{3-x}$Br$_x$ absorber material [25]. They have reported PCE exceeding 20% with 90% retention of initial PCE at 100°C without encapsulation under one sun illumination for 500 h.

Oxide free materials for perovskite solar cells **Chapter | 12** **291**

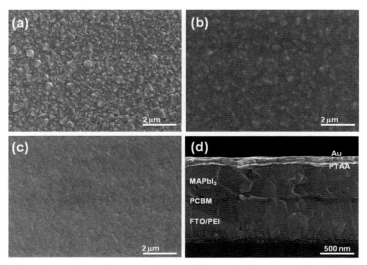

FIG. 3 Scanning electron microscopy images of the various layers coated on FTO substrate (A) PEI, (B) PCBM, (C) MAPbI$_3$, and (D) device cross-section. *(Reproduced from S. Ryu, J. Seo, S.S. Shin, Y.C. Kim, N.J. Jeon, J.H. Noh, S.Il. Seok, J. Mater. Chem. A 3 (2015) 3271–3275. Copyright (2015) The Royal Society of Chemistry.)*

Li et al., have employed amorphous ribboned Bi$_2$S$_3$ with good stability and intrinsic mobility as ETM for low temperature processed inverted PSCs [26]. They have achieved a PCE of 13% using NiO/CH$_3$NH$_3$PbI$_3$/Bi$_2$S$_3$ architecture with improved ambient storage stability than PCBM.

3. Oxide free hole transporting materials (HTMs)

HTMs are crucial for the enhanced photovoltaic performance of PSCs as they act as hole extraction layers, which suppress charge recombination simultaneously blocking the holes. Especially in the n-i-p architecture, they act as a capping layer that prevents ingress of moisture and oxygen into the perovskite layer [27]. The following section highlights the recent advancements in oxide-free HTMs.

3.1 Spiro-OMeTAD

Organic materials are often selected as HTMs due to their easy solution processing, tunable optoelectronic properties by modifying their chemical structure [28–30]. Of all HTMs, Spiro-OMeTAD was regularly used, owing to its suitable energy level matching, simple recipe, high melting temperature, solution processability, and good conductivity leading to high performance [31]. Spiro-OMeTAD was initially reported as HTM replacing the liquid electrolyte used until then in PSCs where a PCE of 9.7% was reported [32]. Since then,

292 Oxide free nanomaterials for energy storage and conversion applications

PSCs based on Spiro-OMeTAD showed a tremendous enhancement in performance and reached PCEs above 20% [33]. The high conductivity of Spiro-OMeTAD can be attributed to the widely used dopants tert-butyl pyridine (TBP) and LiTFSI salt, which increase the free carrier generation. However, its application is limited by the cost factor and dopant penetration into the ETM layer, thus affect its charge extraction leading to hysteresis [30,34]. To address this issue, Zhang et al., have substituted LiTFSI and TBP dopants with tetra butyl ammonium (TBA) salts in Spiro-OMeTAD for efficient and thermally stable PSCs [35]. They have reported a higher PCE of 18.4% using TBATFSI dopant compared to the 18.1% of (bis(trifluoromethane)sulfonimide lithium salt (LiTFSI) and tert-butyl pyridine (TBP) dopants. It has been observed that the anions in the TBA salts such as $TFSI^-$ and PF^{6-} dopant has a critical role in hole conductivity, uniformity of the HTM, and reduced hysteresis of the devices along with improved environmental and thermal stability as shown in Fig. 4. It has been observed that the films with TBA salts with $TFSI^-$ and PF^{6-} anions exhibited better thermal stability when heated at 85°C for 30 min. This enhanced stability has led to their corresponding device stability with 60% of original PCE retention after 30 days of aging in dark and 8 h of aging at 85°C; both at an RH of 50%.

3.2 Poly [bis(4-phenyl)(2,5,6-trimethylphenyl)amine] (PTAA)

Another polymeric material that has gained wide attention in the recent past for its hole conducting properties is PTAA. PTAA has shown improved stability under illumination and higher temperature [36]. PTAA was initially used by Seok et al., where 12% PCE was obtained [37] and the highest certified PCE of 22.1% by Yang et al., for small cells and 19.7% for cells of $1 cm^2$ using mixed anion and mixed cation perovskite [38]. They have incorporated iodide ions into the cation precursor solution through an intramolecular exchange process which was shown to reduce the defect states which negatively affect the V_{oc} and J_{sc}.

3.3 Copper thiocyanate (CuSCN)

CuSCN is the most commonly used inorganic HTM owing to its chemical stability, high transparency, suitable band alignment, solution, and low-temperature processability, making it compatible for flexible PSCs and high hole mobility. CuSCN as HTM was first introduced by Ito et al., in mesoscopic devices, and by Subbiah et al., in planar devices and a PCE of 4.85% and 3.8% was obtained respectively [39,40]. By interface engineering and modifications in the method of deposition, the thickness of perovskite layer, suitable device architecture, etc., their PCEs have risen up to 20.4% [41]. The highest PCE of 20.4% employing CuSCN as HTM has been reported by Arora et al. [42] using $FTO/TiO_2/CsFAMAPbI_{3-x}Br_x/CuSCN/reduced$ graphene oxide (rGO)/Au. Even though they have obtained a slightly higher PCE using Spiro-OMeTAD

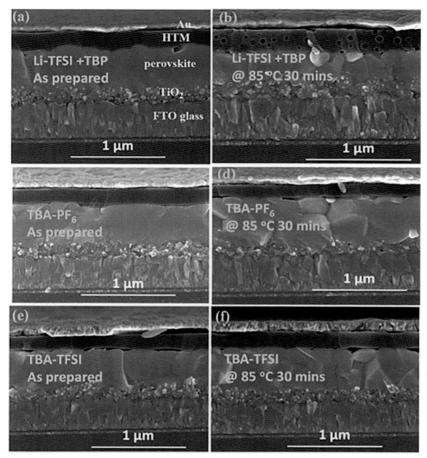

FIG. 4 Cross-sectional images of the as-prepared and heat-treated devices at 85°C for 30 min with different dopants. *(Reproduced from J. Zhang, T. Zhang, L. Jiang, U. Bach, Y.-B. Cheng, ACS Energy Lett. 3 (7) (2018) 1677–1682. Copyright (2018) American Chemical Society.)*

as HTM, the CuSCN based devices were much durable than the organic HTM devices. The CuSCN devices retained 90% of their initial PCE when stored under dark at 85°C in the air compared to only 60% retention by the Spiro-OMeTAD devices. This study proves CuSCN to be an efficient and stable HTM.

3.4 Copper iodide (CuI)

CuI is another potential p-type material that can be used as HTM in PSCs. It has properties similar to that of CuSCN except that it is more compatible with the perovskite layer during solution deposition, unlike CuSCN that uses solvents

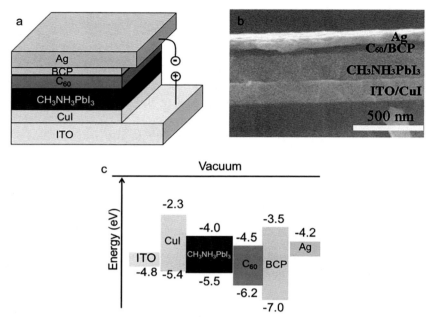

FIG. 5 (A) Device architecture, (B) device cross-section, and (C) device band diagram. *(Reproduced from W. Sun, S. Ye, H. Rao, Y. Li, Z. Liu, L. Xiao, Z. Chen, Z. Bian, C. Huang, Nanoscale 8 (2016) 15954–15960. Copyright (2016) The Royal Society of Chemistry.)*

corrosive to the perovskite layer. In spite of having electrical conductivity two times higher than Spiro-OMeTAD, their usage is low due to the high recombination rate. Chen et al. [43] have fabricated a device using CuI as HTM on two-step sequentially deposited $CH_3NH_3PbI_3$ and obtained a PCE of 13.58%. By replacing the uneven surface of the two-step fabricated films with a smooth surface of the one-step fast deposition-crystallization films, Sun et al., have reported the highest reported efficiency using p-i-n ITO/CuI/perovskite/C_{60}/BCP/Ag planar architecture [44]. They have fabricated the CuI layer in a simple, solution-processed, low-temperature method and obtained an efficiency of 16.8% due to the improvement in carrier extraction and reduced recombination at the HTM-perovskite interface. Device architecture, cross-sectional image, and band energy diagrams are given in Fig. 5A–C respectively. The device maintained 93% of its initial efficiency for 288 h when stored at room temperature and 25% humidity due to the hydrophobic nature of CuI.

3.5 Copper chalcogenides

It is a well-known fact that copper chalcogenides have been employed as absorber layers in second-generation solar cells. However, due to their p-type nature, they have also been explored by researchers as HTM in PSCs. Copper

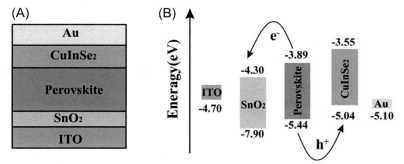

FIG. 6 (A) Device configuration employed and (B) band diagram. *(Reproduced from Y. Zhang, Z. Zhang, Y. Liu, Y. Liu, H. Gao, Y. Mao, An inorganic hole-transport material of CuInSe₂ for stable and efficient perovskite solar cells, 67 (2019) 168–174. Copyright (2019), with permission from Elsevier.)*

chalcogenides such as $CuInS_2$, $Cu(InGa)(SSe)_2$, $Cu_2ZnSnSe_4$, Cu_2ZnSnS_4, $CuInSe_2$ have been reported as HTMs for PSCs [45–47]. Of all, $CuInSe_2$ was found to be a potential candidate among all the copper chalcogenides HTMs due to their high absorption co-efficient, tunable bandgap, low toxicity, ease of processing, and low cost [48]. Using the device architecture shown in Fig. 6A, they have reported the highest PCE of 12.8% with 78% retention in their efficiencies for 96 h whereas the Spiro-OMeTAD device retained only 30%. The charge transfer mechanism is illustrated in Fig. 6B.

3.6 Carbazole based HTMs

Carbazoles are small-molecule organic HTMs that are inexpensive, have good stability both in chemical and environment, and have strategic sites to attach the functional groups [49]. Several works based on carbazole-based HTMs have been reported by various researchers by varying the substituted compounds where efficiencies were comparable to the devices based on Spiro-OMeTAD [50–53]. Lu et al., have used a star-shaped carbazole HTM, with an efficiency of 18.87% using mesoporous device architecture [54]. Due to the simplicity of the synthesis of the star-shaped carbazole, the cost has been reduced to almost three times compared to Spiro-OMeTAD and with better stability. However, most of the carbazole-based HTMs require Li and Co salts for better conductivity and limit the device stability. In this regard Christians et al., have used carbazole based EH44 (9-(2 ethylhexyl)-*N,N,N,N*-tetrakis(4-methoxyphenyl)-9*H*-carbazole-2,7-diamine) as HTM [55]. They eliminated the Li salt of the dopant and replaced it with Ag to obtain EH44-ox and the core fluorine component of Spiro-OMeTAD is replaced with carbazole to obtain hydrophobicity. The device architecture they have used is $SnO_2/(FA_{0.79}MA_{0.16}Cs_{0.05})_{0.97}Pb(I_{0.84}Br_{0.16})_{2.97}/EH44/MoO_x/Al$ and obtained operational stability of 94% up

296 Oxide free nanomaterials for energy storage and conversion applications

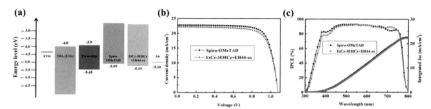

FIG. 7 (A) Band diagram of the PSCs, (B) photocurrent-voltage characteristics of the devices with two HTMs, and (C) IPCE of the highest efficiency devices with two HTMs. *(Reproduced from L. Gao, T.H. Schloemer, F. Zhang, X. Chen, C. Xiao, K. Zhu, A. Sellinger, ACS Appl. Energy Mater. 3 (2020) 4492–4498. Copyright (2020) American Chemical Society.)*

to 1000h of accelerated degradation under continuous illumination (including UV) in ambient conditions at 10%–20% RH; this is among the best-recorded stability without encapsulation till date. However, the performance is not at par with the Li dopant HTMs. An extension to this work, Gao et al., have explored the impact of carbazole-based HTMs on the stability and performance of PSCs [56]. They have synthesized EtCz-3EHCz (9-ethyl-N^2,N^7-bis(9-(2-ethylhexyl)-9H-carbazol-3-yl)-N^2,N^7-bis(4-methoxyphenyl)-9H-carbazole-2,7-diamine)doped with EH44-ox as HTM. The band diagram is given in Fig. 7A and device performance and IPCE curves in Fig. 7B and C respectively. Devices based on this HTM showed 17.75% PCE comparable to the 18.5% of spiroOMeTAD based devices. The carbazole-based devices retained 50% PCE for 500h of continuous illumination at 50°C and 10%–20% RH without encapsulation.

3.7 PEDOT:PSS (PEDOT = poly(3,4-ethylene dioxythiophene) and PSS = polystyrene sulfonate)

PEDOT:PSS is an inexpensive, highly conductive polymer that is employed as HTM in inverted architectures [57]. Even though PEDOT has decent conductivity it lacks solubility in solvents, hence PSS is added to PEDOT to have good solubility along with conductivity. PEDOT:PSS has good conductivity, optical transparency, mechanical flexibility, thermal stability, and good wettability with the perovskite layer that makes it a good choice for HTM. Due to its low processing temperature unlike metal oxides, PEDOT:PSS has also been applied in flexible substrates where a PCE of 9.2% has been reported on flexible ITO substrates and an 11.5% PCE on rigid ITO substrate [58]. Wang et al., have reported fabrication of large grains of perovskite by room temperature water vapor annealing method using PEDOT:PSS HTM and obtained an efficiency of 16.4% [59]. An interesting way to improve the hole conductivity was reported by Hu et al. [60] They have realized a very unique way of improving the PEDOT:PSS layer conductivity by lowering the thickness from 9nm to monolayers of PEDOT:PSS on ITO substrate. On ITO substrate, they

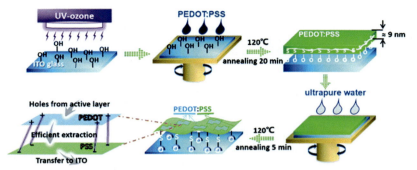

FIG. 8 Schematic illustration of the water rinsing process and hole extraction. *(Reproduced from L. Hu, M. Li, K. Yang, Z. Xiong, B. Yang, M. Wang, X. Tang, Z. Zang, X. Liu, B. Li, Z. Xiao, S. Lu, H. Gong, J. Ouyang, K. Sun, J. Mater. Chem. A 6 (2018) 16583–16589. Copyright (2018) The Royal Society of Chemistry.)*

self-assembled PEDOT:PSS monolayers by a simple water rinsing process. This has improved the performance from 13.4% to 18% due to the orientation of the internal electric field on ITO that helped in better extraction of holes. In Fig. 8, a major portion of PEDOT:PSS has swept away but a monolayer of PEDOT:PSS is attached to the ITO owing to the strong bond between In-O-S. A coulombic interaction between the PEDOT:PSS and ITO lead to a bi-layered structure that orients the electric field which has higher work function and hydrophobicity. This led to improved V_{OC} and better air stability of the water rinsed device compared to the control device.

3.8 Poly (3-hexylthiophene) (P3HT) based HTMs

P3HT is a p-type polymer with good conductivity and stability, which is a suitable HTM for PSCs. Nia et al., have reported a molecular weight (M.W.) dependent performance of P3HT in mesoscopic PSCs [61]. They have observed that at P3HT M.W. of 124 kDa, the performance of the device was 16.2% which is almost double of the PCE at 44 kDa due to increased light absorption, electron lifetime, and decreased charge recombination. However, its poor band alignment with absorber resulting in inefficient carrier extraction and transfer makes it a less popular choice for an HTM. In this regard, doping of P3HT to align its band positions was reported with tetrafluoro-tetracyano-quinodimethane (F4TCNQ), graphdiyne, etc., for better performance [62,63]. One such doping was done by Jung et al., with tris(2-(1*H*-pyrazol-1-yl)pyridine) cobalt(II) di[bis(trifluoromethane)sulfonimide] (Co(II)-TFSI) in P3HT for good performance PSCs, the device side morphology image is illustrated in Fig. 9A [64]. Due to the better charge extraction, they reported efficiency of 16.28% on a glass substrate and 11.84% on flexible substrates with a 1.4 cm^2 active area. This is because of the downward shift of the HOMO levels as shown in Fig. 9B

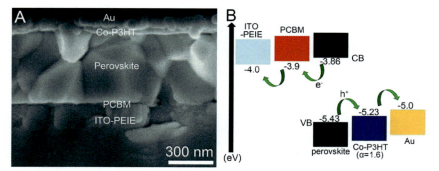

FIG. 9 (A) Device cross-section and (B) band energy illustration of the different layers of the device. *(Reproduced from J.W. Jung, J.-S. Park, I.K. Han, Y. Lee, C. Park, W. Kwon, M. Park, J. Mater. Chem. A 5 (2017) 12158–12167. Copyright (2017) The Royal Society of Chemistry.)*

and enhanced conductivity in the cobalt doped P3HT. The flexible devices showed mechanical stability of up to 600 bending cycles with 93% retention in initial PCE and air stability of 40 days with 72% retention in PCE. This study highlights the improvement in PCE with Co doping in HTMs based on P3HT.

3.9 Lead sulfide (PbS)

Usually, the high-performing solar cells use Spiro-OMeTAD as the HTM [65]. The additives utilized in Spiro-OMeTAD are extremely hygroscopic that leads to the degradation of the solar cell. So, the quest has been for suitable HTMs that is stable and efficient. In this point of view, PbS was found to be a decent substitute due to its bulky exciton Bohr radius of 18 nm which enables modification of the bandgap by changing the PbS nanocrystal size [66–68].

Li et al., have used PbS as HTM on $CH_3NH_3PbI_3$ perovskite and obtained a PCE of 7.88% [69]. The oleic acid capped PbS quantum were synthesized by hot injection method with a bandgap of 1.37 eV and their energy levels matching that of the absorber layer for adequate charge extraction. The colloidal solution of PbS in octane is spin-coated onto one step and two-step processed $CH_3NH_3PbI_3$ layers with TiO_2 as the ETM and Au as back contact. The performance improved from 3.5% to 4.73% in the case of one-step processed perovskite films and from 5.53% to 7.88% in the case of the two-step processed film owing to the improvement in V_{OC} and J_{SC}.

Recently, Sidhik et al. [70] have described the use of PbS quantum dots (QDs) with ~5.4 nm average crystallite size as HTM for the first time. A PCE of 5.83% was reported using mesoporous device configuration with PbS as HTM. The PbS QDs have also been sandwiched as a protective film between the absorber and the Spiro-OMeTAD HTM layer, which caused a decrease in PCE from 17.19% to 14.2% for a PbS colloidal solution of 5 mg mL^{-1}. Nonetheless, when the devices were stored at 27°C and 70%

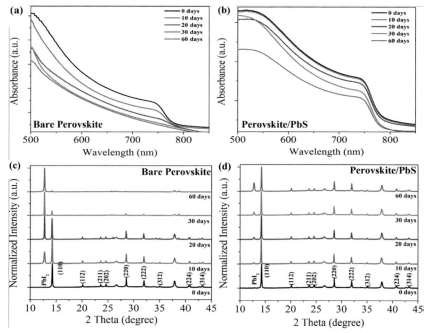

FIG. 10 Absorbance and XRD spectra of the (A, C) bare perovskite and (B, D) PbS QD layer incorporated perovskite. *(Reproduced from S. Sidhik, C. Rosiles Pérez, M.A. Serrano Estrada, T. López-Luke, A. Torres, E. De la Rosa, Improving the stability of perovskite solar cells under harsh environmental conditions, Sol. Energy 202 (2020) 438–445. Copyright (2020), with permission from Elsevier.)*

relative humidity (RH), the bare films degraded within 15 days while the PbS incorporated devices retained 76% of their PCE up to 60 days. It was also found that the films covered by PbS QDs exhibited better tolerance when brought in direct contact with water compared to the bare perovskite films. From the optical and structural results, it was found that the films were stable even after 60 days of exposure to 70% RH at room temperature compared to the bare films that degraded within 20 days as shown in Fig. 10. Hence, PbS QDs can potentially pave a way for water-resistant PSCs.

4. HTM free carbon-based PSCs

Carbon was thought of as an inexpensive alternative to the expensive Au back contact initially. However, its dual functional role as a hole conductor and electrode material has attracted tremendous attention recently. The merits of carbon as an attractive alternative to HTM cum electrode material include its easy processability, low cost, good mechanical strength, good adhesion with the perovskite absorber material thus reducing the interface resistance, suitable work function, to name a few [71,72]. Zhang et al., have used mesoporous graphite/carbon black as an electrode is fully printable mesoscopic PSCs and reported a PCE of 11% [73]. They have observed that the porosity of graphite has an effect on the device performance as it affects the filling of PbI_2 and MAI precursors. Yang et al., have observed that the contact between $MAPbI_3$ perovskite and carbon film plays a vital role in charge extraction and PCE. They have reported a PCE of 10.2% with a combination of mesoscopic carbon black particles and flexible graphite sheet [74]. Wei et al., have extracted soot from the candle, which is a low-cost, stable, and abundantly available HTM in PSCs where a PCE of 11.2% was obtained [75]. The candle soot layer is carefully clamped onto the perovskite photoanode as shown in Fig. 11 directly, with the aid of the rolling transfer process and chemically aided rolling transfer process. The best performance of 11.2% was obtained in the case of third generation

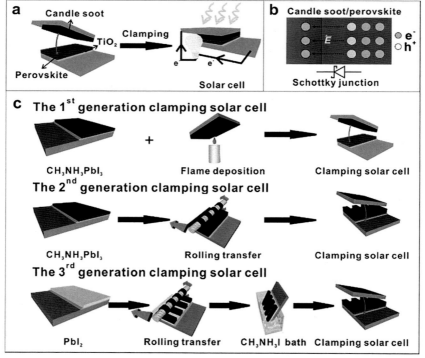

FIG. 11 (A) Concept of clamping FTO substrate with candle soot with perovskite photoanode, (B) Schottky junction formation, and (C) first generation clamping solar cell where the candle soot FTO is clamped on the perovskite absorber layer, second-generation clamping with the help of roll-to-roll processing and third-generation clamping with chemically aided rolling transfer clamping with MAI bath. *(Reprinted with permission from Z. Wei, K. Yan, H. Chen, Y. Yi, T. Zhang, X. Long, J. Li, L. Zhang, J. Wang, S. Yang, Energ. Environ. Sci. 7 (2014) 3326–3333. Copyright (2014) The Royal Society of Chemistry.)*

clamping solar cells since it has a better interface between soot and perovskite. They have revealed that clamping of solar cells is an easy and efficient way to realize high-performance solar cells paving the way towards the commercialization of PSCs. Liu et al., have used carbon-based PSCs with an efficiency of 13.1% where they found the importance of substrate, precursor temperature, spacer layer thickness, the impact of pore filling, and morphology of the absorber are critical parameters for better light-harvesting and lower hole-transporting resistance [76]. Wei et al., have reported the highest PCE in HTM free carbon-based PSCs using carbon layer of free-standing thermoplastic type with good conductivity and flexibility which is hot-pressed onto the perovskite absorber resulting in a PCE of 13.5% [77].

5. Oxide free TCO for PSCs

Usually, TCOs such as FTO or ITO is used as the anodes in PSCs due to their lower sheet resistance ($15\,\Omega$/square) and greater transmittance ($\sim85\%$) [78]. These materials require extensive vacuum processing, high temperature, and low sputtering making them expensive for large-scale manufacturing. Also, ITO is very expensive owing to the rare earth element. It is estimated that $\sim75\%$ of the perovskite solar module cost comes from the TCO [79]. Not only the cost, but TCOs are also rigid which limits their functions in potential flexible applications of PSCs. For flexible PSCs, the p-i-n configuration is usually employed due to its low-temperature processability, which can be used in roll-to-roll processing for large-scale production of PSCs. It has been found that PSCs fabricated on flexible ITO led to cracks owing to their brittleness upon bending by 4 mm curvature for 1000 cycles which makes it unsuitable for deposition on plastic substrates [80]. For flexible PSCs, apart from being mechanically strong and flexible, the materials should display high transmittance and low sheet resistance. Carbon-based materials, conducting polymers, and metallic grids have been employed [81–88]. Sung et al., have reported highly efficient PSCs employing graphene as a substitute for ITO and achieved a PCE of 17.1% but they had to introduce a thin layer of MoO_3 at the interface for better band alignment and conductivity enhancement [89]. Metal grids have high transmittance and low sheet resistance but they need high pressures to flatten out the nanowires and nanofibers to prevent shorting of the device [90]. Even though the individual conductivity of a carbon nanotube is high, it suffers from inefficient carrier transport between the nanotubes and requires to be fabricated with high compactness for the formation of a percolation network [91]. Among all, PEDOT:PSS has gained considerable attention as a conducting polymer owing to its low-temperature solution processability, high transparency, good mechanical strength, thermal stability, good conductivity, and compatibility for roll to roll processing [81,82]. Also, its production on a large scale can give highly uniform, smooth, and transparent films over metallic grids and nanowires which is much suitable for photovoltaic applications. Zhang et al.,

302 Oxide free nanomaterials for energy storage and conversion applications

have described a low temperature, vacuum-free solution processing of TCO-free semitransparent (st) PSCs using nitric acid annealed PEDOT:PSS as conducting electrodes [92]. The TCO-free st-PSCs have shown a PCE of 13.9% on glass and 19.2% on tandem cells. Since it uses a low-temperature process fabrication process (130°C), the PSCs were fabricated on flexible polyethylene terephthalate (PET) and polyimide substrates which showed excellent bending stability. The devices fabricated on PET showed 10.3% PCE with 1000h of bending stability at a 5mm radius. This study highlights the independence of PSCs from TCOs and potential applications in wearable flexible electronics. In most of the studies, the performance is limited to around 12% of the TCO-free PSCs due to the large interface energy barriers owing to the low work function of the HTMs and wettability issues. Hence, it is necessary to fabricate HTMs with suitable work functions and favorable wettability with the anode to prevent large energy barriers. In this context, Wen et al., have reported TCO-free PSCs using poly(3-(4-methylamincarboxylbutyl)thiophene) (P3CT-N) modified PEDOT:PSS anodes where PCE improved from 4.63% to 13.13% [93]. This is due to the improved work function, better hole transport, and good wettability of P3CT-N with the PEDOT:PSS anode. Hence, by modifying the PEDOT:PSS anode, TCO-free PSCs can potentially be used for the commercialization of PSCs owing to their roll-to-roll processability.

6. Conclusion

This chapter focuses on the different oxide-free materials available for PSCs. Oxide-free electron and hole transport materials, as well as transparent conducting oxide free materials for PSCs, have been discussed in detail. The critical factors that influence the performance and stability are elaborated and provide a platform for identifying the loopholes for strategic advancements in efficiency enhancement and stability of PSCs. The oxide-free materials performed at par with the oxide materials available for PSCs in PCE and stability, which can replace the oxide materials for commercialization due to their low processing temperatures, applicability in flexible substrates, compatible with roll-to-roll processing with superior performance and stability.

Acknowledgments

Dr. GV acknowledges the Department of Science and Technology, New Delhi, India for financial support through DST-SYST (SP/YO/012/2017(G)) and SERB Research Scientist (SB/SRS/2019-20/22PS).

References

[1] A. Kojima, K. Teshima, Y. Shirai, T. Miyasaka, J. Am. Chem. Soc. 131 (2009) 6050–6051.
[2] K. Branker, M.J.M. Pathak, J.M. Pearce, Renew. Sustain. Energy Rev. 15 (2011) 4470–4482.

Oxide free materials for perovskite solar cells **Chapter | 12** **303**

[3] Best Research-Cell Efficiency Chart | Photovoltaic Research, NREL, 2021. https://www.nrel. gov/pv/assets/pdfs/best-research-cell-efficiencies-rev211011.pdf.

[4] H.J. Snaith, L. Schmidt-Mende, Adv. Mater. 19 (2007) 3187–3200.

[5] S. Maniarasu, M.K. Rajbhar, R.K. Dileep, E. Ramasamy, G. Veerappan, Mater. Lett. 245 (2019) 226–229.

[6] K. Tanaka, T. Takahashi, T. Kondo, K. Umeda, K. Ema, T. Umebayashi, K. Asai, K. Uchida, N. Miura, Jpn. J. Appl. Phys. 44 (2005) 5923–5932.

[7] J. Am, Chem. Soc. 121 (1999) 8970.

[8] M.M. Lee, J. Teuscher, T. Miyasaka, T.N. Murakami, H.J. Snaith, Science 338 (2012) 643–647.

[9] J.M. Ball, M.M. Lee, A. Hey, H.J. Snaith, Energ. Environ. Sci. 6 (2013) 1739.

[10] J. Burschka, N. Pellet, S.-J. Moon, R. Humphry-Baker, P. Gao, M.K. Nazeeruddin, M. Grätzel, Nature 499 (2013) 316–319.

[11] M. Liu, M.B. Johnston, H.J. Snaith, Nature 501 (2013) 395–398.

[12] G. Xing, N. Mathews, S. Sun, S.S. Lim, Y.M. Lam, M. Gratzel, S. Mhaisalkar, T.C. Sum, Science 342 (2013) 344–347.

[13] Y. Li, Y. Zhao, Q. Chen, Y.(.M.). Yang, Y. Liu, Z. Hong, Z. Liu, Y.-T. Hsieh, L. Meng, Y. Li, Y. Yang, J. Am. Chem. Soc. 137 (2015) 15540–15547.

[14] R. Wang, M. Mujahid, Y. Duan, Z. Wang, J. Xue, Y. Yang, Adv. Funct. Mater. 29 (2019) 1808843.

[15] S. Pfuetzner, J. Meiss, A. Petrich, M. Riede, K. Leo, Appl. Phys. Lett. 94 (2009) 223307.

[16] Q. Wang, Y. Shao, Q. Dong, Z. Xiao, Y. Yuan, J. Huang, Energ. Environ. Sci. 7 (2014) 2359–2365.

[17] H. Yoon, S.M. Kang, J.-K. Lee, M. Choi, Energ. Environ. Sci. 9 (2016) 2262–2266.

[18] X. Meng, Y. Bai, S. Xiao, T. Zhang, C. Hu, Y. Yang, X. Zheng, S. Yang, Nano Energy 30 (2016) 341–346.

[19] M.A. Faist, S. Shoaee, S. Tuladhar, G.F.A. Dibb, S. Foster, W. Gong, T. Kirchartz, D.D.C. Bradley, J.R. Durrant, J. Nelson, Adv. Energy Mater. 3 (2013) 744–752.

[20] Z. Xing, S.-H. Li, Y. Hui, B.-S. Wu, Z.-C. Chen, D.-Q. Yun, L.-L. Deng, M.-L. Zhang, B.-W. Mao, S.-Y. Xie, R.-B. Huang, L.-S. Zheng, Nano Energy 74 (2020) 104859.

[21] J.-Y. Jeng, Y.-F. Chiang, M.-H. Lee, S.-R. Peng, T.-F. Guo, P. Chen, T.-C. Wen, Adv. Mater. 25 (2013) 3727–3732.

[22] C. Tian, K. Kochiss, E. Castro, G. Betancourt-Solis, H. Han, L. Echegoyen, J. Mater. Chem. A 5 (2017) 7326–7332.

[23] S. Ryu, J. Seo, S.S. Shin, Y.C. Kim, N.J. Jeon, J.H. Noh, S. Il Seok, J. Mater. Chem. A 3 (2015) 3271–3275.

[24] O. Malinkiewicz, C. Roldán-Carmona, A. Soriano, E. Bandiello, L. Camacho, M.K. Nazeeruddin, H.J. Bolink, Adv. Energy Mater. 4 (2014) 1400345.

[25] S.-K. Jung, J.H. Heo, D.W. Lee, S.-C. Lee, S.-H. Lee, W. Yoon, H. Yun, S.H. Im, J.H. Kim, O.-P. Kwon, Adv. Funct. Mater. 28 (2018) 1800346.

[26] D.-B. Li, L. Hu, Y. Xie, G. Niu, T. Liu, Y. Zhou, L. Gao, B. Yang, J. Tang, ACS Photonics 3 (2016) 2122–2128.

[27] G.-W. Kim, G. Kang, M. Malekshahi Byranvand, G.-Y. Lee, T. Park, ACS Appl. Mater. Interfaces 9 (2017) 27720–27726.

[28] W. Kong, W. Li, C. Liu, H. Liu, J. Miao, W. Wang, S. Chen, M. Hu, D. Li, A. Amini, S. Yang, J. Wang, B. Xu, C. Cheng, ACS Nano 13 (2) (2019) 1625–1634.

[29] W. Li, C. Liu, Y. Li, W. Kong, X. Wang, H. Chen, B. Xu, C. Cheng, Sol. RRL 2 (2018) 1800173.

304 Oxide free nanomaterials for energy storage and conversion applications

[30] T.H. Schloemer, J.A. Christians, J.M. Luther, A. Sellinger, Chem. Sci. 10 (2019) 1904–1935.

[31] Z. Hawash, L.K. Ono, Y. Qi, Adv. Mater. Interfaces 5 (2018) 1700623.

[32] H.-S. Kim, C.-R. Lee, J.-H. Im, K.-B. Lee, T. Moehl, A. Marchioro, S.-J. Moon, R. Humphry-Baker, J.-H. Yum, J.E. Moser, M. Grätzel, N.-G. Park, Sci. Rep. 2 (2012) 591.

[33] N.J. Jeon, H.G. Lee, Y.C. Kim, J. Seo, J.H. Noh, J. Lee, S. Il Seok, J. Am. Chem. Soc. 136 (2014) 7837–7840.

[34] Z. Li, C. Xiao, Y. Yang, S.P. Harvey, D.H. Kim, J.A. Christians, M. Yang, P. Schulz, S.U. Nanayakkara, C.-S. Jiang, J.M. Luther, J.J. Berry, M.C. Beard, M.M. Al-Jassim, K. Zhu, Energ. Environ. Sci. 10 (2017) 1234–1242.

[35] J. Zhang, T. Zhang, L. Jiang, U. Bach, Y.-B. Cheng, ACS Energy Lett. 3 (2018) 1677–1682.

[36] Z. Wang, Q. Lin, B. Wenger, M.G. Christoforo, Y.-H. Lin, M.T. Klug, M.B. Johnston, L.M. Herz, H.J. Snaith, Nat. Energy 3 (2018) 855–861.

[37] J.H. Heo, S.H. Im, J.H. Noh, T.N. Mandal, C.-S. Lim, J.A. Chang, Y.H. Lee, H. Kim, A. Sarkar, M.K. Nazeeruddin, M. Grätzel, S. Il Seok, Nat. Photonics 7 (2013) 486–491.

[38] W.S. Yang, B.-W. Park, E.H. Jung, N.J. Jeon, Y.C. Kim, D.U. Lee, S.S. Shin, J. Seo, E.K. Kim, J.H. Noh, S. Il Seok, Science 356 (2017) 1376–1379.

[39] A.S. Subbiah, A. Halder, S. Ghosh, N. Mahuli, G. Hodes, S.K. Sarkar, J. Phys. Chem. Lett. 5 (2014) 1748–1753.

[40] S. Ito, S. Tanaka, H. Vahlman, H. Nishino, K. Manabe, P. Lund, ChemPhysChem 15 (2014) 1194–1200.

[41] R. Singh, P.K. Singh, B. Bhattacharya, H.-W. Rhee, Appl. Mater. Today 14 (2019) 175–200.

[42] N. Arora, M.I. Dar, A. Hinderhofer, N. Pellet, F. Schreiber, S.M. Zakeeruddin, M. Grätzel, Science 358 (2017) 768–771.

[43] W.-Y. Chen, L.-L. Deng, S.-M. Dai, X. Wang, C.-B. Tian, X.-X. Zhan, S.-Y. Xie, R.-B. Huang, L.-S. Zheng, J. Mater. Chem. A 3 (2015) 19353–19359.

[44] W. Sun, S. Ye, H. Rao, Y. Li, Z. Liu, L. Xiao, Z. Chen, Z. Bian, C. Huang, Nanoscale 8 (2016) 15954–15960.

[45] M. Lv, J. Zhu, Y. Huang, Y. Li, Z. Shao, Y. Xu, S. Dai, ACS Appl. Mater. Interfaces 7 (2015) 17482–17488.

[46] L. Xu, L.-L. Deng, J. Cao, X. Wang, W.-Y. Chen, Z. Jiang, Nanoscale Res. Lett. 12 (2017) 159.

[47] M. Yuan, X. Zhang, J. Kong, W. Zhou, Z. Zhou, Q. Tian, Y. Meng, S. Wu, D. Kou, Electrochim. Acta 215 (2016) 374–379.

[48] Y. Zhang, Z. Zhang, Y. Liu, Y. Liu, H. Gao, Y. Mao, Org. Electron. 67 (2019) 168–174.

[49] N. Prachumrak, S. Pojanasopa, S. Namuangruk, T. Kaewin, S. Jungsuttiwong, T. Sudyoadsuk, V. Promarak, ACS Appl. Mater. Interfaces 5 (2013) 8694–8703.

[50] S. Do Sung, M.S. Kang, I.T. Choi, H.M. Kim, H. Kim, M. Hong, H.K. Kim, W.I. Lee, Chem. Commun. 50 (2014) 14161–14163.

[51] P. Gratia, A. Magomedov, T. Malinauskas, M. Daskeviciene, A. Abate, S. Ahmad, M. Grätzel, V. Getautis, M.K. Nazeeruddin, Angew. Chem. Int. Ed. 54 (2015) 11409–11413.

[52] M. Daskeviciene, S. Paek, Z. Wang, T. Malinauskas, G. Jokubauskaite, K. Rakstys, K.T. Cho, A. Magomedov, V. Jankauskas, S. Ahmad, H.J. Snaith, V. Getautis, M.K. Nazeeruddin, Nano Energy 32 (2017) 551–557.

[53] A. Magomedov, S. Paek, P. Gratia, E. Kasparavicius, M. Daskeviciene, E. Kamarauskas, A. Gruodis, V. Jankauskas, K. Kantminiene, K.T. Cho, K. Rakstys, T. Malinauskas, V. Getautis, M.K. Nazeeruddin, Adv. Funct. Mater. 28 (2018) 1704351.

[54] C. Lu, I.T. Choi, J. Kim, H.K. Kim, J. Mater. Chem. A 5 (2017) 20263–20276.

[55] J.A. Christians, P. Schulz, J.S. Tinkham, T.H. Schloemer, S.P. Harvey, B.J. Tremolet de Villers, A. Sellinger, J.J. Berry, J.M. Luther, Nat. Energy 3 (2018) 68–74.

Oxide free materials for perovskite solar cells Chapter | 12 **305**

[56] L. Gao, T.H. Schloemer, F. Zhang, X. Chen, C. Xiao, K. Zhu, A. Sellinger, ACS Appl. Energy Mater. 3 (2020) 4492–4498.

[57] A.T. Barrows, A.J. Pearson, C.K. Kwak, A.D.F. Dunbar, A.R. Buckley, D.G. Lidzey, Energ. Environ. Sci. 7 (2014) 2944–2950.

[58] J. You, Z. Hong, Y.(.M.). Yang, Q. Chen, M. Cai, T.-B. Song, C.-C. Chen, S. Lu, Y. Liu, H. Zhou, Y. Yang, ACS Nano 8 (2014) 1674–1680.

[59] B. Wang, Z.-G. Zhang, S. Ye, H. Rao, Z. Bian, C. Huang, Y. Li, J. Mater. Chem. A 4 (2016) 17267–17273.

[60] L. Hu, M. Li, K. Yang, Z. Xiong, B. Yang, M. Wang, X. Tang, Z. Zang, X. Liu, B. Li, Z. Xiao, S. Lu, H. Gong, J. Ouyang, K. Sun, J. Mater. Chem. A 6 (2018) 16583–16589.

[61] N.Y. Nia, F. Matteocci, L. Cina, A. Di Carlo, ChemSusChem 10 (2017) 3854–3860.

[62] Y. Zhang, M. Elawad, Z. Yu, X. Jiang, J. Lai, L. Sun, RSC Adv. 6 (2016) 108888–108895.

[63] J. Xiao, J. Shi, H. Liu, Y. Xu, S. Lv, Y. Luo, D. Li, Q. Meng, Y. Li, Adv. Energy Mater. 5 (2015) 1401943.

[64] J.W. Jung, J.-S. Park, I.K. Han, Y. Lee, C. Park, W. Kwon, M. Park, J. Mater. Chem. A 5 (2017) 12158–12167.

[65] M. Saliba, T. Matsui, J.-Y. Seo, K. Domanski, J.-P. Correa-Baena, M.K. Nazeeruddin, S.M. Zakeeruddin, W. Tress, A. Abate, A. Hagfeldt, M. Grätzel, Energ. Environ. Sci. 9 (2016) 1989–1997.

[66] M.A. Basit, M.A. Abbas, E.S. Jung, J.H. Bang, T.J. Park, Mater. Chem. Phys. 196 (2017) 170–176.

[67] I. Moreels, K. Lambert, D. Smeets, D. De Muynck, T. Nollet, J.C. Martins, F. Vanhaecke, A. Vantomme, C. Delerue, G. Allan, Z. Hens, ACS Nano 3 (2009) 3023–3030.

[68] P.R. Brown, D. Kim, R.R. Lunt, N. Zhao, M.G. Bawendi, J.C. Grossman, V. Bulović, ACS Nano 8 (2014) 5863–5872.

[69] Y. Li, J. Zhu, Y. Huang, J. Wei, F. Liu, Z. Shao, L. Hu, S. Chen, S. Yang, J. Tang, J. Yao, S. Dai, Nanoscale 7 (2015) 9902–9907.

[70] S. Sidhik, C. Rosiles Pérez, M.A. Serrano Estrada, T. López-Luke, A. Torres, E. De la Rosa, Sol. Energy 202 (2020) 438–445.

[71] K. Aitola, K. Sveinbjörnsson, J.-P. Correa-Baena, A. Kaskela, A. Abate, Y. Tian, E.M.J. Johansson, M. Grätzel, E.I. Kauppinen, A. Hagfeldt, G. Boschloo, Energ. Environ. Sci. 9 (2016) 461–466.

[72] R. Dileep, G. Kesavan, V. Reddy, M.K. Rajbhar, S. Shanmugasundaram, E. Ramasamy, G. Veerappan, Sol. Energy 187 (2019) 261–268.

[73] L. Zhang, T. Liu, L. Liu, M. Hu, Y. Yang, A. Mei, H. Han, J. Mater. Chem. A 3 (2015) 9165–9170.

[74] Y. Yang, J. Xiao, H. Wei, L. Zhu, D. Li, Y. Luo, H. Wu, Q. Meng, RSC Adv. 4 (2014) 52825–52830.

[75] Z. Wei, K. Yan, H. Chen, Y. Yi, T. Zhang, X. Long, J. Li, L. Zhang, J. Wang, S. Yang, Energ. Environ. Sci. 7 (2014) 3326–3333.

[76] T. Liu, L. Liu, M. Hu, Y. Yang, L. Zhang, A. Mei, H. Han, J. Power Sources 293 (2015) 533–538.

[77] H. Wei, J. Xiao, Y. Yang, S. Lv, J. Shi, X. Xu, J. Dong, Y. Luo, D. Li, Q. Meng, Carbon N. Y. 93 (2015) 861–868.

[78] C.J.M. Emmott, A. Urbina, J. Nelson, Sol. Energy Mater. Sol. Cells 97 (2012) 14–21.

[79] M. Cai, Y. Wu, H. Chen, X. Yang, Y. Qiang, L. Han, Adv. Sci. 4 (2017) 1600269.

[80] B.J. Kim, D.H. Kim, Y.-Y. Lee, H.-W. Shin, G.S. Han, J.S. Hong, K. Mahmood, T.K. Ahn, Y.-C. Joo, K.S. Hong, N.-G. Park, S. Lee, H.S. Jung, Energ. Environ. Sci. 8 (2015) 916–921.

306 Oxide free nanomaterials for energy storage and conversion applications

[81] Y. Xia, K. Sun, J. Ouyang, Adv. Mater. 24 (2012) 2436–2440.

[82] N. Kim, S. Kee, S.H. Lee, B.H. Lee, Y.H. Kahng, Y.-R. Jo, B.-J. Kim, K. Lee, Adv. Mater. 26 (2014) 2268–2272.

[83] P. Lee, J. Lee, H. Lee, J. Yeo, S. Hong, K.H. Nam, D. Lee, S.S. Lee, S.H. Ko, Adv. Mater. 24 (2012) 3326–3332.

[84] A. Schindler, J. Brill, N. Fruehauf, J.P. Novak, Z. Yaniv, Phys. E: Low-Dimens. Syst. Nanostruct. 37 (2007) 119–123.

[85] M.W. Rowell, M.A. Topinka, M.D. McGehee, H.-J. Prall, G. Dennler, N.S. Sariciftci, L. Hu, G. Gruner, Appl. Phys. Lett. 88 (2006) 233506.

[86] D.-Y. Cho, K. Eun, S.-H. Choa, H.-K. Kim, Carbon N. Y. 66 (2014) 530–538.

[87] X. Fan, W. Nie, H. Tsai, N. Wang, H. Huang, Y. Cheng, R. Wen, L. Ma, F. Yan, Y. Xia, Adv. Sci. 6 (2019) 1900813.

[88] K.-Y. Chun, Y. Oh, J. Rho, J.-H. Ahn, Y.-J. Kim, H.R. Choi, S. Baik, Nat. Nanotechnol. 5 (2010) 853–857.

[89] H. Sung, N. Ahn, M.S. Jang, J.-K. Lee, H. Yoon, N.-G. Park, M. Choi, Adv. Energy Mater. 6 (2016) 1501873.

[90] S. Soltanian, R. Rahmanian, B. Gholamkhass, N.M. Kiasari, F. Ko, P. Servati, Adv. Energy Mater. 3 (2013) 1332–1337.

[91] S. Pang, Y. Hernandez, X. Feng, K. Müllen, Adv. Mater. 23 (2011) 2779–2795.

[92] Y. Zhang, Z. Wu, P. Li, L.K. Ono, Y. Qi, J. Zhou, H. Shen, C. Surya, Z. Zheng, Adv. Energy Mater. 8 (2018) 1701569.

[93] R. Wen, Y. Xia, H. Huang, S. Wen, J. Wang, J. Fang, X. Fan, J. Phys. D Appl. Phys. 53 (2020) 284001.

Chapter 13

Multijunction solar cells based on III–V and II–VI semiconductors

Raja Arumugam Senthil[a], Jayaraman Theerthagiri[b],
S.K. Khadheer Pasha[c], Madhavan Jagannathan[d],
Andrews Nirmala Grace[e], and Sivakumar Manickam[f]

[a]*Faculty of Materials and Manufacturing, Beijing University of Technology, Beijing, PR China,* [b]*Department of Chemistry and Research Institute of Natural Sciences, Gyeongsang National University, Jinju, South Korea,* [c]*Department of Physics, Vellore Institute of Technology (Amaravati Campus), Amaravati, Guntur, Andhra Pradesh, India,* [d]*Solar Energy Lab, Department of Chemistry, Thiruvalluvar University, Vellore, India,* [e]*Centre for Nanotechnology Research, Vellore Institute of Technology, Vellore, Tamil Nadu, India,* [f]*Petroleum and Chemical Engineering, Faculty of Engineering, Universiti Teknologi Brunei, Bandar Seri Begawan, Brunei Darussalam*

Chapter outline

1. Introduction   307
2. Multijunction solar cells   310
3. Multijunction solar cells based on III–V semiconductors   312
4. Multijunction solar cells based on II–VI semiconductors   321
5. Conclusion and future prospects   321
References   322

1. Introduction

Recently, the development of renewable energy technology is essential for human society due to the rapid depletion of fossil fuels and the increasing world population [1–3]. Besides, the fossil fuel burning could emit greenhouse gases such as carbon dioxide, which directly affects human health and the environment. Among the several renewable energies like wind energy, hydro energy, and solar energy (sunlight), the latter has received great attention to solving energy demand issues due to its most abundant, inexhaustible, cleaner, and more sustainable features [4–7]. Solar energy is a mixture of electromagnetic waves comprising infrared (IR), visible and ultraviolet (UV) light regions of the electromagnetic spectrum with a peak at 500 nm. Also, the solar spectrum is comparable with a black body of 5800 K, which is influenced by the

Oxide Free Nanomaterials for Energy Storage and Conversion Applications.
https://doi.org/10.1016/B978-0-12-823936-0.00009-7
Copyright © 2022 Elsevier Inc. All rights reserved.

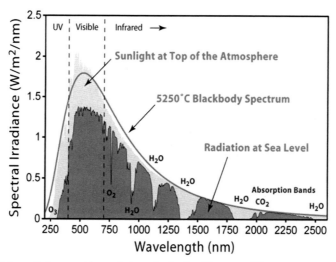

FIG. 1 Solar radiation spectrum and atmospheric absorbing gases [8].

absorption of molecules in the atmosphere including O_3, CO_2, and H_2O, as depicted in Fig. 1 [8]. The sun supplies approximately 120,000 TW of solar energy per annum on the earth's surface, which is 6–7 times higher than that of the world's current energy demands. On the other hand, covering 0.16% of the earth's surface would produce 20 TW of power, with only 10% efficiency [9, 10]. The solar energy reaching the earth's surface mainly depends on the location, atmospheric conditions, time of the day, earth/sun distance, solar rotation, and activity.

Among the different renewable energy technologies, the focus has been directed towards PV technology owing to its excellent property of direct conversion of solar energy to electrical energy without any environmental pollution [11–14]. It is becoming the key source of power generation worldwide. Additionally, there is no mechanical movement or movable parts for PV cells to produce electricity. PV cells are generally based on the charge separation at the single or heterojunction between the two different types (n and p-type) of semiconductors or semiconductor-metal junctions [15, 16]. Therefore, the PV cells can work continuously without any maintenance and longer than other power generation technologies. The power conversion efficiency (PCE) of a solar cell is dependent on the variations in the intensity and energy distribution of the incident light source.

On the other hand, the photovoltaic devices' operational principle is based on the photoelectric effect, which is the direct conversion of incident light into electricity by a p-n junction semiconductor device. Fundamentally, photons' energies are higher than that of the bandgap of the semiconductor, resulting in the generation of electron-hole pairs [17]. The electron-hole pairs generated within the depletion region of the p-n junction are transported in the opposite direction due to the electric field's existence in the depletion region via an external load.

The performance of PV cells is determined by their PCE, stability, and cost [18, 19]. The most important photovoltaic parameter is PCE (η) which is evaluated from the product of short circuit current (J_{sc}), open-circuit voltage (V_{oc}), and fill factor (FF) divided by the intensity of incident light. On the other hand, the PV cells' PCE is defined as the ratio of the maximum electrical energy output to the energy input from the sunlight. The PCE is evaluated from the following Eq. (1) [20, 21]:

$$\eta(\%) = \frac{P_{out}}{P_{in}} = \frac{V_{oc} \times I_{sc} \times FF}{P_{in}} \times 100 \qquad (1)$$

where η is the PCE, P_{in} is the power input from the sunlight. All photovoltaic parameters of PV cells are attained from the current–voltage (I–V) curve, as given in Fig. 2.

The PV cells are divided into four generations based on their performance, cost, and materials utilized to assemble the solar cell, as illustrated below:

(i) The first-generation PV cells are also known as silicon-based solar cells [22]. This solar cell is currently leading in the photovoltaic market owing to its comparatively high PCE (up to 25%) [23, 24]. Nevertheless, some drawbacks such as high cost, low accessibility of silicon, and tedious operation are limited to their widespread application. For this reason, it is essential to develop a solar cell with easily available and low-cost materials.

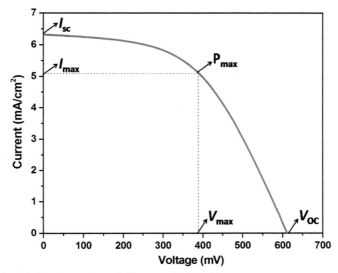

FIG. 2 A typical current–voltage (I–V) curve of solar cells.

(ii) The second-generation PV cells are referred to as thin-film solar cells. Mostly, three kinds of materials such as amorphous silicon, cadmium telluride (GaTe), and copper indium gallium selenide (CIGS) are frequently employed in the second-generation solar cells [25, 26]. The production cost of thin-film solar cells is significantly lower due to the minimal consumption of materials compared with silicon-based solar cells. However, thin-film solar cells exhibit a lower PCE of about 6%–10% than the first-generation solar cells.

(iii) The third generation PV cells are the target in power generation technology [27, 28]. Third-generation PV cells include nanocrystalline solar cells, dye-sensitized solar cells (DSSCs), organic polymer-based solar cells and multijunction solar cells (MJSCs) [29, 30]. These solar cells are considered to be the technical (easy manufacture) and economical (low-cost materials) alternatives for p-n junction solar cells.

(iv) The fourth generation of PV cells, also called as composite photovoltaic technology. This technology combines conducting polymers (organic materials) and stable nanoparticles (inorganic materials) built with a single multispectrum layer [31, 32]. The main advantage of this generation of solar cells is efficient PCE and cost-effectiveness due to the low cost and more flexibility of the polymer thin films with more stability of inorganic nanostructures. Finally, the improvement of best cell efficiencies from 1976 until now is presented in Fig. 3 [33]. It is shown that MJSCs deliver the highest performance as compared with other PV cells. Hence, in this chapter, the recent development and progress of MJSCs have been reviewed.

2. Multijunction solar cells

Multijunction solar cells (MJSCs) are the outstanding PV technology for offering a high PCE for space and terrestrial applications compared to single-junction solar cells [34, 35]. MJSCs can absorb different wavelengths of sunlight due to different semiconductor layers, leading to high PCE from effective utilization of the entire solar spectrum to electricity conversion [36]. Thus, MJSCs are also called tandem solar cells as they are built up of stacked materials (also named sub-cells), which are optimized to absorb different wavelengths of sunlight [37]. The concept of MJSCs was introduced during the 1960s. The first MJSC was reported in the early 1980s with a PCE of 16% [38–40]. It has been reported that the highest theoretical PCE of MJSCs is about 86.8% [41]. Unfortunately, MJSCs have some limitations such as the high cost of manufacturing and low availability of their constituents, which are restricting their practical application. Consequently, it is necessary to focus on reducing the manufacturing cost of MJSCs.

MJSCs are built from multilayers of single-junction solar cells connected on the top. MJSCs are not assembled from silicon as a semiconductor. Mostly,

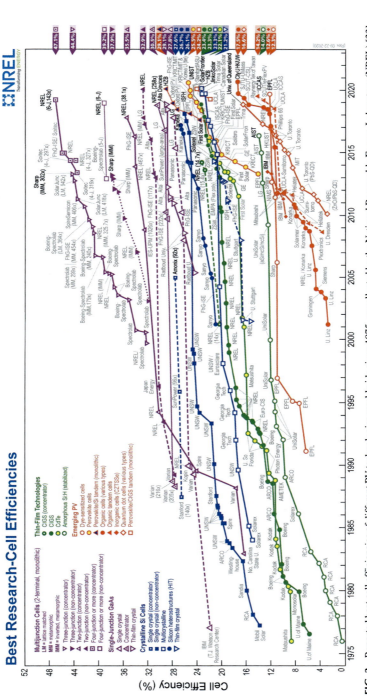

FIG. 3 Reported best cell efficiencies of different PV technologies worldwide since 1975, as collected by the National Renewable Energy Laboratory (NREL) [33].

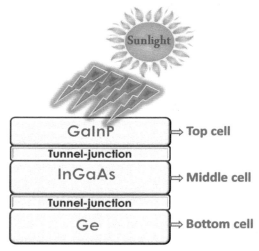

FIG. 4 Schematic showing the structure of GaInP/InGaAs/Ge based MJSC.

MJSCs are assembled based on III–V and II–VI semiconductors due to their broadband gap values, high electron mobility, and complex structures [42–44]. The gallium indium phosphide (GaInP), indium gallium arsenide (InGaAs), and germanium (Ge) are utilized as separate layers of semiconductors to make MJSCs, which react at different wavelengths of sunlight they receive and achieving an enhanced PCE [45, 46]. A schematic structure of GaInP/InGaAs/Ge-based MJSC is displayed in Fig. 4. In this chapter, the recent developments of III–V and II–VI semiconductors-based MJSCs are mainly discussed. Also, the future perspectives and challenges to the development of these MJSCs are considered.

3. Multijunction solar cells based on III–V semiconductors

The MJSCs assembled based on III–V semiconductors (InGaAs, AlInP, AlGaInP, GaInP, InP, GaAs, etc.) are receiving significant interest in the PV technology owing to their highest PCE of over 40% as compared with other solar cells [34, 47]. Therefore, the ongoing development of III–V semiconductors-based MJSCs is summarized below. Bagheri et al. [48] proposed and simulated InGaP/GaAs MJSC along with AlGaAs tunnel junction. They also investigated the PCE of MJSCs via varying window layers and adding back-surface field layers. From the simulation results, it has been observed that the MJSC assembled with six tunnel layers of different thicknesses and utilizing InAlGaP in the back-surface field and window layer can derive a high PCE of 35.5%. Furthermore, the PCE was enhanced to 36.24% by increasing the thickness of the back surface field layer of the top cell. Chee et al. [49]

reported 2-junction InGaP/GaAs solar cell without an unreflective coating through theoretical and numerical simulation studies. The optimized solar cell with back surface field designs showed a PCE of 32.4% under 1-sun AM0 illumination. Cho et al. [50] studied the effect of tunnel junction growth rate on 2-junction InGaP/GaAs solar cells' performance using a metal-organic chemical vapor deposition route. Such an InGaP/GaAs solar cell showed a higher PCE of 14.95% when the tunnel junction was grown at 1.0 Å/s compared with the tunnel junction that was grown at 1.5 Å/s (10.82%). This study postulates that the luminescence properties are also a vital factor in improving the efficiency of MJSCs. Ozen et al. [51] designed a 2-junction GaInP/GaAs solar cell to integrate AlGaAs tunnel junction under a solid-source molecular beam epitaxy method. This GaInP/GaAs solar cell exhibited a PCE of 13.52% under AM1.5G and room temperature conditions, as displayed in Fig. 5.

Bertness et al. [52] proposed a 2-junction GaInP/GaAs solar cell with a maximum PCE of 29.5% at 1-sun concentration AM1.5 illumination by optimizing the grid pattern design, AlInP window layer growth, and back surface field layers. Furthermore, Bertness et al. [53] fabricated 2-junction GaInP/GaAs solar cell with a PCE of 29.5% at 1-sun (AM1.5G) illumination and 30.2% at concentrated 160-suns AM1.5D illumination. It was obtained with a PCE of

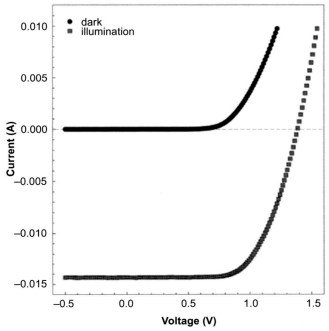

FIG. 5 Current–voltage curves of the GaInP/GaAs solar cell under both dark and AM1.5G conditions [51].

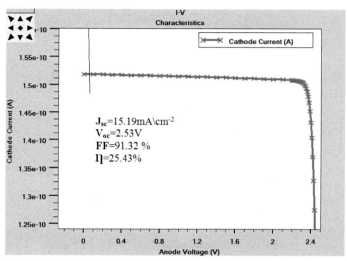

FIG. 6 I–V curve of 2-junction InGaP/GaAs solar cell model at 300 K under AM1.5 illumination [55].

25.7% at 1-sun AM0 illumination. Thus, the proposed GaInP/GaAs solar cell has the potential for both space and terrestrial power applications. Takamoto et al. [54] reported a monolithic 2-junction InGaP/GaAs solar cell and achieved a PCE of 30.28% with a huge area of 4 cm^2 by introducing the InGaP tunnel junction with AlInP barriers at 1-sun AM1.5G illumination. Djaafar et al. [55] demonstrated a high-performance 2-junction InGaP/GaAs solar cell through examining the optimal parameters. The optimized InGaP/GaAs solar cell model as simulated provided a PCE of 25.43% at an optimized temperature of 300 K under AM1.5 illumination, as depicted in Fig. 6. This work demonstrates that the structure simulation makes the fabrication of MJSCs simpler, which also reduces the production cost.

Essig et al. [56] reported a high PCE of 32.45% and 32.82% by assembling mechanically stacked III–V/Si based 2-junction GaInP//Si and GaAs//Si solar cells, respectively. They reached an outstanding PCE of 35.91% for 3-junction GaInP/GaAs//Si solar cell under AM1.5G illumination at 25°C. The attained efficiencies of III–V/Si based MJSCs is higher than both the theoretical efficiency from Si technology (29.4%) and the recorded efficiency of III–V semiconductors based 2-junction cell (32.6%), proving the future application of III–V/Si-based MJSCs in energy applications. Sasaki et al. [57] developed InGaP/GaAs/InGaAs inverted 3-junction solar cell and achieved a PCE of 43.5% with the maximum concentration ratio of around 300-suns illumination. The assembled InGaP/GaAs/InGaAs solar cell with an aperture area of 1.047 cm^2 attained an improved PCE of 37.7% at 1-sun illumination. Meanwhile, King et al. [58] observed that the metamorphic 3-junction

$Ga_{0.44}In_{0.56}P/Ga_{0.92}In_{0.08}As/Ge$ solar cell could achieve 40.7% and 31.3% of PCEs at concentrated 240 suns illumination and 1-sun illumination, respectively. Meanwhile, the lattice-matched $Ga_{0.44}In_{0.56}P/Ga_{0.92}In_{0.08}As/Ge$ solar cell provided the PCE of 40.1% and 32% under the concentrated 135 suns illumination and 1-sun illumination, respectively. This study proved that MJSCs based on III–V semiconductors could be a promising candidate for space and terrestrial power application. Yamaguchi et al. [59] also demonstrated a 3-junction InGaP/InGaP/Ge solar cell with the wide bandgap InGaP double-heterostructure which showed a PCE of 38.9% under concentrated 489-suns, AM1.5G illumination. The 3-junction concentrator cell modules with a large area (5.445 cm^2) exhibited a PCE of 28.9%. Recently, Kotamraju et al. [60] studied the irradiation-induced deep level traps on 3-junction InGaP/InGaAs/Ge solar cell efficiency via 2-D numerical simulations. The simulation results indicate that a 3-junction solar cell provided a PCE of 30% at a trap concentration of $1 \times 10^{16} cm^{-3}$ and surface recombination velocity of $10^4 cm s^{-1}$ in the top, middle, and bottom cells under concentrated 200-suns illumination. Lin et al. [61] reported the 3-junction InGaP/InGaAs/Ge triple-junction solar cell with graphene quantum dots (GQDs). They have observed a significantly enhanced efficiency after deposition of GQDs on the InGaP/InGaAs/Ge triple-junction solar cell surface. Furthermore, the schematic structure, photograph of InGaP/InGaAs/Ge 3-junction solar cell, and a diagram of carrier injection from GQDs in the InGaP/InGaAs/Ge 3-junction solar cell are presented in Fig. 7. The PCE of the 3-junction solar cells was increased with an increase in the concentration of GQD. A maximum PCE of 33.2% was reached for InGaP/InGaAs/Ge 3-junction solar when the concentration of GQDs was at 1.2 mg/mL, which is much higher than the 3-junction solar cell without GQDs (24.6%). Therefore, this study suggested that the carrier injection from GQDs to MJSCs is a potential route for improving cell efficiencies.

Moreover, King et al. [62] reported 3-junction GaInP/GaAs/Ge solar cell with wide-bandgap tunnel junctions and enhanced heterointerfaces. It showed a PCE of 29.3% under 1-sun (AM0, space) illumination. The PCE was further improved to 30.0% under low concentrated 7.6 suns (AM0, space) illumination and reached 32.3% under 440 suns (AM1.5D) illumination. The assembled 3-junction GaInP/GaAs/Ge solar cell is a potential candidate for space and terrestrial applications. Geisz et al. [63] achieved a PCE of 40.8% for inverted 3-junction $Ga_{0.51}In_{0.49}P/In_{0.04}Ga_{0.96}As/In_{0.37}Ga_{0.63}As$ solar cell under concentrated 326-suns illumination, which is greater than the PCE of the same cell under AM1.5G illumination (33.2%). Zhang et al. [64] proposed the 3-junction $Ga_{0.51}In_{0.49}P/In_{0.01}Ga_{0.99}As/Ge$ solar cells grown on 4 in. Ge substrates through the metal-organic chemical vapor deposition technique. The 3-junction solar cell reached a PCE of 31% with an area of 30.15 cm^2 under AM0 condition. Kao et al. [65] fabricated III–V//Si and III–V//InGaAs MJSCs through combined mechanical stacking and wire bonding approaches. The J–V curves of the fabricated MJSCs under 1-sun AM1.5G illumination are shown in Fig. 8.

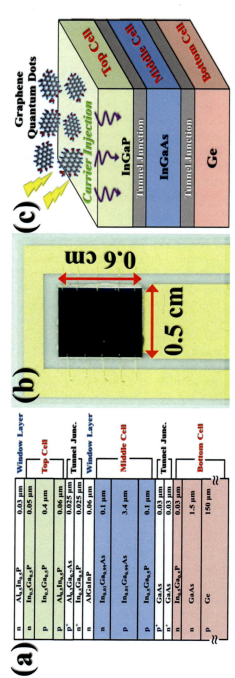

FIG. 7 (A) Schematic structure and (B) photograph of InGaP/InGaAs/Ge 3-junction solar cell; (C) schematic diagram of carrier injection from GQDs to InGaP/InGaAs/Ge 3-junction solar cell [61].

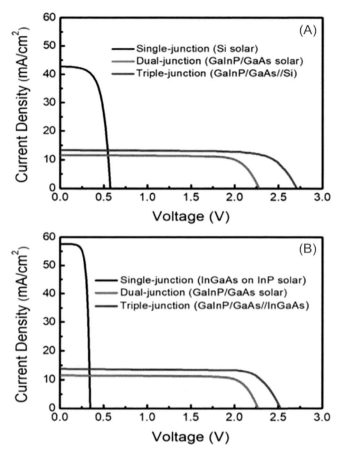

FIG. 8 J–V curves of (A) GaInP/GaAs//Si and (B) GaInP/GaAs//InGaAs TJ solar cells under 1-sun AM1.5G illumination [65].

It is verified that the mechanically stacked GaInp/GaAs//Si and GaInp/GaAs//InGaAs 3-junction solar cells exhibited a significant improvement in the PCE of 27.31% and 26.95%, respectively as compared with GaInP/GaAs 2-junction solar cell (20.60%). Therefore, bonding the GaInP/GaAs 2-junction with Si and InGaAs solar cells would be a promising approach to attain high-performance MJSCs for future power applications.

Wang et al. [66] investigated the performance of heat-pipe cooled 3-junction GaInP/GaInAs/Ge solar cells. They observed a high average output power of 1.52 W/cm^2 and an average PCE of 29.3% under the sunlight illumination of 450 W/m^2. Almonacid et al. [67] evaluated the influence of cell temperature on PCE and observed that the cell's PCE was decreased by increasing the cell temperature. The PCE of 3-junction Ga$_{0.50}$In$_{0.50}$P/Ga$_{0.99}$In$_{0.01}$As/Ge solar cell was reduced from 39% to 31% when the cell temperature was changed from 25°C to 100°C. Sukeerthi et al. [68] investigated the deep level defects in InGaP/InGaAs-GaAsP/InGaAsN quantum well based MJSC using a 2D numerical device simulator. They reached a high PCE of 38% and 44% at 1-sun and 500-sun AM0 illumination, respectively. They also noted that the PCE was dropped by about 4% via introducing a realistic SRV of 1×10^4 cm.s^{-1} and a trap concentration of 1×10^{16} cm^{-3}. It demonstrates the influence of traps and lifetime on the overall cell performance. In 2003, Yamaguchi et al. [69] reported a high-performance concentrated 3-junction InGaP/InGaAs/Ge solar cell with a PCE of 36.5% at 200-suns AM1.5 illumination by using InGaP-Ge heterophase bottom cell and introducing DH-structure tunnel junction. Furthermore, in 2004, they obtained a PCE of 37.4% for concentrator 3-junction InGaP/InGaAs/Ge solar cells at 200-suns AM1.5 illumination. Segev et al. [70] investigated the single-diode and two-diode equivalent circuit semiempirical models for 3-junction InGaP/InGaAs/Ge solar cells. They observed that the two diodes model exhibited the dependence of PCE as compared to the single-diode. Nevertheless, some temperature dependence deviations are still standing for both models. Besides, Yamaguchi et al. [71] realized that the monolithic cascade 3-junction InGaP/InGaAs/Ge solar cell with a double-heterostructure InGaP tunnel junction delivered a high PCE of 31.7% at 1-sun AM1.5 illumination. They also realized that the mechanically stacked 3-junction InGaP/GaAs//InGaAs solar cell supplied an excellent PCE of 33.3% at the same 1-sun AM1.5 illumination. Mizuno et al. [72] introduced a smart stack technology to fabricate integrated III–V/Si based MJSCs with Pd nanoparticle arrays as bonding mediators. The fabricated 3-junction InGaP/GaAs//Si solar cell achieved a PCE of 25.1% under 1-sun illumination (AM1.5G, 100 mW/cm^2). Makita et al. [73] also designed and fabricated III–V//Si based 3-junction InGaP/AlGaAs//Si solar cell from smart stack technology with Pd nanoparticle array. The fabricated III–V//Si-based 3-junction solar cell exhibited a PCE of 30.8% under AM1.5G illumination. This 3-junction solar cell reached the highest PCE of 32.6% under low concentrated 5.5 suns illumination, demonstrating its low concentration PV applications. Therefore, the designed III–V//Si-based MJSCs from smart stack technology is a potential way to develop next-generation solar cells. Walker et al. [74] investigated an inverted metamorphic 3-junction solar cell consisting of GaInP/Ga(In)As/CuInSe$_2$ with GaAs substrate by using epitaxial lift-off technique, which achieved a PCE of 32.6% under 1-sun illumination and also 39.6% under concentrated 750-suns illumination. This study further illustrated that the high absorption coefficient of CuInSe$_2$ is more beneficial in reducing the active material thickness to efficient MJSCs. Sumaryada et al. [75] investigated the influence of spectral irradiance and temperature on the performance of Al$_{0.3}$Ga$_{0.7}$As/InP/Ge MJSCs via simulation studies. From the simulation results, the proposed Al$_{0.3}$Ga$_{0.7}$As/InP/Ge MJSCs showed the highest efficiency under 100-suns illumination at an operating temperature of 25°C. In 2019, Ataser et al. [76] also examined the performance of 3-junction GaInP/GaInAs/Ge solar cells from the analytical solar cell. The PCE of GaInP/GaInAs/Ge solar cell was 35.114% under 1 sun illumination at room temperature. This work also suggested that triple or higher junction structures are needed to construct high-performance MJSCs. King et al. [77] reported a lattice-matched 3-junction GaInP/GaInAs/Ge solar cell with an outstanding PCE of 41.6% at concentrated 364-suns illumination and also showed a PCE of 40.1% under concentrated 822-suns illumination. It is proved that lattice-matched 3-junction solar cell has more potential to fulfill the future space power application. Rauf et al. [78] designed the high-performance MJSCs from simulation studies. They observed that the PCE of 3-junction InGaN/AlInP/AlGaAs solar cell was 37.2% for visible spectrum illumination. Meanwhile, this 3-junction solar cell showed a PCE of 21.95% for UV, visible, and IR spectrum illumination. Also, the PCE of 7-junction solar cells was 62.6% under visible spectrum illumination through the simulation studies. Geisz et al. [79] fabricated the Ge-free 3-junction III-V semiconductors based $Ga_{0.5}In_{0.5}P/GaAs/In_{0.3}Ga_{0.7}As$ solar cell, which offered the PCE of 33.8% for global spectrum (AM1.5G), 30.6% for space spectrum (AM0), and 38.9% for concentrated direct spectrum (AM1.5D, 81 suns) illumination.

To achieve high-efficiency solar cells, Krause et al. [80] developed a wafer bonded 4-junction GaInP/GaAs//GaInAsP/GaInAs solar cell with the bandgap energies of 1.88/1.44//1.11/0.70 eV, respectively. This 4-junction solar cell reached a PCE of 44.7%, considered the world record, at concentrated 297-suns illumination. This study indicates that increasing the number of junctions supports attaining high PCEs. A higher PCE obtained is mainly due to combining lattice-mismatched semiconductors with a stable, electrically conductive, and optically transparent interface from the direct wafer bonding approach. Thus, this work offers a new idea for developing MJSCs with outstanding efficiency. Huang et al. [81] developed a flexible 4-junction AlGaInP/AlGaAs/$In_{0.17}Ga_{0.83}As/In_{0.47}Ga_{0.53}As$ solar cell on a 50 μm thick polyimide film without antireflection coating (ARC). This flexible Al-based 4-junction solar cell exhibited a PCE of 25.76% under AM1.5G illumination, as shown in Fig. 9. Using polyimide film, the 4-junction solar cell's mass density was controlled only at 467 g.m$^{-2,}$ and also the specific power was reached 550 W.kg^{-1}. This study suggests a successful technology to design high-performance MJSCs for future aerospace applications.

Dimroth et al. [82] offered a wafer bonded 4-junction GaInP/GaAs//GaInAsP/GaInAs solar cell with a remarkable PCE of 44.7% under concentrated 297-suns, AM1.5d illumination. This study also provides a potential approach to obtaining high-efficiency III–V MJSCs for space and terrestrial concentrator applications. Dimroth et al. [83] assembled a wafer-bonded 4-junction GaInP/GaAs//GaInAsP/GaInAs solar cell grown on InP-engineered substrate, which

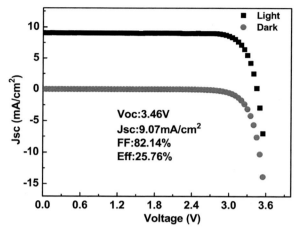

FIG. 9 J–V curves of the flexible 4-junction AlGaInP/AlGaAs/In$_{0.17}$Ga$_{0.83}$As/In$_{0.47}$Ga$_{0.53}$As solar cell at dark and AM1.5G light illumination without ARC [81].

achieved a PCE of up to 46.1% under concentrated 312-suns, AM1.5d illumination. Furthermore, this work indicates that wafer bonding could be a promising route to merge lattice-mismatched III–V compounds to develop efficient MJSCs. France et al. [84] designed an inverted metamorphic 4-junction GaInP/GaAs/GaInAs/GaInAs solar cell with the mismatched bandgap values of 1.8 eV/1.4 eV/1.0 eV/0.7 eV, respectively. The designed 4-junction solar cell achieved the PCE of 35.3% for AM0 spectrum, 37.8% for global spectrum, and 36.6% for direct spectrum under 1-sun illumination at 25°C. The same 4-junction solar cell reached the PCE of 45.6% for concentrated 690-suns illumination and 45.2% at 1000-suns illumination under the direct spectrum, conveying that the as-assembled III–V 4-junction solar cell has more potential to solve future energy demands. Geisz et al. [85] assembled a 6-junction inverted metamorphic solar cell and observed a PCE of 35.8% under 1-sun AM1.5D illumination, which further increased to about 43% by using 200-suns concentrated condition. Geisz et al. [86] also proposed an inverted metamorphic 6-junction solar cell and observed a PCE of over 50% under concentrated AM0 illumination, signifying its application in space power application. Recently, Geisz et al. [87] assembled an inverted metamorphic 6-junction III–V based solar cell with the highest PCE of 47.1% under concentrated 143 suns illumination. Meanwhile, this 6-junction III–V solar cell showed a PCE of 39.2% under 1-sun illumination. Also, they identified that a further decrease in the series resistance within this structure would be practically possible to achieve efficiencies above 50%. These studies confirm the fabrication of MJSCs based III–V semiconductors could be considered the most acceptable approach for future power applications.

4. Multijunction solar cells based on II–VI semiconductors

In the past few decades, the MJSCs assembled based on II–VI semiconductors, including CdS, CdTe, ZnS, CdTe, ZnTe, etc. also have been studied extensively in the PV technology due to their high absorption coefficient and wide band-gap energy [88, 89]. However, to the best of our knowledge, only a few reports are available for the MJSCs based on II–VI semiconductors, as indicated below. Khosroabadi et al. [90] reported the CdS/CdTe solar cell with a maximum PCE of 22.76% under the AM1.5G condition. Also, Britt et al. [91] fabricated the thin-film CdS/CdTe solar cells. The CdS film was attained from the chemical bath deposition method, and the p-CdTe film was deposited via close-spaced sublimation. This thin-film CdS/CdTe solar cell achieved a PCE of 15.8% under AM1.5G illumination. Enam et al. [92] investigated the CdTe/Si 2-junction solar cell's photovoltaic performance through simulation studies. It has been noticed that the optimized CdTe/Si solar cell showed a significantly improved PCE of 28.457% than that of the CdTe solar cell (19.701%). This study proved that combining II–VI and Si solar cells would be a suitable approach to attain high-efficiency solar cells. Olusola et al. [93] reported a multijunction ZnS/CdS/CdTe solar cell with a PCE of 12.8% under AM1.5 illumination at room temperature. In the case of the assembled high-performance solar cell, Garland et al. [94] reported the II–VI/Si-based 2-junction solar cell and II–VI-based 4-junction solar cell under different illumination conditions. The II–VI/Si-based CdZnTe/Si 2-junction solar cell showed the PCE of 40.3%, 43.6% and 44.1% under 1-sun, 200-suns and 500-suns illuminations, respectively. Furthermore, the II–VI based CdZnTe/CdZnTe/CdHgTe/CdHgTe 4-junction solar cell achieved the PCE of 53.1%, 57.2% and 58.2% under 1-sun, 200-suns and 500-suns illuminations, respectively. This work indicates that MJSCs cells could be assembled at low cost using large area, inexpensive, and significantly mechanically stable Si substrates instead of Ge substrates. Xu et al. [95] determined the efficiencies of II–VI/Si-based 2-junction and 3-junction solar cells. They have calculated the theoretical PCE of 2-junction CdZnTe/Si solar cells, which were 38.06% under 1-sun illumination and 43.10% under concentrated 500-suns illumination. The theoretically calculated PCE of 3-junction CdZnTe/CdTe/Si solar cell were 44.08% under 1-sun illumination and 50.19% under 500-suns illumination. Overall, these studies reveal that II–VI semiconductors could also be promising candidates for assembled high-performance MJSCs.

5. Conclusion and future prospects

In recent times, widespread investigations are performed to find out more efficient eco-friendly and renewable energy technology to reduce environmental pollution and control the depletion of fossil energy. Due to the direct conversion

322 Oxide free nanomaterials for energy storage and conversion applications

of solar energy into electricity without any environmental issues, PV technology is one of the most potential candidates to solve future energy needs worldwide. To the best of our knowledge, the MJSCs achieved the highest PCE of over 40% compared to other PV cells. This chapter focused on the fundamentals and recent developments of MJSCs for space and terrestrial power applications. The key focus is directed towards developing different types of MJSCs based on the III–V and II–VI semiconductor materials. Based on the analysis, it has been noted that more studies are reporting on the III–V semiconductors-based MJSCs with significantly improved PCE by utilizing different new technologies. Few reports also centered on the III–V semiconductor- based MJSCs with efficient photovoltaic performances.

This chapter further concludes that the fabrication of MJSCs based on the III–V and II–VI semiconductor materials is promising for developing renewable energy conversion devices. Moreover, this chapter gives some valuable suggestions to construct high-performance MJSCs for space and terrestrial power applications. The MJSCs still have some problems such as high production cost and low availability of their constituents. Hence, it is still a great challenge to reduce the production cost of MJSCs to apply in large-scale power applications. New approaches are necessary for the development of MJSCs in the field of energy conversion devices. The utilization of fewer amount of semiconductor materials by controlling the layer thickness is required to reduce the production cost of MJSCs. Lastly, the production cost of MJSCs is significantly increased by increasing the number of junctions. Controlling the number of junctions has a brilliant prospect to fabricate low-cost, high-performance MJSCs. In the future, it is necessary to find out the more available and inexpensive materials for the utilization of MJSCs.

References

[1] J. Theerthagiri, R.A. Senthil, J. Madhavan, T. Maiyalagan, Recent progress in non-platinum counter electrode materials for dye-sensitized solar cells, ChemElectroChem 2 (2015) 928–945.

[2] B.K. Sahu, A study on global solar PV energy developments and policies with special focus on the top ten solar PV power producing countries, Renew. Sustain. Energy Rev. 43 (2015) 621–634.

[3] Z. Hu, J. Wang, X. Ma, J. Gao, C. Xu, K. Yang, Z. Wang, J. Zhang, F. Zhang, A critical review on semitransparent organic solar cells, Nano Energy 78 (2020) 105376.

[4] S. Bailey, R. Raffaelle, Space solar cells and arrays, in: A. Luque, S. Hegedus (Eds.), Handbook of Photovoltaic Science and Engineering, John Wiley & Sons, Ltd, Chichester, West Sussex, 2011, pp. 365–401.

[5] N. Ali, A. Hussain, R. Ahmed, M.K. Wang, C. Zhao, B. Ul Haq, Y.Q. Fu, Advances in nanostructured thin film materials for solar cell applications, Renew. Sustain. Energy Rev. 59 (2016) 726–737.

[6] R.A. Senthil, J. Theerthagiri, J. Madhavan, Organic dopant added polyvinylidene fluoride based solid polymer electrolytes for dye sensitized solar cells, J. Phys. Chem. Solid 89 (2016) 78–83.

[7] K. Pal, P. Singh, A. Bhaduri, K.B. Thapa, Current challenges and future prospects for a highly efficient (>20%) kesterite CZTS solar cell: a review, Solar Energy Mater. Solar Cells 196 (2019) 138–156.

[8] Solar spectrum, available from https://commons.wikimedia.org/wiki/File:Solar_Spectrum. png.

[9] A. Khare, A critical review on the efficiency improvement of upconversion assisted solar cells, J. Alloys Compd. 821 (2020) 153214.

[10] P.V. Kamat, Meeting the clean energy demand: nanostructure architectures for solar energy conversion, J. Phys. Chem. C 111 (2017) 2834–2860.

[11] B. Boro, B. Gogoi, B.M. Rajbongshi, A. Ramchiary, Nano-structured TiO_2/ZnO nanocomposite for dye-sensitized solar cells application: a review, Renew. Sustain. Energy Rev. 81 (2018) 2264–2270.

[12] R.A. Senthil, J. Theerthagiri, J. Madhavan, A.K.M. Arof, High performance dye-sensitized solar cell based on 2-mercaptobenzimidazole doped poly(vinylidinefluoride-co-hexafluoropropylene) based polymer electrolyte, J. Macromol. Sci., Part A: Pure Appl. Chem. 53 (2016) 245–251.

[13] Z. Xiao, T. Duan, H. Chen, K. Sun, S. Lu, The role of hydrogen bonding in bulk-heterojunction (BHJ) solar cells: a review, Solar Energy Mater. Solar Cells 182 (2018) 1–13.

[14] D. Kishore Kumar, J. Kriz, N. Bennett, B. Chen, H. Upadhayaya, K.R. Reddy, V. Sadhu, Functionalized metal oxide nanoparticles for efficient dye-sensitized solar cells (DSSCs): a review, Mater. Sci. Energy Technol. 3 (2020) 472–481.

[15] N. Asim, K. Sopian, S. Ahmadi, K. Saeedfar, M.A. Alghoul, O. Saadatian, S.H. Zaidi, A review on the role of materials science in solar cells, Renew. Sustain. Energy Rev. 16 (2012) 5834–5847.

[16] W. Feng, R. Geng, D. Liu, T. Wang, T.T. Testoff, W. Li, W. Hu, L. Wang, X. Zhou, A charge-separated interfacial hole transport semiconductor for efficient and stable perovskite solar cells, Organic Electron. 88 (2021) 105988.

[17] J. Day, S. Senthilarasu, T.K. Mallick, Improving spectral modification for applications in solar cells: a review, Renew. Energy 132 (2019) 186–205.

[18] J. Wu, Z. Lan, J. Lin, M. Huang, Y. Huang, L. Fan, G. Luo, Y. Lin, Y. Xie, Y. Wei, Counter electrodes in dye-sensitized solar cells, Chem. Soc. Rev. 46 (2017) 5975–6023.

[19] J. Theerthagiri, R.A. Senthil, M.H. Buraidah, J. Madhavan, A.K. Arof, M. Ashokkumar, One-step electrochemical deposition of $Ni_{1-x}Mo_xS$ ternary sulfides as an efficient counter electrode for dye-sensitized solar cells, J. Mater. Chem. A 4 (2016) 16119–16127.

[20] B. Parida, S. Yoon, S.M. Jeong, J.S. Cho, J.-K. Kim, D.-W. Kang, Recent progress on cesium lead/tin halide-based inorganic perovskites for stable and efficient solar cells: a review, Solar Energy Mater. Solar Cells 204 (2020) 110212.

[21] R.A. Senthil, J. Theerthagiri, J. Madhavan, S. Ganesan, A.K. Arof, Influence of organic additive to PVDF-HFP mixed iodide electrolytes on the photovoltaic performance of dye-sensitized solar cells, J. Phys. Chem. Solid 101 (2017) 18–24.

[22] M.F. Bhopal, D.W. Lee, A. Rehman, S.H. Lee, Past and future of graphene/silicon heterojunction solar cells: a review, J. Mater. Chem. C 5 (2017) 10701–10714.

[23] M. Boubakeur, A. Aissat, M.B. Arbia, H. Maaref, J.P. Vilcot, Enhancement of the efficiency of ultra-thin CIGS/Si structure for solar cell applications, Superlattice. Microst. 138 (2020) 106377.

[24] L. Meng, Z. Yao, L. Cai, W. Wang, L.K. Zhang, K. Qiu, W. Lin, H. Shen, Z. Liang, Indium sulfide-based electron-selective contact and dopant-free heterojunction silicon solar cells, Sol. Energy 211 (2020) 759–766.

324 Oxide free nanomaterials for energy storage and conversion applications

[25] M. Suresh Kumar, S.P. Madhusudanan, S.K. Batabyal, Substitution of Zn in Earth-Abundant $Cu_2ZnSn(S,Se)_4$ based thin film solar cells—a status review, Solar Energy Mater. Solar Cells 185 (2018) 287–299.

[26] J. Ramanujam, D.M. Bishop, T.K. Todorov, O. Gunawan, J. Rath, R. Nekovei, E. Artegiani, A. Romeo, Flexible CIGS, CdTe and a-Si:H based thin film solar cells: a review, Prog. Mater. Sci. 110 (2020) 100619.

[27] J. Yan, B.R. Saunders, Third-generation solar cells: a review and comparison of polymer:fullerene, hybrid polymer and perovskite solar cells, RSC Adv. 4 (2014) 43286–43314.

[28] R.A. Senthil, J. Theerthagiri, J. Madhavan, K. Murugan, P. Arunachalam, A.K. Arof, Enhanced performance of dye-sensitized solar cells based on organic dopant incorporated PVDF-HFP/PEO polymer blend electrolyte with g-C_3N_4/TiO_2 photoanode, J. Solid State Chem. 242 (2016) 199–206.

[29] B. Du, R. Yang, Y. He, F. Wang, S. Huang, Nondestructive inspection, testing and evaluation for Si-based, thin film and multi-junction solar cells: an overview, Renew. Sustain. Energy Rev. 78 (2017) 1117–1151.

[30] J. Ajayan, D. Nirmal, P. Mohankumar, M. Saravanan, M. Jagadesh, L. Arivazhagan, A review of photovoltaic performance of organic/inorganic solar cells for future renewable and sustainable energy technologies, Superlattice. Microst. 143 (2020) 106549.

[31] K.D.G.I. Jayawardena, L.J. Rozanski, C.A. Mills, M.J. Beliatis, N.A. Nismy, S.R.P. Silva, 'Inorganics-in-organics': recent developments and outlook for 4G polymer solar cells, Nanoscale 5 (2013) 8411–8427.

[32] B. Tian, T.J. Kempa, C.M. Lieber, Single nanowire photovoltaics, Chem. Soc. Rev. 38 (2009) 16–24.

[33] National Renewable Energy Laboratory (NREL), Best Research-Cell Efficiencies, available from https://commons.wikimedia.org/wiki/File:Best_Research-Cell_Efficiencies.png.

[34] H. Cotal, C. Fetzer, J. Boisvert, G. Kinsey, R. King, P. Hebert, H. Yoon, N. Karam, III-V multijunction solar cells for concentrating photovoltaics, Energ. Environ. Sci. 2 (2009) 174–192.

[35] S.P. Philipps, A.W. Bett, III-V multi-junction solar cells and concentrating photovoltaic (CPV) systems, Adv. Opt. Techn. 3 (2014) 469–478.

[36] S. Kurtz, J. Geisz, Multijunction solar cells for conversion of concentrated sunlight to electricity, Opt. Express 18 (2010) A73–A78.

[37] O.E. Tereshchenko, V.A. Golyashov, A.A. Rodionov, I.B. Chistokhin, N.V. Kislykh, A.V. Mironov, V.V. Aksenov, Solar energy converters based on multi-junction photoemission solar cells, Sci. Rep. 7 (2017) 16154.

[38] D.J. Friedman, J.M. Olson, S. Kurtz, in: A. Luque, S. Hegedus (Eds.), Handbook of Photovoltaic Science and Engineering, second ed., John Wiley & Sons, West Sussex, UK, 2011, pp. 314–364.

[39] R.R. King, A. Boca, W. Hong, X.-Q. Liu, D. Bhusari, D. Larrabee, K.M. Edmondson, D.C. Law, C.M. Fetzer, S. Mesropian, N.H. Karam, Proceedings of the 24th European Photovoltaic Solar Energy Conference and Exhibition, Hamburg, Germany, 2009, pp. 55–61.

[40] J.M. Olson, D.J. Friedman, S. Kurtz, High-efficiency III-V multijunction solar cells, in: A. Luque, S. Hegedus (Eds.), Handbook of Photovoltaic Science and Engineering, Chapter 9, John Wiley & Sons, Ltd, Chichester, West Sussex, 2003, pp. 359–411.

[41] S. Kurtz, D. Myers, W.E. McMahon, J. Geisz, M. Steiner, A comparison of theoretical efficiencies of multi-junction concentrator solar cells, Prog. Photovolt. Res. Appl. 16 (2008) 537–546.

[42] D.C. Law, R.R. King, H. Yoon, M.J. Archer, A. Boca, C.M. Fetzer, S. Mesropian, T. Isshiki, M. Haddad, K.M. Edmondson, D. Bhusari, J. Yen, R.A. Sherif, H.A. Atwater, N.H. Karam,

Future technology pathways of terrestrial III-V multijunction solar cells for concentrator photovoltaic systems, Solar Energy Mater. Solar Cells 94 (2010) 1314–1318.

[43] H. Lv, F. Sheng, J. Dai, W. Liu, C. Cheng, J. Zhang, Temperature-dependent model of concentrator photovoltaic modules combining optical elements and III-V multi-junction solar cells, Sol. Energy 112 (2015) 351–360.

[44] D. Ding, S.-N. Wu, S. Wang, S.R. Johnson, S.-Q. Yu, X. Liu, J.K. Furdyna, Y.-H. Zhang, Multi-junction solar cells based on the integration of II/VI and III/V semiconductors, LEOS 2008—21st Annual Meeting of the IEEE Lasers and Electro-Optics Society, Acapulco, 2008, pp. 93–94.

[45] M. Yamaguchi, T. Takamoto, K. Araki, M. Imaizumi, N. Kojima, Y. Ohshita, Present and Future of High Efficiency Multi-Junction Solar Cells, CLEO: 2011-Laser Science to Photonic Applications, Baltimore, MD, 2011, pp. 1–3.

[46] R.R. King, D.C. Law, K.M. Edmondson, C.M. Fetzer, G.S. Kinsey, H. Yoon, D.D. Krut, J.H. Ermer, R.A. Sherif, N.H. Karam, Advances in high-efficiency III-V multijunction solar cells, Adv. OptoElectron. (2007) 29523.

[47] M. Yamaguchi, T. Takamoto, K. Araki, N. Ekins-Daukes, Multi-junction III-V solar cells: current status and future potential, Sol. Energy 79 (2005) 78–85.

[48] S. Bagheri, R. Talebzadeh, B. Sardari, F. Mehdizadeh, Design and simulation of a high efficiency InGaP/GaAs multi junction solar cell with AlGaAs tunnel junction, Optik 199 (2019) 163315.

[49] K.W.A. Chee, Y. Hu, Design and optimization of ARC less InGaP/GaAs single −/multijunction solar cells with tunnel junction and back surface field layers, Superlattice. Microst. 119 (2018) 25–39.

[50] I.W. Cho, S.H. Park, T.T. Nguyen, Y. Kim, S.J. Lee, M.-Y. Ryu, Effect of tunnel junction grown at different growth rates on the optical properties and improved efficiency of InGaP/GaAs double-junction solar cells, J. Alloys Compd. 832 (2020) 154989.

[51] Y. Ozen, N. Akın, B. Kinaci, S. Ozcelik, Performance evaluation of a GaInP/GaAs solar cell structure with the integration of AlGaAs tunnel junction, Solar Energy Mater. Solar Cells 137 (2015) 1–5.

[52] K.A. Bertness, S.R. Kurtz, D.J. Friedman, A.E. Kibbler, C. Kramer, J.M. Olson, 29.5%-efficient GaInP/GaAs tandem solar cells, Appl. Phys. Lett. 65 (1994) 989–991.

[53] K.A. Bertness, S.R. Kurtz, D.J. Friedman, A.E. Kibbler, C. Kramer, J.M. Olson, High-efficiency GaInP/GaAs tandem solar cells for space and terrestrial applications, in: "Proceedings of 1994 IEEE 1st World Conference on Photovoltaic Energy Conversion-WCPEC", Waikoloa, HI, 2, 1994, pp. 1671–1678.

[54] T. Takamoto, E. Ikeda, H. Kurita, M. Ohmori, Over 30% efficient InGaP/GaAs tandem solar cells, Appl. Phys. Lett. 70 (1997) 381–383.

[55] F. Djaafar, B. Hadri, G. Bachir, Optimal parameters for performant heterojunction InGaP/GaAs solar cell, Int. J. Hydrogen Energy 42 (2017) 8644–8649.

[56] S. Essig, C. Allebé, T. Remo, J.F. Geisz, M.A. Steiner, K. Horowitz, L. Barraud, J.S. Ward, M. Schnabel, A. Descoeudres, D.L. Young, M. Woodhouse, M. Despeisse, C. Ballif, A. Tamboli, Raising the one-sun conversion efficiency of III-V/Si solar cells to 32.8% for two junctions and 35.9% for three junctions, Nat. Energy 2 (2017) 17144.

[57] K. Sasaki, T. Agui, K. Nakaido, N. Takahashi, R. Onitsuka, T. Takamoto, Development of InGaP/GaAs/InGaAs inverted triple junction concentrator solar cells in '9th international conference on concentrator photovoltaic systems, AIP Conf. Proc. 1556 (2013) 22–25.

[58] R.R. King, D.C. Law, K.M. Edmondson, C.M. Fetzer, G.S. Kinsey, H. Yoon, R.A. Sherif, N.H. Karam, 40% efficient metamorphic GaInP/GaInAs/Ge multijunction solar cells, Appl. Phys. Lett. 90 (2007) 183516.

326 Oxide free nanomaterials for energy storage and conversion applications

[59] M. Yamaguchi, K. Nishimura, T. Sasaki, H. Suzuki, K. Arafune, N. Kojima, Y. Ohsita, Y. Okada, A. Yamamoto, T. Takamoto, K. Araki, Novel materials for high-efficiency III-V multi-junction solar cells, Sol. Energy 82 (2008) 173–180.

[60] S. Kotamraju, M. Sukeerthi, S.E. Puthanveettil, M. Sankaran, Study of degradation in InGaP/InGaAs/Ge multi-junction solar cell characteristics due to irradiation-induced deep level traps using finite element analysis, Sol. Energy 178 (2019) 215–221.

[61] T.N. Lin, S.R.M.S. Santiago, J.A. Zheng, Y.C. Chao, C.T. Yuan, J.L. Shen, C.H. Wu, C.A.J. Lin, W.R. Liu, M.C. Cheng, W.C. Chou, Enhanced conversion efficiency of III-V triple-junction solar cells with graphene quantum dots, Sci. Rep. 6 (2016) 39163.

[62] R.R. King, N.H. Karam, J.H. Ermer, M. Haddad, P. Colter, T. Isshiki, H. Yoon, H.L. Cotal, D. E. Joslin, D.D. Krut, R. Sudharsanan, K. Edmondson, B.T. Cavicchi, D.R. Lillington, Next-generation, high-efficiency III-V multijunction solar cells, in: Conference Record of the Twenty-Eighth IEEE Photovoltaic Specialists Conference-2000 (Cat. No. 00CH37036), Anchorage, AK, USA, 2000, pp. 998–1001.

[63] J.F. Geisz, D.J. Friedman, J.S. Ward, A. Duda, W.J. Olavarria, T.E. Moriarty, J.T. Kiehl, M.J. Romero, A.G. Norman, K.M. Jones, 40.8% efficient inverted triple-junction solar cell with two independently metamorphic junctions, Appl. Phys. Lett. 93 (2008) 123505.

[64] L. Zhang, P. Niu, Y. Li, M. Song, J. Zhang, P. Ning, P. Chen, Investigation on high-efficiency $Ga_{0.51}In_{0.49}P/In_{0.01}Ga_{0.99}As/Ge$ triple-junction solar cells for space applications, AIP Adv. 7 (2017) 125217.

[65] Y.C. Kao, H.M. Chou, S.C. Hsu, A. Lin, C.C. Lin, Z.H. Shih, C.L. Chang, H.F. Hong, R.H. Horng, Performance comparison of III-V//Si and III-V//InGaAs multi-junction solar cells fabricated by the combination of mechanical stacking and wire bonding, Sci. Rep. 9 (2019) 4308.

[66] Z. Wang, H. Zhang, D. Wen, W. Zhao, Z. Zhou, Characterization of the InGaP/InGaAs/Ge triple-junction solar cell with a two-stage dish-style concentration system, Energ. Conver. Manage. 76 (2013) 177–184.

[67] F. Almonacid, P.J. Perez-Higueras, E.F. Ferna'ndez, P. Rodrigo, Relation between the cell temperature of a HCPV module and atmospheric parameters, Sol. Energy Mater. Sol. Cells 105 (2012) 322–327.

[68] M. Sukeerthi, S. Kotamraju, S.E. Puthanveettil, Study of deep level defects in InGaP/InGaAs-GaAsP/InGaAsN quantum well based multi-junction solar cell using finite element analysis, Superlattice. Microst. 130 (2019) 28–37.

[69] M. Yamaguchi, T. Takamoto, K. Araki, Super high-efficiency multi-junction and concentrator solar cells, Solar Energy Mater. Solar Cells 90 (2006) 3068–3077.

[70] G. Segev, G. Mittelman, A. Kribus, Equivalent circuit models for triple-junction concentrator solar cells, Solar Energy Mater. Solar Cells 98 (2012) 57–65.

[71] M. Yamaguchi, Multi-junction solar cells and novel structures for solar cell applications, Physica E: Low Dimens. Syst. Nanostruct. 14 (2002) 84–90.

[72] H. Mizuno, K. Makita, T. Tayagaki, T. Mochizuki, T. Sugaya, H. Takato, High-efficiency III-V//Si tandem solar cells enabled by the Pd nanoparticle array-mediated "smart stack" approach, Appl. Phys. Express 10 (2017), 072301.

[73] K. Makita, H. Mizuno, T. Tayagaki, T. Aihara, R. Oshima, Y. Shoji, H. Sai, H. Takato, R. Müller, P. Beutel, D. Lackner, J. Benick, M. Hermle, F. Dimroth, T. Sugaya, III-V//Si multi-junction solar cells with 30% efficiency using smart stack technology with Pd nanoparticle array, Prog. Photovolt. 28 (2020) 16–24.

[74] A.W. Walker, F. Bouchard, A.H. Trojnar, K. Hinzer, Inverted metamorphic III-V triple-junction solar cell with a 1eV $CuInSe_2$ bottom subcell, Int. J. Photoenergy (2014) 913170.

Multijunction solar cells **Chapter | 13** **327**

[75] T. Sumaryada, S. Rohaeni, N.E. Damayanti, H. Syafutra, H. Hardhienata, Simulating the performance of $Al_{0.3}Ga_{0.7}As/InP/Ge$ multijunction solar cells under variation of spectral irradiance and temperature, Model. Simulat. Eng. (2019) 5090981.

[76] T. Ataser, M.K. Ozturk, O. Zeybek, S. Ozcelik, An examination of the GaInP/GaInAs/Ge triple junction solar cell with the analytical solar cell model, Acta Phys. Pol. A 136 (2019) 21–25.

[77] R.R. King, A. Boca, W. Hong, X.-Q. Liu, D. Bhusari, D. Larrabee, K.M. Edmondson, D.C. Law, C.M. Fetzer, S. Mesropian, N.H. Karam, Band-gap-engineered architectures for high-efficiency multijunction concentrator solar cells, in: Presented at the 24th European Photovoltaic Solar Energy Conference and Exhibition, Hamburg, Germany, 2009, pp. 55–61.

[78] S. Rauf, A.R. Kalair, N. Khan, Simulation design of multi-junction solar cell, MOJ Sol. Photoenergy Syst. 3 (2019) 24–28.

[79] J.F. Geisz, S. Kurtz, M.W. Wanlass, J.S. Ward, A. Duda, D.J. Friedman, J.M. Olson, W.E. McMahon, T.E. Moriarty, J.T. Kiehl, High-efficiency GaInP/GaAs/InGaAs triple-junction solar cells grown inverted with a metamorphic bottom junction, Appl. Phys. Lett. 91 (2007), 023502.

[80] R. Krause, M. Piccin, N. Blanc, M.M. Rico, C. Charles-Alfred, C. Drazek, E. Guiot, et al., Wafer bonded 4-junction GaInP/GaAs//GaInAsP/GaInAs concentrator solar cells, 10th international conference on concentrator photovoltaic systems, AIP Conf. Proc. 1616 (2014) 45–49.

[81] X. Huang, J. Long, D. Wu, S. Ye, X. Li, Q. Sun, Z. Xing, W. Yang, M. Song, Y. Guo, S. Lu, Flexible four-junction inverted metamorphic $AlGaInP/AlGaAs/In_{0.17}Ga_{0.83}As/In0.47Ga_{0.53}As$ solar cell, Solar Energy Mater. Solar Cells 208 (2020) 110398.

[82] F. Dimroth, M. Grave, P. Beutel, U. Fiedeler, C. Karcher, T.N.D. Tibbits, E. Oliva, G. Siefer, M. Schachtner, A. Wekkeli, A.W. Bett, R. Krause, M. Piccin, N. Blanc, C. Drazek, E. Guiot, B. Ghyselen, T. Salvetat, A. Tauzin, T. Signamarcheix, A. Dobrich, T. Hannappel, K. Schwarzburg, Wafer bonded four-junction GaInP/GaAs//GaInAsP/GaInAs concentrator solar cells with 44.7% efficiency, Prog. Photovolt. 22 (2014) 277–282.

[83] F. Dimroth, T.N.D. Tibbits, M. Niemeyer, F. Predan, P. Beutel, C. Karcher, E. Oliva, et al., Four-junction wafer-bonded concentrator solar cells, IEEE J. Photovoltaics 6 (2016) 343–349.

[84] R.M. France, J.F. Geisz, I. Garcia, M.A. Steiner, W.E. McMahon, D.J. Friedman, et al., Design flexibility of ultrahigh efficiency four-junction inverted metamorphic solar cells, IEEE J. Photovolt. 6 (2016) 578–583.

[85] J.F. Geisz, M.A. Steiner, K.L. Schulte, M. Young, R.M. France, D.J. Friedman, Six-junction concentrator solar cells, AIP Conf. Proc. 2012 (2018), 040004.

[86] J.F. Geisz, M.A. Steiner, N. Jain, K.L. Schulte, R.M. France, W.E. McMahon, E.E. Perl, D.J. Friedman, Building a six-junction inverted metamorphic concentrator solar cell, IEEE J. Photovolt. 8 (2018) 626–632.

[87] J.F. Geisz, R.M. France, K.L. Schulte, M.A. Steiner, A.G. Norman, H.L. Guthrey, M.R. Young, T. Song, T. Moriarty, Six-junction III-V solar cells with 47.1% conversion efficiency under 143 Suns concentration, Nat. Energy 5 (2020) 326–335.

[88] S. Girish Kumar, K.S.R. Koteswara Rao, Physics and chemistry of CdTe/CdS thin film heterojunction photovoltaic devices: fundamental and critical aspects, Energ. Environ. Sci. 7 (2014) 45–102.

[89] S.R. Kurtz, J.M. Olson, P. Faine, The difference between standard and average efficiencies of multijunction compared with single-junction concentrator cells, Sol. Energy Mater. Sol. Cells 30 (1991) 501–513.

[90] S. Khosroabadi, S.H. Keshmiri, S. Marjani, Design of a high efficiency CdS/CdTe solar cell with optimized step doping, film thickness, and carrier lifetime of the absorption layer, J. Eur. Opt. Soc. Rap. Public. 9 (2014) 14052.

328 Oxide free nanomaterials for energy storage and conversion applications

[91] J. Britt, C. Ferekides, Thin-film CdS/CdTe solar cell with 15.8% efficiency, Appl. Phys. Lett. 62 (1993) 2851.

[92] F.M.T. Enam, K.S. Rahman, M.I. Kamaruzzaman, K. Sobayel, P. Chelvanathan, B. Bais, M. Akhtaruzzaman, A.R.M. Alamoud, N. Amin, Design prospects of cadmium telluride/silicon (CdTe/Si) tandem solar cells from numerical simulation, Optik 139 (2017) 397–406.

[93] O.I. Olusola, M.L. Madugu, I.M. Dharmadasa, Investigating the electronic properties of multi-junction ZnS/CdS/CdTe graded bandgap solar cells, Mater. Chem. Phys. 191 (2017) 145–150.

[94] J.W. Garland, T. Biegala, M. Carmody, C. Gilmore, S. Sivananthan, Next-generation multi-junction solar cells: the promise of II-VI materials, J. Appl. Phys. 109 (2011) 102423.

[95] D. Xu, T. Biegala, M. Carmody, J.W. Garland, C. Grein, S. Sivananthan, Proposed monolithic triple-junction solar cell structures with the potential for ultrahigh efficiencies using II-VI alloys and silicon substrates, Appl. Phys. Lett. 96 (2010), 073508.

Chapter 14

Recent advances in nanostructured nonoxide materials—Borides, borates, chalcogenides, phosphides, phosphates, nitrides, carbides, alloys, and metal-organic frameworks

Leticia S. Bezerra[a], Bibiana K. Martini[a], Eduardo S.F. Cardoso[a], Guilherme V. Fortunato[b], and Gilberto Maia[a]

[a]*Institute of Chemistry, Federal University of Mato Grosso do Sul, Campo Grande, MS, Brazil,* [b]*São Carlos Institute of Chemistry, University of São Paulo, São Carlos, SP, Brazil*

Chapter outline

1. Introduction	**329**	
2. Existing technology	**330**	
2.1 Alkaline electrolysis cell	331	
2.2 Proton exchange membrane electrolysis cell	331	
2.3 Solid oxide electrolysis cell	331	
3. Methodology or research design to overcome the drawbacks of existing protocols	**331**	
3.1 Borides	332	
3.2 Borates	332	
3.3 Chalcogenides (sulfides and selenides)	333	

3.4 Phosphides	334
3.5 Phosphates	335
3.6 Nitrides	336
3.7 Carbides	337
3.8 Alloys	338
3.9 MOFs	339
4. Findings or results with implications for managers and decision-makers	**341**
5. Discussion, limitations, future research, and conclusion	**350**
References	**353**

1. Introduction

Most of the energy used comes from fossil fuels [1]. However, the use of fossil fuels is resulting in environmental problems, future shortage of this fuel, and consequently future energy crisis [2–5]. Hydrogen-based technology is a

Oxide Free Nanomaterials for Energy Storage and Conversion Applications
https://doi.org/10.1016/B978-0-12-823936-0.00011-5
Copyright © 2022 Elsevier Inc. All rights reserved.

330 Oxide free nanomaterials for energy storage and conversion applications

prominent option to overcome these troubles, representing a sustainable source of energy that has a high energy density and no carbon emission [1,6–10]. A great way to produce hydrogen with high purity and sustainability is through electrochemistry water splitting in an electrolyzer, which occurs by cathodic and anodic reactions—hydrogen evolution reaction (HER) and oxygen evolution reaction (OER), respectively [7,11–14]. The two half-reactions (cathodic and anodic) are presented below, in both acid and alkaline media [2]:

Water splitting reaction in acidic medium:

$$2H_2O\ (l) \rightarrow 4H^+\ (aq) + 4e^- + O_2\ (g)\quad (anode) \tag{1}$$

$$4e^- + 4H^+\ (aq) \rightarrow 2H_2\ (g)\quad (cathode) \tag{2}$$

Water splitting in alkaline medium:

$$4OH^-\ (aq) \rightarrow 2H_2O\ (l) + 4e^- + O_2\ (g)\quad (anode) \tag{3}$$

$$4e^- + 4H_2O\ (l) \rightarrow 4OH^-\ (aq) + 2H_2\ (g)\quad (cathode) \tag{4}$$

At 25°C and 1 atm, the free energy variation (ΔG) for water splitting is $273\,kJ\,mol^{-1}$, which is equal to the potential of 1.23 V vs. Reversible hydrogen electrode (RHE) [13]. However, these reactions present sluggish kinetic demanding overpotential (extra amount of energy) to overcome the high energy barriers. To lower these energy barriers (decrease or cancel the overpotentials for these reactions to take place), highly efficient, economically viable, and durable electrocatalysts are crucial [11,13,15]. The state-of-the-art materials used to catalyze HER and OER reactions comprise precious metals-based catalysts [15], like platinum (Pt/C) for HER [16–20], and ruthenium (RuO_2) and iridium (IrO_2) for OER [21–23]. Nevertheless, the scarcity and high price of these precious metals [7] hinder their use, strongly motivating the research for more Earth-abundant [24] and equally efficient materials.

2. Existing technology

A water-splitting electrolyzer is composed of an anode (electrode where the oxygen evolution takes place) and a cathode (electrode where the hydrogen evolution takes place), placed face-to-face separated by an ion-conducting membrane [25]. A power source is applied to electrodes (both covered by catalysts) to overcome the overpotential of HER and OER half-reactions of water splitting [13,25]. Most of the energy losses are related to the anodic process [26]. The cell is fed with water and the gases (H_2 and O_2) are obtained separately at the exit of the cell [25]. The adjustment in series of an appropriate number of cells can provide the proper rate of gas production of the stack [25]. Proton exchange membrane, alkaline, and solid oxide are the major types of electrolyzers [25].

Oxide free materials for electrochemical water splitting **Chapter | 14 331**

2.1 Alkaline electrolysis cell

Alkaline electrolysis cell (AEC) operates with KOH aqueous solution electrolyte and the cell also uses a diaphragm separator against which the porous grid electrodes are pressed, decreasing distance and ohmic resistance [25]. The main limitation of this type of electrolyzer is the current density limit (typically less than $0.45\,A\,cm^{-2}$), caused by the gas bubbles that block the electrode surface [25]. The actual capital cost of an AEC is approximately 900–1700 € kW^{-1} [25].

2.2 Proton exchange membrane electrolysis cell

The proton exchange membrane electrolysis cell (PEMEC) is constituted by a membrane electrode assembly (MEA), which is typically composed of perfluorosulfonic acid (commercially found as Nafion, for example), located in the middle of the compact unit cell [25]. Right next to this membrane are coated the catalysts (on both sides), which can also be placed onto the next porous transport layers located then right next to MEA [25]. The water is constantly fed in cathodic and anodic sections aiming to supply the anode and remove the excess of heat from both sides, and at the back of porous transport layers, the O_2 and H_2 products are collected [25]. The last layers of PEMEC are the terminal endplates [25].

PEMEC presents many advantages, like the high current densities achieved (more than $10\,A\,cm^{-2}$) and high purity of hydrogen produced [25]. The bubble effect, which is worrying in AEC, in PEMEC does not indicate concern once the gases are collected in the back of the device [25]. The capital cost of this technology is 1700–2500 € kW^{-1}, more expensive than an AEC system [25].

2.3 Solid oxide electrolysis cell

Solid oxide electrolysis cell (SOEC) consists of two electrodes separated by an oxygen-ion-conducting ceramic and interconnects that organize the water vapor and flow of the gases (water is pumped on one side while O_2 is collected in another) [25]. This device requires much energy to supply the cell with water vapor, once the temperature of operation is around 700–900°C [25]. This system presents some disadvantages, such as the extended turn on and turns off proceedings, the elevated temperature that causes a fast degradation, and the O_2 produced, which at such high temperature is corrosive [25]. The SOEC technology is still not commercialized, but its capital costs are more expensive than AEC and PEMEC systems, reaching 2000–4000 € kW^{-1} [25].

3. Methodology or research design to overcome the drawbacks of existing protocols

Earth-abundant transition metal (TM)-based—like borides, borates, chalcogenides (sulfides and selenides), phosphides, phosphates, nitrides, carbides, alloys,

332 Oxide free nanomaterials for energy storage and conversion applications

and MOF-structured nanoparticles—electrocatalysts are an option of very promising materials to replace noble metal-based catalysts [13,27–29]. In sequence, they are discussed some of the main classes of TM-based electrocatalysts for water splitting.

3.1 Borides

Recently, transition metal-borides have emerged as efficient bifunctional electrocatalysts for HER and OER. Ni, Co, and Mo borides have demonstrated the highest catalytic performances in a wide pH range. Compared to other catalyst families (i.e., carbon materials, sulfides), their long-term stability is superior even for OER conditions. Their catalytic performance can be ascribed to the (1) facilitated charge transfer between orbital 3d (or 4d) of TM and orbital 2p of boron that reduces the energetic barrier for the electrochemical reactions, (2) a large number of unsaturated catalytic sites, and (3) common amorphous structure [13,14,30,31].

TM-borides can be produced by different syntheses approaches, but the most common way is by the chemical reduction of transition metal borohydride-precursors, like $LiBH_4$, KBH_4, and $NaBH_4$, which are inexpensive and nontoxic compounds [13,30,32–36]. This method presents the advantage of being simple (typically one-pot) and applicable to a wide range of metals, however, the crystallinity and morphology are parameters that are not controllable by using this method [13,35]. Electrochemical deposition, or electrodeposition, is also used to prepare TM-borides. In this method, an electric current or negative potential is applied to produce nanostructures on the surface of conductive substrates [7,13]. The morphology is also difficult to control by using this process, but it's simple and fast way to nucleate nanomaterials makes this method very attractive [13,37]. TM-borides can be synthesized by the ball-milling method that is a simple and cheap mechanical process where time, pressure, area, and temperature are controlled [13,38,39]. The disadvantage of this method is the inadequate purity and the heterogeneous particle size [13,40].

As recent examples of catalytically successful TM-borides for water splitting, Zhang, et al. deposited amorphous nickel boride on nickel foam (NiB/NF). The combination NiB/NF showed high activity and satisfying stability, with under an overpotential of 41.2 mV for HER takes place [41]. For OER catalysis, Masa et al. produced also onto nickel foam an ultrathin amorphous nickel boride (Ni_xB) nanosheets that exhibited a density current of $20\,mA\,cm^{-2}$ in an overpotential of 280 mV in 1.0 M KOH solution [42].

3.2 Borates

Recently, the borates have attracted attention as OER electrocatalysts due to their stability in neutral and alkaline medium, low cost and earth abundance, and able to reduce the kinetic and thermodynamic barriers during the water splitting. They can work as efficient proton acceptors, becoming a good electrocatalyst candidate toward OER [35,43,44].

The production of robust and electroactive borates can be made through electrodeposition in borate buffer [45], liquid phase chemical reduction [46], solid-state reactions, chemical vapor deposition [47], sol-gel [44] and hydrothermal approaches [48,49]. Additionally, the borates have been extensively used in an intercalated way between layered metal hydroxides, a way that provided them with improved catalytic activity for both HER and OER. It was demonstrated that intercalated borate ions increase the proton transportability and stability of catalysts like $Ni(OH)_2$, and consequently improve the electrocatalytic and photocatalytic HER activity [50]. Also, the introduction of borate in NiFe (oxy)hydroxide structure led to a favorable electronic structure of the Ni active sites, working as an efficient proton accepting species, increasing the charge transferability and stability of the catalyst for OER [44]. For Ni-Fe layered double hydroxide, the interlayer spacing, specific surface area, pore size range, mass transport, charge transfer, and consequently, the catalytic activity, were increased as a result of the borate ion intercalation. According to the authors, the borate ions have the function of proton acceptor, promoting the formation of O—O bonds during the OER [51].

3.3 Chalcogenides (sulfides and selenides)

Transition metal dichalcogenides (TMD), mainly MoS_2, WS_2, $MoSe_2$, and $NiSe_2$ have been extensively studied as promising catalyst materials for HER. This class of materials is attractive due to interesting chemical, optical, thermal, electrical, and mechanical properties. TMD in the form of layered nanomaterials can provide additional characteristics as a result of the quantum size effects [52,53]. They have natural abundance and have been exhibited high chemical stability in alkaline and acidic environments [54].

TMD materials can be produced in different shapes and sizes: the nanospheres and nanodots were classified as zero-dimensional (0D), nanotubes, nanowires, and nanorods as one dimensional (1D), nanoflakes and nanosheets as two dimensional (2D), and mesopores hierarchical structures as three dimensional (3D) materials [55]. Furthermore, mono-, bi-, or multimetallic, as well as doped or heterostructures of metal chalcogenides, can be produced [56]. The use of more than one transition metal can improve the catalytic performance for HER. The catalytic gains typically are ascribed to the smaller energy required to electron transfer between cations, the high number of valences and versatility of the adsorption sites, and/or the fine-tuning on the free energy of hydrogen adsorption due to the synergistic effects [57].

Currently, several methods have been proposed to produce the TMD on a large scale, with controlled structure, shape, size, and defects, and the most common are sputtering [58–60], exfoliation (by mechanical [61–63] or liquid phase method [64–66]), chemical vapor deposition (CVD) [67–70], physical

334 Oxide free nanomaterials for energy storage and conversion applications

vapor deposition (PVD) [71,72], sulfurization [73–75], precipitation [76,77], hydrothermal synthesis [78–81], solvothermal synthesis [82–84], among others.

The sulfides have a sulfur anion bonded to metallic or semimetallic cation (M^{X+}) resulting in mono-metallic compounds as MS, M_2S, M_3S_4, and MS_2; or bi-metallic structures as $A_{1-x}B_xS_y$ [56]. The molybdenum disulfide (MoS_2) has been studied and is known as an efficient catalyst for HER. Its good catalytic performance can be assigned to the hydrogen binding energy on its surface being close to that of Pt-group metals. However, the MoS_2 performance is limited by the concentration of unsaturated S^{-2} atoms and the number of active sites on the edges of MoS_2 layers, since the basal planes are thermodynamically favorable but inert for catalytic reactions [85–87].

.The selenides have attracted consideration mainly due to their catalytic activity, chemical stability, and earth abundance. Considered very promising free-noble metal electrocatalysts, $MoSe_2$ structures offer low Gibbs free energy of hydrogen adsorption on edges, making it a potential catalyst to substitute Pt-based materials for HER. However, it is important to avoid aggregation and increase the number of active sites [79,88]. Several methods have been used to improve the transition metal selenides' performance, like synthesized hierarchical porous structures with improved charge transfer and mass diffusion properties. Heteroatoms doping, increasing the number of defects, tuning the electrical properties, and enrichment of active sites on the surface have also been tentatively used to improve the catalytic performance of TM-selenides [27].

Toward OER application, the metal sulfides present poor stability and weak oxygen adsorption, decreasing the catalytic activity; however, several studies involving doping the sulfides with ions, or high surface area carbon derivatives have pointed out improvements in conductivity and chemical stability and, consequently, enhancements on OER performance [56]. Additionally, an observed conversion of the surface selenides to hydroxide/oxyhydroxide during the OER has shown promising results in terms of catalytic performance [27].

3.4 Phosphides

The transition metal phosphides (TMP) are well known as catalysts to hydrodenitrogenation and hydrodesulfurization reactions, but only in recent years have they received attention as catalysts toward HER and OER reactions [89]. By showing high electronic conductivity, catalytic activity, and long-term stability compared to traditional metal noble catalysts, TMPs have emerged as a good option for the energy industry [90,91]. The most widely studied TMPs for electrochemical water splitting are based on the transition metals Co, Ni, Mo, Fe, Cu, and W [92].

The most common metal sources used for TMP synthesis include metal oxides, metal pyrophosphates, metal nitrates, among others; and the phosphorus sources are usually red or yellow phosphorous, phosphine, or trioctylphosphine

(TOP) [89]. The synthesis methodologies to produce TMP include solution-phase reaction [93], gas-solid reactions [94], hydrothermal and solvothermal synthesis followed or not by annealing [95–98]. Also, approaches like fine surface design, construction of nano-porous structures, adjustments on stoichiometric ratio of M/P, attach or grown TMP in carbon materials, and doping with different elements are used to improve the catalytic performance of the TMP [90,92,99].

The phosphorous and metal sites in TMPs act as proton and hydrides acceptor sites during the HER, resulting in an enhanced electrocatalytic activity. Toward the OER process, the in-situ formation of oxy-hydroxides and phosphates species on the phosphides surface are known as the real active sites, increasing the electrical conductivity and consequently the charge transfer process [99]. Metal phosphides with high contents of P atoms exhibit an improved HER performance since the high electronegativity of the P atoms attracts the electron from the metals and the resultant negatively charged P atom captures the proton as a base [100]. In addition, the existence of P ions on the edges of phosphides can promote the HER, similarly to MoS_2 with a high content of sulfur [101]. It is suggested that P atoms content is directly related to the corrosion-resistance of the metal phosphides and with the HER activity [102].

3.5 Phosphates

Recently, metal phosphides have raised as candidate electrocatalyst to HER and OER in neutral and acidic medium. As promising electrocatalysts, Co, Fe, and Ni-based phosphates (PO_4^{3-}) and pyrophosphates ($P_2O_7^{4-}$) have been attracted attention mainly due to their high catalytic performance [103–106]. The versatility of phosphates to act as electrocatalysts for water splitting is related to their capacity of moving protons freely and helping the metal oxidation during proton-coupled electron transfer process since the phosphate groups can act as proton acceptors [107]. The atomic geometry of phosphates can be distorted to favor the adsorption and oxidation of H_2O molecules [103]. Hierarchically structured Ni pyrophosphates combined with ultra-low Pt-loading (2 wt%) showed demonstrated a synergistic effect that led to high catalytic activity and stability for HER in acid medium [104].

The production of the different types of phosphates or pyrophosphates is generally done by precipitation, hydrothermal, solution combustion, and solid-state methods. Other approaches like sol-gel and chemical vapor deposition also have been used more discreetly [107]. Commonly, in the main phosphate synthesis methods, a calcination step is necessary. In the calcination process, combinations of metals, doping, temperature programming, and atmosphere are crucial variables to adjust the crystallography, porosity, conductivity, and, consequently, catalytic properties of the material [104,108].

As examples of catalytically successful phosphates-based materials for water splitting, Sial et al. synthesized by a single-stage and low-temperature

336 Oxide free nanomaterials for energy storage and conversion applications

hydrothermal method NiCoFe phosphate as a bifunctional catalyst [105]. For HER, NiCoFe phosphate provided a current density of $10\,mA\,cm^{-2}$ at a low overpotential of $-231\,mV$, and for OER, it was capable of reaching $10\,mA\,cm^{-2}$ with a cell voltage of $1.52\,V$ with robust durability. Its high catalytic response was attributed to the ultrafine thickness with a greater number of active sites [105]. Zhong et al. demonstrated that Fe phosphate films supported on Ni foam (Fe-Pi/NF) have notable activity for OER in alkaline medium with low overpotentials around 215 and $257\,mV$ to achieve 10 and $100\,mA\,cm^{-2}$. The authors attributed the catalytic performance and durability to the high hydrophilicity and the facilitated oxidation of Fe^{3+} promoted by the addition of phosphate groups.

3.6 Nitrides

TM-nitrides are considered good candidates as electrocatalysts for water splitting. Their successful application for HER and OER reactions is mainly attributed to their interesting properties, such as high conductivity, synergy, high porosity which allows large numbers of active sites, high corrosion-resistant, and fast gas-diffusion capability for O_2 and H_2. When N atoms are inserted in the metallic system [109], higher nitrogen concentrations (large nitrogen/metal ratio) typically lead to a decrease in charge transfer resistance and an increase in catalytic activity [109,110]. The nature of chemical bonding between metal and nitrogen atoms can reveal important characteristics of the electrocatalyst. Covalent, ionic, and metallic bonds provide hardness and withstand more stress, rapid kinetics of electron transference due to the d-band of metals contracted (electronic structure similar to noble metals), and improved conductivity due to increased corrosion resistance, respectively [109,110]. Mono-, bi-, and tri-metallic nitrides show different electrocatalytic activities because of the modified metal d-banding concentration, synergy, and consequently electrical conductivity [109]. Another advantage of TM-nitrides is their chemical stability under diluted acid and alkaline media [110].

There are a lot of methods to synthesize TM-nitrides, among which are noteworthy sol-gel, solvothermal, urea glass-route, pulsed-laser deposition, and temperature-programmed methods, that are the most used approaches [109–112]. The sol-gel method consists of a synthesis from the metal solution at low temperature which provides solid materials with homogenously distributed nanoparticles, increased surface area, and a low oxygen amount, however, it is necessary the application of harmful substances as nitrogen sources [112]. In the solvothermal method, a mixture of precursor materials in a nonaqueous solution is placed in a Teflon-aligned autoclave and it is submitted to a temperature by a certain period [112]. The facility to synthesize materials with raised vapor pressure and reduced melting points is an advantage of this method, in contrast, the time of synthesis, possibility of particles agglomeration, and risk of explosion are drawbacks of the solvothermal method [112].

The urea glass-route comprises the dissolution in ethanol of a metal precursor presynthesized and, after the formation of metal-ortho-ester, urea is added to the mixture and it is stirred, sequence a thermal treatment with temperatures around 700–800°C under N_2 atmosphere is conducted [111,112]. Through this flexible method nanoparticles with high definition and crystallinity are obtained [112]. Through pulse-lased deposition, thin films of TM-nitrides can be produced onto a conductive substrate surface that is disposed of in a vacuum camera. The TM-nitrides films are deposited by nitrogen radicals' irradiation, being possible to control the deposition rate through laser power. The difficult control of nanoparticles agglomeration is the main disadvantage of this method [110–112]. The nitridation of metal precursor (oxides, halides, sulfides) with N_2 and H_2, or mixed in NH_3, comprises the temperature-programmed method, of which is possible to produce ternary nitrides with high purity but to keep the NH_3 and/or N_2 flow may become trouble [111,112].

Recently, impressive results of electrocatalytic performance for TM-nitrides have been reported in the literature, as the low overpotential of OER acclaimed by Zhang, et al. in their study involving iron-nickel nitride nanostructures ($FeNi_3N/NF$). $FeNi_3N/NF$ catalyst presented 202 mV of the extra energy required to achieve a current density of $10 \, mA \, cm^{-2}$ [113]. Moreover, 15 mV of overpotential for HER was reported by Jia et al. by applying nickel molybdenum nitride nanorods ($Ni_{foam}@Ni-Ni_{0.2}Mo_{0.8}N$) as an electrocatalyst, even more, this material was also active for OER electrocatalysis requiring a cell voltage of 1.49 V for water splitting [114].

3.7 Carbides

Transition metal carbides also have been explored as electrocatalysts mainly for HER. This class of materials has been attracted attention mainly due to their characteristics like electronic behavior similar to Pt, chemical stability in neutral, alkaline, or acidic electrolytes, and high conductivity [115]. Carbide structures are constructed with the insertion of carbon atoms in a metal network and their properties differ from the original metals and metal oxide, the unique Pt-like properties of carbides are due to the links between the transition metals and carbon atoms, where the d-band of the metal is broadened from the hybridization between the d-, s- and p-orbitals of the metal and carbon atoms, respectively [116,117]. Considered the most promising catalyst for HER, W and Mo carbides can adsorb and activate hydrogen. DFT and experimental data have shown that the modification of carbides with few monolayers of Pt can lead to catalytic activity for HER very similar to bulk Pt [115].

Carbides can be synthesized by solid-gas, solid-liquid, or solid-solid reaction systems [118,119]. To replace O with C atoms, metal carbides are produced typically under extremely high temperatures ($>700°C$) [118,119]. Solid-gas reaction commonly employs flammable or toxic gases (i.e., methane, ethane, or carbon monoxide) as carbon precursors make this approach unsafe. In

338 Oxide free nanomaterials for energy storage and conversion applications

addition, the high temperature for preparing carbides can lead to uncontrollable sintering and agglomeration of the nanoparticles, forming a material with extremely low surface area, and consequently less catalytic [103,115,117,120].

Nanocomposites from carbides also have led to the very active structures toward HER. As a successful example, by using spray-drying, and followed by annealing, Wei et al. produced small Mo carbide nanoparticles uniformly embedded into graphene N-doped porous carbon microspheres ($Mo_2C/$ G-NCS) as a nanocomposite with high electrocatalytic performance for HER in alkaline and acidic media. The impressive catalytic activity shown by Mo_2C/G-NCS catalyst was ascribed to the ultrasmall size of carbides which increased the number of exposed active sites, graphene-wrapping which improved the conductivity; high porosity which facilitated the electrolyte access and the charge transfer, as well as to the N-dopant which interacted with protons modifying the electronic structures of adjacent Mo and C atoms [121].

3.8 Alloys

Metal alloys are long-standing applied materials as efficient catalysts to hetero-geneous and electrochemical reactions. There are many studies and applications on energy issues involving this class of materials due to their high versatility properties of acting in different catalytic processes. Combining different metals to produce alloyed structures may lead to the formation of new materials with totally different catalytic properties. The catalytic performance for based on metal allows nanostructures can be modulated due to different effects such as synergism between two or more distinct metals, electronic effect, lattice strain, and ensemble effect [122]. As well as the dependence on the metal com-binations, variations on the ratio between the components also have a strong impact on the formed new material properties [123–127]. Allows can make pos-sible several processes considered expensive and also an improvement of lim-iting factors such as electrocatalytic degradation [123–127]. Structurally organized alloys are becoming the focus of researchers, due to their better elec-trocatalytic activities and also stability in reactions such as the oxygen reduction reaction (ORR) and oxidation reactions [128], when compared to disordered alloys [129]. The improvement of the catalytic performance can be attributed to the control of the geometric structure and defined composition since the size and morphology of the particles also affect the catalytic performance [128,130,131]. The different compositions of alloys with defined morphology enable better exposure of active sites and increased stability in applications for electrochemical reactions, helping the development of metallic electrocatalysts applied to the ORR, HER, and OER reactions [116,132–138].

There are enormous possibilities for the production of different alloys with different physicochemical and electrocatalytic properties. Alloys can be pro-duced by several approaches, but the simplest one is when the metals-precursors are in the form of soluble salts that can generate allowed structures by coprecipitation [139]. This methodology is widely used (mainly for noble metals) due to its great facility to control the composition, dispersion, and morphology of the resulting catalysts [140]. Metal alloys also can be produced by chemical vapor deposition process under high or low pressure which, although energetic dispendious, can produce well-defined structures like core-shell system with optimized thickness [137]. Alloys based on Pt, Ru, and Ir typically show the lowest overpotential for water splitting [116,136], however, as already mentioned, due to their high cost and scarcity, alloys involving nonprecious transition metals have been widely studied in the recent years. Bi- or trimetallic alloys involving Ni (i.e., NiFe, NiCu, NiCo, NiCoFe) have been reported as highly active electrocatalysts toward HER and OER in alkaline medium [103,116,133–138], however, to withstand the drastic reaction conditions of OER both acidic and alkaline media, the surface of these kinds of alloys should be naturally or purposely oxidized as a passivation method [135]. Another way to minimize alloy dissolution is through coating or encapsulation with relatively more stable materials [136–138]. As the pure metallic state may not resist in acidic, alkaline media, or at the strongly oxidative potentials of OER without being corroded, covered, or embedded the metal alloys with carbon materials have demonstrated new nanocomposites with high activity and long-term stability to HER and OER reactions [103,116,133–138].

3.9 MOFs

Metal-organic frameworks (MOFs) have been used in several fields, like chemical sensors, optical, magnetic, and conductive materials, heterogeneous catalysis, and photocatalysis, gas storage, adsorption, separation, and purification; in the field of biomedicine as drug delivery, imaging agents, and therapeutic treatment [141–149].

MOFs present organized structures composed of metals ions connected to organic structures. Their characteristics such as open and porous crystalline composition, flexibility, and adaptable functionalities are attractive for the electrocatalysis point-of-view [5,150–152]. These characteristics bring benefits to MOF-based electrocatalyst allowing to control of the exposure of active sites favoring the mass transport. The control of the metal ions distribution around the organic structure allows the modification of physicochemical properties, which, consequently, can improve the catalytic efficiency [153–157].

Synthesis involving MOFs, besides regulation of porosity and shape, also enables the generation of 0D, 1D, 2D, or 3D structures, as well as the possibility of transforming them into carbon-based materials [158–168]. Furthermore, when carrying a proper synthesis route, heteroatoms can be added to MOF structures, carbon-coated metal nanoparticles, alloys, and also carbides can be obtained from MOF precursors [138,169,170]. Chemical species can be encapsulated inside the pores of MOFs, metals can be atomically dispersed

340 Oxide free nanomaterials for energy storage and conversion applications

at the structure by being anchored on N atoms of MOFs, and metals particles can also be removed from MOFs by undergoing a chemical reaction [171–173].

The most commonly used methods to insert active species inside the pores of MOFs are chemical vapor deposition, solid grinding, solution impregnation, double solvent approach, and one-pot synthesis [174]. In the chemical vapor deposition (CVD) method, the MOF material and the volatile metal precursor are placed together under vacuum, then the part of metal diffuses into MOF pores and part stays on the surface. After, to turn the metal into nanoparticles is performed a thermal treatment or a reduction. The CVD provides a catalyst with small nanoparticles inside the pores of the MOF; however, this synthesis needs high precision and volatile and expensive complexes [174]. In the solid grinding, the MOF precursor and the volatile metal precursor are grounded together, and by a reduction step, the composite is produced. Nevertheless, this method also requires expensive and volatile solvents, once the nonutilization of solvents results in a slow rate of preparation [174–176].

The dispersion followed by mixing a MOF and a metal precursor constitute the solution impregnation method. After this step, a reduction step is employed to form the nanoparticles. The drawback of this method is difficult to control the particle distributions on the surface [174,177]. In the double solvent approach, the MOF is dispersed in a hydrophobic solvent, which under vigorous stirring a metal aqueous solution is incorporated to hydrophobic dispersion. The double solvent approach is a good way to produce precisely the nanoparticle inside the MOF pores [15,174]. The employment of reduction agents in a mixture of MOF and metal precursors is called one-pot synthesis. In this method, the nanoparticle agglomeration is avoided [174]. In a colloidal deposition occurs the deposition of the metal nanoparticles presynthetized on the external surface of the MOF by mixing and stirring in the same solvent [174]. By a templated synthesis, a presynthesized metal nanoparticle is added into the MOF, and frameworks are grown up around it, or MOFs can still be transformed into carbon materials [15,174]. In the spray drying method, the fast evaporation of organic ligands and metal ions provides spherical particles [174,178,179].

Some factors are pointed out as prerequisites to achieve the high catalytic performance of MOF-derived materials toward water splitting. One of these factors is the porosity that yields a large area and shreds of evidence active sites, providing small pathways for ion diffusion, better mass transportation, and also electrocatalytic activation [91–93]. The architecture of which particles of MOF-based electrocatalysts are structured is another factor that contributes to good HER/OER electrocatalysts, as the open architecture containing a hollow inside, analog frame/cage-like structure, of NiCo sulfides deposited on CoFe Prussian blue (NiCo-S@CoFeA-TT) electrocatalysts [180]. An appropriate particle structure can offer a synergy effect, expose active centers, avoid particle agglomeration, and increase the electrochemically active surface area (ECSA) [93–96]. High conductivity can also be achieved by MOF-derived electrocatalysts, properties which can contribute to improve activation and furnish quick mass and charge transference [181,182].

Oxide free materials for electrochemical water splitting **Chapter | 14 341**

Good advances involving MOF-derived materials for activities for OER and HER electrocatalysts in alkaline media were reported by Duan et al. and Zhang, et al. where they found an overpotential of 240 mV for Ni-Fe-MOF catalyst and 144 mV for H-Mo$_2$C@Co, respectively [183,184]. Besides that, Chen, et al. published a study where Co$_9$S$_8$ and MoS$_2$ encapsulated on N, S, O-doped carbon (Co$_9$S/MoS$_2$@NSOC) electrocatalyst proved to be active for HER both in alkaline and acidic media [185].

4. Findings or results with implications for managers and decision-makers

Besides high cost [7], noble metals that compound state-of-art electrocatalysts for water splitting (platinum, ruthenium, and iridium) are also on high foreshadowing of endangered [186]. Furthermore, platinum is present in a reduced number of geological environments [187]. South Africa and Russia are the biggest providers of Pt metal group (Pt, Pd, Rh, Ir, Ru, and Os), with 61% and 27%, respectively [188].

These facts have been encouraging researchers all over the world to seek cheaper and equally active and stable electrocatalysts using Earth-abundant transition metals phosphates, borates, chalcogenides (sulfides and selenides), phosphides, nitrides, borides, carbides, alloys, and MOFs. There are a large number of published studies involving the use of these materials, some of which are reported in Table 1, where low overpotential of HER and OER are noticed. This is the case of Ni$_{foam}$@Ni-Ni$_{0.2}$Mo$_{0.8}$N (nickel molybdenum nitride nanorods grown on nickel foam) for example, impressively this high active electrocatalyst requires only 15 and 218 mV of overpotential for HER and OER, respectively [114]. Nevertheless, these electrocatalysts are also highly stable, presenting stability during 110 h [114].

In the chalcogenides group, several recent studies have shown to be relevant in replacing noble-metal-based catalysts. Theoretical computational studies based on the DFT have contributed to the search and design of new materials and combinations to act as electrocatalysts. Important catalytic characteristics (i.e., adsorption energy for H$_2$, O$_2$, or reaction intermediates) are used as trend descriptors for computational modeling. As an example of a successful combination of theoretical and experimental studies, from theoretical calculations, a stable two-dimensional molybdenum sulfide (Mo$_6$S$_4$) showed energetically favorable basal-planes with accentuated catalytic activity for HER, its catalytic performance was confirmed by electrochemical experiments. This study gives insights into the structure of 2D phase catalysts and shows the possibilities of application for these materials [86]. Covered with MoS$_2$ nanosheets cobalt sulfide cages, produced through solvothermal and annealing approach, were reported as bifunctional catalysts for HER and OER, with satisfactory overpotentials of 136 and 280 mV, respectively, in 1 M KOH solution. Additionally, the Co$_3$S$_4$@MoS$_2$ catalysts showed robust durability and stability, and required only 1.58 V to overall water splitting, producing a current density of

TABLE 1 Summary of catalytic activity reported in the earlier literature HER and OER electrocatalysts.

Material	Reaction	Electrolyte	Overpotential ($10\,mA\,cm^{-2}$) /mV	Tafel slope (mV dec^{-1})	Reference
NENU-500	HER	$0.5\,M\,H_2SO_4$	237	96	[189]
THTNi 2DSP	HER	$0.5\,M\,H_2SO_4$	333	80.5	[190]
NiCo-UMOFNs	OER	$1.0\,M\,KOH$	250	42	[191]
NiFe-MOF array	OER	$0.1\,M\,KOH$	240	34	[184]
NiCo-S@CoFeA-TT	OER	$1.0\,M\,KOH$	268	62	[180]
MIL-53(Co-Fe)/NF	OER	$1.0\,M\,KOH$	262[a]	69	[192]
CoHCF/NC-2	OER	$1.0\,M\,KOH$	357	73.97	[181]
CoP@NC/CF-900	HER	$1.0\,M\,KOH$	151.2	64.8	[182]
Co_9S_8/MoS_2@NSOC	HER	$0.5\,M\,H_2SO_4$	233	96	[185]
Co_9S_8/MoS_2@NSOC	HER	$1.0\,M\,KOH$	194	118	[185]
H-Mo_2C@Co	HER	$0.5\,M\,H_2SO_4$	144	58	[183]
ZnSP/NC	HER	$1.0\,M\,KOH$	171	54.78	[193]
NiB/NF	HER	$1.0\,M\,KOH$	41.2	106.5	[41]
Ni-CMB/CC	HER	$1.0\,M\,KOH$	69	76.3	[194]
NiFeB@NiFeB$_i$	OER	$1.0\,M\,KOH$	237	57.65	[195]
Ni-Fe-P-B	HER	$1.0\,M\,KOH$	220	63	[196]

Ni-Fe-P-B	OER	1.0 M KOH	269	38	[196]
$Ni_xFe_{1-x}B$	HER	1.0 M KOH	63.5	56.3	[197]
$Ni_xFe_{1-x}B$	OER	1.0 M KOH	282	86.7	[197]
Co-Mo-B	HER	Potassium phosphate buffer	95	56	[198]
Co-Mo-B	HER	1.0 M NaOH	67	58	[198]
Ni@B-C 500°C	HER	1.0 M KOH	176	78	[36]
A-CoB/f-CNF	OER	0.1 M KOH	350	173	[199]
$(Ni_{0.75}Co_{0.25})_2B$	OER	1.0 M KOH	224	34.5	[200]
CoNiB-700	OER	1.0 M KOH	262	58	[201]
$Fe_{0.2}Co_{0.8}Se_2/g$-C_3N_4	HER	0.5 M H_2SO_4	83[b]	83	[202]
$Fe_{0.2}Co_{0.8}Se_2/g$-C_3N_4	OER	1.0 M KOH	230	44	[202]
CoN_x/NGA	HER	1.0 M KOH	198	97.3	[203]
CoN_x/NGA	OER	1.0 M KOH	273	82.3	[203]
NiCoN-1H	HER	1.0 M NaOH	145	105.2	[204]
NiCoN-1H	OER	1.0 M NaOH	360	46.9	[204]
$FeNi_3N$/NF	HER	1.0 M KOH	75	98	[113]
$FeNi_3N$/NF	OER	1.0 M KOH	202	40	[113]
Ni_3FeN	HER	1.0 M KOH	45	75	[205]
Ni_3FeN	OER	1.0 M KOH	223	40	[205]

Continued

TABLE 1 Summary of catalytic activity reported in the earlier literature HER and OER electrocatalysts—cont'd

Material	Reaction	Electrolyte	Overpotential ($10\,mA\,cm^{-2}$) /mV	Tafel slope (mV dec^{-1})	Reference
$Ni_{foam}@Ni-Ni_{0.2}Mo_{0.8}N$	HER	1.0 M KOH	15	39	[114]
$Ni_{foam}@Ni-Ni_{0.2}Mo_{0.8}N$	OER	1.0 M KOH	218	55	[114]
$Co_{5.47}N$ NP@N-PC	HER	1.0 M KOH	149	86	[206]
$Co_{5.47}N$ NP@N-PC	OER	1.0 M KOH	248	72	[206]
Ni-Fe-MoN NTs	HER	1.0 M KOH	55	109	[207]
Ni-Fe-MoN NTs	OER	1.0 M KOH	228	41	[207]
CoN-400/CC	HER	1.0 M KOH	97	93.9	[208]
CoN-400/CC	OER	1.0 M KOH	251	75.4	[208]
NC-Nicu-NiCuN	HER	Alkaline media	93	55	[209]
NC-Nicu-NiCuN	OER	1.0 M KOH	232	41	[209]
$V-Ti_4N_3T_x$	HER	$0.5\,M\,H_2SO_4$	330	107	[210]
meso-CoSSe-12h	HER	$0.5\,M\,H_2SO_4$	56	52	[54]
$Ni_{0.5}Mo_{0.5}Se$	HER	$0.5\,M\,H_2SO_4$	197	107	[78]
$Ni_{0.5}Mo_{0.5}Se$	OER	1.0 M KOH	420	–	[78]
$G/MoSe_2-CoSe_2$ NTS	HER	1.0 M KOH	198	79	[79]
$Co_3S_4@MoS_2$	HER	1.0 M KOH	136	74	[80]

$Co_3S_4@MoS_2$	OER	1.0M KOH	280	43	[80]
NiFeS/NF	OER	1.0M KOH	65	119	[81]
$(Ni, Co, Fe)_9S_8$/N-doped carbon	HER	1.0M KOH	137	165	[84]
$(Ni, Co, Fe)_9S_8$/N-doped carbon	OER	1.0M KOH	116	184	[84]
MoS_2/Ni_3S_2	HER	1.0M KOH	110	83	[85]
MoS_2/Ni_3S_2	OER	1.0M KOH	218	88	[85]
MoS_2 QD	HER	pH7 buffer	65	74	[87]
Fe-NiS-(BO)/NF	HER	1.0M KOH	200@50	58	[211]
Fe-NiS-(BO)/NF	OER	1.0M KOH	250@50	55	[211]
O-WS2-1T	HER	0.5M H_2SO_4	~80	47	[212]
NE-CF@WS2 NFs	HER	0.5M H_2SO_4	153	59	[213]
NiSe-PANI	HER	1.0M KOH	120	–	[214]
NiSe-PANI	OER	1.0M KOH	303@50	–	[214]
$Co-MoS_2$/BCCF-21	HER	1.0M KOH	48	52	[215]
$Co-MoS_2$/BCCF-21	OER	1.0M KOH	260	85	[215]
$EG/Ni_3Se_2/Co_9S_8$	HER	1.0M KOH	170^b	82	[216]
Co_3S_4 NCs	HER	0.5M H_2SO_4	250	65	[217]
Ni-Co-P nanoflowers	HER	1.0M KOH	83	47	[90]

Continued

TABLE 1 Summary of catalytic activity reported in the earlier literature HER and OER electrocatalysts—cont'd

Material	Reaction	Electrolyte	Overpotential ($10\,mA\,cm^{-2}$) /mV	Tafel slope (mV dec^{-1})	Reference
Ni-Co-P nanoflowers	HER	$0.5\,M\ H_2SO_4$	92	50	[90]
NiCoFe$_x$P/CC	HER	1.0 M KOH	39	50	[95]
NiCoFe$_x$P/CC	OER	1.0 M KOH	275	56	[95]
Ni$_x$P$_y$-325	HER	$0.5\,M\ H_2SO_4$	62^b	46	[218]
Ni$_x$P$_y$-325	OER	1.0 M KOH	320	72	[218]
Ni-P-Pt$_{0.5}$/NF	HER	1.0 M KOH	34	31	[219]
Ni-P-Pt$_{0.5}$/NF	HER	$0.5\,M\ H_2SO_4$	48	36	[219]
S-Ni5P4 NPA/CP	HER	$0.5\,M\ H_2SO_4$	56	44	[220]
NiFeP	OER	1.0 M KOH	330^b	39	[221]
Ni-Co-P/C$_{60}$	HER	$0.5\,M\ H_2SO_4$	97	48	[222]
C-(Fe-Ni)P@PC/(Ni-Co)P@CC	HER	1.0 M KOH	142	98	[223]
C-(Fe-Ni)P@PC/(Ni-Co)P@CC	OER	1.0 M KOH	251	56	[223]
CNTs@NiCoP/C	OER	1.0 M KOH	297	57	[224]
FeMnP/GNF	HER	0.1 M KOH	84	78	[225]
FeMnP/GNF	OER	0.1 M KOH	230	35	[225]

MoP@HCC	HER	0.5M H_2SO_4	129	48	[226]
Co-Fe-Bi/NF	OER	1.0M KOH	307	69	[43]
NiFe-borate	OER	1.0M KOH	230	32	[44]
$Co_3(BO_3)_2$@CNT	OER	pH7 buffer	487	67	[49]
B_x/GCS	HER	0.5M H_2SO_4	52	47	[227]
B_x/GCS	OER	0.5M H_2SO_4	170	67	[227]
Co-B/$Ti_3C_2T_x$	OER	1.0M KOH	250	53	[46]
0.05-SB-$Ni_{0.8}$-$Fe_{0.2}(OH)_2$	OER	1.0M KOH	301	42	[51]
np-AlNiFeCoMo	OER	1.0M KOH	240	46	[135]
NiMo-FG/NF	OER	1.0M KOH	338	67	[136]
FeCo	HER	0.5M H_2SO_4	262	74	[138]
$Ni_2P_2O_7$/Pt	HER	0.5M H_2SO_4	28	32	[104]
NiCoFe phosphate NSs/NF	OER	1.0M KOH	240	58	[105]
NiCoFe phosphate NSs/NF	HER	1.0M KOH	231	86	[105]
$Fe(PO_3)_2$	OER	1.0M KOH	177	51.9	[106]
CoP/CC	HER	0.5M H_2SO_4	45	49	[115]
Mo_2C	HER	0.5M H_2SO_4	124	43	[117]
Mo_2C	OER	1.0M KOH	77	50	[117]

[a]Value reached at $100\,mA\,cm^{-2}$.
[b]Value reached at $20\,mA\,cm^{-2}$.

348 Oxide free nanomaterials for energy storage and conversion applications

$10\,mA\,cm^{-2}$ [80]. A Co-Ni/MoS$_2$ nanocomposite synthesized in a microwave oven exhibited outstanding HER performance, close to the commercial Pt/C catalysts, with an overpotential close of $18\,mV$ in $0.1\,M\,HClO_4$ solution, associated to the synergistic effects and interactions between Co and Ni with the MoS$_2$ nanosheets [228].

Theoretical studies for a Ni-MoS$_2$ monolayer showed better catalytic performance to HER than MoS$_2$, with good conductivity, low-cost, activity in both alkaline and acidic medium, and suitable for high-performance HER applications [229]. Doped with borate and Fe ions Nickel sulfides were hydrothermally synthesized and showed remarkable HER and OER activity as well as long-term stability in alkaline medium, producing $50\,mA\,cm^{-2}$ with a low overpotential of 200 and $250\,mV$, respectively. The authors attributed the great performance of the catalysts to the use of anionic and cationic doping agents (Borate and Fe, respectively), improving elementary steps in HER and OER mechanisms. The potential of $1.59\,V$ to Fe-NiS-(BO)/NF overcome the noble metal catalysts ($1.64\,V$) operating in an alkaline water electrolyzer [211].

Produced via chemical exfoliation and ultrasonication processes tungsten sulfide nanoclusters showed unique properties of the metallic 1T phase and oxygen incorporation, like tailored edges sites and conductivity. The O-WS$_2$-1 T catalyst exhibited catalytic activity to HER with an overpotential close to $80\,mV$ in an acidic medium [212]. Likewise, using the thermal evaporation approach, WS$_2$ nanoflakes were grown on N-enriched carbon foam exhibiting excellent electrocatalytic activity to HER under an overpotential of $153\,mV$ at $10\,mA\,cm^{-2}$, their stability and durability were attributed to fast electron transfer and the content of active sites in the 3D porous structure [213]. Produced via solvothermal process, MoS$_2$ and WS$_2$ combined nanostructures were supported in carbon nanotubes (MoS$_2$-WS$_2$-CNTs), achieving enlarged interlayer spacing to MoS$_2$, enhanced edge sites, and charge transfer between catalyst/substrate. The MoS$_2$-WS$_2$-CNTs catalyst exhibited favorable performance to HER with a low overpotential of $212\,mV$ at $10\,mA\,cm^{-2}$ [230].

MoSe$_2$-CoSe$_2$ nanotubes attached to graphene nanosheets, synthesized via the hydrothermal method, exhibited high performance to HER with an overpotential of $198\,mV$, its catalytic performance was associated with the homogenous distribution of MoSe$_2$ nanosheets and CoSe$_2$ nanoparticles, in addition to a large number of active sites or edges [79].

A nickel selenide functionalized with polyaniline, synthesized through electrodeposition, hydrothermal and electropolymerization approaches presented a surface structure rich in Se species attractive toward HER and Ni^{3+}/Ni^{4+} species formation during the OER. These features provided an excellent activity for OER and HER requiring only $1.53\,V$ to produce $10\,mA\,cm^{-2}$, which surpass the benchmarking Pt and IrO_2 electrocatalysts [214]. Likewise, produced via hydrothermal method nickel molybdenum selenide composite proved to be an alternative catalyst to water splitting, with outstanding catalytic activity to HER and OER, demanding overpotentials close to 190 and $340\,mV$, respectively, to produce $10\,mA\,cm^{-2}$ [78].

Oxide free materials for electrochemical water splitting **Chapter | 14 349**

In the phosphide group, Ni_2P was one of the first transition metal phosphides (TMP) investigated for HER, with catalytic activity and stability in acidic medium attributed to the increased surface area and high density of exposed (001) facets [231]. Nickel phosphides also can present stability in acidic medium; it is cost-effective and efficient for HER electrocatalysis. The catalytic mechanism of the Ni_xP_y is similar to the hydrogenase enzymes, giving exceptional catalytic performance [218]. A nickel phosphide decorated with Pt in a < 0.5 wt% concentration (Ni-P-$Pt_{0.5}$/NF) was synthesized through solvothermal treatment followed by chemical adsorption and exhibited a low overpotential of 34 mV and 48 mV in both alkaline and acidic medium (at 10 mA cm^{-2}), chemical stability and fast reaction kinetics for HER. The authors attributed the outstanding performance of Ni-P-$Pt_{0.5}$/NF to the synergistic effects between the nickel phosphide and Pt, resulting in a low-cost electrocatalyst with an activity comparable to noble metal catalysts [219].

Produced via hydrothermal and calcination approaches sulfur-doped nickel phosphide also presented high performance to HER with a low overpotential of 56 mV to achieve a current density of 10 mA cm^{-2}, a catalytic performance which was associated with the presence of S atoms in Ni_5P_4 structure, increasing the surface area and active sites number, and hindering the dissolution and catalyst surface oxidation [220]. Similarly, Fe-doped Ni phosphide was synthesized via a mixed solvent stabilization approach, resulting in nanoparticles with hcp structure without relevant structural distortion, and notable electrocatalytic activity comparable or even superior to RuO_2 and IrO_2 catalysts. The overpotential to NiFeP was 330 mV at a current density of 20 mA cm^{-2}, and the authors described the Fe-O sites as responsible for the OH adsorption while the Ni sites were responsible for the oxygen evolution in the OER [221].

A nickel-cobalt phosphide coupled with C_{60} molecules also presented remarkable electrocatalytic activity toward HER, with an overpotential of 97 mV at 10 mA cm^{-2} and great stability. The Ni-Co-P/C_{60} synthesized through precipitation and phosphidation approaches exhibited more exposed active sites and a nanosized structure with defects which were beneficial for the HER according to the authors [222]. Likewise, nanocubes containing Ni-Co phosphides on the interior carbon layer and Ni-Fe phosphides on the exterior carbon layer (C-(Fe-Ni)P@PC/(Ni-Co)P@CC) exhibited remarkable electrocatalytic activity for overall water splitting in alkaline medium, with overpotentials of 142 and 251 mV at 10 mA cm^{-2} for HER and OER, respectively. The C-(Fe-Ni)P@PC/(Ni-Co)P@CC catalyst was synthesized in several steps, including precipitation, ammonia carving, conformal polydopamine coating, precursor metal ion coordination, and thermal phosphorization [223].

Hollow nanocages and nanosheets of nickel cobalt phosphide and carbon were perpendicularly anchored onto carbon nanotubes via precipitation and phosphorization reactions and exhibited excellent catalytic performance to OER with an overpotential of 297 mV at 10 mA cm^{-2} and long-term operation stability. According to the authors, the superior OER performance of CNTs@NiCoP/C is related to the unique hierarchical superstructure and multiple compositions [224].

350 Oxide free nanomaterials for energy storage and conversion applications

A synthesized via hydrothermal and phosphorization methods cobalt copper phosphide presented both electrocatalytic activity to HER in acidic, basic, and neutral media and OER in alkaline media, with low overpotentials of 47, 70, 120, and 221 mV, respectively, to reach $10\,mA\,cm^{-2}$. Also, the $Cu_{0.075}Co_{0.925}P$ catalyst overcame the noble metal-based Pt and IrO_2 requiring only 1.55 V to produce a current density of $10\,mA\,cm^{-2}$ to overall water splitting [232]. Using the CVD method, an iron-manganese phosphide was deposited on a graphene-protected nickel foam that demonstrated attractive electrocatalytic activity toward HER and OER, with overpotentials of 57 and 230 mV at a current density of $10\,mA\,cm^{-2}$, and required only 1.55 V to overall water splitting with great stability. Theoretical data suggested that Fe and Mn sites exposed on the facets are essential to achieve high HER activity [225].

Produced through mechanical mixing and pyrolysis, supported on porous honeycomb carbon molybdenum phosphide (MoP@HCC) showed outstanding electrocatalytic activity and stability to HER in acid medium, with a low over-potential of 120 mV at $10\,mA\,cm^{-2}$. According to the authors, the small size of the nanoparticles, the large surface area of the honeycomb structure, and the intrinsic activity of the MoP contributed to the catalytic performance showed by MoP@HCC electrocatalyst [226].

Among the monometallic borates, a crosslinked to carbon-based structure borate showed the excellent performance to OER and HER, with an overpotential of 170 and 52 mV, respectively, related to the rich BO_3 planes, where the B atoms were identified as the major active sites. When 1.45 V was applied to an electrolyzer a current density of $10\,mA\,cm^{-2}$ was achieve [227]. A CNTs cobalt borate-based electrocatalyst was explored in pH 7 phosphate buffer and exhibited high stability and catalytic activity at an overpotential of 487 mV [49].

Bimetallic borate-based OER electrocatalysts demonstrated an excellent catalytic OER activity as well. Amorphous Co-Fe-B_i/NF grown on a nickel foam with a three-dimensional porous structure required only 307 mV to produce a current density of $10\,mA\,cm^{-2}$, related to the number of catalytic sites and the conductivity [43]. Cobalt borates/$Ti_3C_2T_x$ were produced by chemical approach and exhibited low overpotential (250 mV) which was related to the interaction between Co-Bi and $Ti_3C_2T_x$, the charge transfer, and the attraction between the intermediates of the reaction [46]. The most notable reported water-splitting electrocatalysts in recent years are listed in Table 1.

5. Discussion, limitations, future research, and conclusion

Transition metal-borides typically generate a huge quantity of bubbles during the HER and OER reaction, which leads to limitations of the electrocatalytic performance. Moreover, loss of activity is also observed in TM-borides due to the binders applied in the nanostructured electrocatalysts, pH, and the increased energy of the surface [13]. Others challenges for TM-borides materials involve surface oxidation when stocked for an elongated time in ambient conditions, the no improvement of OER catalytic performance in acidic electrolytes, and the essential better exposition of active sites [13].

The metal borates exhibited favorable catalytic activity toward OER, promoting faster proton-coupled electron transfer pathways, however, the catalytic mechanisms are still not well understood. Also, the intrinsic activity of the pure borates hinder the application on a large scale, which may be solved through doping with different elements. The effects of borate doping in carbon materials and transition metal oxides/hydroxides to HER and OER applications have been studied but still need improvements [35,233].

Sulfides and selenides can catalyze the HER and OER in acidic, alkaline, or neutral solutions and present advantages as low cost, availability, efficiency, and are easily constructed. However, despite their catalytic properties, they are still inferior when compared to precious metal-based catalysts. To enhance their performance several strategies such as structure and composition optimization, controlled heteroatom doping, and incorporation with conductive materials have been adopted. Nonetheless, some challenges still exist as the absence of complete understanding of OER mechanisms to metal sulfides and selenides as well as the surface changes during the reaction [57,101].

Transition metal phosphides exhibited high performance for HER and OER reaction, but some efforts must be done to achieve a substantial performance like hierarchically micro/nanostructured interfaces with high roughness, exposed active sites, facilitated charge, and mass transfer, and distinct components on the interface. Likewise, the formation of the oxides on the TMPs surface during the catalytic process does not affect the OER activity but can decrease the conductivity and block the HER active sites. These challenges may be solved through surface doping with others elements improving the stability of HER [99]. The development of efficient OER catalysts in an acidic medium is also needed since the TMPs are generally effective in an alkaline medium. Finally, the use of TM-phosphides as precatalysts to OER has been proved to be more catalytic active to OER than analogous oxide/hydroxide containing the same metals [234].

Despite the great advances achieved by the academic community, for a successful application of phosphates as electrocatalysts for water splitting some obstacles still need to be faced, such as a deep understanding of reaction mechanisms as well as the role of each group in the catalytic process. The activity and catalytic stability of the phosphates that are maintained even after the structure undergo drastic surface reconstructions during the reactions also constitute an issue to be resolved [107].

The great electrocatalytic activity for water splitting of TM-nitrides is attributed to the bond with density highly electronic between metal and nitrogen atoms providing a less difficult charge transfer [109]. As HER and OER succeed on the catalyst surface, a way to improve even more its performance is deepening the investigation around the active sites. Furthermore, it is necessary to prevent TM-nitrides from corrosion during long-term stability aiming to keep the high electrocatalyst activity [110].

Although metal alloys are a potentially active catalyst to HER and OER, problems involving corrosion are still unsolved which make this class of materials a challenge as practical electrocatalysts mainly for OER [103]. In this way, new synthesis approaches to improve the electrochemical stability and to increase the exposure active sites (to maximize the metal utilization) still should be kept in focus. In addition, more efforts toward a better knowledge about the HER mechanism in alkaline media onto metal alloy surfaces also should be addressed.

Regarding the carbides, although this class of materials has shown impressive catalytic performance especially for HER, there are some points to be well understood and improved manly for their application in alkaline medium. One of them is the deep understanding of the mechanism reaction in alkaline electrolytes where the role of surface oxides during the HER still keep contradictory. Another necessary point is the improvement of catalytic properties (i.e., hydrogen binding energy, selectivity), which can be done through the use of hybrid structures and the improvement of synthetic conditions [117]. Efforts must be done to seek synthetic approaches that are capable of controlling the crystallography, are large-scale viable, and that are not dangerous.

The utilization of integrated 1D/2D MOFs offers more actives sites, but despite this, during thermal treatment, they present decreased porosity, also they are not able to show high stability in terms of morphology, which are characteristics of 3D MOFs. In this scenario, more research involving multidimensional MOFs is desired [5]. The use of an appropriate temperature during the thermal treatment of the synthesis of electrocatalysts involving MOFs is still a challenge that must be studied. When the high temperature is applied to synthesize MOFs, problems like aggregation, fusion, and the breakdown of the structure can happen, which result in low electrocatalytic performance. In contrast, if a low temperature is used, the carbonatization of the organic ligands does not happen to a proper degree, and the catalyst will present low conductivity, which leads to poor electrocatalysis performance too. Therefore, more efforts should be directed to evaluate this parameter [5].

In general, despite the great advances achieved in recent years, systems exclusively composed of nonnoble transition metal-based materials still show limitations like inferior electrolytic performance and stability in comparison to Pt/RuO_2 or IrO_2 catalysts. To overcome the challenges, future investigations should be done aiming at the improvements in the intrinsic activity, the interfacial charge transfer, conductivity, and stability of nonprecious materials [13]. Strategies to tune the materials like heteroatom-doping, use of high specific surface area support, construction of hierarchically 3D morphologies, and the development of small nanoclusters and single atom-based catalysts are strongly encouraged to face these challenges. In addition, the OER and HER mechanisms on the transition metal-based catalysts need to be better understood.

References

[1] H. Nazir, C. Louis, S. Jose, J. Prakash, N. Muthuswamy, M.E.M. Buan, C. Flox, S. Chavan, X. Shi, P. Kauranen, T. Kallio, G. Maia, K. Tammeveski, N. Lymperopoulos, E. Carcadea, E. Veziroglu, A. Iranzo, A.M. Kannan, Is the H2 economy realizable in the foreseeable future? Part I: H2 production methods, Int. J. Hydrog. Energy 45 (2020) 13777–13788,- https://doi.org/10.1016/j.ijhydene.2020.03.092.

[2] S. Anantharaj, S.R. Ede, K. Karthick, S. Sam Sankar, K. Sangeetha, P.E. Karthik, S. Kundu, Precision and correctness in the evaluation of electrocatalytic water splitting: revisiting activity parameters with a critical assessment, Energy Environ. Sci. 11 (2018) 744–771, https://doi.org/10.1039/C7EE03457A.

[3] K.G. Reddy, T.G. Deepak, G.S. Anjusree, S. Thomas, S. Vadukumpully, K.R.V. Subramanian, S.V. Nair, A.S. Nair, On global energy scenario, dye-sensitized solar cells and the promise of nanotechnologt, Phys. Chem. Chem. Phys. 16 (2014) 6838–6858, https://doi.org/10.1039/C3CP55448A.

[4] B.K. Bose, Global energy scenario and impact of power electronics in 21st century, IEEE Trans. Ind. Electron. 60 (2013) 2638–2651, https://doi.org/10.1109/TIE.2012.2203771.

[5] Z. Chen, H. Qing, K. Zhou, D. Sun, R. Wu, Metal-organic framework-derived nanocomposites for electrocatalytic hydrogen evolution reaction, Prog. Mater. Sci. 108 (2020) 100618, https://doi.org/10.1016/j.pmatsci.2019.100618.

[6] J.O. Abe, A.P.I. Popoola, E. Ajenifuja, O.M. Popoola, Hydrogen energy, economy and storage: review and recommendation, Int. J. Hydrog. Energy 44 (2019) 15072–15086, https://doi.org/10.1016/j.ijhydene.2019.04.068.

[7] Z. Chen, X. Duan, W. Wei, S. Wang, Z. Zhang, B.J. Ni, Boride-based electrocatalysts: emerging candidates for water splitting, Nano Res. 13 (2020) 293–314, https://doi.org/10.1007/s12274-020-2618-y.

[8] X. Huang, M. Leng, W. Xiao, M. Li, J. Ding, T.L. Tan, W.S.V. Lee, J. Xue, Activating basal planes and S-terminated edges of MoS2 toward more efficient hydrogen evolution, Adv. Funct. Mater. 27 (2017) 1604943, https://doi.org/10.1002/adfm.201604943.

[9] Y. Wang, L. Chen, X. Yu, Y. Wang, G. Zheng, Superb alkaline hydrogen evolution and simultaneous electricity generation by Pt-decorated Ni3N nanosheets, Adv. Energy Mater. 7 (2017) 1–7, https://doi.org/10.1002/aenm.201601390.

[10] Z. Chen, X. Duan, W. Wei, S. Wang, B.J. Ni, Recent advances in transition metal-based electrocatalysts for alkaline hydrogen evolution, J. Mater. Chem. A 7 (2019) 14971–15005, https://doi.org/10.1039/c9ta03220g.

[11] N. Yao, T. Tan, F. Yang, G. Cheng, W. Luo, Well-aligned metal-organic framework array-derived CoS2 nanosheets toward robust electrochemical water splitting, Mater. Chem. Front. 2 (2018) 1732–1738, https://doi.org/10.1039/c8qm00259b.

[12] Q. Wang, Z. Zhang, X. Zhao, J. Xiao, D. Manoj, F. Wei, F. Xiao, H. Wang, S. Wang, MOF-derived copper nitride/phosphide heterostructure coated by multi-doped carbon as electrocatalyst for efficient water splitting and neutral-pH hydrogen evolution reaction, ChemElectroChem 7 (2020) 289–298, https://doi.org/10.1002/celc.201901860.

[13] Y. Jiang, Y. Lu, Designing transition-metal-boride-based electrocatalysts for applications in electrochemical water splitting, Nanoscale 12 (2020) 9327–9351, https://doi.org/10.1039/d0nr01279c.

[14] D. Wang, Y. Song, H. Zhang, X. Yan, J. Guo, Recent advances in transition metal borides for electrocatalytic oxygen evolution reaction, J. Electroanal. Chem. 861 (2020) 113953, https://doi.org/10.1016/j.jelechem.2020.113953.

354 Oxide free nanomaterials for energy storage and conversion applications

[15] H.F. Wang, L. Chen, H. Pang, S. Kaskel, Q. Xu, MOF-derived electrocatalysts for oxygen reduction, oxygen evolution and hydrogen evolution reactions, Chem. Soc. Rev. 49 (2020) 1414–1448, https://doi.org/10.1039/c9cs00906j.

[16] C.G. Morales-Guio, L.A. Stern, X. Hu, Nanostructured hydrotreating catalysts for electrochemical hydrogen evolution, Chem. Soc. Rev. 43 (2014) 6555–6569, https://doi.org/10.1039/c3cs60468c.

[17] B. Reuillard, M. Blanco, L. Calvillo, N. Coutard, A. Ghedjatti, P. Chenevier, S. Agnoli, M. Otyepka, G. Granozzi, V. Artero, Non-covalent integration of a bio-inspired Ni catalyst to graphene acid for reversible electrocatalytic hydrogen oxidation, ACS Appl. Mater. Interfaces 12 (2020) 5805–5811, https://doi.org/10.1021/acsami.9b18922.

[18] M. David, C. Ocampo-Martínez, R. Sánchez-Peña, Advances in alkaline water electrolyzers: a review, J. Energy Storage 23 (2019) 392–403, https://doi.org/10.1016/j.est.2019.03.001.

[19] A.J. Medford, A. Vojvodic, J.S. Hummelshøj, J. Voss, F. Abild-Pedersen, F. Studt, T. Bligaard, A. Nilsson, J.K. Nørskov, From the Sabatier principle to a predictive theory of transition-metal heterogeneous catalysis, J. Catal. 328 (2015) 36–42, https://doi.org/10.1016/j.jcat.2014.12.033.

[20] J. Mohammed-Ibrahim, S. Xiaoming, Recent progress on earth abundant electrocatalysts for hydrogen evolution reaction (HER) in alkaline medium to achieve efficient water splitting—a review, J. Energy Chem. 34 (2019) 111–160, https://doi.org/10.1016/j.jechem.2018.09.016.

[21] L.J. Yang, Y.Q. Deng, X.F. Zhang, H. Liu, W.J. Zhou, MoSe2 nanosheet/MoO2 nanobelt/carbon nanotube membrane as flexible and multifunctional electrodes for full water splitting in acidic electrolyte, Nanoscale 10 (2018) 9268–9275, https://doi.org/10.1039/c8nr01572d.

[22] B. Zhang, Z. Qi, Z. Wu, Y.H. Lui, T.H. Kim, X. Tang, L. Zhou, W. Huang, S. Hu, Defect-rich 2D material networks for advanced oxygen evolution catalysts, ACS Energy Lett. 4 (2019) 328–336, https://doi.org/10.1021/acsenergylett.8b02343.

[23] M. Ma, D. Liu, S. Hao, R. Kong, G. Du, A.M. Asiri, Y. Yao, X. Sun, A nickel-borate-phosphate nanoarray for efficient and durable water oxidation under benign conditions, Inorg. Chem. Front. 4 (2017) 840–844, https://doi.org/10.1039/c6qi00594b.

[24] L.S. Bezerra, G. Maia, Developing efficient catalysts for the OER and ORR using a combination of Co, Ni, and Pt oxides along with graphene nanoribbons and $NiCo_2O_4$, J. Mater. Chem. A 8 (2020) 17691–17705, https://doi.org/10.1039/D0TA05908K.

[25] S.A. Grigoriev, V.N. Fateev, D.G. Bessarabov, P. Millet, Current status, research trends, and challenges in water electrolysis science and technology, Int. J. Hydrog. Energy (2020), https://doi.org/10.1016/j.ijhydene.2020.03.109.

[26] J.O. Bockris, Hydrogen economy in the future, Int. J. Hydrog. Energy 24 (1999) 1–15, https://doi.org/10.1016/S0360-3199(98)00115-3.

[27] X. Xia, L. Wang, N. Sui, V.L. Colvin, W.W. Yu, Recent progress in transition metal selenide electrocatalysts for water splitting, Nanoscale 12 (2020) 12249–12262, https://doi.org/10.1039/D0NR02939D.

[28] H. Jin, X. Liu, S. Chen, A. Vasileff, L. Li, Y. Jiao, L. Song, Y. Zheng, S.Z. Qiao, Heteroatom-doped transition metal electrocatalysts for hydrogen evolution reaction, ACS Energy Lett. 4 (2019) 805–810, https://doi.org/10.1021/acsenergylett.9b00348.

[29] J. Deng, P. Ren, D. Deng, L. Yu, F. Yang, X. Bao, Highly active and durable non-precious-metal catalysts encapsulated in carbon nanotubes for hydrogen evolution reaction, Energy Environ. Sci. 7 (2014) 1919–1923, https://doi.org/10.1039/c4ee00370e.

[30] J.M.V. Nsanzimana, Y. Peng, Y.Y. Xu, L. Thia, C. Wang, B.Y. Xia, X. Wang, An efficient and earth-abundant oxygen-evolving electrocatalyst based on amorphous metal borides, Adv. Energy Mater. 8 (2018) 1–7, https://doi.org/10.1002/aenm.201701475.

[31] Y. Li, B. Huang, Y. Sun, M. Luo, Y. Yang, Y. Qin, L. Wang, C. Li, F. Lv, W. Zhang, S. Guo, Multimetal borides nanochains as efficient electrocatalysts for overall water splitting, Small 15 (2019) 1–8, https://doi.org/10.1002/smll.201804212.

[32] K. Elumeeva, J. Masa, D. Medina, E. Ventosa, S. Seisel, Y.U. Kayran, A. Genç, T. Bobrowski, P. Weide, J. Arbiol, M. Muhler, W. Schuhmann, Cobalt boride modified with N-doped carbon nanotubes as a high-performance bifunctional oxygen electrocatalyst, J. Mater. Chem. A 5 (2017) 21122–21129, https://doi.org/10.1039/c7ta06995b.

[33] H. Li, P. Wen, Q. Li, C. Dun, J. Xing, C. Lu, S. Adhikari, L. Jiang, D.L. Carroll, S.M. Geyer, Earth-abundant iron diboride (FeB2) nanoparticles as highly active bifunctional electrocatalysts for overall water splitting, Adv. Energy Mater. 7 (2017) 1–12, https://doi.org/10.1002/aenm.201700513.

[34] V.I. Simagina, A.M. Ozerova, O.V. Komova, G.V. Odegova, D.G. Kellerman, R.V. Fursenko, E.S. Odintsov, O.V. Netskina, Cobalt boride catalysts for small-scale energy application, Catal. Today 242 (2015) 221–229, https://doi.org/10.1016/j.cattod.2014.06.030.

[35] L. Cui, W. Zhang, R. Zheng, J. Liu, Electrocatalysts based on transition metal borides and borates for the oxygen evolution reaction, Chem. – A Eur. J. 26 (2020) 11661–11672, https://doi.org/10.1002/chem.202000880.

[36] B.A. Yusuf, Y. Xu, N. Ullah, M. Xie, C.J. Oluigbo, W. Yaseen, J.K. Alagarasan, K. Rajalakshmi, J. Xie, B-doped carbon enclosed Ni nanoparticles: a robust, stable and efficient electrocatalyst for hydrogen evolution reaction, J. Electroanal. Chem. 869 (2020) 114085, https://doi.org/10.1016/j.jelechem.2020.114085.

[37] J. Kim, H. Kim, S.K. Kim, S.H. Ahn, Electrodeposited amorphous Co-P-B ternary catalyst for hydrogen evolution reaction, J. Mater. Chem. A 6 (2018) 6282–6288, https://doi.org/10.1039/c7ta11033b.

[38] S.J. Sitler, K.S. Raja, I. Charit, Metal-rich transition metal diborides as electrocatalysts for hydrogen evolution reactions in a wide range of pH, J. Electrochem. Soc. 163 (2016) H1069–H1075, https://doi.org/10.1149/2.0201613jes.

[39] X. Ma, J. Wen, S. Zhang, H. Yuan, K. Li, F. Yan, X. Zhang, Y. Chen, Crystal CoxB (x = 1-3) synthesized by a ball-milling method as high-performance Electrocatalysts for the oxygen evolution reaction, ACS Sustain. Chem. Eng. 5 (2017) 10266–10274, https://doi.org/10.1021/acssuschemeng.7b02281.

[40] S. Carenco, D. Portehault, C. Boissière, N. Mézailles, C. Sanchez, Nanoscaled metal borides and phosphides: recent developments and perspectives, Chem. Rev. 113 (2013) 7981–8065, https://doi.org/10.1021/cr400020d.

[41] R. Zhang, H. Liu, C. Wang, L. Wang, Y. Yang, Y. Guo, Electroless plating of transition metal boride with high boron content as superior HER electrocatalyst, ChemCatChem 12 (2020) 3068–3075, https://doi.org/10.1002/cctc.202000315.

[42] J. Masa, I. Sinev, H. Mistry, E. Ventosa, M. de la Mata, J. Arbiol, M. Muhler, B. Roldan Cuenya, W. Schuhmann, Ultrathin high surface area nickel boride (Ni x B) nanosheets as highly efficient electrocatalyst for oxygen evolution, Adv. Energy Mater. 7 (2017) 1700381, https://doi.org/10.1002/aenm.201700381.

[43] U.P. Suryawanshi, M.P. Suryawanshi, U.V. Ghorpade, S.W. Shin, J.H.J. Kim, J.H.J. Kim, An earth-abundant, amorphous cobalt-iron-borate (Co-Fe-Bi) prepared on Ni foam as highly efficient and durable electrocatalysts for oxygen evolution, Appl. Surf. Sci. 495 (2019) 143462, https://doi.org/10.1016/j.apsusc.2019.07.204.

[44] N. Wang, Z. Cao, X. Kong, J. Liang, Q. Zhang, L. Zheng, C. Wei, X. Chen, Y. Zhao, L. Cavallo, B. Zhang, X. Zhang, Activity enhancement via borate incorporation into a NiFe

(oxy)hydroxide catalyst for electrocatalytic oxygen evolution, J. Mater. Chem. A 6 (2018) 16959–16964, https://doi.org/10.1039/C8TA04762F.

[45] Y. Surendranath, M. Dincă, D.G. Nocera, Electrolyte-dependent electrosynthesis and activity of cobalt-based water oxidation catalysts, J. Am. Chem. Soc 131 (2009) 2615–2620, https://doi.org/10.1021/ja807769r.

[46] J. Liu, T. Chen, P. Juan, W. Peng, Y. Li, F. Zhang, X. Fan, Hierarchical cobalt borate/MXenes hybrid with extraordinary electrocatalytic performance in oxygen evolution reaction, ChemSusChem 11 (2018) 3758–3765, https://doi.org/10.1002/cssc.201802098.

[47] Z. Sun, L. Lin, C. Nan, H. Li, G. Sun, X. Yang, Amorphous boron oxide coated NiCo layered double hydroxide nanoarrays for highly efficient oxygen evolution reaction, ACS Sustain. Chem. Eng. 6 (2018) 14257–14263, https://doi.org/10.1021/acssuschemeng.8b02893.

[48] Z. Zhang, T. Zhang, J.Y. Lee, Enhancement effect of borate doping on the oxygen evolution activity of α-nickel hydroxide, ACS Appl. Nano Mater. 1 (2018) 751–758, https://doi.org/10.1021/acsanm.7b00210.

[49] E.A. Turhan, S.V.K. Nune, E. Ülker, U. Şahin, Y. Dede, F. Karadas, Water oxidation electrocatalysis with a cobalt-borate-based hybrid system under neutral conditions, Chem. – A Eur. J. 24 (2018) 10372–10382, https://doi.org/10.1002/chem.201801412.

[50] Y. Li, W.W. Zhang, H. Li, T. Yang, S. Peng, C. Kao, W.W. Zhang, Ni-B coupled with borate-intercalated $Ni(OH)_2$ for efficient and stable electrocatalytic and photocatalytic hydrogen evolution under low alkalinity, Chem. Eng. J. 394 (2020) 124928, https://doi.org/10.1016/j.cej.2020.124928.

[51] L. Su, H. Du, C. Tang, K. Nan, J. Wu, C. Ming Li, Borate-ion intercalated Ni Fe layered double hydroxide to simultaneously boost mass transport and charge transfer for catalysis of water oxidation, J. Colloid Interface Sci. 528 (2018) 36–44, https://doi.org/10.1016/j.jcis.2018.05.075.

[52] M. Chhowalla, H.S. Shin, G. Eda, L.-J. Li, K.P. Loh, H. Zhang, The chemistry of two-dimensional layered transition metal dichalcogenide nanosheets, Nat. Chem. 5 (2013) 263–275, https://doi.org/10.1038/nchem.1589.

[53] M. Ahmadi, O. Zabihi, S. Jeon, M. Yoonessi, A. Dasari, S. Ramakrishna, M. Naebe, 2D transition metal dichalcogenide nanomaterials: advances, opportunities, and challenges in multifunctional polymer nanocomposites, J. Mater. Chem. A 8 (2020) 845–883, https://doi.org/10.1039/c9ta10130f.

[54] B. Dutta, Y. Wu, J. Chen, J. Wang, J. He, M. Sharafeldin, P. Kerns, L. Jin, A.M. Dongare, J. Rusling, S.L. Suib, Partial surface selenization of cobalt Sulfide microspheres for enhancing the hydrogen evolution reaction, ACS Catal. 9 (2019) 456–465, https://doi.org/10.1021/acscatal.8b02904.

[55] V.V. Pokropivny, V.V. Skorokhod, Classification of nanostructures by dimensionality and concept of surface forms engineering in nanomaterial science, Mater. Sci. Eng. C 27 (2007) 990–993, https://doi.org/10.1016/j.msec.2006.09.023.

[56] S. Chandrasekaran, L. Yao, L. Deng, C. Bowen, Y. Zhang, S. Chen, Z. Lin, F. Peng, P. Zhang, Recent advances in metal sulfides: from controlled fabrication to electrocatalytic, photocatalytic and photoelectrochemical water splitting and beyond, Chem. Soc. Rev. 48 (2019) 4178–4280, https://doi.org/10.1039/C8CS00664D.

[57] G. Fu, J.-M. Lee, Ternary metal sulfides for electrocatalytic energy conversion, J. Mater. Chem. A 7 (2019) 9386–9405, https://doi.org/10.1039/C9TA01438A.

[58] J. Tao, J. Chai, X. Lu, L.M. Wong, T.I. Wong, J. Pan, Q. Xiong, D. Chi, S. Wang, Growth of wafer-scale MoS_2 monolayer by magnetron sputtering, Nanoscale 7 (2015) 2497–2503, https://doi.org/10.1039/C4NR06411A.

Oxide free materials for electrochemical water splitting **Chapter | 14** **357**

[59] F. Xi, P. Bogdanoff, K. Harbauer, P. Plate, C. Höhn, J. Rappich, B. Wang, X. Han, R. van de Krol, S. Fiechter, Structural transformation identification of sputtered amorphous MoSx as an efficient hydrogen-evolving catalyst during electrochemical activation, ACS Catal. 9 (2019) 2368–2380, https://doi.org/10.1021/acscatal.8b04884.

[60] K. Ellmer, Preparation routes based on magnetron sputtering for tungsten disulfide (WS2) films for thin-film solar cells, Phys. Status Solidi 245 (2008) 1745–1760, https://doi.org/10.1002/pssb.200879545.

[61] H. Li, J. Wu, Z. Yin, H. Zhang, Preparation and applications of mechanically exfoliated single-layer and multilayer MoS2 and WSe2 nanosheets, Acc. Chem. Res. 47 (2014) 1067–1075, https://doi.org/10.1021/ar4002312.

[62] L. Ottaviano, S. Palleschi, F. Perrozzi, G. D'Olimpio, F. Priante, M. Donarelli, P. Benassi, M. Nardone, M. Gonchigsuren, M. Gombosuren, A. Lucia, G. Moccia, O.A. Cacioppo, Mechanical exfoliation and layer number identification of MoS2 revisited, 2D Mater. 4 (2017) 045013, https://doi.org/10.1088/2053-1583/aa8764.

[63] G.Z. Magda, J. Pető, G. Dobrik, C. Hwang, L.P. Biró, L. Tapasztó, Exfoliation of large-area transition metal chalcogenide single layers, Sci. Rep. 5 (2015) 14714, https://doi.org/10.1038/srep14714.

[64] Z. Gholamvand, D. McAteer, C. Backes, N. McEvoy, A. Harvey, N.C. Berner, D. Hanlon, C. Bradley, I. Godwin, A. Rovetta, M.E.G.G. Lyons, G.S. Duesberg, J.N. Coleman, Comparison of liquid exfoliated transition metal dichalcogenides reveals $MoSe_2$ to be the most effective hydrogen evolution catalyst, Nanoscale 8 (2016) 5737–5749, https://doi.org/10.1039/C5NR08553E.

[65] J. Zheng, H. Zhang, S. Dong, Y. Liu, C. Tai Nai, H. Suk Shin, H. Young Jeong, B. Liu, K. Ping Loh, High yield exfoliation of two-dimensional chalcogenides using sodium naphthalenide, Nat. Commun. 5 (2014) 2995, https://doi.org/10.1038/ncomms3995.

[66] C.J. Zhang, S.-H. Park, O. Ronan, A. Harvey, A. Seral-Ascaso, Z. Lin, N. McEvoy, C.S. Boland, N.C. Berner, G.S. Duesberg, P. Rozier, J.N. Coleman, V. Nicolosi, Enabling flexible heterostructures for Li-ion battery anodes based on nanotube and liquid-phase exfoliated 2D gallium chalcogenide nanosheet colloidal solutions, Small 13 (2017) 1701677, https://doi.org/10.1002/smll.201701677.

[67] J.-G. Song, G. Hee Ryu, G. Kim, W. Je Woo, K. Yong Ko, Y. Kim, C. Lee, I.-K. Oh, J. Park, Z. Lee, H. Kim, Catalytic chemical vapor deposition of large-area uniform two-dimensional molybdenum disulfide using sodium chloride, Nanotechnology 28 (2017) 465103, https://doi.org/10.1088/1361-6528/aa8f15.

[68] J.C. Shaw, H. Zhou, Y. Chen, N.O. Weiss, Y. Liu, Y. Huang, X. Duan, Chemical vapor deposition growth of monolayer $MoSe_2$ nanosheets, Nano Res. 7 (2014) 511–517, https://doi.org/10.1007/s12274-014-0417-z.

[69] J. Yuan, J. Wu, W.J. Hardy, P. Loya, M. Lou, Y. Yang, S. Najmaei, M. Jiang, F. Qin, K. Keyshar, H. Ji, W. Gao, J. Bao, J. Kono, D. Natelson, P.M. Ajayan, J. Lou, Facile synthesis of single crystal vanadium disulfide nanosheets by chemical vapor deposition for efficient hydrogen evolution reaction, Adv. Mater. 27 (2015) 5605–5609, https://doi.org/10.1002/adma.201502075.

[70] W. Xu, S. Li, S. Zhou, J.K. Lee, S. Wang, S.G. Sarwat, X. Wang, H. Bhaskaran, M. Pasta, J.H. Warner, Large dendritic monolayer MoS2 grown by atmospheric pressure chemical vapor deposition for electrocatalysis, ACS Appl. Mater. Interfaces 10 (2018) 4630–4639, https://doi.org/10.1021/acsami.7b14861.

[71] Q. Feng, N. Mao, J. Wu, H. Xu, C. Wang, J. Zhang, L. Xie, Growth of $MoS_{2(1-x)}Se_{2x}$ (x = 0.41–1.00) monolayer alloys with controlled morphology by physical vapor deposition, ACS Nano 9 (2015) 7450–7455, https://doi.org/10.1021/acsnano.5b02506.

[72] P.M.R. Kumar, T.T. John, C.S. Kartha, K.P. Vijayakumar, T. Abe, Y. Kashiwaba, Effects of thickness and post deposition annealing on the properties of evaporated In_2S_3 thin films, J. Mater. Sci. 41 (2006) 5519–5525, https://doi.org/10.1007/s10853-006-0307-1.

[73] B. Ma, Z. Yang, Z. Yuan, Y. Chen, Effective surface roughening of three-dimensional copper foam via sulfurization treatment as a bifunctional electrocatalyst for water splitting, Int. J. Hydrog. Energy 44 (2019) 1620–1626, https://doi.org/10.1016/j.ijhydene.2018.11.115.

[74] Z. Dai, H. Geng, J. Wang, Y. Luo, B. Li, Y. Zong, J. Yang, Y. Guo, Y. Zheng, X. Wang, Q. Yan, Hexagonal-phase cobalt monophosphosulfide for highly efficient overall water splitting, ACS Nano 11 (2017) 11031–11040, https://doi.org/10.1021/acsnano.7b05050.

[75] H. Liu, F.-X. Ma, C.-Y. Xu, L. Yang, Y. Du, P.-P. Wang, S. Yang, L. Zhen, Sulfurizing-induced hollowing of Co_9S_8 microplates with nanosheet units for highly efficient water oxidation, ACS Appl. Mater. Interfaces 9 (2017) 11634–11641, https://doi.org/10.1021/acsami.7b00899.

[76] S.V.P. Vattikuti, C. Byon, C.V. Reddy, J. Shim, B. Venkatesh, Co-precipitation synthesis and characterization of faceted MoS_2 nanorods with controllable morphologies, Appl. Phys. A Mater. Sci. Process. 119 (2015) 813–823, https://doi.org/10.1007/s00339-015-9163-7.

[77] F. Zhan, Q. Wang, Y. Li, X. Bo, Q. Wang, F. Gao, C. Zhao, Low-temperature synthesis of cuboid silver tetrathiotungstate (Ag_2WS_4) as electrocatalyst for hydrogen evolution reaction, Inorg. Chem. 57 (2018) 5791–5800, https://doi.org/10.1021/acs.inorgchem.8b00108.

[78] K. Premnath, P. Arunachalam, M.S. Amer, J. Madhavan, A.M. Al-Mayouf, Hydrothermally synthesized nickel molybdenum selenide composites as cost-effective and efficient trifunctional electrocatalysts for water splitting reactions, Int. J. Hydrog. Energy 44 (2019) 22796–22805, https://doi.org/10.1016/j.ijhydene.2019.07.034.

[79] X. Wang, B. Zheng, B. Wang, H. Wang, B. Sun, J. He, W. Zhang, Y. Chen, Hierarchical $MoSe_2$-$CoSe_2$ nanotubes anchored on graphene nanosheets: a highly efficient and stable electrocatalyst for hydrogen evolution in alkaline medium, Electrochim. Acta 299 (2019) 197–205, https://doi.org/10.1016/j.electacta.2018.12.101.

[80] Y. Guo, J. Tang, Z. Wang, Y.-M. Kang, Y. Bando, Y. Yamauchi, Elaborately assembled core-shell structured metal sulfides as a bifunctional catalyst for highly efficient electrochemical overall water splitting, Nano Energy 47 (2018) 494–502, https://doi.org/10.1016/j.nanoen.2018.03.012.

[81] B. Dong, B. Zhao, G.-Q. Han, X. Li, X. Shang, Y.-R. Liu, W.-H. Hu, Y.-M. Chai, H. Zhao, C.-G. Liu, Two-step synthesis of binary Ni–Fe sulfides supported on nickel foam as highly efficient electrocatalysts for the oxygen evolution reaction, J. Mater. Chem. A 4 (2016) 13499–13508, https://doi.org/10.1039/C6TA03177C.

[82] X. Wang, Y. Zheng, J. Yuan, J. Shen, J. Hu, A. Wang, L. Wu, L. Niu, Porous NiCo diselenide nanosheets arrayed on carbon cloth as promising advanced catalysts used in water splitting, Electrochim. Acta 225 (2017) 503–513, https://doi.org/10.1016/j.electacta.2016.12.162.

[83] R.J. Deokate, S.H. Mujawar, H.S. Chavan, S.S. Mali, C.K. Hong, H. Im, A.I. Inamdar, Chalcogenide nanocomposite electrodes grown by chemical etching of Ni-foam as electrocatalyst for efficient oxygen evolution reaction, Int. J. Energy Res. 44 (2020) 1233–1243, https://doi.org/10.1002/er.5018.

[84] F. Wang, K. Li, J. Li, L.M. Wolf, K. Liu, H. Zhang, A bifunctional electrode engineered by sulfur vacancies for efficient electrocatalysis, Nanoscale 11 (2019) 16658–16666, https://doi.org/10.1039/C9NR05484G.

[85] J. Zhang, T. Wang, D. Pohl, B. Rellinghaus, R. Dong, S. Liu, X. Zhuang, X. Feng, Interface engineering of MoS_2/Ni_3S_2 heterostructures for highly enhanced electrochemical overall-water-splitting activity, Angew. Chem. Int. Ed. 55 (2016) 6702–6707, https://doi.org/10.1002/anie.201602237.

[86] T. Yang, Y. Bao, W. Xiao, J. Zhou, J. Ding, Y.P. Feng, K.P. Loh, M. Yang, S.J. Wang, Hydrogen evolution catalyzed by a molybdenum sulfide two-dimensional structure with active basal planes, ACS Appl. Mater. Interfaces 10 (2018) 22042–22049, https://doi.org/10.1021/acsami.8b03977.

[87] D. Dinda, M.E. Ahmed, S. Mandal, B. Mondal, S.K. Saha, Amorphous molybdenum sulfide quantum dots: an efficient hydrogen evolution electrocatalyst in neutral medium, J. Mater. Chem. A 4 (2016) 15486–15493, https://doi.org/10.1039/c6ta06101j.

[88] C. Tsai, K. Chan, F. Abild-Pedersen, J.K. Nørskov, Active edge sites in $MoSe_2$ and WSe_2 catalysts for the hydrogen evolution reaction: a density functional study, Phys. Chem. Chem. Phys. 16 (2014) 13156–13164, https://doi.org/10.1039/c4cp01237b.

[89] A. Parra-Puerto, K.L. Ng, K. Fahy, A.E. Goode, M.P. Ryan, A. Kucernak, Supported transition metal phosphides: activity survey for HER, ORR, OER, and corrosion resistance in acid and alkaline electrolytes, ACS Catal. 9 (2019) 11515–11529, https://doi.org/10.1021/acscatal.9b03359.

[90] J. Mu, J. Li, E.-C. Yang, X.-J. Zhao, Three-dimensional hierarchical nickel cobalt phosphide nanoflowers as an efficient electrocatalyst for the hydrogen evolution reaction under both acidic and alkaline conditions, ACS Appl. Energy Mater. 1 (2018) 3742–3751, https://doi.org/10.1021/acsaem.8b00540.

[91] M. Sun, H. Liu, J. Qu, J. Li, Earth-rich transition metal phosphide for energy conversion and storage, Adv. Energy Mater. 6 (2016) 1600087, https://doi.org/10.1002/aenm.201600087.

[92] Y. Shi, B. Zhang, Recent advances in transition metal phosphide nanomaterials: synthesis and applications in hydrogen evolution reaction, Chem. Soc. Rev. 45 (2016) 1529–1541, https://doi.org/10.1039/c5cs00434a.

[93] Y. Xu, R. Wu, J. Zhang, Y. Shi, B. Zhang, Anion-exchange synthesis of nanoporous FeP nanosheets as electrocatalysts for hydrogen evolution reaction, Chem. Commun. 49 (2013) 6656, https://doi.org/10.1039/c3cc43107j.

[94] J. Kibsgaard, T.F. Jaramillo, Molybdenum phosphosulfide: an active, acid-stable, earth-abundant catalyst for the hydrogen evolution reaction, Angew. Chem. Int. Ed. 53 (2014) 14433–14437, https://doi.org/10.1002/anie.201408222.

[95] C. Ray, S.C. Lee, B. Jin, A. Kundu, J.H. Park, S.C. Jun, Stacked porous iron-doped nickel cobalt phosphide nanoparticle: an efficient and stable water splitting electrocatalyst, ACS Sustain. Chem. Eng. 6 (2018) 6146–6156, https://doi.org/10.1021/acssuschemeng.7b04808.

[96] Z. Pu, S. Wei, Z. Chen, S. Mu, Flexible molybdenum phosphide nanosheet array electrodes for hydrogen evolution reaction in a wide pH range, Appl. Catal. B Environ. 196 (2016) 193–198, https://doi.org/10.1016/j.apcatb.2016.05.027.

[97] X. Wang, Y.V. Kolen'ko, L. Liu, Direct solvothermal phosphorization of nickel foam to fabricate integrated Ni_2P-nanorods/Ni electrodes for efficient electrocatalytic hydrogen evolution, Chem. Commun. 51 (2015) 6738–6741, https://doi.org/10.1039/C5CC00370A.

[98] M.A.R. Anjum, J.S. Lee, Sulfur and nitrogen dual-doped molybdenum phosphide nanocrystallites as an active and stable hydrogen evolution reaction electrocatalyst in acidic and alkaline media, ACS Catal. 7 (2017) 3030–3038, https://doi.org/10.1021/acscatal.7b00555.

[99] Y. Wang, B. Kong, D. Zhao, H. Wang, C. Selomulya, Strategies for developing transition metal phosphides as heterogeneous electrocatalysts for water splitting, Nano Today 15 (2017) 26–55, https://doi.org/10.1016/j.nantod.2017.06.006.

[100] J. Joo, T. Kim, J. Lee, S. Choi, K. Lee, Morphology-controlled metal sulfides and phosphides for electrochemical water splitting, Adv. Mater. 31 (2019) 1806682, https://doi.org/10.1002/adma.201806682.

[101] S. Anantharaj, S.R. Ede, K. Sakthikumar, K. Karthick, S. Mishra, S. Kundu, Recent trends and perspectives in electrochemical water splitting with an emphasis on sulfide, selenide,

360 Oxide free nanomaterials for energy storage and conversion applications

and phosphide catalysts of Fe, Co, and Ni: a review, ACS Catal. 6 (2016) 8069–8097, https://doi.org/10.1021/acscatal.6b02479.

[102] A.R.J.J. Kucernak, V.N. Naranammalpuram Sundaram, Nickel phosphide: the effect of phosphorus content on hydrogen evolution activity and corrosion resistance in acidic medium, J. Mater. Chem. A 2 (2014) 17435–17445, https://doi.org/10.1039/C4TA03468F.

[103] Z. Wu, X.F. Lu, S. Zang, X.W. (David) Lou, Non-noble-metal-based electrocatalysts toward the oxygen evolution reaction, Adv. Funct. Mater. 30 (2020) 1910274, https://doi.org/10.1002/adfm.201910274.

[104] J. Theerthagiri, E.S.F. Cardoso, G.V. Fortunato, G.A. Casagrande, B. Senthilkumar, J. Madhavan, G. Maia, Highly electroactive Ni pyrophosphate/Pt catalyst toward hydrogen evolution reaction, ACS Appl. Mater. Interfaces 11 (2019) 4969–4982, https://doi.org/10.1021/acsami.8b18153.

[105] M.A.Z.G. Sial, H. Lin, X. Wang, Microporous 2D NiCoFe phosphate nanosheets supported on Ni foam for efficient overall water splitting in alkaline media, Nanoscale 10 (2018) 12975–12980, https://doi.org/10.1039/c8nr03350a.

[106] H. Zhou, F. Yu, J. Sun, R. He, S. Chen, C.W. Chu, Z. Ren, Highly active catalyst derived from a 3D foam of $Fe(PO_3)_2/Ni_2P$ for extremely efficient water oxidation, Proc. Natl. Acad. Sci. U. S. A. 114 (2017) 5607–5611, https://doi.org/10.1073/pnas.1701562114.

[107] R. Guo, X. Lai, J. Huang, X. Du, Y. Yan, Y. Sun, G. Zou, J. Xiong, Phosphate-based electrocatalysts for water splitting: recent progress, ChemElectroChem 5 (2018) 3822–3834, https://doi.org/10.1002/celc.201800996.

[108] Q. Li, Z. Xing, A.M. Asiri, P. Jiang, X. Sun, Cobalt phosphide nanoparticles film growth on carbon cloth: a high-performance cathode for electrochemical hydrogen evolution, Int. J. Hydrog. Energy 39 (2014) 16806–16811, https://doi.org/10.1016/j.ijhydene.2014.08.099.

[109] K. Karthick, S. Sam Sankar, S. Kundu, Chapter 16—Transition Metal–Based Nitrides for Energy Applications, Elsevier Inc., 2020, pp. 493–515, https://doi.org/10.1016/b978-0-12-819552-9.00016-6.

[110] X. Peng, C. Pi, X. Zhang, S. Li, K. Huo, P.K. Chu, Recent progress of transition metal nitrides for efficient electrocatalytic water splitting, Sustain. Energy Fuels 3 (2019) 366–381, https://doi.org/10.1039/c8se00525g.

[111] J. Theerthagiri, S.J. Lee, A.P. Murthy, J. Madhavan, M.Y. Choi, Fundamental aspects and recent advances in transition metal nitrides as electrocatalysts for hydrogen evolution reaction: a review, Curr. Opin. Solid State Mater. Sci. 24 (2020) 100805, https://doi.org/10.1016/j.cossms.2020.100805.

[112] A.K. Tareen, G.S. Priyanga, S. Behara, T. Thomas, M. Yang, Mixed ternary transition metal nitrides: a comprehensive review of synthesis, electronic structure, and properties of engineering relevance, Prog. Solid State Chem. 53 (2019) 1–26, https://doi.org/10.1016/j.progsolidstchem.2018.11.001.

[113] B. Zhang, C. Xiao, S. Xie, J. Liang, X. Chen, Y. Tang, Iron-nickel nitride nanostructures in situ grown on surface-redox-etching nickel foam: efficient and ultrasustainable electrocatalysts for overall water splitting, Chem. Mater. 28 (2016) 6934–6941, https://doi.org/10.1021/acs.chemmater.6b02610.

[114] J. Jia, M. Zhai, J. Lv, B. Zhao, H. Du, J. Zhu, Nickel molybdenum nitride nanorods grown on Ni foam as efficient and stable bifunctional electrocatalysts for overall water splitting, ACS Appl. Mater. Interfaces 10 (2018) 30400–30408, https://doi.org/10.1021/acsami.8b09854.

[115] H. Sun, Z. Yan, F. Liu, W. Xu, F. Cheng, J. Chen, Self-supported transition-metal-based electrocatalysts for hydrogen and oxygen evolution, Adv. Mater. 32 (2020) 1806326, https://doi.org/10.1002/adma.201806326.

Oxide free materials for electrochemical water splitting **Chapter | 14** **361**

[116] A. Eftekhari, Electrocatalysts for hydrogen evolution reaction, Int. J. Hydrog. Energy 42 (2017) 11053–11077, https://doi.org/10.1016/j.ijhydene.2017.02.125.

[117] Q. Gao, W. Zhang, Z. Shi, L. Yang, Y. Tang, Structural design and electronic modulation of transition-metal-carbide electrocatalysts toward efficient hydrogen evolution, Adv. Mater. 31 (2019) 1802880, https://doi.org/10.1002/adma.201802880.

[118] Y. Ma, G. Guan, X. Hao, J. Cao, A. Abudula, Molybdenum carbide as alternative catalyst for hydrogen production – a review, Renew. Sust. Energ. Rev. (2016) 1–29, https://doi.org/10.1016/j.rser.2016.11.092.

[119] Y.J. Tang, C.H. Liu, W. Huang, X.L. Wang, L.Z. Dong, S.L. Li, Y.Q. Lan, Bimetallic carbides-based nanocomposite as superior electrocatalyst for oxygen evolution reaction, ACS Appl. Mater. Interfaces 9 (2017) 16977–16985, https://doi.org/10.1021/acsami.7b01096.

[120] D.V. Esposito, J.G. Chen, Monolayer platinum supported on tungsten carbides as low-cost electrocatalysts: opportunities and limitations, Energy Environ. Sci. 4 (2011) 3900, https://doi.org/10.1039/c1ee01851e.

[121] H. Wei, Q. Xi, X. Chen, D. Guo, F. Ding, Z. Yang, S. Wang, J. Li, S. Huang, Molybdenum carbide nanoparticles coated into the graphene wrapping N-doped porous carbon micro-spheres for highly efficient electrocatalytic hydrogen evolution both in acidic and alkaline media, Adv. Sci. 5 (2018) 1700733, https://doi.org/10.1002/advs.201700733.

[122] Z.-P. Zhang, Y.-W. Zhang, Introduction of Bimetallic Nanostructures, in: Bimetallic Nanostructures, John Wiley & Sons, Ltd., Chichester, UK, 2018, pp. 1–22, https://doi.org/10.1002/9781119214618.ch1.

[123] C. Zhang, R. Hao, H. Yin, F. Liu, Y. Hou, Iron phthalocyanine and nitrogen-doped graphene composite as a novel non-precious catalyst for the oxygen reduction reaction, Nanoscale 4 (2012) 7326–7329, https://doi.org/10.1039/c2nr32612d.

[124] L.B. Venarusso, C.V. Boone, G. Maia, Carbon-supported metal nanodendrites as efficient, stable catalysts for the oxygen reduction reaction, J. Mater. Chem. A (2018) 1714–1726, https://doi.org/10.1039/c7ta08964c.

[125] K. Zhang, Q. Yue, G. Chen, Y. Zhai, L. Wang, H. Wang, J. Zhao, J. Liu, J. Jia, H. Li, Effects of acid treatment of Pt-Ni alloy nanoparticles @ graphene on the kinetics of the oxygen reduction reaction in acidic and alkaline solutions, J. Phys. Chem. C 115 (2011) 379–389, https://doi.org/10.1021/jp108305v.

[126] M. Zeng, Y. Li, Recent advances in heterogeneous electrocatalysts for the hydrogen evolution reaction, J. Mater. Chem. A 3 (2015) 14942–14962, https://doi.org/10.1039/c5ta02974k.

[127] H. Tabassum, A. Mahmood, B. Zhu, Z. Liang, R. Zhong, S. Guo, R. Zou, Recent advances in confining metal-based nanoparticles into carbon nanotubes for electrochemical energy conversion and storage devices, Energy Environ. Sci. 12 (2019) 2924–2956, https://doi.org/10.1039/c9ee00315k.

[128] C. Kim, F. Dionigi, V. Beermann, X. Wang, T. Möller, P. Strasser, Alloy nanocatalysts for the electrochemical oxygen reduction (ORR) and the direct electrochemical carbon dioxide reduction reaction (CO2RR), Adv. Mater. 31 (2019) 1–19, https://doi.org/10.1002/adma.201805617.

[129] W. Xiao, W. Lei, M. Gong, H.L. Xin, D. Wang, Recent advances of structurally ordered intermetallic nanoparticles for electrocatalysis, ACS Catal. 8 (2018) 3237–3256, https://doi.org/10.1021/acscatal.7b04420.

[130] J. Liu, H. Zhang, M. Qiu, Z. Peng, M.K.H. Leung, W.F. Lin, J. Xuan, A review of non-precious metal single atom confined nanomaterials in different structural dimensions (1D-3D) as highly active oxygen redox reaction electrocatalysts, J. Mater. Chem. A 8 (2020) 2222–2245, https://doi.org/10.1039/c9ta11852g.

362 Oxide free nanomaterials for energy storage and conversion applications

[131] C. Zhang, X. Shen, Y. Pan, Z. Peng, A review of Pt-based electrocatalysts for oxygen reduction reaction, Front. Energy. 11 (2017) 268–285, https://doi.org/10.1007/s11708-017-0466-6.

[132] G.V. Fortunato, E.S.F. Cardoso, B.K. Martini, G. Maia, Ti/Pt−Pd-based nanocomposite: effects of metal oxides on the oxygen reduction reaction, ChemElectroChem 7 (2020) 1610–1618, https://doi.org/10.1002/celc.202000268.

[133] D. Lim, E. Oh, C. Lim, S.E. Shim, S.-H. Baeck, Bimetallic NiFe alloys as highly efficient electrocatalysts for the oxygen evolution reaction, Catal. Today 352 (2020) 27–33, https://doi.org/10.1016/j.cattod.2019.09.046.

[134] A. Wang, Z. Zhao, D. Hu, J. Niu, M. Zhang, K. Yan, G. Lu, Tuning the oxygen evolution reaction on a nickel–iron alloy via active straining, Nanoscale 11 (2019) 426–430, https://doi.org/10.1039/C8NR08879A.

[135] H.-J. Qiu, G. Fang, J. Gao, Y. Wen, J. Lv, H. Li, G. Xie, X. Liu, S. Sun, Noble metal-free nanoporous high-entropy alloys as highly efficient electrocatalysts for oxygen evolution reaction, ACS Mater. Lett. 1 (2019) 526–533, https://doi.org/10.1021/acsmaterialslett.9b00414.

[136] S. Jeong, K. Hu, T. Ohto, Y. Nagata, H. Masuda, J. Fujita, Y. Ito, Effect of graphene encapsulation of NiMo alloys on oxygen evolution reaction, ACS Catal. 10 (2020) 792–799, https://doi.org/10.1021/acscatal.9b04134.

[137] Y. Shen, Y. Zhou, D. Wang, X. Wu, J. Li, J. Xi, Nickel-copper alloy encapsulated in graphitic carbon shells as electrocatalysts for hydrogen evolution reaction, Adv. Energy Mater. 8 (2018) 1701759, https://doi.org/10.1002/aenm.201701759.

[138] Y. Yang, Z. Lun, G. Xia, F. Zheng, M. He, Q. Chen, Non-precious alloy encapsulated in nitrogen-doped graphene layers derived from MOFs as an active and durable hydrogen evolution reaction catalyst, Energy Environ. Sci. 8 (2015) 3563–3571, https://doi.org/10.1039/C5EE02460A.

[139] M. Gao, J. Liang, Y. Zheng, Y. Xu, J. Jiang, Q. Gao, J. Li, S. Yu, An efficient molybdenum disulfide/cobalt diselenide hybrid catalyst for electrochemical hydrogen generation, Nat. Commun. 6 (2015) 5982, https://doi.org/10.1038/ncomms6982.

[140] C. Xu, Q. Li, J. Shen, Z. Yuan, J. Ning, Y. Zhong, Z. Zhang, Y. Hu, A facile sequential ion exchange strategy to synthesize $CoSe_2/FeSe_2$ double-shelled hollow nanocuboids for the highly active and stable oxygen evolution reaction, Nanoscale 11 (2019) 10738–10745, https://doi.org/10.1039/c9nr02599e.

[141] J.R. Li, R.J. Kuppler, H.C. Zhou, Selective gas adsorption and separation in metal-organic frameworks, Chem. Soc. Rev. 38 (2009) 1477–1504, https://doi.org/10.1039/b802426j.

[142] A. Corma, H. García, F.X. Llabrés, I. Xamena, Engineering metal organic frameworks for heterogeneous catalysis, Chem. Rev. 110 (2010) 4606–4655, https://doi.org/10.1021/cr9003924.

[143] Y. Cui, Y. Yue, G. Qian, B. Chen, Luminescent functional metal-organic frameworks, Chem. Rev. 112 (2012) 1126–1162, https://doi.org/10.1021/cr200101d.

[144] K. Lu, T. Aung, N. Guo, R. Weichselbaum, W. Lin, Nanoscale metal–organic frameworks for therapeutic, imaging, and sensing applications, Adv. Mater. 30 (2018) 1–20, https://doi.org/10.1002/adma.201707634.

[145] P. Ramaswamy, N.E. Wong, G.K.H. Shimizu, MOFs as proton conductors-challenges and opportunities, Chem. Soc. Rev. 43 (2014) 5913–5932, https://doi.org/10.1039/c4cs00093e.

[146] L.E. Kreno, K. Leong, O.K. Farha, M. Allendorf, R.P. Van Duyne, J.T. Hupp, Metal-organic framework materials as chemical sensors, Chem. Rev. 112 (2012) 1105–1125, https://doi.org/10.1021/cr200324t.

[147] B. Li, H.M. Wen, Y. Cui, W. Zhou, G. Qian, B. Chen, Emerging multifunctional metal–organic framework materials, Adv. Mater. 28 (2016) 8819–8860, https://doi.org/10.1002/adma.201601133.

Oxide free materials for electrochemical water splitting **Chapter | 14** **363**

[148] F.Y. Yi, D. Chen, M.K. Wu, L. Han, H.L. Jiang, Chemical sensors based on metal–organic frameworks, Chem. Aust. 81 (2016) 675–690, https://doi.org/10.1002/cplu.201600137.

[149] J. Ren, X. Dyosiba, N.M. Musyoka, H.W. Langmi, M. Mathe, S. Liao, Review on the current practices and efforts towards pilot-scale production of metal-organic frameworks (MOFs), Coord. Chem. Rev. 352 (2017) 187–219, https://doi.org/10.1016/j.ccr.2017.09.005.

[150] D. Li, H.-Q. Xu, L. Jiao, H.-L. Jiang, Metal-organic frameworks for catalysis: state of the art, challenges, and opportunities, EnergyChem 1 (2019) 100005, https://doi.org/10.1016/j.enchem.2019.100005.

[151] L. Chen, R. Luque, Y. Li, Controllable design of tunable nanostructures inside metal-organic frameworks, Chem. Soc. Rev. 46 (2017) 4614–4630, https://doi.org/10.1039/c6cs00537c.

[152] H. Furukawa, K.E. Cordova, M. O'Keeffe, O.M. Yaghi, The chemistry and applications of metal-organic frameworks, Science 341 (2013), https://doi.org/10.1126/science.1230444.

[153] K. Shen, X. Chen, J. Chen, Y. Li, Development of MOF-derived carbon-based nanomaterials for efficient catalysis, ACS Catal. 6 (2016) 5887–5903, https://doi.org/10.1021/acscatal.6b01222.

[154] A. Mahmood, W. Guo, H. Tabassum, R. Zou, Metal-organic framework-based nanomaterials for electrocatalysis, Adv. Energy Mater. 6 (2016), https://doi.org/10.1002/aenm.201600423.

[155] J. Liu, D.D. Zhu, C.X. Guo, A. Vasileff, S.Z. Qiao, Design strategies toward advanced mof-derived electrocatalysts for energy-conversion reactions, Adv. Energy Mater. 7 (2017) 1–26, https://doi.org/10.1002/aenm.201700518.

[156] W. Yang, X. Li, Y. Li, R. Zhu, H. Pang, Applications of metal-organic-framework-derived carbon materials, Adv. Mater. 31 (2019) 1–35, https://doi.org/10.1002/adma.201804740.

[157] Z. Liang, R. Zhao, T. Qiu, R. Zou, Q. Xu, Metal-organic framework-derived materials for electrochemical energy applications, EnergyChem 1 (2019) 100001, https://doi.org/10.1016/j.enchem.2019.100001.

[158] Y.V. Kaneti, J. Tang, R.R. Salunkhe, X. Jiang, A. Yu, K.C.W. Wu, Y. Yamauchi, Nanoarchitectured design of porous materials and nanocomposites from metal-organic frameworks, Adv. Mater. 29 (2017) 1604898, https://doi.org/10.1002/adma.201604898.

[159] S.L. Zhang, B.Y. Guan, X.W. (David) Lou, Co–Fe alloy/N-doped carbon hollow spheres derived from dual metal–organic frameworks for enhanced electrocatalytic oxygen reduction, Small 15 (2019) 1–6, https://doi.org/10.1002/smll.201805324.

[160] B. Liu, H. Shioyama, T. Akita, Q. Xu, Metal-organic framework as a template for porous carbon synthesis, J. Am. Chem. Soc. 130 (2008) 5390–5391, https://doi.org/10.1021/ja7106146.

[161] S. Sorribas, B. Zornoza, C. Téllez, J. Coronas, Ordered mesoporous silica-(ZIF-8) core-shell spheres, Chem. Commun. 48 (2012) 9388–9390, https://doi.org/10.1039/c2cc34893d.

[162] F. Zhang, Y. Wei, X. Wu, H. Jiang, W. Wang, H. Li, Hollow zeolitic imidazolate framework nanospheres as highly efficient cooperative catalysts for [3+3] cycloaddition reactions, J. Am. Chem. Soc. 136 (2014) 13963–13966, https://doi.org/10.1021/ja506372z.

[163] Y. Chen, S. Ji, Y. Wang, J. Dong, W. Chen, Z. Li, R. Shen, L. Zheng, Z. Zhuang, D. Wang, Y. Li, Isolated single iron atoms anchored on N-doped porous carbon as an efficient electrocatalyst for the oxygen reduction reaction, Angew. Chem. – Int. Ed. 56 (2017) 6937–6941, https://doi.org/10.1002/anie.201702473.

[164] B. You, N. Jiang, M. Sheng, S. Gul, J. Yano, Y. Sun, High-performance overall water splitting electrocatalysts derived from cobalt-based metal-organic frameworks, Chem. Mater. 27 (2015) 7636–7642, https://doi.org/10.1021/acs.chemmater.5b02877.

[165] L. Zhou, J. Meng, P. Li, Z. Tao, L. Mai, J. Chen, Ultrasmall cobalt nanoparticles supported on nitrogen-doped porous carbon nanowires for hydrogen evolution from ammonia borane, Mater. Horizons 4 (2017) 268–273, https://doi.org/10.1039/c6mh00534a.

364 Oxide free nanomaterials for energy storage and conversion applications

[166] Y. Lin, G. Chen, H. Wan, F. Chen, X. Liu, R. Ma, 2D free-standing nitrogen-doped Ni-Ni_3S_2 @carbon nanoplates derived from metal–organic frameworks for enhanced oxygen evolution reaction, Small 15 (2019) 1–11, https://doi.org/10.1002/smll.201900348.

[167] L. Zou, M. Kitta, J. Hong, K. Suenaga, N. Tsumori, Z. Liu, Q. Xu, Fabrication of a spherical superstructure of carbon nanorods, Adv. Mater. 31 (2019) 1–7, https://doi.org/10.1002/adma.201900440.

[168] G. Cai, W. Zhang, L. Jiao, S.H. Yu, H.L. Jiang, Template-directed growth of well-aligned MOF arrays and derived self-supporting electrodes for water splitting, Chem. 2 (2017) 791–802, https://doi.org/10.1016/j.chempr.2017.04.016.

[169] R. Wang, X.Y. Dong, J. Du, J.Y. Zhao, S.Q. Zang, MOF-derived bifunctional Cu3P nanoparticles coated by a N,P-codoped carbon shell for hydrogen evolution and oxygen reduction, Adv. Mater. 30 (2018) 1–10, https://doi.org/10.1002/adma.201703711.

[170] M.Q. Wang, C. Ye, M. Wang, T.H. Li, Y.N. Yu, S.J. Bao, Synthesis of M (Fe$_3$C, Co, Ni)-porous carbon frameworks as high-efficient ORR catalysts, Energy Storage Mater. 11 (2018) 112–117, https://doi.org/10.1016/j.ensm.2017.10.003.

[171] C. Chen, A. Wu, H. Yan, Y. Xiao, C. Tian, H. Fu, Trapping [PMo$_{12}$O$_{40}$]$_3$-clusters into pre-synthesized ZIF-67 toward Mo: XCoxC particles confined in uniform carbon polyhedrons for efficient overall water splitting, Chem. Sci. 9 (2018) 4746–4755, https://doi.org/10.1039/c8sc01454j.

[172] Z. Liang, C. Qu, D. Xia, R. Zou, Q. Xu, Atomically dispersed metal sites in MOF-based materials for electrocatalytic and photocatalytic energy conversion, Angew. Chem. – Int. Ed. 57 (2018) 9604–9633, https://doi.org/10.1002/anie.201800269.

[173] X. Han, X. Ling, Y. Wang, T. Ma, C. Zhong, W. Hu, Y. Deng, Generation of nanoparticle, atomic-cluster, and single-atom cobalt catalysts from zeolitic imidazole frameworks by spatial isolation and their use in zinc–air batteries, Angew. Chem. – Int. Ed. 58 (2019) 5359–5364, https://doi.org/10.1002/anie.201901109.

[174] A. Bavykina, N. Kolobov, I.S. Khan, J.A. Bau, A. Ramirez, J. Gascon, Metal-organic frameworks in heterogeneous catalysis: recent progress, new trends, and future perspectives, Chem. Rev. 120 (2020) 8468–8535, https://doi.org/10.1021/acs.chemrev.9b00685.

[175] C.W. Huang, V.H. Nguyen, S.R. Zhou, S.Y. Hsu, J.X. Tan, K.C.W. Wu, Metal-organic frameworks: preparation and applications in highly efficient heterogeneous photocatalysis, Sustain. Energy Fuels 4 (2020) 504–521, https://doi.org/10.1039/c9se00972h.

[176] M. Rubio-Martinez, C. Avci-Camur, A.W. Thornton, I. Imaz, D. Maspoch, M.R. Hill, New synthetic routes towards MOF production at scale, Chem. Soc. Rev. 46 (2017) 3453–3480, https://doi.org/10.1039/c7cs00109f.

[177] B.Y. Guan, X.Y. Yu, H. Bin Wu, X.W.D. Lou, Complex nanostructures from materials based on metal–organic frameworks for electrochemical energy storage and conversion, Adv. Mater. 29 (2017) 1–20, https://doi.org/10.1002/adma.201703614.

[178] A. Carné-Sánchez, I. Imaz, M. Cano-Sarabia, D. Maspoch, A spray-drying strategy for synthesis of nanoscale metal-organic frameworks and their assembly into hollow superstructures, Nat. Chem. 5 (2013) 203–211, https://doi.org/10.1038/nchem.1569.

[179] L. Garzón-Tovar, S. Rodríguez-Hermida, I. Imaz, D. Maspoch, Spray drying for making covalent chemistry: postsynthetic modification of metal-organic frameworks, J. Am. Chem. Soc. 139 (2017) 897–903, https://doi.org/10.1021/jacs.6b11240.

[180] M. Hafezi Kahnamouei, S. Shahrokhian, Mesoporous nanostructured composite derived from thermal treatment CoFe Prussian blue analogue cages and electrodeposited NiCo-S as an efficient Electrocatalyst for an oxygen evolution reaction, ACS Appl. Mater. Interfaces 12 (2020) 16250–16263, https://doi.org/10.1021/acsami.9b21403.

Oxide free materials for electrochemical water splitting **Chapter | 14** **365**

[181] X. Zhang, Y. Chen, W. Zhang, D. Yang, Coral-like hierarchical architecture self-assembled by cobalt hexacyanoferrate nanocrystals and N-doped carbon nanoplatelets as efficient electrocatalyst for oxygen evolution reaction, J. Colloid Interface Sci. 558 (2019) 190–199, https://doi.org/10.1016/j.jcis.2019.09.108.

[182] Y. Wang, S. Li, Y. Chen, X. Shi, C. Wang, L. Guo, 3D hierarchical MOF-derived CoP@N-doped carbon composite foam for efficient hydrogen evolution reaction, Appl. Surf. Sci. 505 (2020) 144503, https://doi.org/10.1016/j.apsusc.2019.144503.

[183] L. Zhang, J. Zhu, Y. Shi, Z. Wang, W. Zhang, Defect-induced nucleation and epitaxial growth of a MOF-derived hierarchical Mo2C@Co architecture for an efficient hydrogen evolution reaction, RSC Adv. 10 (2020) 13838–13847, https://doi.org/10.1039/d0ra01197e.

[184] J. Duan, S. Chen, C. Zhao, Ultrathin metal-organic framework array for efficient electrocatalytic water splitting, Nat. Commun. 8 (2017) 1–7, https://doi.org/10.1038/ncomms15341.

[185] T.T. Chen, R. Wang, L.K. Li, Z.J. Li, S.Q. Zang, MOF-derived Co_9S_8/MoS_2 embedded in tri-doped carbon hybrids for efficient electrocatalytic hydrogen evolution, J. Energy Chem. 44 (2020) 90–96, https://doi.org/10.1016/j.jechem.2019.09.018.

[186] A.S. Fajardo, P. Westerhoff, C.M. Sanchez-Sanchez, S. Garcia-Segura, Earth-abundant elements a sustainable solution for electrocatalytic reduction of nitrate, Appl. Catal. B Environ. 281 (2021) 119465, https://doi.org/10.1016/j.apcatb.2020.119465.

[187] R. Jaffe, J. Price, G. Ceder, R. Eggert, T. Graedel, K. Gschneidner, M. Hitzman, F. Houle, A. Hurd, R. Kelley, A. King, D. Milliron, B. Skinner, F. Slakey, Energy Critical Elements: Securing Materials for Emerging Technologies, 2011.

[188] European Commission, Report on Critical Raw Materials for the EU, Report of the Ad hoc Working Group on Defining Critical Raw Materials, 2014. doi:Ref. Ares(2015)1819595-29/04/2015.

[189] J.S. Qin, D.Y. Du, W. Guan, X.J. Bo, Y.F. Li, L.P. Guo, Z.M. Su, Y.Y. Wang, Y.Q. Lan, H.C. Zhou, Ultrastable polymolybdate-based metal-organic frameworks as highly active electrocatalysts for hydrogen generation from water, J. Am. Chem. Soc. 137 (2015) 7169–7177, https://doi.org/10.1021/jacs.5b02688.

[190] R. Dong, M. Pfeffermann, H. Liang, Z. Zheng, X. Zhu, J. Zhang, X. Feng, Large-area, free-standing, two-dimensional supramolecular polymer single-layer sheets for highly efficient Electrocatalytic hydrogen evolution, Angew. Chem. – Int. Ed. 54 (2015) 12058–12063, https://doi.org/10.1002/anie.201506048.

[191] S. Zhao, Y. Wang, J. Dong, C.T. He, H. Yin, P. An, K. Zhao, X. Zhang, C. Gao, L. Zhang, J. Lv, J. Wang, J. Zhang, A.M. Khattak, N.A. Khan, Z. Wei, J. Zhang, S. Liu, H. Zhao, Z. Tang, Ultrathin metal-organic framework nanosheets for electrocatalytic oxygen evolution, Nat. Energy 1 (2016) 1–10, https://doi.org/10.1038/nenergy.2016.184.

[192] M. Xie, Y. Ma, D. Lin, C. Xu, F. Xie, W. Zeng, Bimetal-organic framework MIL-53(Co-Fe): An efficient and robust electrocatalyst for the oxygen evolution reaction, Nanoscale 12 (2020) 67–71, https://doi.org/10.1039/c9nr06883j.

[193] Y. Jing, H. Yin, H. Zhang, B. Yu, MOF-derived Zn, S, and P co-doped nitrogen-enriched carbon as an efficient electrocatalyst for hydrogen evolution reaction, Int. J. Hydrog. Energy 45 (2020) 19174–19180, https://doi.org/10.1016/j.ijhydene.2020.05.035.

[194] S. Dutta, H.S. Han, M. Je, H. Choi, J. Kwon, K. Park, A. Indra, K.M. Kim, U. Paik, T. Song, Chemical and structural engineering of transition metal boride towards excellent and sustainable hydrogen evolution reaction, Nano Energy 67 (2020) 104245, https://doi.org/10.1016/j.nanoen.2019.104245.

366 Oxide free nanomaterials for energy storage and conversion applications

[195] P. Han, T. Tan, F. Wu, P. Cai, G. Cheng, W. Luo, Nickel-iron borate coated nickel-iron boride hybrid for highly stable and active oxygen evolution electrocatalysis, Chinese Chem. Lett. 31 (2020) 2469–2472, https://doi.org/10.1016/j.cclet.2020.03.009.

[196] W. Tang, X. Liu, Y. Li, Y. Pu, Y. Lu, Z. Song, Q. Wang, R. Yu, J. Shui, Boosting electrocatalytic water splitting via metal-metalloid combined modulation in quaternary Ni-Fe-P-B amorphous compound, Nano Res. 13 (2020) 447–454, https://doi.org/10.1007/s12274-020-2627-x.

[197] W. Hong, S. Sun, Y. Kong, Y. Hu, G. Chen, Ni: $xFe_{1-x}B$ nanoparticle self-modified nanosheets as efficient bifunctional electrocatalysts for water splitting: experiments and theories, J. Mater. Chem. A 8 (2020) 7360–7367, https://doi.org/10.1039/c9ta14058a.

[198] R. Fernandes, A. Chunduri, S. Gupta, R. Kadrekar, A. Arya, A. Miotello, N. Patel, Exploring the hydrogen evolution capabilities of earth-abundant ternary metal borides for neutral and alkaline water-splitting, Electrochim. Acta 354 (2020) 136738, https://doi.org/10.1016/j.electacta.2020.136738.

[199] R. Sukanya, S. Chen, Amorphous cobalt boride nanosheets anchored surface-functionalized carbon nanofiber: an bifunctional and efficient catalyst for electrochemical sensing and oxygen evolution reaction, J. Colloid Interface Sci. 580 (2020) 318–331, https://doi.org/10.1016/j.jcis.2020.07.037.

[200] X. Ma, K. Zhao, Y. Sun, Y. Wang, F. Yan, X. Zhang, Y. Chen, Direct observation of chemical origins in crystalline (Ni: $XCo_{1-x})_2B$ oxygen evolution electrocatalysts, Catal. Sci. Technol. 10 (2020) 2165–2172, https://doi.org/10.1039/d0cy00099j.

[201] H. Yuan, S. Wei, B. Tang, Z. Ma, J. Li, M. Kundu, X. Wang, Self-supported 3D ultrathin cobalt–nickel–boron nanoflakes as an efficient electrocatalyst for the oxygen evolution reaction, ChemSusChem 13 (2020) 3662–3670, https://doi.org/10.1002/cssc.202000784.

[202] M. Zulqarnain, A. Shah, M.A. Khan, F. Jan Iftikhar, J. Nisar, $FeCoSe_2$ nanoparticles embedded in g-C3N4: a highly active and stable bifunctional electrocatalyst for overall water splitting, Sci. Rep. 10 (2020) 1–9, https://doi.org/10.1038/s41598-020-63319-7.

[203] H. Zou, G. Li, L. Duan, Z. Kou, J. Wang, In situ coupled amorphous cobalt nitride with nitrogen-doped graphene aerogel as a trifunctional electrocatalyst towards Zn-air battery deriven full water splitting, Appl. Catal. B Environ. 259 (2019) 118100, https://doi.org/10.1016/j.apcatb.2019.118100.

[204] L. Han, K. Feng, Z. Chen, Self-supported cobalt nickel nitride nanowires electrode for overall electrochemical water splitting, Energy Technol. 5 (2017) 1908–1911, https://doi.org/10.1002/ente.201700108.

[205] Y. Wang, C. Xie, D. Liu, X. Huang, J. Huo, S. Wang, Nanoparticle-stacked porous nickel-iron nitride nanosheet: a highly efficient bifunctional electrocatalyst for overall water splitting, ACS Appl. Mater. Interfaces 8 (2016) 18652–18657, https://doi.org/10.1021/acsami.6b05811.

[206] Z. Chen, Y. Ha, Y. Liu, H. Wang, H. Yang, H. Xu, Y. Li, R. Wu, In situ formation of cobalt nitrides/graphitic carbon composites as efficient bifunctional electrocatalysts for overall water splitting, ACS Appl. Mater. Interfaces 10 (2018) 7134–7144, https://doi.org/10.1021/acsami.7b18858.

[207] C. Zhu, Z. Yin, W. Lai, Y. Sun, L. Liu, X. Zhang, Y. Chen, S.L. Chou, Fe-Ni-Mo nitride porous nanotubes for full water splitting and Zn-air batteries, Adv. Energy Mater. 8 (2018) 1–12, https://doi.org/10.1002/aenm.201802327.

[208] Z. Xue, J. Kang, D. Guo, C. Zhu, C. Li, X. Zhang, Y. Chen, Self-supported cobalt nitride porous nanowire arrays as bifunctional electrocatalyst for overall water splitting, Electrochim. Acta 273 (2018) 229–238, https://doi.org/10.1016/j.electacta.2018.04.056.

Oxide free materials for electrochemical water splitting **Chapter | 14** **367**

[209] J. Hou, Y. Sun, Z. Li, B. Zhang, S. Cao, Y. Wu, Z. Gao, L. Sun, Electrical behavior and electron transfer modulation of nickel–copper nanoalloys confined in nickel–copper nitrides nanowires array encapsulated in nitrogen-doped carbon framework as robust bifunctional Electrocatalyst for overall water splitting, Adv. Funct. Mater. 28 (2018) 23–25, https://doi.org/10.1002/adfm.201803278.

[210] A. Djire, X. Wang, C. Xiao, O.C. Nwamba, M.V. Mirkin, N.R. Neale, Basal plane hydrogen evolution activity from mixed metal nitride MXenes measured by scanning electrochemical microscopy, Adv. Funct. Mater. 30 (2020) 2001136, https://doi.org/10.1002/adfm.202001136.

[211] Z. Zhang, T. Zhang, J.Y. Lee, 110th anniversary: a total water splitting electrocatalyst based on borate/Fe co-doping of nickel sulfide, Ind. Eng. Chem. Res. 58 (2019) 13053–13063, https://doi.org/10.1021/acs.iecr.9b01976.

[212] P.V. Sarma, C.S. Tiwary, S. Radhakrishnan, P.M. Ajayan, M.M. Shaijumon, Oxygen incorporated WS2 nanoclusters with superior electrocatalytic properties for hydrogen evolution reaction, Nanoscale 10 (2018) 9516–9524, https://doi.org/10.1039/c8nr00253c.

[213] H. Li, A. Li, Z. Peng, X. Fu, Free-standing N-enriched C foam@WS$_2$ nanoflakes for efficient electrocatalytic hydrogen evolution, Appl. Surf. Sci. 487 (2019) 972–980, https://doi.org/10.1016/j.apsusc.2019.05.185.

[214] P.F. Liu, L. Zhang, L.R. Zheng, H.G. Yang, Surface engineering of nickel selenide for an enhanced intrinsic overall water splitting ability, Mater. Chem. Front. 2 (2018) 1725–1731, https://doi.org/10.1039/C8QM00292D.

[215] Q. Xiong, Y. Wang, P.-F. Liu, L.-R. Zheng, G. Wang, H.-G. Yang, P.-K. Wong, H. Zhang, H. Zhao, Cobalt covalent doping in MoS$_2$ to induce bifunctionality of overall water splitting, Adv. Mater. 30 (2018) 1801450, https://doi.org/10.1002/adma.201801450.

[216] Y. Hou, M. Qiu, G. Nam, M.G. Kim, T. Zhang, K. Liu, X. Zhuang, J. Cho, C. Yuan, X. Feng, Integrated hierarchical cobalt sulfide/nickel selenide hybrid nanosheets as an efficient three-dimensional electrode for electrochemical and photoelectrochemical water splitting, Nano Lett. 17 (2017) 4202–4209, https://doi.org/10.1021/acs.nanolett.7b01030.

[217] Y. Pan, Y. Liu, C. Liu, Phase- and morphology-controlled synthesis of cobalt sulfide nanocrystals and comparison of their catalytic activities for hydrogen evolution, Appl. Surf. Sci. 357 (2015) 1133–1140, https://doi.org/10.1016/j.apsusc.2015.09.125.

[218] J. Li, J. Li, X. Zhou, Z. Xia, W. Gao, Y. Ma, Y. Qu, Highly efficient and robust nickel phosphides as bifunctional electrocatalysts for overall water-splitting, ACS Appl. Mater. Interfaces 8 (2016) 10826–10834, https://doi.org/10.1021/acsami.6b00731.

[219] J. Xia, K. Dhaka, M. Volokh, G. Peng, Z. Wu, Y. Fu, M. Caspary Toroker, X. Wang, M. Shalom, Nickel phosphide decorated with trace amount of platinum as an efficient electrocatalyst for the alkaline hydrogen evolution reaction, Sustain. Energy Fuels 3 (2019) 2006–2014, https://doi.org/10.1039/C9SE00221A.

[220] J. Chang, K. Li, Z. Wu, J. Ge, C. Liu, W. Xing, Sulfur-doped nickel phosphide nanoplates arrays: a monolithic electrocatalyst for efficient hydrogen evolution reactions, ACS Appl. Mater. Interfaces 10 (2018) 26303–26311, https://doi.org/10.1021/acsami.8b08068.

[221] H.-W. Man, C.-S. Tsang, M.M.-J. Li, J. Mo, B. Huang, L.Y.S. Lee, Y. Leung, K.-Y. Wong, S.C.E. Tsang, Tailored transition metal-doped nickel phosphide nanoparticles for the electrochemical oxygen evolution reaction (OER), Chem. Commun. 54 (2018) 8630–8633, https://doi.org/10.1039/C8CC03870H.

[222] Z. Du, N. Jannatun, D. Yu, J. Ren, W. Huang, X. Lu, C 60-decorated nickel–cobalt phosphide as an efficient and robust electrocatalyst for hydrogen evolution reaction, Nanoscale 10 (2018) 23070–23079, https://doi.org/10.1039/C8NR07472K.

368 Oxide free nanomaterials for energy storage and conversion applications

[223] C.-N. Lv, L. Zhang, X.-H. Huang, Y.-X. Zhu, X. Zhang, J.-S. Hu, S.-Y. Lu, Double functionalization of N-doped carbon carved hollow nanocubes with mixed metal phosphides as efficient bifunctional catalysts for electrochemical overall water splitting, Nano Energy 65 (2019) 103995, https://doi.org/10.1016/j.nanoen.2019.103995.

[224] Y. Zhao, G. Fan, L. Yang, Y. Lin, F. Li, Assembling Ni–Co phosphides/carbon hollow nanocages and nanosheets with carbon nanotubes into a hierarchical necklace-like nanohybrid for electrocatalytic oxygen evolution reaction, Nanoscale 10 (2018) 13555–13564, https://doi.org/10.1039/C8NR04776F.

[225] Z. Zhao, D.E. Schipper, A.P. Leitner, H. Thirumalai, J.-H. Chen, L. Xie, F. Qin, M.K. Alam, L.C. Grabow, S. Chen, D. Wang, Z. Ren, Z. Wang, K.H. Whitmire, J. Bao, Bifunctional metal phosphide FeMnP films from single source metal organic chemical vapor deposition for efficient overall water splitting, Nano Energy 39 (2017) 444–453, https://doi.org/10.1016/j.nanoen.2017.07.027.

[226] M. Hou, X. Teng, J. Wang, Y. Liu, L. Guo, L. Ji, C. Cheng, Z. Chen, Multiscale porous molybdenum phosphide of honeycomb structure for highly efficient hydrogen evolution, Nanoscale 10 (2018) 14594–14599, https://doi.org/10.1039/C8NR04246B.

[227] Y. Cheng, K. Pang, X. Xu, P. Yuan, Z. Zhang, X. Wu, L. Zheng, J. Zhang, R. Song, Borate crosslinking synthesis of structure tailored carbon-based bifunctional electrocatalysts directly from guar gum hydrogels for efficient overall water splitting, Carbon N. Y. 157 (2020) 153–163, https://doi.org/10.1016/j.carbon.2019.10.024.

[228] A. Bin Yousaf, M. Imran, M. Farooq, P. Kasak, Synergistic effect of co-Ni co-bridging with MoS2 nanosheets for enhanced electrocatalytic hydrogen evolution reactions, RSC Adv. 8 (2018) 3374–3380, https://doi.org/10.1039/c7ra12692a.

[229] D. Liang, Y.-W. Zhang, P. Lu, Z.G. Yu, Strain and defect engineered monolayer Ni-MoS2 for pH-universal hydrogen evolution catalysis, Nanoscale 11 (2019) 18329–18337, https://doi.org/10.1039/C9NR06541E.

[230] P. Thangasamy, S. Oh, S. Nam, I.-K. Oh, Rose-like MoS_2 nanostructures with a large interlayer spacing of ~ 9.9 Å and exfoliated WS2 nanosheets supported on carbon nanotubes for hydrogen evolution reaction, Carbon N. Y. 158 (2020) 216–225, https://doi.org/10.1016/j.carbon.2019.12.019.

[231] E.J. Popczun, J.R. McKone, C.G. Read, A.J. Biacchi, A.M. Wiltrout, N.S. Lewis, R.E. Schaak, Nanostructured nickel phosphide as an electrocatalyst for the hydrogen evolution reaction, J. Am. Chem. Soc. 135 (2013) 9267–9270, https://doi.org/10.1021/ja403440e.

[232] L. Yan, B. Zhang, J. Zhu, S. Zhao, Y. Li, B. Zhang, J. Jiang, X. Ji, H. Zhang, P.K. Shen, Chestnut-like copper cobalt phosphide catalyst for all-pH hydrogen evolution reaction and alkaline water electrolysis, J. Mater. Chem. A 7 (2019) 14271–14279, https://doi.org/10.1039/C9TA03686E.

[233] T. Zhang, Y. Zhu, J.Y. Lee, Unconventional noble metal-free catalysts for oxygen evolution in aqueous systems, J. Mater. Chem. A 6 (2018) 8147–8158, https://doi.org/10.1039/c8ta01363b.

[234] J. Xu, J. Li, D. Xiong, B. Zhang, Y. Liu, K.-H. Wu, I. Amorim, W. Li, L. Liu, Trends in activity for the oxygen evolution reaction on transition metal (M = Fe, co, Ni) phosphide precatalysts, Chem. Sci. 9 (2018) 3470–3476, https://doi.org/10.1039/C7SC05033J.

Chapter 15

Oxides free nanomaterials for (photo)electrochemical water splitting

Lakshmana Reddy Nagappagari[a,b], Santosh S. Patil[a,b], Kiyoung Lee[a,b], and Shankar Muthukonda Venkatakrishnan[c]

[a]*School of Nano & Materials Science and Engineering, Kyungpook National University, Sangju, Gyeongbuk, South Korea,* [b]*Research Institute of Environmental Science & Technology, Kyungpook National University, Daegu, South Korea,* [c]*Nanocatalysis and Solar Fuels Research Laboratory, Department of Materials Science & Nanotechnology, Yogi Vemana University, Kadapa, Andhra Pradesh, India*

Chapter outline

1. **General introduction** — 369
2. **Metal nitrides** — 370
   2.1 General synthesis methods of metal nitrides — 372
   2.2 Mono metal nitrides for PEC water splitting — 373
   2.3 Binary metal nitrides for PEC water splitting — 374
3. **Metal phosphides** — 378
   3.1 Synthesis methods of metal phosphides — 379
   3.2 Mono metal phosphides for PEC water splitting — 383

   3.3 Binary metal phosphides for PEC water splitting — 383
4. **Metal sulfides** — 386
   4.1 Synthesis methods of metal sulfides — 386
   4.2 Mono metal sulfides for PEC water splitting — 391
   4.3 Binary metal sulfides for PEC water splitting — 391
5. **Summary and future prospects** — 393
   Acknowledgment — 396
   References — 396

1. General introduction

Owing to the rapid end of fossil fuels and environmental threats, the development of clean and renewable alternatives to fossil fuels has become an important task. Among the various energy sources, hydrogen (H_2) has attracted significant attention because of its high gravimetric energy density beyond that of known fuels, compatibility with electrochemical processes, and energy

---

Oxide Free Nanomaterials for Energy Storage and Conversion Applications.
https://doi.org/10.1016/B978-0-12-823936-0.00018-8
Copyright © 2022 Elsevier Inc. All rights reserved.

370 Oxide free nanomaterials for energy storage and conversion applications

conversion without CO_2 emission [1, 2]. The effective and economical production of high-purity H_2 gas has become the key issue. In the industrial process, H_2 is generally produced by expensive and energy-demanding steam reformation of hydrocarbons derived from fossil fuels. This process inevitably releases a lot of CO_2 and therefore, alternative green techniques to produce H_2 on a large scale are attractive for the realization of practical applications. Photoelectrochemical water splitting (PEC-WS) is an environmentally friendly and carbon-free alternative if the power can be derived from renewable energy sources [3–5]. To date, a wide variety of semiconducting electrode materials (anode, cathode) has been reported for solar PEC-WS applications, including noble metals based [6, 7], metal oxides [8–10], metal chalcogenides [11]. Pt and RuO_2 are known as benchmark materials for HER and OER, respectively since they all have excellent onset potential for water splitting [12, 13]. For example, the Pt has an onset potential very close to that for ideal HER (0 V vs. RHE), and a cell using such electrode is generally used as a standard to evaluate the performance of a new catalyst [14]. Though these materials perform a superior activity, the high cost of them severely influences the massive production, and researchers have steered their focus to the searching and development of novel nonnoble metal [15] or even nonmetal catalyst [16] for oxygen evolution and hydrogen evolution [17]. In this connection, the transition metal compounds have attracted increasing interest in electrochemistry especially electrocatalysis for solar energy production. Numerous nonnoble-metal and oxide-free materials such as nitrides, carbides, sulfides, phosphides have been exemplified as electrocatalysts for HER and OER in water splitting because of their multivalence states, special electronic structure, unique metal-like physical and chemical characteristics [18–23] (Fig. 1).

Thus, in this book chapter, a detailed discussion was evaluated on various oxide-free electrocatalysts more importantly metal nitrides (MNs), metal phosphides (MPs), and metal sulfides (MSs) for solar PEC water splitting. Materials with different combinations, structures, and other important properties will be discussed that indeed led to remarkably improved PEC performances over those of metal oxide-based materials.

2. Metal nitrides

Metal nitrides (MNs) have attracted increasing interest in the photo and electrochemistry and energy fields, for example, metal-ion batteries [24, 25], supercapacitors [26], solar cells [27], sensors [28], and photo/electrocatalysts [29–31] because of the multivalence states, special electronic structure, and physical properties [32]. On the one hand, the inclusion of nitrogen atoms modifies the nature of the d-band of the parent metal, resulting in a contraction of the metal d-band. This makes the electronic structure of MNs more similar to noble metals (e.g., Pd and Pt) [33]. On the other hand, nitrogen can nest in the

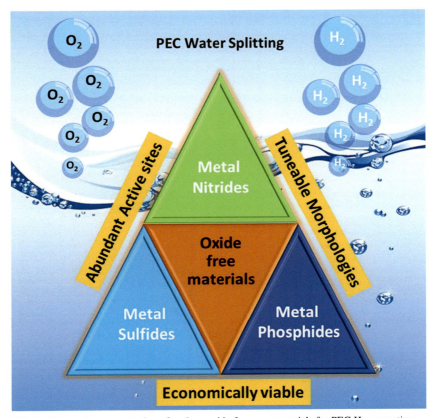

FIG. 1 Schematic representation of various oxide-free nanomaterials for PEC H$_2$ generation.

interstices of the lattices due to the small atomic radius, so that the arrangement of metal atoms always maintains close-packed or near close-packed. Besides the metallic conductivity, MNs have high corrosion resistance boding well for electrocatalytic water splitting into acidic and alkaline solutions. This endows MNs with an attractive electronic conductivity [34]. These promising features, coupled with their high resistance against corrosion, make this kind of material more reliable relative to metal or metal alloys [35]. Fig. 2 shows various applications of MNs in energy storage devices and electrochemical water splitting and the relationship between the theoretical capacities, experimental capacities, and free energy of formation G_f^o (kJ mol^{-1}) of some selected MNs.

In brief, the MNs can be classified into (i) mono metal nitrides and (ii) binary metal nitrides. Both are acquiring of their own importance that is widely used in

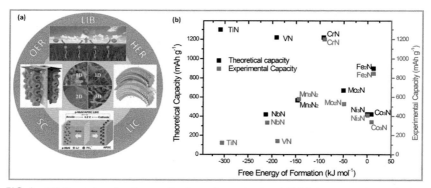

FIG. 2 (A) Examples of various applications of nanostructured TMNs in energy storage devices and electrochemical water splitting. The summarized topics include lithium-ion batteries, supercapacitors, lithium-ion capacitors, oxygen, and hydrogen evolution reactions. (B) Relationship between the theoretical capacities, experimental capacities, and G_f^o (kJ mol^{-1}) of some selected TMNs. *(Reproduced with permission from reference M.-S. Balogun, Y. Huang, W. Qiu, H. Yang, H. Ji, Y. Tong, Updates on the development of nanostructured transition metal nitrides for electrochemical energy storage and water splitting, Mater. Today 20 (2017) 425–451. https://doi.org/10.1016/j.mattod.2017.03.019. Copyright 2017, Elsevier.)*

energy applications. The important synthesis methods, various structures of binary, ternary metal nitrides used in PEC water splitting are discussed in detail as follows.

2.1 General synthesis methods of metal nitrides

The general synthesis strategies of MNs can be classified into (i) physical methods and (ii) chemical methods. The physical methods include laser ablation, arc discharge, evaporation, and pulsed-laser deposition [36–39]. The chemical synthesis is more preferred for MNs which can be divided into five categories: (1) direct nitridation of the transition metals [40], (2) nitridation of transition metal oxides [41, 42], (3) ammonolysis of metal chlorides [43], (4) solvothermal method [44] and (5) thermal decomposition of polymeric precursors [45]. MNs are generally prepared by topochemical processes by using metal oxides as precursors and N_2, NH_3, urea, and nitrogen-containing organic chemicals as nitrogen sources. Typically, annealing under an N_2 gaseous environment to produce MNs requires a high temperature (>1200°C) [46]. For example, as demonstrated by Xu et al., ultrathin metallic Ni_3N nanosheets were fabricated with a relatively low-temperature reaction where nanosheet-like NiO was submitted to heat treatment in NH_3 atmosphere [47]. According to the Atomic force microscopy (AFM), the nanosheets scanning heights ranged from 2.15 to 2.95 nm, indicating that the Ni_3N nanosheets comprise about 5–7-unit cells. Another approach is using supramolecular complexes as a nitrogen source but that needs several kinds of nitrogen-containing organic chemicals and moreover, the MNs were formed via a slow solid-state process. But compared to the aforementioned preparation protocols, producing MNs by thermal ammonia reduction (gas–solid reaction) at a moderate temperature (normally 300–800°C) by a simple operation is preferred. All these mentioned methods are basically used for the synthesis of (i) mono metal nitrides (ii) binary metal nitrides and (iii) ternary metal nitrides. Let us discuss these metal nitrides in detail for PEC water splitting.

2.2 Mono metal nitrides for PEC water splitting

The mono metal nitrides such as MoN, VN, WN, NbN, TiN, Ni_3N, etc., have been extensively studied in PEC water splitting and show promising catalytic ability. The comparison and summary of recently reported MNs-based catalysts for HER and OER reactions are presented in Table 1. Xie et al. [48] reported metallic MoN nanosheets with atomic thickness as efficient electrocatalyst for HER reactions and Mo atoms on the surface act as active sites (Fig. 3) to reduce protons to H_2. This work provides fundamental understanding and useful guidance to optimize the surface structure to achieve high HER activity.

The Tungsten-based nitrides have recognized important metal nitrides since they exhibit similar behavior as molybdenum-based nitrides and have better corrosion resistance and stability under neutral and higher pH conditions compared to tungsten carbides [57]. Chakrapani et al. [58] prepared a W_2N nanowire array by nitridation of WO_3 in an NH_3 atmosphere and evaluated the PEC performances for hydrogen generation. Significant cathodic current attributable to hydrogen evolution was observed with the W_2N nanowires as in Fig. 4A, the onset potential for the hydrogen evolution reaction on W_2N shifted to a less negative potential compared to WO_3, suggesting that W_2N is electrocatalytic toward the hydrogen evolution reaction. The prepared material showed well-defined nanowire arrays morphology as seen from the SEM images of W_2N in Fig. 4B. GaN is also one of the important metal nitrides widely used in PEC reactions. Especially the heterojunction composites consist of GaN showed improved PEC activities for OER and HER reactions. Recently Santosh and co-workers have studied the anchoring the MWCNTs to 3D honeycomb ZnO/GaN heterostructures for enhanced photoelectrochemical water oxidation and concluded that [59] the fabricating ZnO porous nanostructures on underlying GaN is beneficial for trapping light through their voids which in turn help to improve the photocurrent density and in another study, they reported the enhanced photoelectrocatalytic water oxidation using CoPi modified GaN/MWCNTs composite photoanodes [60].

Apart from this, many other MNs have been synthesized and demonstrated for efficient electrochemical OER and HER reactions. For example Ni_3N [47], TiN, different cobalt nitrides with different valence states (Co_2N, Co_3N, and Co_4N) prepared by Chen et al. [61] for OER reactions. Thus, the mono MNs have become one of the promising materials for PEC water splitting applications with improved performances.

374 Oxide free nanomaterials for energy storage and conversion applications

TABLE 1 Comparison and summary of the recently reported MNs with overpotentials, Tafel slopes for HER, OER reactions in alkaline (1 M KOH) media.

MNs	Electrolyte	Type of reaction	Overpotential (@ 10 mA cm^{-2}) mV	Tafel slope (mV dec^{-1})	Ref.
Ni$_3$N nanoparticles	1 M KOH	OER	430	64	[49]
CoN nanowires	1 M KOH	OER	290	70	[50]
Co$_4$N nanowires	1 M KOH	OER	257	44	[51]
FeNi$_3$N nanoparticles	1 M KOH	OER	280	46	[49]
TiN@Ni$_3$N NANOWIRES	1 M KOH	OER	350	94	[52]
Co$_3$FeN$_x$	1 M KOH	OER	222	46	[53]
Ni$_3$N nanoparticles	1 M KOH	HER	–	67	[49]
Ni$_3$N nanostructures	1 M KOH	HER	121	109	[54]
Ni$_3$FeN nanoparticles	1 M KOH	HER	158	42	[49]
TiN nanowires	1 M HClO$_4$	HER	–	54	[46]
γ-Mo$_2$N	0.5 M H$_2$SO$_4$	HER	381	100	[55]
CoNiNx	1 M KOH	HER	–	130	[55]
Co$_{0.6}$Mo$_{1.4}$N$_2$	0.1 M HClO$_4$	HER	190	–	[56]

2.3 Binary metal nitrides for PEC water splitting

Although single-phase metal nitrides (MNs) show promising electrical conductivity and PEC activity toward water splitting, single-phase MNs have steady M—H bonding strength which is either higher or lower than that of Pt-based catalysts resulting in inferior conversion efficiency compared to Pt-group metals [62]. More importantly, the long-term stability of some single-phase

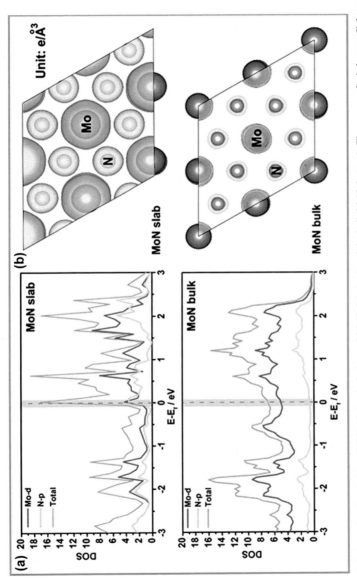

FIG. 3 (A) Calculated density of states (DOS) of the single-layered MoN slab (top) and the bulk MoN (bottom). The *orange shaded areas* (light gray in print versions) highlight the DOS contribution near the Fermi level. (B) Charge density distributions of the single-layered MoN slab (top) and the bulk MoN (bottom) near the Fermi level based on a specific equal value, respectively. *(Reproduced with permission from reference J. Xie, S. Li, X. Zhang, J. Zhang, R. Wang, H. Zhang, B. Pan, Y. Xie, Atomically-thin molybdenum nitride nanosheets with exposed active surface sites for efficient hydrogen evolution, Chem. Sci. 5 (2014) 4615–4620. https://doi.org/10.1039/C4SC02019G. Copyright 2014, Royal Society of Chemistry.)*

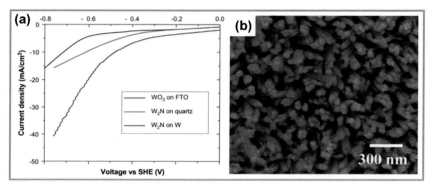

FIG. 4 (A) Polarization curves of various electrodes recorded during cathodic polarization in 0.5 M H_2SO_4 at a scan rate of 25 mV s^{-1} and (B) SEM images of W_2N/quartz nanowire arrays. *(Reprinted with permission from reference V. Chakrapani, J. Thangala, M.K. Sunkara, WO3 and W2N nanowire arrays for photoelectrochemical hydrogen production, Int. J. Hydrogen Energy 34 (2009) 9050–9059. https://doi.org/10.1016/j.ijhydene.2009.09.031. Copyright 2009, Elsevier.)*

nitrides is not satisfactory under extreme electrochemical conditions. A simple way to improve the PEC activity of single-phase MNs is to alter the electronic properties by introducing another metal to weaken or strengthen M—H bonding. The "volcano plot" of the MNs elucidates the HER theory of single transition metals [57, 63]. Navarro-Flores et al. have proposed a synergetic mechanism to account for the enhanced kinetics in hydrogen evolution observed from NiMo, NiW, and NiFe bimetallic carbide alloys [64]. The binary metal nitrides show similar electrocatalytic behavior. Chen et al. have synthesized NiMoN nanosheets on a carbon substrate (NiMoNx/C) and demonstrated high HER activity with a low overpotential and small Tafel slope [65]. The nanostructured cobalt molybdenum nitride ($Co_{0.6}Mo_{1.4}N_2$) reported by P.G. Khalifah et al. shows high HER activity and stability under acidic conditions with a small overpotential of 200 mV accomplished at a current density of 10 mA cm^{-2} for a small catalyst loading of 0.24 mg cm^{-2}, as shown in Fig. 5 [56]. Wei et al. have prepared bimetallic vanadium–molybdenum nitride (MoVN) thin films for PEC activity for HER and the materials show superior electrocatalytic activity in an alkaline electrolyte in contrast to pristine VN and Mo_2N in addition to excellent long-term durability. This study reveals the potential of bimetallic MNs is superior for HER activities [66].

Moreover, the binary MNs showed better catalytic activities toward overall PEC water splitting. Wang et al. have fabricated a 3-D porous $NiCo_2N$ on nickel foam which showed improved OER and HER performance with small overpotentials of 0.29 and 0.18 V at 10 mA cm^{-2}, respectively [67]. Similarly, the Yan and co-workers have designed and prepared porous interconnected iron-nickel nitride nanosheets on a carbon fiber cloth with a low mass loading of 0.25 mg cm^{-2} which exhibited excellent catalytic activity in OER with an overpotential of 232 mV at 20 mA cm^{-2} and HER with an overpotential of 106 mV at

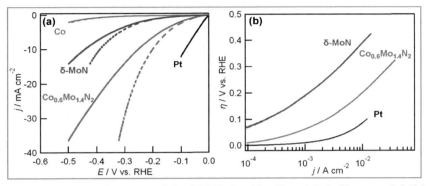

FIG. 5 (A) Polarization curves of Co, δ-MoN, Co$_{0.6}$Mo$_{1.4}$N$_2$ and Pt in H$_2$-saturated 0.1 M HClO$_4$ with (dashed line) and without (solid line) *iR* correction. (B) Corresponding Tafel plots at low overpotentials after *iR* correction. *(Reproduced with permission from reference B. Cao, G. M. Veith, J.C. Neuefeind, R.R. Adzic, P.G. Khalifah, Mixed close-packed cobalt molybdenum nitrides as non-noble metal electrocatalysts for the hydrogen evolution reaction, J. Am. Chem. Soc. 135 (2013) 19186–19192. https://doi.org/10.1021/ja4081056. Copyrighted 2013, American Chemical Society.)*

10 mA cm^{-2} [68]. The results suggesting a new route to prepare low-cost and efficient bifunctional catalysts for overall water splitting.

Most of the TMNs with excellent catalytic activities for HER is bimetallic nitrides of Mo, Ni, and Fe because they display more excellent conductivity than the single MNs. For example, Zhang et al. synthesized a novel bi-metallic NiMoN catalyst for HER reactions on self-supporting carbon cloth and compared it with Ni$_3$N, NiMo alloy, and MoN in 1 M KOH [69]. This ternary composite delivered an advanced onset with low overpotential roughly 50 mV and 109 mV at 10 mA cm^{-2}. Furthermore, the FeNi$_3$N has emerged as promising bimetallic MNs due to the low charge transfer resistance, favorable kinetics in the Ni$_3$N species, high availability, and cost-effectiveness when compared to Mo compounds [70, 71]. According to the reports of Jia et al. [72], the FeNi$_3$N nanoparticles prepared by annealing of ultrathin NiFe-LDH precursor in NH$_3$ gas environment exhibited excellent catalytic performance with considerably low overpotential for both HER and OER. Meanwhile, the performance of most reported TMN catalysts for HER is usually tested in alkaline electrolytes such as KOH and NaOH. TMNs have been reported to catalyze the evolution of hydrogen in acidic electrolytes such as HClO$_4$ and H$_2$SO$_4$ [54,65,73]. Furthermore, there are no reports yet on acidic and neutral electrolyzers, especially for TMNs. Research on acidic and neutral electrolyzers is another major challenge and prospect for electrochemical water splitting. The comparison and summary of recently reported MNs-based catalysts for HER and OER reactions are presented in Table 1. Therefore, from the above discussions, it was concluded that the binary MNs can be promising future electrocatalysts for stable PEC performance under extreme experimental conditions.

378 Oxide free nanomaterials for energy storage and conversion applications

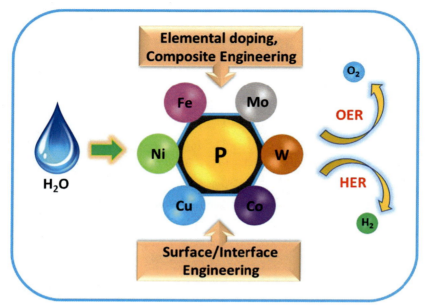

FIG. 6 Schematic representation of various engineering strategies of different metal phosphides for PEC water splitting reactions.

3. Metal phosphides

Recently, metal phosphides (MPs) have attracted great attention due to their admirable catalytic properties, long stability, and bifunctional properties as both HER and OER catalysts in alkaline solution. Like metal sulfides, the conjecture about their possible application in this field also originates from the imitation of hydrogenase. To date, six metal phosphides (MPs), FeP_x, Ni_xP_y, Co_xP_y, MoP_x, WP_x, and Cu_xP have shown promising HER or OER activities [74] as shown in Fig. 6. Their application for HER is inspired by their proven performance as excellent catalysts for hydrodesulfurization, which shares a catalytic mechanism analogous to HER. According to theoretical studies by Rodriguez's group in 2005, the first time brought forward a standpoint that Ni_2P might be the best practical catalyst for the HER, which indicated the excellent catalytic behavior of $Ni_2P(001)$ toward the HER [75]. In their study, they found that in Ni_2P the Ni concentration was diluted by the introduction of P elements, which made Ni_2P (001) behave somewhat like the hydrogenase rather than the pure metal surface. The negatively charged nonmetal atoms functioned as proton-acceptor sites and isolated metal atoms as hydride-acceptor sites. In other words, the proton-acceptor sites and hydride-acceptor sites co-exist on the surface of $Ni_2P(001)$, and this so-called "ensemble effect" would facilitate the HER. Furthermore, they found that during the HER progress hydrogen strongly bound with the Ni hollow sites, but with the assistance of P, the bonded hydrogen could be easily removed from the $Ni_2P(001)$ surface. This important theoretical prediction stimulates enormously the investigation on metal phosphides as HER catalysts.

Further, the Fe-, Co-, Ni-based phosphides also have excellent HER and OER performance. It is proposed that the real active sites in these phosphides are in situ formed surface oxy/hydroxides (M—OOH, M = Fe, Co, Ni) and phosphates during the OER process. Many research groups have studied the development of MPs catalysts for water splitting. For example, Xiao et al. and Callejas et al. reviewed various synthesis methods to prepare MPs, characterization techniques, and applications as HER electrocatalysts [76, 77]. Shi et al. [78] also reported the basic principles, evaluation approaches, the new trends, and challenges in the use of MPs as HER electrocatalysts. Kundo group [79] reviewed the Fe, Co, Ni-based phosphides, sulfides, and selenides as catalysts for water splitting. Therefore, the MPs have provided a lot of room for the advancement to effectively utilize it for PEC water splitting reactions.

3.1 Synthesis methods of metal phosphides

The MPs electrocatalysts can be prepared via various synthesis routes. Similar to the nitridation process of preparing MNs, the post phosphidation at a lower temperature, such as at about 300°C using NaH_2PO_2 enables fabrication of MPs. Moreover, wet chemical methods can be used to prepare MPs [80]. Metal phosphides in various forms can be synthesized using different methods, including the metal-organic precursor decomposition method [81], solvothermal [20], solid-phase reaction [82], electrodeposition [83], and several other synthesis methods. Let us discuss in detail important synthesis methods of metal phosphides.

3.1.1 Metal-organic precursor decomposition method

Metal-organic precursor decomposition is one of the emerging methods which is operated at moderate temperatures and minimizes the problems present in traditional phosphide synthesis methods, such as handling and storage of highly pyrophoric reagents of phosphine gas or white phosphorus. In this method, metalorganic and organic phosphorus are mixed and subjected to a certain reaction to form an organometallic complex, followed by thermal decomposition to get the final product of transition-metal phosphides. Trioctylphosphine (TOP) has been widely used as a phosphorus source for the colloidal synthesis of metal phosphides via the thermal decomposition method. Read et al., [81] transformed many metal foils (Fe, Co, Ni, Cu, and NiFe) into phosphides by reacting with TOP or tributylphosphine (TBP) as a phosphorus source. In this method, the vapor of organic phosphorus source was injected into a quartz tube by a syringe pump (Fig. 7A), where the corresponding metal foils were placed and heated at a required temperature. At the end of the reaction, high purity metal phosphides can be obtained.

380 Oxide free nanomaterials for energy storage and conversion applications

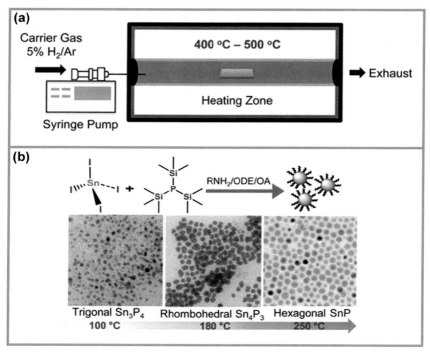

FIG. 7 (A) Schematic representation of metal-organic precursor decomposition method for preparation of metal phosphides, (B) solvothermal synthesis of various structures of SnP by varying reaction temperature and Sn/P molar ratio. *(Panel A reprinted with permission from reference C.G. Read, J.F. Callejas, C.F. Holder, R.E. Schaak, General strategy for the synthesis of transition metal phosphide films for electrocatalytic hydrogen and oxygen evolution, ACS Appl. Mater. Interfaces 8 (2016) 12798–12803. 10.1021/acsami.6b02352. Copyrighted 2016, American Chemical Society, Panel B Reproduced with permission from reference V. Tallapally, R.J.A. Esteves, L. Nahar, I.U. Arachchige, Multivariate synthesis of tin phosphide nanoparticles: temperature, time, and ligand control of size, shape, and crystal structure, Chem. Mater. 28 (2016) 5406–5414. https://doi.org/10.1021/acs.chemmater.6b01749. Copyright 2016, American Chemical Society.)*

3.1.2 Solvothermal method

Solvothermal synthesis is one of the important approaches to prepare MPs. In this method organic solvent e.g., 1-octadecene and oleylamine are used as reaction mediums. MPs with various morphologies and crystal structures can be controlled by tuning the nucleation/growth process with different temperatures, Sn/P molar ratios, or incorporation of additional coordinating solvents [84]. Arachchige and his group reported the colloidal synthesis of size-, shape-, and phase-controlled, narrowly disperse rhombohedral Sn_4P_3, hexagonal SnP, and trigonal Sn_3P_4 nanoparticles (NPs) displaying tunable morphologies and size-dependent physical properties [84]. The control over NP morphology and crystal phase was achieved by tuning the nucleation/growth temperature,

Sn/P molar ratio, and incorporation of additional coordinating solvents (alkylphosphines) as shown in Fig. 7B. Solvothermal methods using another solvent e.g., N-dimethylformamide (DMF) or ethylenediamine tetra acetic acid (EDTA) were also reported [85]. Therefore, this method has been attracting much interest to synthesize many MPs nanostructures.

3.1.3 Solid-state reaction method

The solid-state reaction involves mixing solid metal source and phosphorus source followed with thermal treatment in an inert atmosphere or under vacuum. By these methods, MPs can be achieved by phosphidation of metal oxides, hydroxides, or other solid metal compound sources by solid phosphorus species and by adjusting the temperature and the reactant molar ratio. Li et al. [82] prepared a series of Ni_xP_y catalysts by solid-phase reaction of β-Ni(OH)$_2$ and $NaH_2PO_2 \cdot H_2O$ with a molar ratio of 1:5 under Ar atmosphere. This approach can be extendable to synthesize binary MPs For example CoMoP catalysts were prepared by Wang et al. [86] in which stoichiometric mixture of cobalt nitrate, ammonium molybdate tetrahydrate, and ammonium hydrogen phosphate is calcined in air at 500°C for approximately 10 h and followed with hydrogenation treatment at 400–650°C for 2.5 h.

3.1.4 Gas-solid reaction method

In this method, the metal phosphorous precursor or red phosphorous is put on the upstream of carrier gas while the metal precursors on the downstream are added to the quartz tube in the furnace to synthesize MPs with a high P percentage, such as the synthesis of MoP_2 [87] or CoP_3 [88]. Phosphine gas (PH$_3$) gas is also used to directly transform the precursors into MPs, such as FeP at 350°C, FeP_2 at 400°C [89].

3.1.5 Electrodeposition method

By using this method, MPs can be grown on the conductive substrates like FTO/ITO or metal foils directly by reducing the metal ions and $H_2PO_2^-$. Zhu et al. electrodeposited the mesoporous CoP nanorods on Ni foam directly [90]. This method can avoid phosphidation at high temperatures, which alleviates the energy use and the environmental issues of using phosphorous sources. However, MPs obtained from this method are accompanied by a high percentage of corresponding phosphates.

3.1.6 Other miscellaneous synthetic methods

There are many other methods available to synthesize various MPs nanostructures. For example, a high-energy mechanical milling route can be used [91] to synthesize Sn_4P_3 powders. Also, the Sn_4P_3 thin films can be fabricated by pulsed laser deposition (PLD) method [92], from the mixed targets of Sn and P powders with elemental molar ratios of Sn:P = 1:3. A solid-vapor reaction

382 Oxide free nanomaterials for energy storage and conversion applications

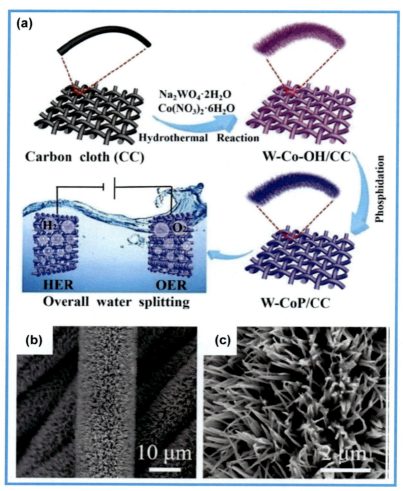

FIG. 8 (A) Illustration of the fabrication procedure of the W-CoP electrocatalyst on CC and two-electrode configuration of W-CoP/CC in overall water splitting in alkaline medium (B and C) SEM images of W-CoP/CC. *(Reproduced with permission from reference Z. Ren, X. Ren, L. Zhang, C. Fu, X. Li, Y. Zhang, B. Gao, L. Yang, P.K. Chu, K. Huo, Tungsten-doped CoP nanoneedle arrays grown on carbon cloth as efficient bifunctional electrocatalysts for overall water splitting, ChemElectroChem 6 (2019) 5229–5236. https://doi.org/10.1002/celc.201901417. Copyright 2019, Wiley VCH.)*

method was reported by Barry and Gillan [93], where they used metal chlorides to react directly with P_4 vapors to produce the corresponding MPs with different types such as white phosphorus, red phosphorus, or yellow phosphorus and resulted in different particle sizes of the products. As shown in Fig. 8, Ren et al. [94] reported the tungsten doped CoP nanoneedle arrays grown on carbon

Oxides free nanomaterials **Chapter | 15** **383**

cloth by hydrothermal assisted phosphidation process. Briefly, the precursor of tungsten-doped cobalt-carbonate-hydroxide $(W-Co(CO_3)_{0.5}(OH)\cdot0.11H_2O)$ is hydrothermally prepared on CC (denoted as W-Co-OH/CC). The W content in W-Co-OH/CC is controlled by adjusting the concentration of $Na_2WO_4\cdot2H_2O$ in the solution during the hydrothermal process. W-Co-OH/CC is then thermally phosphatized to produce W-Co-OH/CC with $NaH_2PO_2\cdot H_2O$ as the phosphorus source. A direct phosphidation of commercially available nickel foam using phosphorus vapor was also reported by Wang et al. [95], this approach can be used to synthesize self-supported MPs as HER cathodes. Therefore, there are several methods available to synthesize MPs of various structures, in which some are selectively prepared to deserve for OER and HER applications.

3.2 Mono metal phosphides for PEC water splitting

In this section, we discuss the different types of mono metal phosphides and their properties suitable for PEC water splitting reactions. Many mono MPs have been synthesized by many methods as discussed in the previous sections for HER and OER applications. Cobalt and nickel phosphides are the most common phosphides. Many interesting morphologies for Co- or Ni-based phosphides have been reported [96, 97]. Fang et al. [98] synthesized the CoP phosphide nanorods and CoP nanoparticles and compared the PEC activity. Here the CoP nanorods show a higher OER catalytic activity compared to the CoP nanoparticle. The overpotential of CoP nanoparticles was $340\,mV/10\,mA\,cm^{-2}$, while that of CoP nanorods was $320\,mV/10\,mA\,cm^{-2}$. The improved catalytic activity for CoP nanorods was ascribed to the larger specific surface area, as well as the higher surface O content of CoP nanorods relative to the nanoparticle samples. Shalom and his group explored the OER performance of Ni_5P_4 in alkaline media, which was even superior to metal Ni. They studied the OER activity of Ni_5P_4 to the fast electrochemical formation of highly OER active NiOOH [99]. Fu and co-workers have also found that both Ni_2P and Co_2P can be used as efficient cocatalysts for hydrogen evolution in a system with CdS nanorods as photosensitizers [100].

In addition, several other MPs have been successfully explored as efficient electrocatalysts for OER and HER applications of water splitting. Table 2 summarizes PEC water splitting performances on various types of MPs electrodes.

3.3 Binary metal phosphides for PEC water splitting

The binary MPs show unique physicochemical and enhanced catalytic properties compared to single MPs. For example, the Co, Ni, Fe based binary phosphides not only show higher HER performance but also act as excellent OER catalysts in alkaline solution [114–117]. Several previous works revealed that the real active sites of these phosphides were surface oxy/hydroxide and phosphate in situ formed during the OER catalysts in alkaline solution [118].

TABLE 2 Comparison of HER and OER properties for MPs with various M/P ratios, overpotential, and Tafel slope.

MPs	Electrolytes	Type of reaction	Tafel slope (mV dec^{-1})	Overpotential (@ 10mAcm^{-2})	Ref.
$Ni_{12}P_5$ nanoparticles	0.5 M H_2SO_4	HER	63	110	[101]
Ni_2P nanoparticles	0.5 M H_2SO_4	HER	62	172	[102]
Mo_3P	0.5 M H_2SO_4	HER	147	500	[103]
MoP nanoparticles	0.5 M H_2SO_4	HER	45	90	[104]
MoP_2 nanoparticles	0.5 M H_2SO_4	HER	52	121	[87]
Co_2P nanorods	0.5 M H_2SO_4	HER	51.7	167	[105]
FeP nanowires	0.5 M H_2SO_4	HER	39	96	[89]
WP nanoparticles	0.5 M H_2SO_4	HER	54	120	[80]
WP_2 nanorods	0.5 M H_2SO_4	HER	52	148	[106]
3D Cu_3P/copper foam	0.1 M KOH	HER	148	222	[107]
Co_xFeP_{1-x} nanocubes	0.5 M H_2SO_4	HER	52	72	[106]
Mo-W-P nanosheet	0.5 M H_2SO_4	HER	52	138	[108]
NiCoP Nanoparticles	0.5 M H_2SO_4	HER	50	97	[109]
Ni-Fe-P nanosheets	0.5 M H_2SO_4	HER	64.6	89	[110]
$Ni-Co_2P$/N CNTs	0.5 M H_2SO_4	HER	105	–	[111]
$Cu-Co_2P$/NCNTs	0.5 M H_2SO_4	HER	112	–	[111]
Co-MoP	0.5 M H_2SO_4	HER	50	215	[86]
Fe-MoP-0.10	0.5 M H_2SO_4	HER	49	195	[112]
$MoS_{0.60}P_{0.70}$	0.5 M H_2SO_4	HER	57	190	[113]

Oxides free nanomaterials **Chapter | 15** **385**

Due to their bi-functionality, these phosphides can even be assembled as an alkaline electrolyzer for overall water splitting. The Liang et al. used PH$_3$ plasma to successfully transform NiCo hydroxide (NiCo-OH) precursor into NiCoP at a low temperature (250°C) which avoided the phosphidation of Ni foam as the substrate [119]. Fig. 9 depicts the synthetic route, structural characterization, and electrocatalytic overall water splitting of the NiCoP nanostructure in 1 M KOH as demonstrated by Liang and his research group [119]. The Qiao group prepared for the first time Fe and O anion and cation co-doped Co$_2$P on Ni foam (CoFePO) which displayed outstanding HER and OER

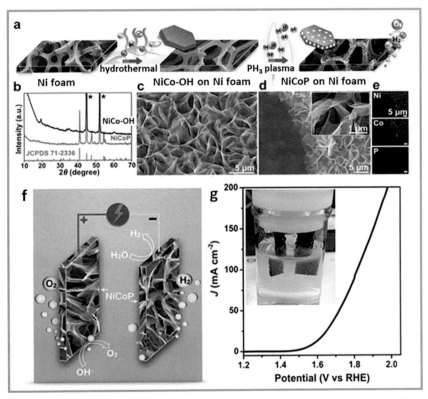

FIG. 9 (A) Schematic illustration of the synthetic route for NiCoP nanostructure on Ni foam. (B) PXRD patterns of NiCo-OH and the converted NiCoP. The asterisks mark the diffraction peaks from Ni foam. (C) SEM image of NiCo-OH. (D) SEM images and (E) corresponding EDS elemental maps of the NiCoP. (F) NiCoP/NF electrocatalyst for overall water splitting in 1 M KOH, (G) Schematic illustration of the two-electrode cell using NiCoP/NF for both anode and cathode for water splitting. *(Reproduced with permission from reference H. Liang, A.N. Gandi, D.H. Anjum, X. Wang, U. Schwingenschlögl, H.N. Alshareef, Plasma-assisted synthesis of NiCoP for efficient overall water splitting, Nano Lett. 16 (2016) 7718–7725. https://doi.org/10.1021/acs.nanolett. 6b03803. Copyright 2016, American Chemical Society.)*

performance with an overpotential of 335.5 mV lower than the state-of-the-art Pt/CIrO$_2$ counterparts (491.5 mV) for overall water splitting at a current density of 10 mAcm^{-2} [120]. Many other combinations of MPs have been studied to explore the PEC performance for water splitting (Table 2).

4. Metal sulfides

Metal sulfides are compounds in which the sulfur anion is combined with metal or semimetal cations to form M$_x$S$_y$ giving the mono-metal sulfides of MS, M$_2$S, M$_3$S$_4$, and MS$_2$ stoichiometries. Similarly, the formation of bimetal sulfides leads to A$_{1-x}$B$_x$S$_y$, where x and y are integers [121]. Transition metal sulfides have attracted significant attention, due to their narrow band gaps and valence bands (VB) being relatively negative potentials compared to oxides, therefore can be good candidates for visible-light-driven photocatalysts. The VB of metal sulfides usually consists of S 3p orbitals, this level is more negative than O 2p, which generally implied narrower band gaps [122]. In addition, most of the metal sulfides are a major group of minerals that provide the crystal chemist a rich field for investigation due to their diverse structural types. They are abundant and cheap since they usually exist in nature as minerals [123]. Transition metal sulfides have many different compositions with various lattice structures and unique electronic structures [79]. Based on those superior properties, the transition metal sulfides show promising application in many energy applications [124], such as electrochemical catalysis, photocatalysis, metal-air batteries, and other energy conversion reactions. Especially, for their abundant defect sites [122, 124], tunable electronic structure [125–127], and various morphology [128, 129], the transition metal chalcogenides exhibit boosting performance for water splitting [130]. Fig. 10 depicts the various synthesis methods and structural, morphological engineering strategies of metal sulfides for HER and OER applications.

4.1 Synthesis methods of metal sulfides

In the realm of functional MS, the convenient, low cost, and large-scale synthesis has still remained a daunting challenge. Hydro/solvothermal synthesis is a commonly used approach to prepare TMS-based materials because of its numerous advantages over other preparative methods such as simplicity, low temperature ($\sim$80–200°C), modest cost, and environmental benignity. In the hydro/solvothermal synthesis, the water/organic solvents are used as reaction media, preferentially the boiling point of these solvents is lower than the reaction temperature thereby led to building up a high pressure in closed system promoting the strong interaction among metal precursors, surfactant and sulfur (source: thioacetamide, thiocarbamide, thiourea, carbon disulfide, sodium sulfide, L-cysteine, etc.) allowing chemical reactions which are generally inaccessible in solution assisted methods.

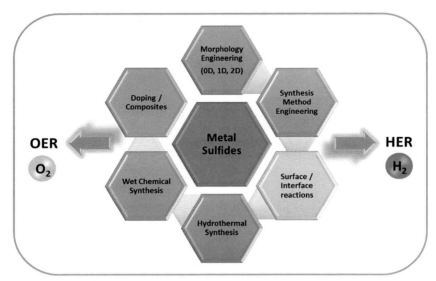

FIG. 10 Schematic representation of various engineering strategies of metal sulfides for PEC water splitting reactions.

The preparative parameters such as concentration of precursors, nature of the complexing agent, pH value, temperature, etc. are easily adjustable to control the morphology and compositions in hydro/solvothermal synthesis. For example, Sun et al. [131] prepared morphology tunable bismuth sulfide (Bi_2S_3) nanostructures in H_2WO_4 aqueous solution via surfactant-free green hydrothermal approach. In a typical procedure, Na_2WO_4 (with different molar concentrations) was dissolved in distilled water followed by adjusting to pH 2 with concentric nitric acid to afford H_2WO_4 aqueous solution, that could provide suitable acidic condition and suppresses the hydrolysis of Bi^{3+} ions to tune the growth dynamics. Bismuth nitrate and sodium sulfide as Bi and S sources were gradually added into a prepared H_2WO_4 aqueous solution and transferred Teflon-lined autoclave followed by heat treatment. As seen in Fig. 11 (ii), by simply varying the concentration of H_2WO_4 solution (0–0.1 M), Bi_2S_3 morphology can be modulated from a nanoparticle, nanorod to the nanotube, presumably when higher concentrations of H_2WO_4 solution was used because of strong interlayer interaction and preferential growth in [001] direction.

Chandrasekaran et al. [124] reviewed number of MS with hierarchical dendrites PbS, CdS, and microsphere like ZnS, CoS, Cu_2S nanostructures through hydrothermal method using metal acetate and ionic liquid 1-butyl-3-methylimidazole thiocyanate ([BMIM][SCN]) acting as both sulfur source and capping ligand. Wang et al. [132] have synthesized $CdIn_2S_4$ nanostructures by facile hydrothermal method wherein cadmium nitrate, indium nitrate, and thioacetamide were dissolved into distilled water and transferred to Teflon

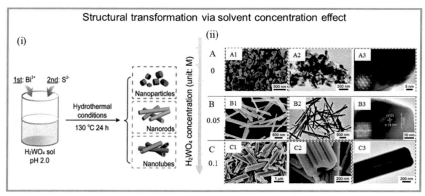

FIG. 11 (i) Schematic of synthesizing Bi_2S_3 nanostructures. (ii) Morphology evolution of Bi_2S_3 nanostructures synthesized in aqueous solution with different concentrations of H_2WO_4: (A) without H_2WO_4, (B) 0.05 M, (C) 0.1 M. SEM images are summarized in the left dashed-line (*blue*; dark gray in print versions) area, TEM images in the right dashed-line (*magenta*; light gray in print versions) area. Inset figures in (A3), (B3), (C3): FFT electron diffraction patterns. *(Reproduced with permission from reference B. Sun, T. Feng, J. Dong, X. Li, X. Liu, J. Wu, S. Ai, Green synthesis of bismuth sulfide nanostructures with tunable morphologies and robust photoelectrochemical performance, CrstEngComm 21 (2019) 1474–1481. https://doi.org/10.1039/C8CE02089B. Copyright 2018, Royal Society of Chemistry.)*

autoclave for heat treatment. Before this, the conducting FTO glass substrate was placed vertically aligned to the wall of the Teflon reactor enabling the growth of $CdIn_2S_4$ thin film photoelectrodes. It was extensively investigated (Fig. 12) theoretically and experimentally that high-temperature reductive annealing under H_2/Ar gaseous environment introduces sulfur vacancies leading to the transformation of surface-active sites beneficial for photoelectrochemical water splitting. Yanming Fu et al. [133]. have prepared CdS/SnS_x 1D/2D heterostructure photoanodes by a two-step hydrothermal and solvothermal process in which thiourea and thioacetamide were used as a sulfur source, respectively. First, the 1D CdS nanorods were grown on FTO by hydrothermal method following synthesis of three different crystal phase 2D SnS_x nanosheets such as hexagonal, orthorhombic, and combined phase of orthorhombic and zinc blende was achieved by simply varying precursor concentrations and solvothermal reaction time. These important findings purports an effective unique approach to a phase modulated band structure engineering which brings forth the twin benefits of improving the photoelectrochemical properties and restrains the photo corrosion of CdS owing to the heterojunction effect.

Moreover, MoS_2/montmorillonite (MMT) composite nanosheets have been prepared by Peng et al. [134] by hydrothermal method. Initially, a natural clay mineral (MMT) sheet was well dispersed in precursor solutions of sodium molybdate and thiourea. The thiourea decomposed and adsorbed onto the MMT surface through amidogen. During hydrothermal treatment, in-situ

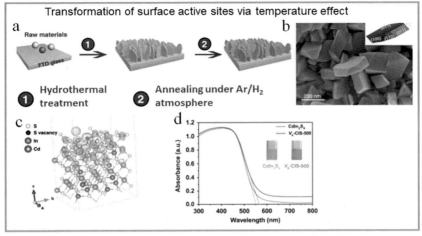

FIG. 12 (A) The schematic diagram for the synthetic procedure of the CdIn$_2$S$_4$ nanocrystals with sulfur vacancies. (B) The SEM image of Vs-CIS-500. Inset is the reconstructed shape based on the BFDH theory. (C) Different charge density images for the CdIn$_2$S$_4$ (011) with surface sulfur vacancies (isosurface set at 0.04e Bohr^{-3}); *yellow* (light gray in print versions) and *cyan* (dark gray in print versions) regions represent electron accumulation and depletion, respectively. (D) UV-vis diffuse reflectance spectra. *(Reproduced with permission from reference H. Wang, Y. Xia, H. Li, X. Wang, Y. Yu, X. Jiao, D. Chen, Highly active deficient ternary sulfide photoanode for photoelectrochemical water splitting, Nat. Commun. 11 (2020) 3078. https://doi.org/10.1038/s41467-020-16800-w. Copyright 2020, Nature.)*

nucleation, and growth of MoS$_2$ nanosheets take place where MMT acting as a template. The prepared MoS$_2$/MMT composite nanosheets exhibit a high BET surface area of ~72.10 m^2/g which is advantageous for photocatalytic and photoelectrochemical applications.

Along the line, direct sulphidation approach have been extensively used to transform the metal oxide materials to corresponding metal sulfides. It can be executed by introducing the sulfide ions in the lattice of host materials using H$_2$S, a mixture of H$_2$S and hydrogen (10%–11%), sodium sulfide, or sulfur powder by inducing high-temperature reactions under a reductive gaseous environment. For example, Su et al. [135] have synthesized yolk-shell CdS microcubes using the reaction between Cd-Fe Prussian blue (Cd-Fe-PBA) and Na$_2$S via microwave-assisted sulfidation at 150°C. The time-dependent sulfidation reaction (0.5–2 h) enabled morphology evolution from solid cube to yolk-shell CdS microcubes (Fig. 13) exhibiting bandgap energy of 2.24 eV and a large specific surface area of ~94.0 m^2/g beneficial for enhancing catalytic and photoelectrochemical properties. In addition to these strategies, numerous synthesis approaches such as top-down (sputtering [136], electrospinning [137], lithography [138], ball milling [139] and exfoliation [140], etc.) and bottom-up (chemical vapor deposition [141], atomic layer deposition [142], spray pyrolysis [143], combustion, thermal decomposition [144] and coprecipitation [144]) have been

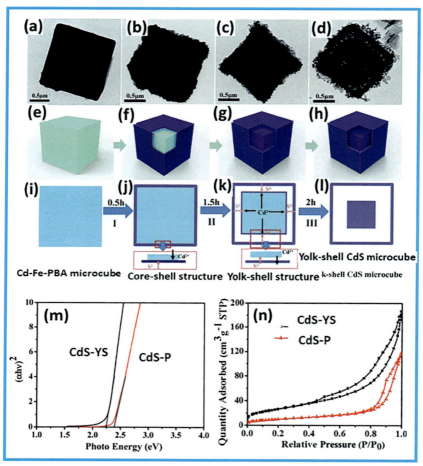

FIG. 13 FE-TEM images of the yolk-shell CdS microcubes products after reactions for (A) 0 h, (B) 0.5 h, (C) 1.5 h and (D) 2 h, and (E–L) related schematic illustration of the formation process of the yolk-shell CdS microcubes, (M, N) estimated band gap values and N_2 adsorption-desorption isotherms of yolk-shell CdS microcubes and CdS nanoparticle. *(Reproduced with permission from reference Y. Su, D. Ao, H. Liu, Y. Wang, MOF-derived yolk–shell CdS microcubes with enhanced visible-light photocatalytic activity and stability for hydrogen evolution, J. Mater. Chem. A 5 (2017) 8680–8689. https://doi.org/10.1039/C7TA00855D. Copyright 2017, Royal Society of Chemistry.)*

explored to prepare high-quality TMS nanostructures including mono-metal sulfides, bi-metal sulfide nanocomposites, and heterostructures. From the method's perceptive viewpoint, template-based controllable synthesis of TMS has proven to be a promising approach to prepare 2D MoS_2, WS_2 layered nanostructures and its heterostructures and/or 3D nanostructures with complex unique morphologies such as hollow spheres, nano-frames, cubes, yolk-shell structures, etc.

4.2 Mono metal sulfides for PEC water splitting

Recent advances in material research have identified that mono metal sulfides are one of the important classes of photoelectrode materials well suitable for PEC reactions. Particularly ZnS [145], CdS [146], Bi_2S_3 [147], PbS [148], SnS [149], Sb_2S [150], MoS_2 [151, 152], WS_2 [153], CoS [154], FeS [155] and NiS [156] are recognized as the best and low-cost alternatives to Pt for the generation of H_2 from water by electrochemical reactions. In this section, we discuss the important properties of these electrocatalysts that helped for improved PEC performance. Due to the quantum confinement effect, nanosized CdS and ZnS were chosen as a prototype system for several PEC systems. For example, CdS-QDs sensitized TiO_2 nanotube array photoelectrode attained a PEC efficiency of $\sim 4.15\%$ [157]. Similarly, in our previous studies, we have reported high quantum efficiency of 8.78% for CdS/ZnS core/shell nanostructured photocatalysts for solar hydrogen generation. The morphologies and different lattice fringes of CdS/ZnS confirm heterojunction formation as shown in Fig. 14. Moreover, the optimized catalysts showed excellent stability under solar and visible light [145]. Another important mono MS is Ag_2S, since it has a large absorption coefficient of $(Eg = 0.9\text{--}1.05\,eV)$ that helps to improve the optical absorption of $CdS/TiO_2/ITO$ photoelectrode, and also efficiently reduce charge recombination between the TiO_2 and electrolyte, which could increase the overall efficiency of CdS-sensitized TiO_2/ITO photoelectrode [158].

The MoS_2 is another important MS, extensively investigated owing to its layered structure, excellent stability, favorable electronic arrangement, and light absorption properties. For example, the 2D nanosheets of MoS_2 with a direct bandgap of $\sim 1.96\,eV$ due to a quantum confinement effect provide MoS_2 nanosheets with suitable band positions for visible-light absorption [159]. The MoS_2 layered silicon photoelectrode showed excellent photocurrent for 100 h of continuous operation without loss of photocurrent performance under $0.5\,M\ H_2SO_4$ environments [160]. FeS_2 (Pyrite) is another important economically viable MS electrode widely used as a potential photoanode for PEC water splitting with visible light absorption. Barawi et al. [161] fabricated n-type FeS_2 on titanium substrates and reported the flat band potential of $-0.75 \pm 0.05\,V$ vs. Ag/AgCl, which resulted in high efficiency of 8% for H_2 generation. Therefore, mono metal sulfides (MSs) have become an important class of photoelectrode materials for water splitting applications.

4.3 Binary metal sulfides for PEC water splitting

In addition to mono metal sulfides, the binary metal sulfides have been extensively investigated for PEC applications, due to their high stability and high performance compared to mono MSs. Various combinations of bimetal sulfides such as $ZnIn_2S_4$ [162], $CdIn_2S_4$ [163], $Fe_{1-x}Co_xS_2$ [164], Ag_3CuS_2 [165], and $CuInS_2$ [166], which absorb visible light have reported as PEC catalysts,

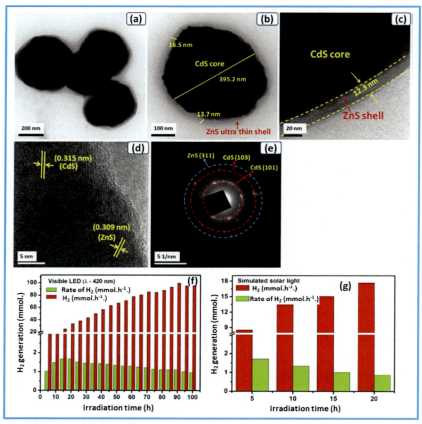

FIG. 14 (A) TEM images of CdS/ZnS core/shell nanoparticles, (B and C), HR-TEM conforming CdS core and ultra-thin shell of ZnS, (D) Lattice fringes confirming CdS core and ZnS shell, (E) SAED pattern of CdS/ZnS core/shell catalyst and (F and G) photocatalytic H_2 generation stability under visible LED and simulated solar light respectively. *(Reproduced with permission from reference N. Lakshmana Reddy, V.N. Rao, M.M. Kumari, M. Sathish, S. Muthukonda Venkatakrishnan, N.L. Reddy, V.N. Rao, M.M. Kumari, M. Sathish, S. Muthukonda Venkatakrishnan, S.M. Venkatakrishnan, S. Muthukonda Venkatakrishnan, Development of high quantum efficiency CdS/ZnS core/shell structured photocatalyst for the enhanced solar hydrogen evolution, Int. J. Hydrogen Energy 43 (2018) 22315–22328. https://doi.org/10.1016/j.ijhydene.2018.10.054. Copyright 2018, Elsevier.)*

and even can act as sensitizers that are capable of transferring electrons to the wide bandgap semiconductors such as ZnO, SnO_2 or TiO_2, and thereby increase PEC water splitting performance. Similarly, the $CuInS_2$ thin film was formed between the TiO_2 and CdS photoelectrodes to suppress charge recombination in $CuInS_2$-sensitized TiO_2 photoelectrodes [167]. The $ZnIn_2S_4$ is one of the widely studied bimetal sulfide semiconductor photocatalysts that has an energy

Oxides free nanomaterials **Chapter | 15** **393**

gap consistent with the absorption of visible-light with chemical stability. In addition, $ZnIn_2S_4$ is an eco-friendly photocatalyst for photocatalytic H_2 generation and thus several morphologies such as nanoparticles, microspheres, nanoribbons, and nanotubes have been fabricated and used for photocatalysis [168]. Further, the valence band hole (h^+_{VB}) of $ZnIn_2S_4$ photocatalyst powder oxidizes the $H\bar{S}$ ion to a disulfide (S_2^{2-}) ion, releasing a proton from the $H\bar{S}$ ion, where the electron of the conduction band electron (e^-_{CB}) of the photocatalyst reduces H^+ into H_2 molecules. An HER rate of $5287\,\mu mol\,h^{-1}$ was achieved under visible light. The highest quantum yield of 18.4% at 420 nm wavelength with a hydrogen production rate of $112.45\,mmol\,h^{-1}$ was achieved for the $ZnIn_2S_4$ photocatalyst prepared by the 9.6 mol CTAB-assisted hydrothermal method [169]. In another study, the highly crystalline $CdIn_2S_4$ photocatalyst with a marigold-like morphology and a diameter of 25 nm was prepared by Kale et al. [170] which showed a quantum yield of 16.8% at 500 nm light irradiation for photocatalytic HER H_2 production.

Zhao et al. [171] reported cubic zinc blende $Zn_{0.9}Cd_{0.1}S$ via a hydrothermal technique using L-cystine. Interestingly it performed photocatalytic H_2 activity of $\sim 4400\,\mu mol\,h^{-1}$, without a co-catalyst, which was higher than CdS. In addition, the other bimetal sulfides have been studied by many researchers. Recently, Takayama and co-workers prepared a series of sulvanite structured Cu_3VS_4, Cu_3NbS_4, and Cu_3TaS_4 photocatalysts by a solid-state reaction method [172]. The sulvanite structured Cu_3MS_4 (M = V, Ta, and Nb) contained edge and corner assimilated MS_4 and CuS_4 tetrahedral. Particularly, the Cu_3VS_4 photocatalyst reacted to near-IR light, up to 800 nm, and performed excellent photocatalytic HER. Xu et al. [173] reported the room-temperature synthesis of aqueous $CuInS_2$ (CIS) and Cu_2SnS_3 (CTS) nanocrystals supported by ZnO for efficient photoelectrochemical water splitting as shown in Fig. 15. Here the photocurrent density of ZnO photoanode increases from $0.46\,mA\,cm^{-2}$ to $5.98\,mA\,cm^{-2}$ for the ZnO/CIS photoanode and $4.85\,mA\,cm^{-2}$ for the ZnO/CTS photoanode. From the above discussion, it was concluded that the binary metal sulfide photoelectrodes and heterostructures are promising over mono metal sulfides toward PEC-WS.

5. Summary and future prospects

1. In summary, this book chapter highlighted the synthetic techniques and importance of oxide-free nanostructured materials, especially for PEC water splitting. The most important three oxide-free materials viz., metal nitrides, metal phosphides, and metal sulfides have been discussed. Mono and binary metals -nitrides, -phosphides, -sulfides nanostructures have been discussed with emphasis on photoelectrodes physicochemical properties in correlation with PEC-WS performances.

2. First the metal nitrides nanostructures and their important characteristics suitable for PEC water splitting are discussed. For example, the

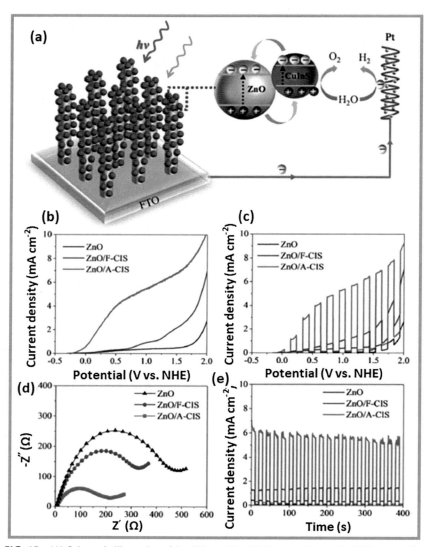

FIG. 15 (A) Schematic illustration of the CIS sensitized ZnO nanorod array for PEC water splitting; (B) LSV curves, (C) chopped LSV curves, (D) Nyquist plots, and (E) time-dependent photo response at a bias of 1.23 V vs. NHE of the pristine and optimized photoanodes. *(Reproduced with permission from reference J. Xu, W. Sun, Z. Hu, Y. You, Y. Lu, R. Zhou, J. Zhang, Green and room-temperature synthesis of aqueous $CuInS_2$ and Cu_2SnS_3 nanocrystals for efficient photoelectrochemical water splitting, Mater. Today Energy 10 (2018) 200–207. https://doi.org/10.1016/j.mtener.2018.09.010. Copyright 2018, Elsevier.)*

multivalence states, special electronic structure, and physical properties of metal nitrides are discussed. Then other diverged properties, like high corrosion resistance boding for electrocatalytic water splitting into acidic and alkaline solutions, the inclusion of nitrogen atoms that modifies the nature of the d-band of the parent metal, resulting in a contraction of the metal d-band, besides the metallic conductivity. A discussion on mono and binary metal nitrides, especially the synthesis strategies of metal nitrides like (i) physical methods and (ii) chemical methods. The chemical synthesis is more preferred for MNs which can be divided into five categories: (1) direct nitridation of the transition metals, (2) nitridation of transition metal oxides, (3) ammonolysis of metal chlorides, (4) solvothermal method, and (5) thermal decomposition of polymeric precursors. Later the physical and chemical properties of various types of metal nitrides were explained and summarized for photocurrent densities, Tafel slopes, etc.

3. Various types of metal phosphides have been discussed on account of brief introduction of key features and different combinations of metals. The admirable catalytic properties, long stability, and bifunctional properties as both HER and OER catalysts in alkaline solution have been discussed. Important synthetic methodologies for metal phosphides preparation have been discussed viz., (a) metal-organic precursor decomposition method, (b) solvothermal method, (c) solid-state reaction method, (d) gas-solid reaction method, (e) electrodeposition method, and other synthetic methods have explored as efficient PEC catalysts for HER and OER reactions. At last, the potential utility of mono metal and binary metal phosphides as efficient OER as well as HER catalysts have been reviewed and summarized.

4. At last important physicochemical and catalytic properties of metal sulfides have been discussed on account of their narrow bandgaps and negative valence bands (VB) beneficial for visible-light-driven photoelectrocatalysis. The unique properties of metal sulfides have been highlighted such as the VB of metal sulfides usually consists of S 3p orbitals, this level is more negative than O 2p, which generally implied narrower band gaps. Depicts the various synthesis methods and structural, morphological engineering strategies of metal sulfides for HER and OER applications. Recent advances in material research have addressed that mono metal sulfides are one of the important classes of electrode materials well suitable for PEC reactions. Apart from mono metal sulfides, the binary metal sulfides have attracted great attention for PEC applications, due to their high stability and high performance compared to mono metal sulfides. The important synthesis methods of metals sulfides (i) solvothermal, (ii) hydrothermal, and (iii) sol-gel methods have been discussed well. Hydrothermal/solvothermal strategies are found to be best choice to prepare mono and bi-metal sulfide and composite nanostructures owing to low cost, simplicity, and reaction controllability over morphology and compositions. Thus, in this book chapter, we discussed the important oxide-free

396 Oxide free nanomaterials for energy storage and conversion applications

nanomaterials as efficient electrocatalysts for PEC water splitting. Thus present book chapter obviously provides important information for a wide range of readerships in material science especially in the fields of renewable energy production and conversion systems.

Acknowledgment

Authors acknowledge the National Research Foundation of Korea Grant funded by the Korean Government (NRF-2019R111A3A01041454).

References

[1] C.G. Morales-Guio, L.-A. Stern, X. Hu, Nanostructured hydrotreating catalysts for electrochemical hydrogen evolution, Chem. Soc. Rev. 43 (2014) 6555–6569, https://doi.org/10.1039/C3CS60468C.

[2] A. Eftekhari, V.J. Babu, S. Ramakrishna, Photoelectrode nanomaterials for photoelectrochemical water splitting, Int. J. Hydrogen Energy 42 (2017) 11078–11109, https://doi.org/10.1016/j.ijhydene.2017.03.029.

[3] Z. Liu, X. Zhang, P. Li, X. Gao, Recent development of efficient A-D-A type fused-ring electron acceptors for organic solar, Sol. Energy 174 (2018) 171–188, https://doi.org/10.1016/j.solener.2018.09.008.

[4] N. Lakshmana Reddy, K.K. Cheralathan, V. Durga Kumari, B. Neppolian, S. Muthukonda Venkatakrishnan, Photocatalytic reforming of biomass derived crude glycerol in water: a sustainable approach for improved hydrogen generation using $Ni(OH)_2$ decorated TiO_2 nanotubes under solar light irradiation, ACS Sustain. Chem. Eng. 6 (2018) 3754–3764, https://doi.org/10.1021/acssuschemeng.7b04118.

[5] Y. Ma, X.L. Wang, Y.S. Jia, X.B. Chen, H.X. Han, C. Li, Titanium dioxide-based nanomaterials for photocatalytic fuel generations, Chem. Rev. 114 (2014) 9987–10043.

[6] A. Galińska, J. Walendziewski, Photocatalytic water splitting over $Pt–TiO_2$ in the presence of sacrificial reagents, Energy Fuel 19 (2005) 1143–1147, https://doi.org/10.1021/ef0400619.

[7] Z. Lian, W. Wang, S. Xiao, X. Li, Y. Cui, D. Zhang, G. Li, H. Li, Plasmonic silver quantum dots coupled with hierarchical TiO_2 nanotube arrays photoelectrodes for efficient visible-light photoelectrocatalytic hydrogen evolution, Sci. Rep. 5 (2015) 10461, https://doi.org/10.1038/srep10461.

[8] P.S. Shinde, A. Annamalai, J.H. Kim, S.H. Choi, J.S. Lee, J.S. Jang, Photoelectrochemical, impedance and optical data for self Sn-diffusion doped Fe_2O_3 photoanodes fabricated at high temperature by one and two-step annealing methods, Data Brief 5 (2015) 796–804, https://doi.org/10.1016/j.dib.2015.10.031.

[9] K. Sivula, F. Le Formal, M. Grätzel, Solar water splitting: progress using hematite (α-Fe_2O_3) photoelectrodes, ChemSusChem 4 (2011) 432–449, https://doi.org/10.1002/cssc.201000416.

[10] W. Yang, R.R. Prabhakar, J. Tan, S.D. Tilley, J. Moon, Strategies for enhancing the photocurrent, photovoltage, and stability of photoelectrodes for photoelectrochemical water splitting, Chem. Soc. Rev. 48 (2019) 4979–5015, https://doi.org/10.1039/C8CS00997J.

[11] S.V. Kershaw, A.S. Susha, A.L. Rogach, Narrow bandgap colloidal metal chalcogenide quantum dots: synthetic methods, heterostructures, assemblies, electronic and infrared optical properties, Chem. Soc. Rev. 42 (2013) 3033–3087, https://doi.org/10.1039/C2CS35331H.

Oxides free nanomaterials **Chapter | 15** **397**

[12] Y. Xu, M. Kraft, R. Xu, Metal-free carbonaceous electrocatalysts and photocatalysts for water splitting, Chem. Soc. Rev. 45 (2016) 3039–3052, https://doi.org/10.1039/C5CS00729A.

[13] A.-M. Alexander, J.S.J. Hargreaves, Alternative catalytic materials: carbides, nitrides, phosphides and amorphous boron alloys, Chem. Soc. Rev. 39 (2010) 4388–4401, https://doi.org/10.1039/B916787K.

[14] S. Bai, C. Wang, M. Deng, M. Gong, Y. Bai, J. Jiang, Y. Xiong, Surface polarization matters: enhancing the hydrogen-evolution reaction by shrinking Pt shells in Pt–Pd–graphene stack structures, Angew. Chem. Int. Ed. 53 (2014) 12120–12124, https://doi.org/10.1002/anie.201406468.

[15] X.-F. Lu, P.-Q. Liao, J.-W. Wang, J.-X. Wu, X.-W. Chen, C.-T. He, J.-P. Zhang, G.-R. Li, X.-M. Chen, An alkaline-stable, metal hydroxide mimicking metal–organic framework for efficient electrocatalytic oxygen evolution, J. Am. Chem. Soc. 138 (2016) 8336–8339, https://doi.org/10.1021/jacs.6b03125.

[16] S. Chen, J. Duan, M. Jaroniec, S.-Z. Qiao, Nitrogen and oxygen dual-doped carbon hydrogel film as a substrate-free electrode for highly efficient oxygen evolution reaction, Adv. Mater. 26 (2014) 2925–2930, https://doi.org/10.1002/adma.201305608.

[17] Y. Jiao, Y. Zheng, M. Jaroniec, S.Z. Qiao, Design of electrocatalysts for oxygen- and hydrogen-involving energy conversion reactions, Chem. Soc. Rev. 44 (2015) 2060–2086, https://doi.org/10.1039/C4CS00470A.

[18] X. Zou, Y. Zhang, Noble metal-free hydrogen evolution catalysts for water splitting, Chem. Soc. Rev. 44 (2015) 5148–5180, https://doi.org/10.1039/C4CS00448E.

[19] P.D. Tran, M. Nguyen, S.S. Pramana, A. Bhattacharjee, S.Y. Chiam, J. Fize, M.J. Field, V. Artero, L.H. Wong, J. Loo, J. Barber, Copper molybdenum sulfide: a new efficient electrocatalyst for hydrogen production from water, Energ. Environ. Sci. 5 (2012) 8912–8916, https://doi.org/10.1039/C2EE22611A.

[20] E.J. Popczun, J.R. McKone, C.G. Read, A.J. Biacchi, A.M. Wiltrout, N.S. Lewis, R.E. Schaak, Nanostructured nickel phosphide as an electrocatalyst for the hydrogen evolution reaction, J. Am. Chem. Soc. 135 (2013) 9267–9270, https://doi.org/10.1021/ja403440e.

[21] J. Xu, Y. Qi, C. Wang, L. Wang, NH2-MIL-101(Fe)/Ni(OH)$_2$-derived C,N-codoped Fe$_2$P/Ni$_2$P cocatalyst modified g-C$_3$N$_4$ for enhanced photocatalytic hydrogen evolution from water splitting, Appl. Catal. Environ. 241 (2019) 178–186, https://doi.org/10.1016/j.apcatb.2018.09.035.

[22] V. Navakoteswara Rao, N. Lakshmana Reddy, M. Mamatha Kumari, P. Ravi, M. Sathish, K.M. Kuruvilla, V. Preethi, K.R. Reddy, N.P. Shetti, T.M. Aminabhavi, M.V. Shankar, Photocatalytic recovery of H$_2$ from H$_2$S containing wastewater: surface and interface control of photo-excitons in Cu$_2$S@TiO$_2$ core-shell nanostructures, Appl. Catal. Environ. 254 (2019) 174–185, https://doi.org/10.1016/j.apcatb.2019.04.090.

[23] S.-W. Cao, Y.-P. Yuan, J. Fang, M.M. Shahjamali, F.Y.C. Boey, J. Barber, S.C. Joachim Loo, C. Xue, In-situ growth of CdS quantum dots on g-C$_3$N$_4$ nanosheets for highly efficient photocatalytic hydrogen generation under visible light irradiation, Int. J. Hydrogen Energy 38 (2013) 1258–1266, https://doi.org/10.1016/j.ijhydene.2012.10.116.

[24] M. Hong, J. Li, W. Zhang, S. Liu, H. Chang, Semimetallic 1T′ WTe$_2$ nanorods as anode material for the sodium ion battery, Energy Fuel 32 (2018) 6371–6377, https://doi.org/10.1021/acs.energyfuels.8b00454.

[25] G. Xu, L. Yang, X. Wei, J. Ding, J. Zhong, P.K. Chu, MoS$_2$-quantum-dot-interspersed Li$_4$Ti$_5$O$_{12}$ nanosheets with enhanced performance for Li- and Na-ion batteries, Adv. Funct. Mater. 26 (2016) 3349–3358, https://doi.org/10.1002/adfm.201505435.

398 Oxide free nanomaterials for energy storage and conversion applications

[26] N. Cai, J. Fu, H. Zeng, X. Luo, C. Han, F. Yu, Reduced graphene oxide-silver nanoparticles/ nitrogen-doped carbon nanofiber composites with meso-microporous structure for high-performance symmetric supercapacitor application, J. Alloys Compd. 742 (2018) 769–779, https://doi.org/10.1016/j.jallcom.2018.01.011.

[27] P. Qin, G. Fang, W. Ke, F. Cheng, Q. Zheng, J. Wan, H. Lei, X. Zhao, In situ growth of double-layer MoO_3/MoS_2 film from MoS_2 for hole-transport layers in organic solar cell, J. Mater. Chem. A 2 (2014) 2742–2756, https://doi.org/10.1039/C3TA13579A.

[28] C. Zhu, F. Du, P. Fu, S. Wang, Z. Lin, Highly sensitive sensor based on $\{NaBi\}(\{MoO\}_4)_2/\{MWCNT\}$ composites, Mater. Res. Express 5 (2018) 125016, https://doi.org/10.1088/2053-1591/aae17a.

[29] R. Wang, G. Cheng, Z. Dai, J. Ding, Y. Liu, R. Chen, Ionic liquid-employed synthesis of Bi_2E_3 (E = S, Se, and Te) hierarchitectures: the case of Bi_2S_3 with superior visible-light-driven Cr(VI) photoreduction capacity, Chem. Eng. J. 327 (2017) 371–386, https://doi.org/10.1016/j.cej.2017.06.119.

[30] S. Luo, F. Qin, Y. Ming, H. Zhao, Y. Liu, R. Chen, Fabrication uniform hollow Bi_2S_3 nanospheres via Kirkendall effect for photocatalytic reduction of Cr(VI) in electroplating industry wastewater, J. Hazard. Mater. 340 (2017) 253–262, https://doi.org/10.1016/j.jhazmat.2017.06.044.

[31] X. Zou, Y. Dong, S. Li, J. Ke, Y. Cui, Facile anion exchange to construct uniform AgX (X = Cl, Br, I)/Ag_2CrO_4 NR hybrids for efficient visible light driven photocatalytic activity, Sol. Energy 169 (2018) 392–400, https://doi.org/10.1016/j.solener.2018.05.017.

[32] S. Dong, X. Chen, X. Zhang, G. Cui, Nanostructured transition metal nitrides for energy storage and fuel cells, Coord. Chem. Rev. 257 (2013) 1946–1956, https://doi.org/10.1016/j.ccr.2012.12.012.

[33] D.J. Ham, J.S. Lee, Transition metal carbides and nitrides as electrode materials for low temperature fuel cells, Energies 2 (2009) 873–899, https://doi.org/10.3390/en20400873.

[34] E. Furimsky, Metal carbides and nitrides as potential catalysts for hydroprocessing, Appl. Catal. A. Gen. 240 (2003) 1–28, https://doi.org/10.1016/S0926-860X(02)00428-3.

[35] X. Peng, C. Pi, X. Zhang, S. Li, K. Huo, P.K. Chu, Recent progress of transition metal nitrides for efficient electrocatalytic water splitting, Sustain. Energy Fuels. 3 (2019) 366–381, https://doi.org/10.1039/C8SE00525G.

[36] S. Xu, J. Xu, P. Munroe, Z.-H. Xie, Nanoporosity improves the damage tolerance of nanostructured tantalum nitride coatings, Scr. Mater. 133 (2017) 86–91, https://doi.org/10.1016/j.scriptamat.2017.02.008.

[37] J.J. Ma, J. Xu, S. Jiang, P. Munroe, Z.-H. Xie, Effects of pH value and temperature on the corrosion behavior of a Ta_2N nanoceramic coating in simulated polymer electrolyte membrane fuel cell environment, Ceram. Int. 42 (2016) 16833–16851, https://doi.org/10.1016/j.ceramint.2016.07.175.

[38] J. Cheng, J. Xu, L.L. Liu, S. Jiang, Electrochemical corrosion behavior of Ta_2N nanoceramic coating in simulated body fluid, Materials (Basel) 9 (2016), https://doi.org/10.3390/ma9090772.

[39] J. Xu, W. Hu, Z.-H. Xie, P. Munroe, Reactive-sputter-deposited β-Ta_2O_5 and TaON nanoceramic coatings on Ti–6Al–4V alloy against wear and corrosion damage, Surf. Coat. Technol. 296 (2016) 171–184, https://doi.org/10.1016/j.surfcoat.2016.04.004.

[40] H. Deno, T. Kamemoto, S. Nemoto, A. Koshio, F. Kokai, Formation of TiN–Ir particle films using pulsed-laser deposition and their electrolytic properties in producing hypochlorous acid, Appl. Surf. Sci. 254 (2008) 2776–2782, https://doi.org/10.1016/j.apsusc.2007.10.019.

[41] X. Peng, W. Li, L. Wang, L. Hu, W. Jin, A. Gao, X. Zhang, K. Huo, P.K. Chu, Lithiation kinetics in high-performance porous vanadium nitride nanosheet anode, Electrochim. Acta 214 (2016) 201–207, https://doi.org/10.1016/j.electacta.2016.08.023.

[42] X. Xiao, X. Peng, H. Jin, T. Li, C. Zhang, B. Gao, B. Hu, K. Huo, J. Zhou, Freestanding mesoporous VN/CNT hybrid electrodes for flexible all-solid-state supercapacitors, Adv. Mater. 25 (2013) 5091–5097, https://doi.org/10.1002/adma.201301465.

[43] C. Giordano, C. Erpen, W. Yao, M. Antonietti, Synthesis of Mo and W carbide and nitride nanoparticles via a simple "urea glass" route, Nano Lett. 8 (2008) 4659–4663, https://doi.org/10.1021/nl8018593.

[44] X.Z. Chen, J.L. Dye, H.A. Eick, S.H. Elder, K.-L. Tsai, Synthesis of transition-metal nitrides from nanoscale metal particles prepared by homogeneous reduction of metal halides with an alkalide, Chem. Mater. 9 (1997) 1172–1176, https://doi.org/10.1021/cm960565n.

[45] D. Choi, P.N. Kumta, Synthesis, structure, and electrochemical characterization of nanocrystalline tantalum and tungsten nitrides, J. Am. Ceram. Soc. 90 (2007) 3113–3120, https://doi.org/10.1111/j.1551-2916.2007.01873.x.

[46] Y. Han, X. Yue, Y. Jin, X. Huang, P.K. Shen, Hydrogen evolution reaction in acidic media on single-crystalline titanium nitride nanowires as an efficient non-noble metal electrocatalyst, J. Mater. Chem. A 4 (2016) 3673–3677, https://doi.org/10.1039/C5TA09976E.

[47] K. Xu, P. Chen, X. Li, Y. Tong, H. Ding, X. Wu, W. Chu, Z. Peng, C. Wu, Y. Xie, Metallic nickel nitride nanosheets realizing enhanced electrochemical water oxidation, J. Am. Chem. Soc. 137 (2015) 4119–4125, https://doi.org/10.1021/ja5119495.

[48] J. Xie, S. Li, X. Zhang, J. Zhang, R. Wang, H. Zhang, B. Pan, Y. Xie, Atomically-thin molybdenum nitride nanosheets with exposed active surface sites for efficient hydrogen evolution, Chem. Sci. 5 (2014) 4615–4620, https://doi.org/10.1039/C4SC02019G.

[49] X. Jia, Y. Zhao, G. Chen, L. Shang, R. Shi, X. Kang, G.I.N. Waterhouse, L.-Z. Wu, C.-H. Tung, T. Zhang, Water splitting: Ni3FeN nanoparticles derived from ultrathin NiFe-layered double hydroxide nanosheets: an efficient overall water splitting electrocatalyst (Adv. Energy Mater. 10/2016), Adv. Energy Mater. 6 (2016), https://doi.org/10.1002/aenm.201502585, 1502585.

[50] Y. Zhang, B. Ouyang, J. Xu, G. Jia, S. Chen, R.S. Rawat, H.J. Fan, Rapid synthesis of cobalt nitride nanowires: highly efficient and low-cost catalysts for oxygen evolution, Angew. Chem. Int. Ed. 55 (2016) 8670–8674, https://doi.org/10.1002/anie.201604372.

[51] P. Chen, K. Xu, Z. Fang, Y. Tong, J. Wu, X. Lu, X. Peng, H. Ding, C. Wu, Y. Xie, Metallic Co_4N porous nanowire arrays activated by surface oxidation as electrocatalysts for the oxygen evolution reaction, Angew. Chem. Int. Ed. 54 (2015) 14710–14714, https://doi.org/10.1002/anie.201506480.

[52] Q. Zhang, Y. Wang, Y. Wang, A.M. Al-Enizi, A.A. Elzatahry, G. Zheng, Myriophyllum-like hierarchical TiN@Ni_3N nanowire arrays for bifunctional water splitting catalysts, J. Mater. Chem. A 4 (2016) 5713–5718, https://doi.org/10.1039/C6TA00356G.

[53] Y. Wang, D. Liu, Z. Liu, C. Xie, J. Huo, S. Wang, Porous cobalt–iron nitride nanowires as excellent bifunctional electrocatalysts for overall water splitting, Chem. Commun. 52 (2016) 12614–12617, https://doi.org/10.1039/C6CC06608A.

[54] Z. Xing, Q. Li, D. Wang, X. Yang, X. Sun, Self-supported nickel nitride as an efficient high-performance three-dimensional cathode for the alkaline hydrogen evolution reaction, Electrochim. Acta 191 (2016) 841–845, https://doi.org/10.1016/j.electacta.2015.12.174.

[55] L. Ma, L.R.L. Ting, V. Molinari, C. Giordano, B.S. Yeo, Efficient hydrogen evolution reaction catalyzed by molybdenum carbide and molybdenum nitride nanocatalysts synthesized

400 Oxide free nanomaterials for energy storage and conversion applications

via the urea glass route, J. Mater. Chem. A 3 (2015) 8361–8368, https://doi.org/10.1039/C5TA00139K.

[56] B. Cao, G.M. Veith, J.C. Neuefeind, R.R. Adzic, P.G. Khalifah, Mixed close-packed cobalt molybdenum nitrides as non-noble metal electrocatalysts for the hydrogen evolution reaction, J. Am. Chem. Soc. 135 (2013) 19186–19192, https://doi.org/10.1021/ja4081056.

[57] W.-F. Chen, J.T. Muckerman, E. Fujita, Recent developments in transition metal carbides and nitrides as hydrogen evolution electrocatalysts, Chem. Commun. 49 (2013) 8896–8909, https://doi.org/10.1039/C3CC44076A.

[58] V. Chakrapani, J. Thangala, M.K. Sunkara, WO3 and W2N nanowire arrays for photoelectrochemical hydrogen production, Int. J. Hydrogen Energy 34 (2009) 9050–9059, https://doi.org/10.1016/j.ijhydene.2009.09.031.

[59] S.S. Patil, M.A. Johar, M.A. Hassan, D.R. Patil, S.-W. Ryu, Anchoring MWCNTs to 3D honeycomb ZnO/GaN heterostructures to enhancing photoelectrochemical water oxidation, Appl. Catal. Environ. 237 (2018) 791–801, https://doi.org/10.1016/j.apcatb.2018.06.047.

[60] S.S. Patil, M.A. Johar, M.A. Hassan, D.R. Patil, S.-W. Ryu, Enhanced photoelectrocatalytic water oxidation using CoPi modified GaN/MWCNTs composite photoanodes, Sol. Energy 178 (2019) 125–132, https://doi.org/10.1016/j.solener.2018.12.028.

[61] P. Chen, K. Xu, Y. Tong, X. Li, S. Tao, Z. Fang, W. Chu, X. Wu, C. Wu, Cobalt nitrides as a class of metallic electrocatalysts for the oxygen evolution reaction, Inorg. Chem. Front. 3 (2016) 236–242, https://doi.org/10.1039/C5QI00197H.

[62] J.K. Nørskov, T. Bligaard, A. Logadottir, J.R. Kitchin, J.G. Chen, S. Pandelov, U. Stimming, Trends in the exchange current for hydrogen evolution, J. Electrochem. Soc. 152 (2005) J23, https://doi.org/10.1149/1.1856988.

[63] J. Deng, H. Li, J. Xiao, Y. Tu, D. Deng, H. Yang, H. Tian, J. Li, P. Ren, X. Bao, Triggering the electrocatalytic hydrogen evolution activity of the inert two-dimensional MoS_2 surface via single-atom metal doping, Energ. Environ. Sci. 8 (2015) 1594–1601, https://doi.org/10.1039/C5EE00751H.

[64] E. Navarro-Flores, Z. Chong, S. Omanovic, Characterization of Ni, NiMo, NiW and NiFe electroactive coatings as electrocatalysts for hydrogen evolution in an acidic medium, J. Mol. Catal. A Chem. 226 (2005) 179–197, https://doi.org/10.1016/j.molcata.2004.10.029.

[65] W.-F. Chen, K. Sasaki, C. Ma, A.I. Frenkel, N. Marinkovic, J.T. Muckerman, Y. Zhu, R.R. Adzic, Hydrogen-evolution catalysts based on non-noble metal nickel–molybdenum nitride nanosheets, Angew. Chem. Int. Ed. 51 (2012) 6131–6135, https://doi.org/10.1002/anie.201200699.

[66] B. Wei, G. Tang, H. Liang, Z. Qi, D. Zhang, W. Hu, H. Shen, Z. Wang, Bimetallic vanadium-molybdenum nitrides using magnetron co-sputtering as alkaline hydrogen evolution catalyst, Electrochem. Commun. 93 (2018) 166–170, https://doi.org/10.1016/j.elecom.2018.07.012.

[67] Y. Wang, B. Zhang, W. Pan, H. Ma, J. Zhang, 3 D porous nickel–cobalt nitrides supported on nickel foam as efficient electrocatalysts for overall water splitting, ChemSusChem 10 (2017) 4170–4177, https://doi.org/10.1002/cssc.201701456.

[68] F. Yan, Y. Wang, K. Li, C. Zhu, P. Gao, C. Li, X. Zhang, Y. Chen, Highly stable three-dimensional porous nickel-Iron nitride nanosheets for full water splitting at high current densities, Chem. A Eur. J. 23 (2017) 10187–10194, https://doi.org/10.1002/chem.201701662.

[69] Y. Zhang, B. Ouyang, J. Xu, S. Chen, R.S. Rawat, H.J. Fan, 3D porous hierarchical nickel–molybdenum nitrides synthesized by RF plasma as highly active and stable hydrogen-evolution-reaction electrocatalysts, Adv. Energy Mater. 6 (2016) 1600221, https://doi.org/10.1002/aenm.201600221.

[70] Y. Wang, C. Xie, D. Liu, X. Huang, J. Huo, S. Wang, Nanoparticle-stacked porous nickel–iron nitride nanosheet: a highly efficient bifunctional electrocatalyst for overall water splitting, ACS Appl. Mater. Interfaces 8 (2016) 18652–18657, https://doi.org/10.1021/acsami.6b05811.

[71] B. Zhang, C. Xiao, S. Xie, J. Liang, X. Chen, Y. Tang, Iron–nickel nitride nanostructures in situ grown on surface-redox-etching nickel foam: efficient and ultrasustainable electrocatalysts for overall water splitting, Chem. Mater. 28 (2016) 6934–6941, https://doi.org/10.1021/acs.chemmater.6b02610.

[72] X. Jia, Y. Zhao, G. Chen, L. Shang, R. Shi, X. Kang, G.I.N. Waterhouse, L.-Z. Wu, C.-H. Tung, T. Zhang, Ni_3FeN nanoparticles derived from ultrathin NiFe-layered double hydroxide nanosheets: an efficient overall water splitting electrocatalyst, Adv. Energy Mater. 6 (2016) 1502585, https://doi.org/10.1002/aenm.201502585.

[73] J. Shi, Z. Pu, Q. Liu, A.M. Asiri, J. Hu, X. Sun, Tungsten nitride nanorods array grown on carbon cloth as an efficient hydrogen evolution cathode at all pH values, Electrochim. Acta 154 (2015) 345–351, https://doi.org/10.1016/j.electacta.2014.12.096.

[74] Y. Wang, B. Kong, D. Zhao, H. Wang, C. Selomulya, Strategies for developing transition metal phosphides as heterogeneous electrocatalysts for water splitting, Nano Today 15 (2017) 26–55, https://doi.org/10.1016/j.nantod.2017.06.006.

[75] P. Liu, J.A. Rodriguez, Catalysts for hydrogen evolution from the [NiFe] hydrogenase to the Ni2P(001) surface: the importance of ensemble effect, J. Am. Chem. Soc. 127 (2005) 14871–14878, https://doi.org/10.1021/ja0540019.

[76] P. Xiao, W. Chen, X. Wang, A review of phosphide-based materials for electrocatalytic hydrogen evolution, Adv. Energy Mater. 5 (2015) 1500985, https://doi.org/10.1002/aenm.201500985.

[77] J.F. Callejas, C.G. Read, C.W. Roske, N.S. Lewis, R.E. Schaak, Synthesis, characterization, and properties of metal phosphide catalysts for the hydrogen-evolution reaction, Chem. Mater. 28 (2016) 6017–6044, https://doi.org/10.1021/acs.chemmater.6b02148.

[78] Y. Shi, B. Zhang, Recent advances in transition metal phosphide nanomaterials: synthesis and applications in hydrogen evolution reaction, Chem. Soc. Rev. 45 (2016) 1529–1541, https://doi.org/10.1039/C5CS00434A.

[79] S. Anantharaj, S.R. Ede, K. Sakthikumar, K. Karthick, S. Mishra, S. Kundu, Recent trends and perspectives in electrochemical water splitting with an emphasis on sulfide, selenide, and phosphide catalysts of Fe, Co, and Ni: a review, ACS Catal. 6 (2016) 8069–8097, https://doi.org/10.1021/acscatal.6b02479.

[80] J.M. McEnaney, J. Chance Crompton, J.F. Callejas, E.J. Popczun, C.G. Read, N.S. Lewis, R. E. Schaak, Electrocatalytic hydrogen evolution using amorphous tungsten phosphide nanoparticles, Chem. Commun. 50 (2014) 11026–11028, https://doi.org/10.1039/C4CC04709E.

[81] C.G. Read, J.F. Callejas, C.F. Holder, R.E. Schaak, General strategy for the synthesis of transition metal phosphide films for electrocatalytic hydrogen and oxygen evolution, ACS Appl. Mater. Interfaces 8 (2016) 12798–12803, https://doi.org/10.1021/acsami.6b02352.

[82] J. Li, J. Li, X. Zhou, Z. Xia, W. Gao, Y. Ma, Y. Qu, Highly efficient and robust nickel phosphides as bifunctional electrocatalysts for overall water-splitting, ACS Appl. Mater. Interfaces 8 (2016) 10826–10834, https://doi.org/10.1021/acsami.6b00731.

[83] N. Jiang, B. You, M. Sheng, Y. Sun, Electrodeposited cobalt-phosphorous-derived films as competent bifunctional catalysts for overall water splitting, Angew. Chem. Int. Ed. 54 (2015) 6251–6254, https://doi.org/10.1002/anie.201501616.

402 Oxide free nanomaterials for energy storage and conversion applications

[84] V. Tallapally, R.J.A. Esteves, L. Nahar, I.U. Arachchige, Multivariate synthesis of tin phosphide nanoparticles: temperature, time, and ligand control of size, shape, and crystal structure, Chem. Mater. 28 (2016) 5406–5414, https://doi.org/10.1021/acs.chemmater.6b01749.

[85] S. Liu, H. Zhang, L. Xu, L. Ma, Synthesis of hollow spherical tin phosphides (Sn_4P_3) and their high adsorptive and electrochemical performance, J. Cryst. Growth 438 (2016) 31–37, https://doi.org/10.1016/j.jcrysgro.2015.12.018.

[86] D. Wang, X. Zhang, D. Zhang, Y. Shen, Z. Wu, Influence of Mo/P ratio on CoMoP nanoparticles as highly efficient HER catalysts, Appl. Catal. A. Gen. 511 (2016) 11–15, https://doi.org/10.1016/j.apcata.2015.11.029.

[87] T. Wu, M. Pi, D. Zhang, S. Chen, Three-dimensional porous structural MoP_2 nanoparticles as a novel and superior catalyst for electrochemical hydrogen evolution, J. Power Sources 328 (2016) 551–557, https://doi.org/10.1016/j.jpowsour.2016.08.050.

[88] T. Wu, M. Pi, D. Zhang, S. Chen, 3D structured porous CoP3 nanoneedle arrays as an efficient bifunctional electrocatalyst for the evolution reaction of hydrogen and oxygen, J. Mater. Chem. A 4 (2016) 14539–14544, https://doi.org/10.1039/C6TA05838H.

[89] C.Y. Son, I.H. Kwak, Y.R. Lim, J. Park, FeP and FeP2 nanowires for efficient electrocatalytic hydrogen evolution reaction, Chem. Commun. 52 (2016) 2819–2822, https://doi.org/10.1039/C5CC09832G.

[90] Y.-P. Zhu, Y.-P. Liu, T.-Z. Ren, Z.-Y. Yuan, Self-supported cobalt phosphide mesoporous nanorod arrays: a flexible and bifunctional electrode for highly active electrocatalytic water reduction and oxidation, Adv. Funct. Mater. 25 (2015) 7337–7347, https://doi.org/10.1002/adfm.201503666.

[91] W.C. Zhou, H.X. Yang, S.Y. Shao, X.P. Ai, Y.L. Cao, Superior high rate capability of tin phosphide used as high capacity anode for aqueous primary batteries, Electrochem. Commun. 8 (2006) 55–59, https://doi.org/10.1016/j.elecom.2005.10.016.

[92] J.-J. Wu, Z.-W. Fu, Pulsed-laser-deposited Sn_4P_3 electrodes for lithium-ion batteries, J. Electrochem. Soc. 156 (2009) A22, https://doi.org/10.1149/1.3005960.

[93] B.M. Barry, E.G. Gillan, A general and flexible synthesis of transition-metal polyphosphides via PCl3 elimination, Chem. Mater. 21 (2009) 4454–4461, https://doi.org/10.1021/cm9010663.

[94] Z. Ren, X. Ren, L. Zhang, C. Fu, X. Li, Y. Zhang, B. Gao, L. Yang, P.K. Chu, K. Huo, Tungsten-doped CoP nanoneedle arrays grown on carbon cloth as efficient bifunctional electrocatalysts for overall water splitting, ChemElectroChem 6 (2019) 5229–5236, https://doi.org/10.1002/celc.201901417.

[95] X. Wang, Y.V. Kolen'ko, X.-Q. Bao, K. Kovnir, L. Liu, One-step synthesis of self-supported nickel phosphide nanosheet array cathodes for efficient electrocatalytic hydrogen generation, Angew. Chem. Int. Ed. 54 (2015) 8188–8192, https://doi.org/10.1002/anie.201502577.

[96] S.H. Yu, D.H.C. Chua, Toward high-performance and low-cost hydrogen evolution reaction electrocatalysts: nanostructuring cobalt phosphide (CoP) particles on carbon fiber paper, ACS Appl. Mater. Interfaces 10 (2018) 14777–14785, https://doi.org/10.1021/acsami.8b02755.

[97] B. Owens-Baird, J.P.S. Sousa, Y. Ziouani, D.Y. Petrovykh, N.A. Zarkevich, D.D. Johnson, Y.V. Kolen'ko, K. Kovnir, Crystallographic facet selective HER catalysis: exemplified in FeP and NiP_2 single crystals, Chem. Sci. 11 (2020) 5007–5016, https://doi.org/10.1039/D0SC00676A.

[98] J. Chang, Y. Xiao, M. Xiao, J. Ge, C. Liu, W. Xing, Surface oxidized cobalt-phosphide nanorods as an advanced oxygen evolution catalyst in alkaline solution, ACS Catal. 5 (2015) 6874–6878, https://doi.org/10.1021/acscatal.5b02076.

Oxides free nanomaterials **Chapter | 15** **403**

[99] M. Ledendecker, S. Krick Calderón, C. Papp, H.-P. Steinrück, M. Antonietti, M. Shalom, The synthesis of nanostructured Ni_5P_4 films and their use as a non-noble bifunctional electrocatalyst for full water splitting, Angew. Chem. Int. Ed. 54 (2015) 12361–12365, https://doi.org/10.1002/anie.201502438.

[100] S. Cao, Y. Chen, C.-C. Hou, X.-J. Lv, W.-F. Fu, Cobalt phosphide as a highly active non-precious metal cocatalyst for photocatalytic hydrogen production under visible light irradiation, J. Mater. Chem. A 3 (2015) 6096–6101, https://doi.org/10.1039/C4TA07149B.

[101] Z. Huang, Z. Chen, Z. Chen, C. Lv, H. Meng, C. Zhang, $Ni_{12}P_5$ nanoparticles as an efficient catalyst for hydrogen generation via electrolysis and photoelectrolysis, ACS Nano 8 (2014) 8121–8129, https://doi.org/10.1021/nn5022204.

[102] T. Tian, L. Ai, J. Jiang, Metal–organic framework-derived nickel phosphides as efficient electrocatalysts toward sustainable hydrogen generation from water splitting, RSC Adv. 5 (2015) 10290–10295, https://doi.org/10.1039/C4RA15680C.

[103] P. Xiao, M.A. Sk, L. Thia, X. Ge, R.J. Lim, J.-Y. Wang, K.H. Lim, X. Wang, Molybdenum phosphide as an efficient electrocatalyst for the hydrogen evolution reaction, Energ. Environ. Sci. 7 (2014) 2624–2629, https://doi.org/10.1039/C4EE00957F.

[104] J.M. McEnaney, J.C. Crompton, J.F. Callejas, E.J. Popczun, A.J. Biacchi, N.S. Lewis, R.E. Schaak, Amorphous molybdenum phosphide nanoparticles for electrocatalytic hydrogen evolution, Chem. Mater. 26 (2014) 4826–4831, https://doi.org/10.1021/cm502035s.

[105] Z. Huang, Z. Chen, Z. Chen, C. Lv, M.G. Humphrey, C. Zhang, Cobalt phosphide nanorods as an efficient electrocatalyst for the hydrogen evolution reaction, Nano Energy 9 (2014) 373–382, https://doi.org/10.1016/j.nanoen.2014.08.013.

[106] H. Du, S. Gu, R. Liu, C.M. Li, Tungsten diphosphide nanorods as an efficient catalyst for electrochemical hydrogen evolution, J. Power Sources 278 (2015) 540–545, https://doi.org/10.1016/j.jpowsour.2014.12.095.

[107] C.-C. Hou, Q.-Q. Chen, C.-J. Wang, F. Liang, Z. Lin, W.-F. Fu, Y. Chen, Self-supported Cedarlike semimetallic Cu_3P nanoarrays as a 3D high-performance Janus electrode for both oxygen and hydrogen evolution under basic conditions, ACS Appl. Mater. Interfaces 8 (2016) 23037–23048, https://doi.org/10.1021/acsami.6b06251.

[108] X.-D. Wang, Y.-F. Xu, H.-S. Rao, W.-J. Xu, H.-Y. Chen, W.-X. Zhang, D.-B. Kuang, C.-Y. Su, Novel porous molybdenum tungsten phosphide hybrid nanosheets on carbon cloth for efficient hydrogen evolution, Energ. Environ. Sci. 9 (2016) 1468–1475, https://doi.org/10.1039/C5EE03801D.

[109] C. Wang, J. Jiang, T. Ding, G. Chen, W. Xu, Q. Yang, Monodisperse ternary NiCoP nanostructures as a bifunctional electrocatalyst for both hydrogen and oxygen evolution reactions with excellent performance, Adv. Mater. Interfaces 3 (2016) 1500454, https://doi.org/10.1002/admi.201500454.

[110] Z. Ma, R. Li, M. Wang, H. Meng, F. Zhang, X.-Q. Bao, B. Tang, X. Wang, Self-supported porous Ni-Fe-P composite as an efficient electrocatalyst for hydrogen evolution reaction in both acidic and alkaline medium, Electrochim. Acta 219 (2016) 194–203, https://doi.org/10.1016/j.electacta.2016.10.004.

[111] Y. Pan, Y. Liu, Y. Lin, C. Liu, Metal doping effect of the M–Co_2P/nitrogen-doped carbon nanotubes (M = Fe, Ni, Cu) hydrogen evolution hybrid catalysts, ACS Appl. Mater. Interfaces 8 (2016) 13890–13901, https://doi.org/10.1021/acsami.6b02023.

[112] X. Liang, D. Zhang, Z. Wu, D. Wang, The Fe-promoted MoP catalyst with high activity for water splitting, Appl. Catal. A. Gen. 524 (2016) 134–138, https://doi.org/10.1016/j.apcata.2016.06.029.

404 Oxide free nanomaterials for energy storage and conversion applications

[113] R. Ye, P. del Angel-Vicente, Y. Liu, M.J. Arellano-Jimenez, Z. Peng, T. Wang, Y. Li, B.I. Yakobson, S.-H. Wei, M.J. Yacaman, J.M. Tour, High-performance hydrogen evolution from $MoS_{2(1-x)}P_x$ solid solution, Adv. Mater. 28 (2016) 1427–1432, https://doi.org/10.1002/adma.201504866.

[114] J. Chang, L. Liang, C. Li, M. Wang, J. Ge, C. Liu, W. Xing, Ultrathin cobalt phosphide nanosheets as efficient bifunctional catalysts for a water electrolysis cell and the origin for cell performance degradation, Green Chem. 18 (2016) 2287–2295, https://doi.org/10.1039/C5GC02899J.

[115] G.-Q. Han, X. Li, Y.-R. Liu, B. Dong, W.-H. Hu, X. Shang, X. Zhao, Y.-M. Chai, Y.-Q. Liu, C.-G. Liu, Controllable synthesis of three dimensional electrodeposited Co–P nanosphere arrays as efficient electrocatalysts for overall water splitting, RSC Adv. 6 (2016) 52761–52771, https://doi.org/10.1039/C6RA04478F.

[116] J. Masa, S. Barwe, C. Andronescu, I. Sinev, A. Ruff, K. Jayaramulu, K. Elumeeva, B. Konkena, B. Roldan Cuenya, W. Schuhmann, Low overpotential water splitting using cobalt–cobalt phosphide nanoparticles supported on nickel foam, ACS Energy Lett. 1 (2016) 1192–1198, https://doi.org/10.1021/acsenergylett.6b00532.

[117] M. Liu, J. Li, Cobalt phosphide hollow polyhedron as efficient bifunctional electrocatalysts for the evolution reaction of hydrogen and oxygen, ACS Appl. Mater. Interfaces 8 (2016) 2158–2165, https://doi.org/10.1021/acsami.5b10727.

[118] J. Ryu, N. Jung, J.H. Jang, H.-J. Kim, S.J. Yoo, In situ transformation of hydrogen-evolving CoP nanoparticles: toward efficient oxygen evolution catalysts bearing dispersed morphologies with co-oxo/hydroxo molecular units, ACS Catal. 5 (2015) 4066–4074, https://doi.org/10.1021/acscatal.5b00349.

[119] H. Liang, A.N. Gandi, D.H. Anjum, X. Wang, U. Schwingenschlögl, H.N. Alshareef, Plasma-assisted synthesis of NiCoP for efficient overall water splitting, Nano Lett. 16 (2016) 7718–7725, https://doi.org/10.1021/acs.nanolett.6b03803.

[120] J. Duan, S. Chen, A. Vasileff, S.Z. Qiao, Anion and cation modulation in metal compounds for bifunctional overall water splitting, ACS Nano 10 (2016) 8738–8745, https://doi.org/10.1021/acsnano.6b04252.

[121] Q. Zhao, Z. Yan, C. Chen, J. Chen, Spinels: controlled preparation, oxygen reduction/evolution reaction application, and beyond, Chem. Rev. 117 (2017) 10121–10211, https://doi.org/10.1021/acs.chemrev.7b00051.

[122] X. Huang, Z. Zeng, H. Zhang, Metal dichalcogenide nanosheets: preparation, properties and applications, Chem. Soc. Rev. 42 (2013) 1934–1946, https://doi.org/10.1039/C2CS35387C.

[123] Y. Guo, T. Park, J.W. Yi, J. Henzie, J. Kim, Z. Wang, B. Jiang, Y. Bando, Y. Sugahara, J. Tang, Y. Yamauchi, Nanoarchitectonics for transition-metal-sulfide-based electrocatalysts for water splitting, Adv. Mater. 31 (2019) 1807134, https://doi.org/10.1002/adma.201807134.

[124] S. Chandrasekaran, L. Yao, L. Deng, C. Bowen, Y. Zhang, S. Chen, Z. Lin, F. Peng, P. Zhang, Recent advances in metal sulfides: from controlled fabrication to electrocatalytic, photocatalytic and photoelectrochemical water splitting and beyond, Chem. Soc. Rev. 48 (2019) 4178–4280, https://doi.org/10.1039/C8CS00664D.

[125] C. Tan, H. Zhang, Two-dimensional transition metal dichalcogenide nanosheet-based composites, Chem. Soc. Rev. 44 (2015) 2713–2731, https://doi.org/10.1039/C4CS00182F.

[126] Y. Li, H. Zhang, M. Jiang, Y. Kuang, X. Sun, X. Duan, Ternary NiCoP nanosheet arrays: an excellent bifunctional catalyst for alkaline overall water splitting, Nano Res. 9 (2016) 2251–2259, https://doi.org/10.1007/s12274-016-1112-z.

[127] K. Tang, X. Wang, Q. Li, C. Yan, High edge selectivity of in situ electrochemical Pt deposition on edge-rich layered WS2 nanosheets, Adv. Mater. 30 (2018) 1704779, https://doi.org/10.1002/adma.201704779.

[128] T.-T. Zhuang, Y. Liu, Y. Li, Y. Zhao, L. Wu, J. Jiang, S.-H. Yu, Integration of semiconducting sulfides for full-spectrum solar energy absorption and efficient charge separation, Angew. Chem. Int. Ed. 55 (2016) 6396–6400, https://doi.org/10.1002/anie.201601865.

[129] S.-K. Han, C. Gu, S. Zhao, S. Xu, M. Gong, Z. Li, S.-H. Yu, Precursor triggering synthesis of self-coupled sulfide polymorphs with enhanced photoelectrochemical properties, J. Am. Chem. Soc. 138 (2016) 12913–12919, https://doi.org/10.1021/jacs.6b06609.

[130] J. Yin, J. Jin, H. Lin, Z. Yin, J. Li, M. Lu, L. Guo, P. Xi, Y. Tang, C.-H. Yan, Optimized metal chalcogenides for boosting water splitting, Adv. Sci. 7 (2020) 1903070, https://doi.org/10.1002/advs.201903070.

[131] B. Sun, T. Feng, J. Dong, X. Li, X. Liu, J. Wu, S. Ai, Green synthesis of bismuth sulfide nanostructures with tunable morphologies and robust photoelectrochemical performance, CrstEngComm 21 (2019) 1474–1481, https://doi.org/10.1039/C8CE02089B.

[132] H. Wang, Y. Xia, H. Li, X. Wang, Y. Yu, X. Jiao, D. Chen, Highly active deficient ternary sulfide photoanode for photoelectrochemical water splitting, Nat. Commun. 11 (2020) 3078, https://doi.org/10.1038/s41467-020-16800-w.

[133] Y. Fu, F. Cao, F. Wu, Z. Diao, J. Chen, S. Shen, L. Li, Phase-modulated band alignment in CdS nanorod/SnSx nanosheet hierarchical heterojunctions toward efficient water splitting, Adv. Funct. Mater. 28 (2018) 1706785, https://doi.org/10.1002/adfm.201706785.

[134] K. Peng, L. Fu, J. Ouyang, H. Yang, Emerging parallel dual 2D composites: natural clay mineral hybridizing MoS_2 and interfacial structure, Adv. Funct. Mater. 26 (2016) 2666–2675, https://doi.org/10.1002/adfm.201504942.

[135] Y. Su, D. Ao, H. Liu, Y. Wang, MOF-derived yolk–shell CdS microcubes with enhanced visible-light photocatalytic activity and stability for hydrogen evolution, J. Mater. Chem. A 5 (2017) 8680–8689, https://doi.org/10.1039/C7TA00855D.

[136] L. Zeng, L. Tao, C. Tang, B. Zhou, H. Long, Y. Chai, S.P. Lau, Y.H. Tsang, High-responsivity UV-vis photodetector based on transferable WS2 film deposited by magnetron sputtering, Sci. Rep. 6 (2016) 20343, https://doi.org/10.1038/srep20343.

[137] Y. Zhong, X. Qiu, D. Chen, N. Li, Q. Xu, H. Li, J. He, J. Lu, Flexible electrospun carbon nanofiber/tin(IV) sulfide core/sheath membranes for photocatalytically treating chromium (VI)-containing wastewater, ACS Appl. Mater. Interfaces 8 (2016) 28671–28677, https://doi.org/10.1021/acsami.6b10241.

[138] M.S.M. Saifullah, M. Asbahi, M. Binti-Kamran Kiyani, S. Tripathy, E.A.H. Ong, A. Ibn Saifullah, H.R. Tan, T. Dutta, R. Ganesan, S. Valiyaveettil, K.S.L. Chong, Direct patterning of zinc sulfide on a Sub-10 nanometer scale via electron beam lithography, ACS Nano 11 (2017) 9920–9929, https://doi.org/10.1021/acsnano.7b03951.

[139] A. Ambrosi, X. Chia, Z. Sofer, M. Pumera, Enhancement of electrochemical and catalytic properties of MoS_2 through ball-milling, Electrochem. Commun. 54 (2015) 36–40, https://doi.org/10.1016/j.elecom.2015.02.017.

[140] D. Voiry, H. Yamaguchi, J. Li, R. Silva, D.C.B. Alves, T. Fujita, M. Chen, T. Asefa, V.B. Shenoy, G. Eda, M. Chhowalla, Enhanced catalytic activity in strained chemically exfoliated WS2 nanosheets for hydrogen evolution, Nat. Mater. 12 (2013) 850–855, https://doi.org/10.1038/nmat3700.

[141] A. Gurarslan, Y. Yu, L. Su, Y. Yu, F. Suarez, S. Yao, Y. Zhu, M. Ozturk, Y. Zhang, L. Cao, Surface-energy-assisted perfect transfer of centimeter-scale monolayer and few-layer MoS2

406 Oxide free nanomaterials for energy storage and conversion applications

films onto arbitrary substrates, ACS Nano 8 (2014) 11522–11528, https://doi.org/10.1021/nn5057673.

[142] E. Thimsen, S.C. Riha, S.V. Baryshev, A.B.F. Martinson, J.W. Elam, M.J. Pellin, Atomic layer deposition of the quaternary chalcogenide Cu_2ZnSnS_4, Chem. Mater. 24 (2012) 3188–3196, https://doi.org/10.1021/cm3015463.

[143] E. Veena, K.V. Bangera, G.K. Shivakumar, Effect of annealing on the properties of spray-pyrolysed lead sulphide thin films for solar cell application, Appl. Phys. A 123 (2017) 366, https://doi.org/10.1007/s00339-017-0982-6.

[144] N.M. Hosny, A. Dahshan, Synthesis, structure and optical properties of SnS2, CdS and HgS nanoparticles from thioacetate precursor, J. Mol. Struct. 1085 (2015) 78–83, https://doi.org/10.1016/j.molstruc.2014.11.074.

[145] N. Lakshmana Reddy, V.N. Rao, M.M. Kumari, M. Sathish, S. Muthukonda Venkatakrishnan, N.L. Reddy, V.N. Rao, M.M. Kumari, M. Sathish, S. Muthukonda Venkatakrishnan, S.M. Venkatakrishnan, S. Muthukonda Venkatakrishnan, Development of high quantum efficiency CdS/ZnS core/shell structured photocatalyst for the enhanced solar hydrogen evolution, Int. J. Hydrogen Energy 43 (2018) 22315–22328, https://doi.org/10.1016/j.ijhydene.2018.10.054.

[146] Z. Zhang, X. Li, C. Gao, F. Teng, Y. Wang, L. Chen, W. Han, Z. Zhang, E. Xie, Synthesis of cadmium sulfide quantum dot-decorated barium stannate nanowires for photoelectrochemical water splitting, J. Mater. Chem. A 3 (2015) 12769–12776, https://doi.org/10.1039/C5TA01948F.

[147] Q. Zeng, J. Bai, J. Li, Y. Li, X. Li, B. Zhou, Combined nanostructured Bi_2S_3/TNA photoanode and Pt/SiPVC photocathode for efficient self-biasing photoelectrochemical hydrogen and electricity generation, Nano Energy 9 (2014) 152–160, https://doi.org/10.1016/j.nanoen.2014.06.023.

[148] R. Trevisan, P. Rodenas, V. Gonzalez-Pedro, C. Sima, R.S. Sanchez, E.M. Barea, I. Mora-Sero, F. Fabregat-Santiago, S. Gimenez, Harnessing infrared photons for photoelectrochemical hydrogen generation. A PbS quantum dot based "Quasi-Artificial Leaf", J. Phys. Chem. Lett. 4 (2013) 141–146, https://doi.org/10.1021/jz301890m.

[149] W. Gao, C. Wu, M. Cao, J. Huang, L. Wang, Y. Shen, Thickness tunable SnS nanosheets for photoelectrochemical water splitting, J. Alloys Compd. 688 (2016) 668–674, https://doi.org/10.1016/j.jallcom.2016.07.083.

[150] J. Zhang, Z. Liu, Z. Liu, Novel WO_3/Sb_2S_3 heterojunction photocatalyst based on WO_3 of different morphologies for enhanced efficiency in photoelectrochemical water splitting, ACS Appl. Mater. Interfaces 8 (2016) 9684–9691, https://doi.org/10.1021/acsami.6b00429.

[151] T.F. Jaramillo, K.P. Jørgensen, J. Bonde, J.H. Nielsen, S. Horch, I. Chorkendorff, Identification of active edge sites for electrochemical H_2 evolution from MoS_2 nanocatalysts, Science (80-.) 317 (2007) 100–102, https://doi.org/10.1126/science.1141483.

[152] H.I. Karunadasa, E. Montalvo, Y. Sun, M. Majda, J.R. Long, C.J. Chang, A molecular MoS_2 edge site mimic for catalytic hydrogen generation, Science (80-.) 335 (2012) 698–702, https://doi.org/10.1126/science.1215868.

[153] Z. Wu, B. Fang, A. Bonakdarpour, A. Sun, D.P. Wilkinson, D. Wang, WS_2 nanosheets as a highly efficient electrocatalyst for hydrogen evolution reaction, Appl. Catal. Environ. 125 (2012) 59–66, https://doi.org/10.1016/j.apcatb.2012.05.013.

[154] M. Cabán-Acevedo, M.L. Stone, J.R. Schmidt, J.G. Thomas, Q. Ding, H.-C. Chang, M.-L. Tsai, J.-H. He, S. Jin, Efficient hydrogen evolution catalysis using ternary pyrite-type cobalt phosphosulphide, Nat. Mater. 14 (2015) 1245–1251, https://doi.org/10.1038/nmat4410.

Oxides free nanomaterials **Chapter | 15 407**

[155] M.S. Faber, M.A. Lukowski, Q. Ding, N.S. Kaiser, S. Jin, Earth-abundant metal pyrites (FeS_2, CoS_2, NiS_2, and their alloys) for highly efficient hydrogen evolution and polysulfide reduction electrocatalysis, J. Phys. Chem. C 118 (2014) 21347–21356, https://doi.org/10.1021/jp506288w.

[156] D. Kong, J.J. Cha, H. Wang, H.R. Lee, Y. Cui, First-row transition metal dichalcogenide catalysts for hydrogen evolution reaction, Energ. Environ. Sci. 6 (2013) 3553–3558, https://doi.org/10.1039/C3EE42413H.

[157] W.-T. Sun, Y. Yu, H.-Y. Pan, X.-F. Gao, Q. Chen, L.-M. Peng, CdS quantum dots sensitized TiO_2 nanotube-array photoelectrodes, J. Am. Chem. Soc. 130 (2008) 1124–1125, https://doi.org/10.1021/ja0777741.

[158] C. Chen, Y. Zhai, C. Li, F. Li, Improving the efficiency of cadmium sulfide-sensitized titanium dioxide/indium tin oxide glass photoelectrodes using silver sulfide as an energy barrier layer and a light absorber, Nanoscale Res. Lett. 9 (2014) 605, https://doi.org/10.1186/1556-276X-9-605.

[159] K. He, C. Poole, K.F. Mak, J. Shan, Experimental demonstration of continuous electronic structure tuning via strain in atomically thin MoS_2, Nano Lett. 13 (2013) 2931–2936, https://doi.org/10.1021/nl4013166.

[160] J.D. Benck, S.C. Lee, K.D. Fong, J. Kibsgaard, R. Sinclair, T.F. Jaramillo, Designing active and stable silicon photocathodes for solar hydrogen production using molybdenum sulfide nanomaterials, Adv. Energy Mater. 4 (2014) 1400739, https://doi.org/10.1002/aenm.201400739.

[161] M. Barawi, I.J. Ferrer, E. Flores, S. Yoda, J.R. Ares, C. Sánchez, Hydrogen photoassisted generation by visible light and an earth abundant photocatalyst: pyrite (FeS_2), J. Phys. Chem. C 120 (2016) 9547–9552, https://doi.org/10.1021/acs.jpcc.5b11482.

[162] L.R. Nagappagari, S. Samanta, N. Sharma, V.R. Battula, K. Kailasam, Synergistic effect of a noble metal free $Ni(OH)_2$ co-catalyst and a ternary $ZnIn_2S_4/g$-C_3N_4 heterojunction for enhanced visible light photocatalytic hydrogen evolution, Sustain. Energy Fuels. 4 (2020) 750–759, https://doi.org/10.1039/c9se00704k.

[163] J.-P. Song, P.-F. Yin, J. Mao, S.-Z. Qiao, X.-W. Du, Catalytically active and chemically inert $CdIn_2S_4$ coating on a CdS photoanode for efficient and stable water splitting, Nanoscale 9 (2017) 6296–6301, https://doi.org/10.1039/C7NR01170A.

[164] D.-Y. Wang, M. Gong, H.-L. Chou, C.-J. Pan, H.-A. Chen, Y. Wu, M.-C. Lin, M. Guan, J. Yang, C.-W. Chen, Y.-L. Wang, B.-J. Hwang, C.-C. Chen, H. Dai, Highly active and stable hybrid catalyst of cobalt-doped FeS_2 nanosheets–carbon nanotubes for hydrogen evolution reaction, J. Am. Chem. Soc. 137 (2015) 1587–1592, https://doi.org/10.1021/ja511572q.

[165] K. Guo, Z. Liu, J. Han, X. Zhang, Y. Li, T. Hong, C. Zhou, Higher-efficiency photoelectrochemical electrodes of titanium dioxide-based nanoarrays sensitized simultaneously with plasmonic silver nanoparticles and multiple metal sulfides photosensitizers, J. Power Sources 285 (2015) 185–194, https://doi.org/10.1016/j.jpowsour.2015.03.112.

[166] J. Luo, S.D. Tilley, L. Steier, M. Schreier, M.T. Mayer, H.J. Fan, M. Grätzel, Solution transformation of Cu_2O into $CuInS_2$ for solar water splitting, Nano Lett. 15 (2015) 1395–1402, https://doi.org/10.1021/nl504746b.

[167] C. Chen, G. Ali, S.H. Yoo, J.M. Kum, S.O. Cho, Improved conversion efficiency of CdS quantum dot-sensitized TiO_2 nanotube-arrays using $CuInS_2$ as a co-sensitizer and an energy barrier layer, J. Mater. Chem. 21 (2011) 16430–16435, https://doi.org/10.1039/C1JM13616J.

[168] X. Gou, F. Cheng, Y. Shi, L. Zhang, S. Peng, J. Chen, P. Shen, Shape-controlled synthesis of ternary chalcogenide $ZnIn_2S_4$ and $CuIn(S,Se)_2$ nano-/microstructures via facile solution route, J. Am. Chem. Soc. 128 (2006) 7222–7229, https://doi.org/10.1021/ja0580845.

408 Oxide free nanomaterials for energy storage and conversion applications

[169] S. Shen, L. Zhao, L. Guo, Crystallite, optical and photocatalytic properties of visible-light-driven $ZnIn_2S_4$ photocatalysts synthesized via a surfactant-assisted hydrothermal method, Mater. Res. Bull. 44 (2009) 100–105, https://doi.org/10.1016/j.materresbull.2008.03.027.

[170] B.B. Kale, J.-O. Baeg, S.M. Lee, H. Chang, S.-J. Moon, C.W. Lee, $CdIn_2S_4$ nanotubes and "Marigold" nanostructures: a visible-light photocatalyst, Adv. Funct. Mater. 16 (2006) 1349–1354, https://doi.org/10.1002/adfm.200500525.

[171] H. Zhao, Y. He, M. Liu, R. Wang, Y. Li, W. You, Biomolecule-assisted, cost-effective synthesis of a $Zn_{0.9}Cd_{0.1}S$ solid solution for efficient photocatalytic hydrogen production under visible light, Chinese J. Catal. 39 (2018) 495–501, https://doi.org/10.1016/S1872-2067(17)62946-2.

[172] T. Takayama, I. Tsuji, N. Aono, M. Harada, T. Okuda, A. Iwase, H. Kato, A. Kudo, Development of various metal sulfide photocatalysts consisting of d0, d5, and d10 metal ions for sacrificial H_2 evolution under visible light irradiation, Chem. Lett. 46 (2017) 616–619, https://doi.org/10.1246/cl.161192.

[173] J. Xu, W. Sun, Z. Hu, Y. You, Y. Lu, R. Zhou, J. Zhang, Green and room-temperature synthesis of aqueous $CuInS_2$ and Cu_2SnS_3 nanocrystals for efficient photoelectrochemical water splitting, Mater. Today Energy 10 (2018) 200–207, https://doi.org/10.1016/j.mtener.2018.09.010.

Chapter 16

Oxides free materials for photocatalytic water splitting

M.L. Aruna Kumari

Department of Chemistry, Ramaiah College of Arts, Science and Commerce, Bengaluru, India

Chapter outline

1. Introduction	409	4.1 Metal chalcogenides	414
2. Fundamental aspects of photocatalytic water splitting	412	4.2 Metal pnictides	419
		4.3 Metal carbides	424
3. Experimental needs for photocatalytic water splitting	413	5. Conclusion and future aspects	427
		References	427
4. Oxide-free materials for photocatalytic water splitting	414		

1. Introduction

Our dependency on sustainable energy is expanded owing to the rapid population growth and rise in living standards. The development of an inexpensive alternate approach over the traditional use of fossil fuel for energy production is the main challenge for the scientific community. This crucial problem can be alleviated by the production of green fuel hydrogen (H_2) by efficient utilization of two renewable resources such as water and solar energy. H_2 production by Photocatalytic water splitting (PWS) is emerged as a fascinating subject by resolving energy and environmental issues and stood as a benchmark in bringing an energy revolution. The PWS process converts photonic energy into chemical energy and stores it within the H_2 molecules. Thus, the stored energy can be then converted into electricity as per demand through the oxidation of this H_2 fuel, which releases water molecules as a byproduct.

Even though in 1874, Jules Verne mentioned the possibility of H_2 generation by water electrolysis with a comment that "water will be the coal of future," the research on water splitting was triggered by the discovery of PEC water splitting by Fujishima and Honda in 1972 using TiO_2 photoanode [1,2]. From then researchers published several research articles on developing new photocatalyst and on experimental design for PWS. Inspired by the photosynthesis process, in 1979 Bard designed a new approach instead of a single photocatalyst system where he used two photocatalyst systems with redox mediators and called as

---

Oxide Free Nanomaterials for Energy Storage and Conversion Applications.
https://doi.org/10.1016/B978-0-12-823936-0.00003-6
Copyright © 2022 Elsevier Inc. All rights reserved.

Z-Scheme [3]. Even with long research history is H_2 production from PWS is a challenging process.

The PWS process using a photocatalyst is similar in a mechanistic way to photosynthesis, hence it is also termed artificial photosynthesis (Fig. 1). Both the process exhibits three similar reaction processes i.e., light-harvesting, charge generation, followed by separation and catalytic reactions [4]. Hence the overall efficiency is relayed on thermodynamic and kinetic consideration of the aforementioned processes. The catalytic splitting of water into H_2 and O_2 involves a 4-electron process and requires a positive change in Gibb's free energy ($\Delta G° = +237\,kJ\,mol^{-1}$). This indicates that 1.3 eV of photon energy corresponds to ~1000 nm wavelength is thermodynamically required to drive the PWS [5].

There are three main pathways are available for H_2 production from PWS such as (Fig. 2), (i) Photocatalytic overall water splitting (POWS), which yields a stoichiometric amount of H_2 and O_2 (Eq. 1); (ii) Photocatalytic partial water splitting (PPWS), where sacrificial reagents are used for efficient charge separation by abstracting photogenerated holes (Eq. 2); (iii) Photocatalytic intermediate water splitting (PIWS) is a promising approach since it produces H_2O_2 (along with H_2O), which completely avoids the reversible water formation from H_2 and O_2 (Eq. 3) [6].

$$2H_2O \rightarrow 2H_2 + O_2 \quad (1)$$

$$2H_2O + \text{Reagent} \rightarrow 2H_2 + \begin{array}{c}\text{oxidized form}\\\text{of reagent}\end{array} \quad (2)$$

$$2H_2O \rightarrow H_2 + H_2O_2 \quad (3)$$

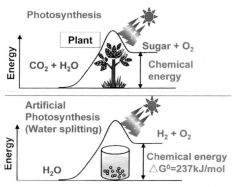

FIG. 1 Represents conceptual similarities between natural and artificial photosynthesis processes. *(Reproduced with permission from A. Kudo, Y. Miseki, Heterogeneous photocatalyst materials for water splitting, Chem. Soc. Rev. 38 (2009) 253–278.)*

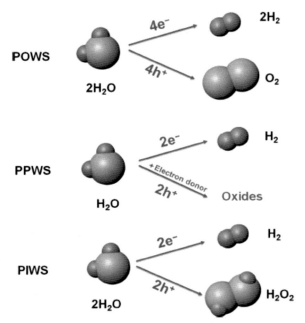

FIG. 2 Schematic representation of POWS, PPWS, and PIWS processes. *(Reproduced with permission from S. Cao, L. Piao, X. Chen, Emerging photocatalysts for hydrogen evolution, Trends Chem. 2 (2020) 57–70.)*

A huge number of photocatalysts like metal oxides, nitrides, carbides, chalcogenides, phosphides, and metal halides are reported for H_2 generation via PWS. To increase the efficiency of above-said materials various modifications are adopted namely; metal/nonmetal doping, surface deposition, heterojunction formation, composites preparations, using co-catalysts by designing their hierarchical structures. These modifications alleviate many problems by tuning light absorption properties, desired band edge position, morphology, phase purity, surface area, and crystallographic quality, etc. Even after extensive research upon photocatalysts design and development, it's difficult to analyze which physicochemical factors dominate the net activity of PWS.

To date, more research work has been carried out on metal oxides owing to their stability, but its large bandgap limits its work under UV-light illumination hence they cannot operate under visible light, this issue is addressed with emerged oxide-free materials. To the best of the author's knowledge, no comprehensive reports are available on metal oxide-free material mediated PWS. This chapter contains important aspects such as principles and processes involved in PWS and a brief discussion on experimental methods and criteria for evaluating photocatalytic activity and various oxide-free nanomaterials used in PWS.

2. Fundamental aspects of photocatalytic water splitting

A schematic illustration of PWS is shown in Fig. 3. The PWS involves three sequential competing steps [6].

(1) Charge carrier generation: When a photocatalyst is illuminated with photon energy higher than its bandgap it generates charge carriers i.e., e^- in the conduction band (CB) and h^+ in the valence band (VB) further these e^--h^+ pairs are involved in redox reactions. For efficient PWS, photocatalyst should possess suitable alignment between the band edge position of photocatalyst and the redox potentials of the water. Precisely, CB minimum must be located at a negative potential than H^+/H_2 energy level (0 V vs NHE at pH 0) and VB maxima must be at a positive potential than O_2/H_2O (1.23 V vs NHE at pH 0).

(2) Charge carrier separation: This step involves the separation and migration of charge carriers towards the surface. The lifetime of photogenerated charge carriers is in few nanoseconds time scales but they easily recombine in femtosecond timescale with the influence of electronic structure and effective masses. The inhibition of the charge carrier's recombination is a key step in PWS since it competes with desired catalytic reaction.

(3) Surface catalytic reaction: After the migration of charge carriers towards the surface they participate in redox reactions. The e^- and h^+ after the reaction with water yields H_2 and O_2 respectively.

The first two steps depend upon various properties of photocatalyst like band gap, band edge positions, crystal structure, the density of defects, surface area, etc. The highly crystalline material increases the charge carrier generation and

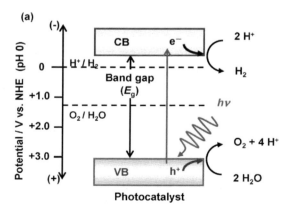

FIG. 3 Schematic representation of desired energy levels for photocatalytic water splitting. *(Reproduced with permission from T. Hisatomi, J. Kubota, K. Domen, Recent advances in semiconductors for photocatalytic and photoelectrochemical water splitting, Chem. Soc. Rev. 42 (2014) 7520–7535.)*

more defects enhance the recombination. The third step efficiency can be further enhanced by utilizing co-catalysts (usually noble metals) to produce an active site and to reduce the activation energy for gas evolution [7]. Many informative review articles are published on PWS on photocatalyst development, kinetic process, and reaction system [1,3,5–18].

3. Experimental needs for photocatalytic water splitting

(a) Apparatus

Reactor: The PWS process is usually carried out either in a vacuum using closed-circulation or at atmospheric pressure using a flow system. In a closed-circulation system, the photocatalyst is dispersed in a reaction medium and evacuated adequately to remove the dissolved air then inert gas is circulated to keep the homogeneous gas composition and avoid boiling of the reaction solution. Whereas in an open flow system, there will be a continuous inert gas flow will be there through the headspace of a reactor. These two systems' gaseous products are analyzed using a reaction system integrated gas chromatography to avoid air contamination. The pressure generated during the reaction affects the water-splitting process by reversible water formation which is termed as a surface back reaction. At high pressure desorption of generated H_2 and O_2 will takes place which induces surface back reaction, which should be inhibited for practical application.

Light source: For the illumination of photocatalyst various light sources like Xe lamp, halogen lamp Hg lamp, and light-emitting diodes are used. The appropriate filters are attached to the light source to get better efficiency. Further, the efficiency of the light source depends upon distance, position, and photon flux. To check the proper photo response some control experiments such as in absence of photocatalysts/irradiation must be carried out to confirm the photocatalytic reaction and neglect the possibility of the mechano-catalytic water splitting. There are many other points that researchers must pay attention to. The details of experiments and quantitative approach are described in the literature [12–14].

(b) Efficiency calculation

The activities of photocatalysts are usually assessed by the rates of gas (O_2 and H_2) evolution [mol h^{-1}] per catalyst amount[g] under similar reaction conditions. However, it is practically difficult to compare the activities measured in different reaction systems because of variations in light sources. Therefore, it is necessary to compare how much photons contribute to the photocatalytic reactions hence, two parameters i.e., quantum yield (QY) or apparent quantum yield (AQY) are used as standard measures of activity [8,10].

$$QY\,(\%) = \frac{\text{Number of reacted electrons}}{\text{Number of absorbed photon}} \times 100$$

$$AQY = \frac{\text{No of electrons involved} \times \text{rate of gas evolution}}{\text{Rate of incident photons}}$$

4. Oxide-free materials for photocatalytic water splitting

The oxide-free materials such as metal chalcogenides (sulfides, selenides, tellurides, oxy sulfides), metal pnictides (nitrides, oxynitrides, phosphides), and metal carbides, received considerable attention in PWS. In these materials oxygen in oxide material is replaced with different anions like S, Se, Te, N, P, and C partially or completely, which constitute the VBM instead of oxygen. This replacement leads to narrowing of the bandgap and leads to visible light activity.

The doping of anion in metal oxide is a conventional method used for bandgap tuning in photocatalytic material. In such cases, anion dopant acts as a visible light absorption center and recombination center based on its concentration [19]. This creates a discrete forbidden band which is disadvantageous for photogenerated h^+ migration, which is overcome by oxide-free materials.

4.1 Metal chalcogenides

To date, numerous types of metal chalcogenides like metal sulfides, metal tellurides, metal selenides, and oxysulfides were used as a suitable photocatalyst in PWS, because of their suitable band edge position to harvest visible spectrum. But photo-corrosion is a key issue on their performance; hence sacrificial reagents are used to trap the photogenerated hole, which avoids corrosion.

4.1.1 Metal sulfides

Among these metal chalcogenides, metal sulfides received more attention due to their suitable band edge position, small effective mass, robust quantum size effect. The empty valence band and small effective mass of metal sulfides have a great contribution to charge mobility and visible light activity. When compared to the metal oxide, metal sulfides have a narrow bandgap since their VB consists of S 3p orbital instead of O 2p orbital [20]. The replacement of sulfur with oxygen laid a route for the generation of sensitive, selective, and stable photocatalytic materials for energy conversion applications but is widely used in optoelectronic devices because of their promising optical and electric properties.

To date, several hundreds of metal sulfide materials have been prepared by a variety of routes in which the sulfur anion is combined with a metal cation to form mono-metal sulfides of type MS, M_2S, M_3S_4, and MS_2 or bimetal sulfides leads to $A_{1-x}B_xS_y$, where x and y are integers or tri-metal sulfides. Among the various mono-metal sulfides CdS, ZnS, CuS, SnS, MOS_2, NiS_2, CoS_2, Bi_2S_3,

IN_2S_3, and bimetallic metal sulfides $ZnIn_2S_4$, $Mn_{1-x}Cd_xS$, Cu_3TaS_4 are widely employed in PWS. However, the H_2 production performance of metal sulfide photocatalysts is still limited because of the fast recombination of photoexcited carriers. Besides, photo corrosion is also a major problem, therefore, many researchers have focused on solving the problems of photo corrosion and charge carrier recombination [21]. Several strategies have been using for enhancing the PWS activity like doping with metal/nonmetal atoms, noble metal deposition, heterojunction formation, morphology controlled and graphene-based structure, etc.

Doping with metal/nonmetal will alter the band edge position and enhance the activity under visible light. When ZnS doped with low-cost metal cations such as Cu, Ni, and Pb exhibits an outstanding photocatalytic activity [22–25] and bandgap shrinkage was observed for Ni-doped ZnS photocatalysts [26]. When rare earth dopants such as La^{3+}, Ce^{3+}, Gd^{3+}, Er^{3+}, and Y^{3+} doped in $ZnIn_2S_4$ exhibited improved efficiency in H_2 production in terms of $\sim46\%$, $\sim53\%$, $\sim61\%$, $\sim69\%$, and $\sim106\%$, respectively [27]. When Shen et al. prepared a series of Ca, Sr, and Ba doped $ZnIn_2S_4$ materials among them Ca-doped $ZnIn_2S_4$ displayed a higher PWS performance. Later, the same group doped Cr, Mn, Fe, and Co into the $ZnIn_2S_4$ to reduce the energy gap. Due to their narrow bandgap, the Mn-doped $ZnIn_2S_4$ enhanced the activity whereas Cr, Fe, and Co dopants retards the activity, the decreased activity is attributed to dopant-related impurity energy levels which act as recombination centers for photoexcited charge carriers of electrons and holes [28,29]. Shi et al. observed that after doping phosphor in CdS, S vacancies act as electron trapping centers thereby reduces the charge carrier recombination and improve the H_2 evolution [30]. Many noble metals like Au, Pd, Pt, and Ni have been deposited on metal sulfides, which also shows enhancement in PWS activity [31–37].

It has been reported that the formation of a heterojunction interface with other semiconductors or coupling with other cocatalysts (metal oxide or metal sulfide) can overcome the carrier recombination and photo corrosion problems. The binary heterojunction such as TiO_2-CdS core-shell is prepared by Zubair et al., via the hydrothermal method with varying shell thickness (Fig. 4). With optimized sample TiO_2-CdS (3:2) this showed better H_2 production by PWS under visible region 954 $\mu mol\ g^{-1}\ h^{-1}$ of hydrogen which is ~1.4 and ~1.7 times higher than pure CdS nanoparticles and pure TiO_2, respectively [38].

Qin et al. prepared Nitrogen-doped hydrogenated TiO_2 modified with CdS nanorods ($CdS/Ti^{3+}/N-TiO_2$) photocatalyst which shows efficient H_2 evolution under visible light irradiation. The higher activity is ascribed to bandgap tuning via Ti^{3+}/O_V defects. i.e., Ti^{3+}/O_V and doped N energy level locates below the CB and above the VB of TiO_2 respectively, which shorten the bandgap from two inverse directions and enhanced optical absorption, charge separation, and thereby PWS. The $10\%CdS/Ti^{3+}/N-TiO_2$ has the best photocatalytic H_2 evolution activity [39]. Wang et al. reported that CuS/ZnS nanomaterials exhibit high visible-light-induced H_2 generation activity. The H_2 generation

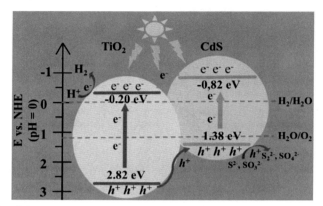

FIG. 4 Pictorial representation desired energy levels for H_2 generation via PWS method utilizing core-shell TiO_2-CdS nanocomposite. *(Reproduced with permission from M. Zubair, I.-H. Svenum, M. Rønning, J. Yang, Facile synthesis approach for core-shell TiO₂–CdS nanoparticles for enhanced photocatalytic H₂ generation from water, Catal. Today 328 (2019) 15–20.)*

rate increases with increasing Cu^{2+} ions. However, as with other cocatalysts, when the maximum amount of Cu^{2+} is reached (above 7 mol%), the hydrogen evolution rate decreases significantly. This is due to light-shielding by excess CuS, which reduces the number of active sites on the surface [40]. By one-pot hydrothermal strategy, Qiu et al. prepared cobalt sulfide (Co_9S_8) hollow cubes decorated by CdS QDs. This exhibits higher activity in PWS in presence of sacrificial electron donor Na_2S and Na_2SO_3, which prevent the backward reaction. The hollow structure of Co_9S_8 leads to multiple reflections of light inside the cavity, thereby leads to higher photocatalytic activity than the heterojunction [41]. Yao et al. prepared a ternary hollow nano-shell TiO_2-Au-CdS photocatalyst is fabricated by a hardcore template method with surface modification, cation exchange, and sulfidation process. They studied the effect of matching the Mie scattering peak with the absorption peak of semiconductors by varying the inner diameters of TiO_2 nano-shells from 150, 185, 225, to 255 nm. When the Mie scattering peak of TiO_2 nano-shell (inner diameter = 185 nm) matches the absorption band of CdS, it yields 669.7 $\mu mol\ h^{-1}\ g^{-1}$ of H_2 under visible light [42]. Jiang et al. prepared a ternary composite of metal sulfides (In_2S_3/MoS_2/CdS composite) via a simple hydrothermal method which showed higher activity and stability under visible light irradiation. The matched band edge position facilitates the hole transfer from VB of CdS to that of In_2S_3, which prevents the hole accumulation and inhibits photo corrosion, this dramatically enhanced stability of the composite. Thus, prepared composite promotes the interfacial charge transfer process and minimizes the charge carrier recombination which leads to efficient PWS [43].

Silver sulfide was used as a cocatalyst of ZnS photocatalyst to enhance the H_2 evolution rate. For instance, Hsu et al. reported the use of Ag_2S-coupled ZnO@ZnS core-shell nanorods to achieve efficient H_2 production [44]. Iwashina et al. demonstrated Z-schematic water splitting using various metal sulfides and RGO-TiO_2 (rutile) composite as an H_2-evolving and O_2-evolving photocatalyst respectively. When ZnS, $AgGaS_2$, $AgInS_2$, Ag_2ZnGeS_2, and Ag_2ZnSnS_2 were employed Only H_2 evolved due to photo corrosion, whereas when $CuGaS_2$, $CuInS_2$, Cu_2ZnGeS_2, and Cu_2ZnSnS_2 were employed H_2 and O_2 evolved in a stoichiometric amount [45]. Liu et al. reviewed the application of metal sulfide/MOF or its derivative composites in H_2evolution. These systems take advantage of the merits of both metal sulfide and MOF by avoiding the drawbacks of each component. The metal sulfides have an excellent capability to harvest solar energy, while the MOFs facilitate the metal sulfides dispersion and improve the activity and stability of metal sulfides [46]. Lim et al. prepared a bimetallic sulfite ($ZnIn_2S_4$ microspheres) by hydrothermal method and co-catalysts like CuS, Ag_2S, and MoS_2 are photo deposited on these microspheres to optimize the contact interfacial charge transfer. This photocatalyst works efficiently under both UV and visible light irradiation. MoS_2 photo deposited $ZnIn_2S_4$ shows profound improvement with maximum photocatalytic H_2 evolution yields, 47.71 μ mol h^{-1} and 30.56 μ mol h^{-1} under UV and visible light irradiation respectively [47].

4.1.2 Metal selenides

Metal selenides having narrow bandgaps for photocatalytic overall pure water splitting have not yet been reported due to the severe self-photooxidation. Chen et al. demonstrated that the solid solutions of zinc selenide and copper gallium selenide (ZnSe:CGSe), with absorption edges ranging from 480 to 750 nm can be employed for PWS [48]. Many theoretical studies have been carried out on the use of metal selenides as photocatalytic material in PWS, Zhuang et al. perform a systematic theoretical study of the single-layer transition metal dichalcogenides MX_2 (M = Nb, Mo, Ta, W, Ti, V, Zr, Hf, and Pt; X = S, Se, and Te) and demonstrated to be potential photocatalysts for PWS using a first-principles design approach [49].

Recently Silva et al. carried out the theoretical studies on Metal Chalcogenides Janus Monolayers of type Ga_2XY and In_2XY (X = S, Se, Te; Y = S, Se, Te). They found that all materials, Ga_2SSe, Ga_2STe, Ga_2SeTe, In_2SSe, In_2STe, and In_2SeTe, are mechanically stable and have a similar structure as their common counterparts (GaS, GaSe, GaTe, InS, InSe, and InTe) but their electronic properties are distinct, due to the difference in the electronegativities of the chalcogens in different sides of the materials. The lower exciton binding energies of these catalysts arise from the internal field which fulfills all required conditions for hydrogen generation by PWS (Fig. 5) [50].

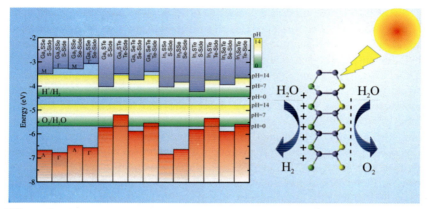

FIG. 5 The electronic band alignment of metal chalcogenides janus monolayers of type Ga$_2$XY and In$_2$XY (X=S, Se, Te; Y=S, Se, Te). *(Reproduced with permission from R. da Silva, R. Barbosa, R.R. Mançano, N. Durães, R.B. Pontes, R.H. Miwa, A. Fazzio, J.E. Padilha, Metal chalcogenides janus monolayers for efficient hydrogen generation by photocatalytic water splitting, ACS Appl. Nano Mater. 2 (2019) 890–897.)*

4.1.3 Metal tellurides

Metal tellurides have been explored extensively just like sulfides and selenides due to their surface chemistry, quantum confinement, and physical properties. But because of its instability its widely used in thermoelectric materials, only a few articles are published in the field of PWS and Xu et al. accounted in their review article [51]. The metal ion (Ni^{2+}, Fe^{2+}, and Co^{2+}) incorporated CdTe QDs were examined for H$_2$ production by the PWS method. Ni^{2+} ions incorporated CdTe effectively produce shallow trapping states to capture photoexcited electrons. These dopants can serve as catalytic sites and substantially improve the photocatalytic activity of QDs [52].

Wang et al. developed an artificial water-soluble [FeFe]-H$_2$ase mimic by incorporating a cyanide group onto three hydrophilic ether chains to the active site of the [FeFe]-H$_2$ase to improve its solubility in water. The generated the H$_2$ in an aqueous solution containing [FeFe]-H$_2$ase mimic which is coupled with CdTe QDs in presence of ascorbic acid as an electron donor. This system produces 786 mmol (17.6 mL) H$_2$ upon the irradiation >400 nm photons with 505 and 50 h^{-1} values TON and TOF respectively even after 10 h [53].

4.1.4 Metal oxysulfides

In oxysulfide's surface S^{2-} ions are stabilized by hybridization of the S-3p and O-2p orbitals and are expected to be more resistant for self-oxidation. To date, only fewer oxysulfide materials were prepared and used in PWS applications. The lack of metal oxysulfide photocatalysts is mainly based on the strong differences between the size and electronegativity of oxygen and sulfur [54].

Wang et al., demonstrated promising photocatalytic water-splitting activity for $Y_2Ti_2O_5S_2$ under solar light owing to its narrow bandgap (1.9 eV) and directly decomposes water into H_2 and O_2 in a stoichiometric ratio. This result opens a new opportunity of using oxy-sulfides in both one-step and Z-scheme water-splitting systems [55].

4.2 Metal pnictides

In metal pnictides, oxyanion of metal oxide has been replaced with nitrides and phosphides and they found huge application in the field of H_2 Storage, Li-batteries, thermo-electrics, superconductivity, and catalysis, etc. Among them, some metal nitrides, oxynitrides, and metal phosphides act as efficient photocatalytic materials in PWS.

4.2.1 Metal nitrides

In metal nitrides O^{2-} is replaced by N^{3-} to main, the charge balance two N^{3-} substitutes three O^{2-} in comparison to metal oxides. Many metal nitrides are synthesized to date but only a few of them like GaN, Ta_3N_5, VN, InGaN, Ni_3N, Co_3N, Ge_3N_4, etc. are utilized in PWS [56]. Among these metal nitrides, GaN is most widely used even its inactive. Domen et al. showed the higher activity of GaN by using along with cocatalyst and by metal ion doping. In 2006, they demonstrated the effect of divalent metal ions (Zn^{2+} and Mg^{2+}) on water splitting in presence of RuO_3 as a promoter and the Mg^{2+} doped GaN showed better performance in PWS [57]. In 2007, they prepared the GaN and modified it with $Rh_{2-y}Cr_yO_3$ nanoparticles as a co-catalyst and showed the H_2 evolution under UV light illumination [58]. They also prepared GaN and ZnO solid solutions with various modifications and achieved water splitting under visible light illumination [59–64]. Wang et al. prepared GaN nanowire by molecular beam epitaxy with the incorporation of Rh/Cr_2O_3, which acts as a co-catalyst and prevents the back reaction in PWS and showed a higher turnover number per unit time (as shown in Fig. 6) [65].

Another choice of metal nitride is Ta_3N_5 which has an orthorhombic structure and is one of the most common photocatalysts employed for water splitting. It generates hydrogen from water by absorbing light up to 600 nm. Domen et al. [66,67] prepared Ta_3N_5 which has a bandgap of 2.1 eV and used it as photocatalyst along with the sacrificial electron donor and acceptor in PWS. Later in 2012 to overcome the drawback of insufficient crystallization upon thermal nitridation which enhances undesirable charge recombination and reduces the photocatalytic efficiency. This drawback was overcome by modifying the surface of the starting Ta_2N_3 with a small amount of alkaline metal. This demonstrated a sixfold improvement in photocatalytic activity for O_2 evolution under visible light by PWS [68]. They further prepared Ta_3N_5 nanoparticles from a mesoporous carbon nitride template, this shows higher activity than

420 Oxide free nanomaterials for energy storage and conversion applications

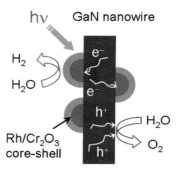

FIG. 6 Schematic illustration of water splitting on GaN nanowires by incorporating an Rh/Cr$_2$O$_3$-core-shell nanostructure as a cocatalyst. *(Reproduced with permission from D. Wang, A. Pierre, M. G. Kibria, K. Cui, X. Han, K.H. Bevan, H. Guo, S. Paradis, A.-R. Hakima. Z. Mi, Wafer-level photocatalytic water splitting on GaN nanowire arrays grown by molecular beam epitaxy, Nano Lett. 11 (2011) 2353–2357.)*

the bulk Ta$_3$N$_5$ for photocatalytic hydrogen evolution under visible light irradiation [69]. Tian et al. reported that metallic vanadium nitride (VN) along with CdS as a co-catalyst yield PWS with an H$_2$ production rate of 6.24 mmol h^{-1} g^{-1} [70] (Fig. 7).

Another known metal-nitride photocatalyst is InGaN, its bandgap can be tuned to cover the entire solar spectrum by varying the ratio between In and Ga, higher the in content enhances the H$_2$O splitting [71,72]. Kibria et al. reported the InGaN/GaN nanowire hetero-structure decorated with Rh/Cr$_2$O$_3$ core-shell nanoparticles produce H$_2$ from PWS under UV light, blue, and green-light irradiation [73]. Recently Liu et al. prepared cobalt zinc nitride coupled with carbon black (Co$_3$ZnN/C) which yields H$_2$ in an Eosin Y-sensitized system under visible light with the rate of 15.4 μmol mg^{-1} h^{-1} as shown in Fig. 8 [74].

4.2.2 Metal oxynitrides

Oxynitride is an emerging material in the field of PWS since it inherits the benefits of both oxides and nitrides. In oxy nitrides the 2p orbital of O is partially replaced with 2p orbital of N, this lowers the bandgap and activates the material efficiency under visible light. The bandgap can be tuned by changing the transition material and ratio of N/O as shown in Fig. 9. The oxynitride material exhibit greater stability and corrosion resistance compared to metal nitrides [75,76]. To date very few oxynitrides are used in PWS TaON, Ti$_3$O$_3$N$_2$, Zr$_3$O$_3$N$_2$, and Mixed Metal oxynitrides (Ga$_{1-x}$Zn$_x$)(N$_{1-x}$O$_x$) were explored Domen et al. prepared the TaON which has a suitable band edge position for water splitting. In presence of sacrificial electron donor (Ag$^+$) TaON generates H$_2$ under visible light [77].

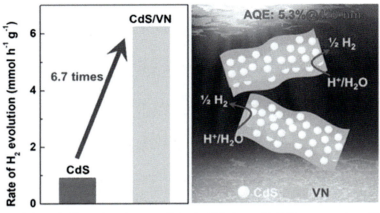

FIG. 7 Schematic illustration of water splitting by VN along with CdS as a cocatalyst. *(Reproduced with permission from L. Tian, S. Min, F. Wang, Z. Zhang, Metallic vanadium nitride as a noble-metal-free cocatalyst efficiently catalyzes photocatalytic hydrogen production with CdS nanoparticles under visible light irradiation, J. Phys. Chem. C 123 (2019) 28640–28650.)*

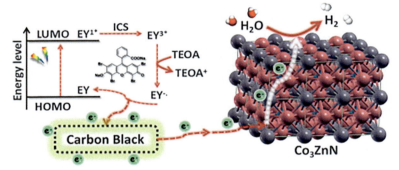

FIG. 8 Schematic representation of the mechanistic aspect of H_2 evolution in Eosin Y sensitized Co_3ZnN/C. *(Reproduced with permission from S. Liu, X. Meng, S. Adimi, H. Guo, W. Qi, J.P. Attfield, M. Yang, Efficient photocatalytic hydrogen evolution over carbon supported antiperovskite cobalt zinc nitride, Chem. Eng. J. 408, (2020) 127307.)*

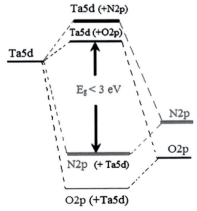

FIG. 9 Schematic band structure of tantalum oxynitride. *(Reproduced with permission from M. Ahmed, G. Xinxin, A review of metal oxynitrides for photocatalysis, Inorg. Chem. Front. 3 (2016) 578–590.)*

422 Oxide free nanomaterials for energy storage and conversion applications

Mixed metal oxynitrides $(Ga_{1-x}Zn_x)(N_{1-x}O_x)$ were prepared by solid solution method which has an absorption edge around 500 nm yields PWS with a quantum yield of 52% [78]. Madea et al. [79] prepared GaN and ZnO solid solution with various modifications and achieved water splitting under visible light illumination. Among these, $(Ga_{1-x}Zn_x)(N_{1-x}O_x)$ with $x = 0.12$ loaded with a 5 wt.% RuO_2 showed higher PWS efficiency. Similarly, when its surface is loaded with mixed oxide of Rh and Cr as cocatalyst it enhances the PWS. The function of the Rh-Cr-O mixed oxide nanoparticles is to trap the excited electron and holes to harness them for the PWS. The Rh^{3+} are the H_2 evolution sites and the Cr^{3+} oxides shell prevents the back reaction [79–81]. Kamata et al. prepared the new photocatalytic material Ga-Zn-In-O-N mixed oxynitride by forming a solid solution, which has the potential for PWS under visible light irradiation (near 600 nm) [82].

4.2.3 Metal phosphides

In PWS, noble metals play an important role as co-catalyst in developing a semiconductor-metalloid interface, which is necessary for electron transfer as well as H_2 reduction. Recently researchers found that, transition metal phosphides (TMP's) such as MoP, Ni_2P, CoP, FeP, NiCoP as an alternative to noble metal. The phosphorus in TMP's will abstract the proton and promotes H_2 production. TMP's have diversely been used in PWS because of their high electrical conductivity, low overpotential, better reduction ability, high stability, and photocatalytic activity. In comparison with nitrides and sulfides, metal phosphides perform well for catalytic applications such as redox reactions, aromatization, hydrogenolysis, etc. [83,84].

Among various TMP's Ni_2P is widely investigated as a cocatalyst by various research groups. Du et al. [85] deposit Ni_2P nanoparticles on CdS nanorods by one-pot hydrothermal method. Ni_2P traps the photo-generated electrons and holes by sacrificial agent, this process reduces the charge carrier recombination and improves the photocatalytic H_2 evolution activity under visible light irradiation. Indra et al. [86] prepared an integrated Ni_2 sP-C_3N_4 catalyst and tested its efficiency for PWS, they found that the photo-generated electrons from g-C_3N_4 are trapped by Ni_2P since unoccupied molecular orbital (LUMO) of Ni_2P has lower energy than the CB of g-C_3N_4 but higher energy than the H^+/H_2 reduction potential. The scheme of Ni_2P as cocatalysts can be summarized in Fig. 10.

When CoP is anchored on the surface of $Cd_xZn_{1-x}Se$ nanobelts, the optimized molar ratio of Cd/Zn showed better PWS activity. The $Cd_{0.25}Zn_{0.75}Se/CoP$ system achieves the highest photocatalytic H_2 evolution rate at 36.6 mmol $g^{-1} h^{-1}$ in artificial seawater [87]. Zhang et al. reported the preparation of CoP/TiO_2 hybrid at 350 nm for PWS application. For 0.5 wt.% CoP/TiO_2, AQE is found to be 3.8%. This demonstrated the highly stable activity of CoP/TiO_2 composite in long-standing photocatalytic

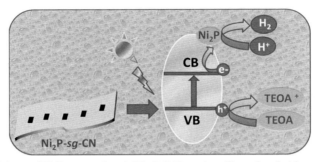

FIG. 10 Scheme of Ni$_2$P as cocatalysts in Ni$_2$sP-C$_3$N$_4$ catalyst. *(Reproduced with permission from A. Indra, A. Acharjya, P.W. Menezes, C. Merschjann, D. Hollmann, M. Schwarze, M. Aktas, A. Friedrich, S. Lochbrunner, A. Thomas, M. Driess, Boosting visible-light-driven photocatalytic hydrogen evolution with an integrated nickel phosphide-carbon nitride system, Angew. Chem. Int. Ed. 56 (2017) 1653–1657.)*

experiments [88]. Yue et al. used Cu$_3$P as a cocatalyst to couple with TiO$_2$ and demonstrated that the optimum Cu$_3$P/TiO$_2$ exhibits 11 times higher activity than that of bare TiO$_2$ in H$_2$ evolution. The enhanced activity is attributed to effective charge separation arises due to the strong interaction between Cu$_3$P and TiO$_2$ [89]. Cao et al. developed a CdS heterojunction with various TMPs nanoparticles (CoP, Ni$_2$P, and Cu$_3$P) as cocatalysts and L(+)-lactic acid as an electron donor (Fig. 11). Among these 3 TMPs/CdS, the CoP/CdS hybrid catalyst yields H$_2$ production of 254,000 μmol h^{-1} g^{-1} during 4.5 h of sunlight irradiation. The system exhibited excellent stability which could maintain its high photocatalytic activity even after 100 h of light irradiation [90].

Sun et al. prepared CoP QDs/CdS binary photocatalyst using ZIF-67 as a template. With optimal CoP concentration, the binary photocatalyst exhibits super high photocatalytic activity (55.2 times higher) compared to bare CdS [84]. Choi et al. prepared CdS/MoS$_2$@Ni$_2$P ternary nanohybrid which shows higher activity 69-fold and 6-folds than bare CdS and CdS-Pt nanocomposites respectively. This proved the use of the Ni$_2$P as a better alternative cocatalyst than Pt in PWS [91]. WP is also used as a cocatalyst with Zn$_x$Cd$_{1-x}$S solid solution, this system yields 15,028.6 μmol g^{-1} h^{-1} H$_2$ under visible light irradiation by PWS [92].

Besides cocatalyst, some of the TMP's are also used as photocatalytic material in PWS, for example, Tian et al. used CoP nanowires in the presence of the probe of the human immunodeficiency virus (PHIV) via the dye-sensitized process and demonstrated the H$_2$ evolution. Under light illumination, this system retards the charge carrier recombination by transferring the electron from PHIV into the CB of CoP and holes scavenged by sacrificial reagent (TEOA) [93]. The CoP and WP were also investigated as a photocatalyst for H$_2$ evolution in presence of Eosin dye via dye sensitization mechanism [94].

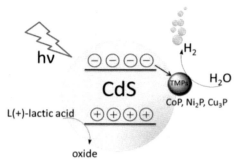

FIG. 11 A schematic representation TMPs (CoP, Ni$_2$P, and Cu$_3$P) as cocatalysts in TMPs/CdS hybrid photocatalysts for visible-light-driven H$_2$ production. *(Reproduced with permission from S. Cao, Y. Chen, C. Wang, X. Lv, W.F. Fu, Spectacular photocatalytic hydrogen evolution using metal-phosphide/CdS hybrid catalysts under sunlight irradiation, Chem. Commun. 51 (2015) 8708–8711.)*

4.3 Metal carbides

Transition metal carbides attracted considerable attention as catalysts because of their availability with high specific areas, high conductivity, high chemical stability, and thermal stability and have been prepared and investigated in various areas. When compared to bulk metal carbide, 2D metal carbides like Ti$_2$C, Ti$_3$C$_2$, Nb$_4$C$_3$, Ta$_4$C$_3$, V$_2$C, etc., are widely used in PWS and commonly termed as MXenes. These 2D materials have some diverse properties than those of other-dimensional and structural materials. The morphological anisotropy, large surface area, mechanical stability, and nearly atomic layer thickness are the promising properties of 2D materials.

In metal carbide-dependent PWS, MXenes are widely used compared to normal metal carbides, to date only Mo$_2$C and WC metal carbides are explored. Dong et al., prepared a molybdenum-rich molybdenum carbide (Mo-Mo$_2$C) loaded onto graphitic carbon nitride (g-C$_3$N$_4$) and tested it in PWS application. They found that the cocatalyst loaded 2.0 wt.% Mo-Mo$_2$C/g-C$_3$N$_4$ composite has excellent photocatalytic performance with H$_2$ evolution as high as 219.7 µmol g^{-1} h^{-1}, which is 440 times higher than that of g-C$_3$N$_4$ and 90% higher than 0.5 wt% Pt/g-C$_3$N$_4$ photocatalyst [95]. Ma et al. have shown that Mo$_2$C can act as a cocatalyst instead of noble metals, they prepared a novel noble-metal-free Mo$_2$C-In$_2$S$_3$ heterojunction that was synthesized by a simple hydrothermal method. The optimized Mo$_2$C-In$_2$S$_3$ sample has the highest photocatalytic performance with lactic acid as a sacrificial agent. This system yields 535.58 L mol g^{-1} h^{-1} of H$_2$, which was 175.6 times higher than the bare In$_2$S$_3$ [96]. Pan et al. showed the ability of tungsten carbide (WC) as a co-catalyst with CdS QDs and TiO$_2$ as a photocatalyst. The CdS/WC/TiO$_2$ photocatalyst has better PWS activity under visible light in presence of lactic acid as an electron donor. The optimal H$_2$ evolution rate on

CdS/WC/TiO$_2$ (624.9 μmol h^{-1}) is comparable to the rate on CdS/Pt/TiO$_2$ (636.2 μmol h^{-1}), indicating that WC is a good candidate to substitute Pt as the co-catalyst [97]. Wang et al. used Rod-like molybdenum carbide (Mo$_2$C) microcrystals in PWS as a photocatalyst via dye sensitization mechanism using Erythrosin B (ErB) as the photosensitizer and triethanolamine (TEOA) as the sacrificial reagent. This system exhibited promising H$_2$ evolution activity even under light irradiation with a long wavelength. The high quantum efficiency of 29.7% was obtained at 480 nm [98].

MXenes has recently attracted intense attention which is produced by etching out the A layers from MAX phases, where M stands for early transition metal (Ti, V, Nb, Ta), A represents an A-group (mostly groups 13 and 14) element, and X corresponds to carbon. MXene is gaining attention as a suitable alternative for promoting photocatalytic performance because of its flexible adjustability of elemental composition, regular layered structure, and excellent electrical conductivity. Coupling MXene with other semiconductors is regarded as a potential photocatalyst for H$_2$ production (Fig. 12) [99,100]. Since it's an emerging field only a few reports were there on in PWS application of MXene as a cocatalyst coupled with benchmark metal oxide TiO$_2$ and transition metal sulfates such as CdS and WS$_2$.

Li et al. studied the effect of two photocatalysts namely TiO$_2$ and octahedral (1T) phase WS$_2$ along with Ti$_3$C$_2$ MXene co-catalyst in photocatalytic H$_2$ evolution. The 1T-WS$_2$@TiO$_2$@Ti$_3$C$_2$ composite with optimized 15 wt.% WS$_2$ shows 50 times higher photocatalytic H$_2$ evolution than that of TiO$_2$ nanosheets. The enhanced activity arises due to an increase of electron transfer efficiency by conductive Ti$_3$C$_2$ MXene and 1T-WS$_2$ [101] (Fig. 13).

Su et al. prepared a series of Ti$_3$C$_2$T$_x$/TiO$_2$ composite photocatalysts with a monolayer and multilayers Ti$_3$C$_2$T$_x$ as the cocatalyst and observed the enhanced

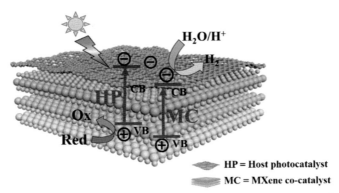

FIG. 12 A schematic representation MXenes as cocatalysts in a hybrid photocatalyst system. *(Reproduced with permission from L. Cheng, X. Li, H. Zhang, Q. Xiang, Two-dimensional transition metal MXene-based photocatalysts for solar fuel generation, J. Phys. Chem. Lett. 10 (2019) 3488–3494.)*

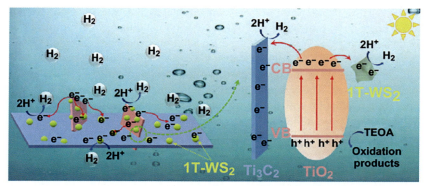

FIG. 13 Pictorial representation of the photocatalytic mechanism of octahedral phase WS$_2$@TiO$_2$@Ti$_3$C$_2$ composite system. *(Reproduced with permission from Y. Li, L. Ding, S. Yin, Z. Liang, Y. Xue, X. Wang, H. Cui, J. Tian, Photocatalytic H$_2$ evolution on TiO$_2$ assembled with Ti$_3$C$_2$ MXene and metallic 1T-WS$_2$ as co-catalysts, Nano-Micro Lett. 12 (2020) 6.)*

activity in monolayers over multilayer Ti$_3$C$_2$T$_x$ MXenes [102]. Very recently, Li et al. had successfully designed g-C$_3$N$_4$@Ti$_3$C$_2$T$_x$ QDs composites by a self-assembly method. As expected, the composite showed better activity compared to individual counterparts with the best photocatalytic activity (5111.8 molgcat^{-1}h^{-1}) under artificial sunlight [103].

Ran et al. coupled Ti$_3$C$_2$T$_x$ with CdS via a hydrothermal method exhibits better performance with the optimized composition (2.5 wt.% Ti$_3$C$_2$T$_x$) [104]. Similarly, Xiao et al. coupled Ti$_3$C$_2$T$_x$ with CdS nanorod to construct a Schottky heterojunction which showed sevenfold better performance than that of pristine CdS (Fig. 14) [105]. Tie et al. decorated ZnS nanoparticles with Ti$_3$C$_2$T$_x$ which demonstrated a better H$_2$ production rate of 502.6 molgcat^{-1}h^{-1} under optimal conditions, which is almost fourfold higher than pure ZnS [106].

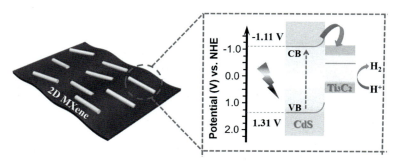

FIG. 14 Pictorial representation of 1D CdS/2D MXene heterojunction for photocatalytic H$_2$ production. *(Reproduced with permission from R. Xiao, C. Zhao, Z. Zou, Z. Chen, L. Tian, H. Xu, H. Tang, Q. Liu, Z. Lin, X. Yang, In situ fabrication of 1D CdS nanorod/2D Ti$_3$C$_2$ MXene nanosheet Schottky heterojunction toward enhanced photocatalytic hydrogen evolution, Appl. Catal. B 268 (2019) 118382.)*

5. Conclusion and future aspects

This chapter presented the fundamental aspects and experimental needs of photocatalytic water splitting. To achieve unassisted water splitting the bandgap of photocatalyst should be in well match with reduction and oxidation potentials of water so that photoexcited electrons and holes have enough overpotential to drive hydrogen and oxygen evolution reactions. In PWS metal oxides are bloomed as active photocatalytic materials because of their photostability however they suffer from the drawback of wide bandgap and fast charge carrier recombination. To overcome the aforementioned problem, researchers fabricated the VB edge positions by replacing oxygen from S, Se, Te, N, P, and C partially or completely to synthesize oxide-free materials such as sulfides, selenides, tellurides, oxy sulfides, nitrides, oxynitrides, phosphides, and metal carbides. These oxide-free materials are potential candidates for solar water splitting but stability and self-corrosion are the major hurdles to ineffective utilization. However, most of them are used as co-catalyst in PWS since they suffer from photo instability which limits long-term operational durability this can be abide using sacrificial agents to improve performance.

Among metal chalcogenides, metal sulfides are extensively used as photocatalysts and as cocatalysts in the form of composite or heterojunctions. Whereas minimal work has been done on selenides and tellurides, but theoretical studies showed that metal selenide and telluride Janus monolayers are efficient photocatalysts for H_2 generation. In the case of oxysulfides difference in size and electronegativity of oxygen and sulfur inhibits the stable material preparation, this can be overcome by using two metals to synthesize bimetallic oxysulfides. In metal pnictides, metal nitrides and oxynitrides are widely used as cocatalyst whereas metal phosphides are found as an alternative to the noble metal which will abstract the proton and increases H_2 production in PWS. In carbides, only Mo_2C and WC metal carbides are explored as photocatalysts in PWS but recently, 2D metal carbides (MXenes) have emerged as new photocatalysts. These MXenes are a suitable alternative for promoting photocatalytic performance because of their flexible adjustability of elemental composition, regular layered structure, and excellent electrical conductivity. Coupling MXene with other semiconductors is regarded as a potential photocatalyst for H_2 production.

References

[1] F.E. Osterloh, B.A. Parkinson, Recent developments in solar water-splitting photocatalysis, MRS Bull. 36 (2011) 17–22.

[2] A. Fujishima, K. Honda, Electrochemical photolysis of water at a semiconductor electrode, Nature 238 (1972) 37–38.

[3] A.J. Bard, Photoelectrochemistry and heterogeneous photo-catalysis at semiconductors, J. Photochem. 10 (1979) 59–75.

428 Oxide free nanomaterials for energy storage and conversion applications

[4] A. Kudo, Y. Miseki, Heterogeneous photocatalyst materials for water splitting, Chem. Soc. Rev. 38 (2009) 253–278.

[5] A.A. Ismail, D.W. Bahnemann, Photochemical splitting of water for hydrogen production by photocatalysis: a review, Sol. Energy Mater. Sol. Cells 128 (2014) 85–101.

[6] S. Cao, L. Piao, X. Chen, Emerging photocatalysts for hydrogen evolution, Trends Chem. 2 (2020) 57–70.

[7] K. Maeda, Photocatalytic water splitting using semiconductor particles: history and recent developments, J. Photochem. Photobiol. C 12 (2011) 237–268.

[8] F.E. Osterloh, Inorganic nanostructures for photoelectrochemical and photocatalytic water splitting, Chem. Soc. Rev. 42 (2013) 2294–2320.

[9] T. Hisatomi, T. Minegishi, K. Domen, Kinetic assessment and numerical modeling of photocatalytic water splitting toward efficient solar hydrogen production, Bull. Chem. Soc. Jpn. 85 (2012) 647–655.

[10] T. Hisatomi, J. Kubota, K. Domen, Recent advances in semiconductors for photocatalytic and photoelectrochemical water splitting, Chem. Soc. Rev. 42 (2014) 7520–7535.

[11] K. Maeda, K. Domen, Photocatalytic water splitting: recent progress and future challenges, J. Phys. Chem. Lett. 1 (2010) 2655–2661.

[12] D. Jing, L. Guo, L. Zhao, X. Zhang, H. Liu, M. Li, S. Shen, G. Liu, X. Hu, X. Zhang, Efficient solar hydrogen production by photocatalytic water splitting: from fundamental study to pilot demonstration, Int. J. Hydrogen Energy 35 (2010) 7087–7097.

[13] K. Takanabe, Photocatalytic water splitting: quantitative approaches toward photocatalyst by design, ACS Catal. 7 (2017) 8006–8022.

[14] B. Ohtani, Preparing articles on photocatalysis-beyond the illusions, misconceptions, and speculation, Chem. Lett. 37 (2008) 216–229.

[15] L. Lin, T. Hisatomi, S. Chen, T. Takata, K. Domen, Visible-light-driven photocatalytic water splitting: recent progress and challenges, Trends Chem. 2 (2020) 813–824.

[16] K. Maeda, Z-scheme water splitting using two different semiconductor photocatalysts, ACS Catal. 3 (2013) 1486–1503.

[17] K. Takanabe, K. Domen, Preparation of inorganic photocatalytic materials for overall water splitting, ChemCatChem 4 (2012) 1485–1497.

[18] K. Maeda, K. Domen, New non-oxide photocatalysts designed for overall water splitting under visible light, J. Phys. Chem. C 111 (2007) 7851–7861.

[19] L.G. Devi, R. Kavitha, A review on non-metal ion doped titania for the photocatalytic degradation of organic pollutants under UV/solar light: role of photogenerated charge carrier dynamics in enhancing the activity, Appl. Catal. Environ. 140-141 (2013) 559–587.

[20] Y. Shiga, N. Umezawa, N. Srinivasan, S. Koyasu, E. Sakai, M. Miyauchi, A metal sulfide photocatalyst composed of ubiquitous elements for solar hydrogen production, Chem. Commun. 52 (2016) 7470–7473.

[21] (a) S. Chandrasekaran, L. Yao, L. Deng, C. Bowen, Y. Zhang, S. Chen, Z. Lin, F. Peng, P. Zhang, Recent advances in metal sulfides: from controlled fabrication to electrocatalytic, photocatalytic and photoelectrochemical water splitting and beyond, Chem. Soc. Rev. 48 (2019) 4178–4280. (b) S.L. Lee, C.-J. Chang, Recent progress on metal sulfide composite nanomaterials for photocatalytic hydrogen production, Catalysts 9 (2019) 457.

[22] A. Kudo, M. Sekizawa, Photocatalytic H_2 evolution under visible light irradiation on Ni-doped ZnS photocatalyst, Chem. Commun. (2000) 1371–1372.

[23] I. Tsuji, A. Kudo, H_2 evolution from aqueous sulfite under visible-light irradiation over Pb and halogen-codoped ZnS photocatalysts, J. Photochem. Photobiol. A 156 (2003) 249–252.

Oxides free materials **Chapter | 16   429**

[24] J. Zhang, S. Liu, J. Yu, M. Jaroniec, A simple cation exchange approach to Bi-doped ZnS hollow spheres with enhanced UV and visible-light photocatalytic H_2-production activity, J. Mater. Chem. 21 (2011) 14655–14662.

[25] T. Arai, S.I. Senda, Y. Sato, H. Takahashi, K. Shinoda, B. Jeyadevan, K. Tohji, Cu-doped ZnS hollow particle with high activity for hydrogen generation from alkaline sulfide solution under visible light, Chem. Mater. 20 (2008) 1997–2000.

[26] L.A. Silva, S.Y. Ryu, J. Choi, W. Choi, M.R. Hoffmann, Photocatalytic hydrogen production with visible light over Pt-interlinked hybrid composites of cubic-phase and hexagonal-phase CdS, J. Phys. Chem. C 112 (2008) 12069–12073.

[27] F. Tian, R. Zhu, Y. He, F. Ouyang, Improving photocatalytic activity for hydrogen evolution over $ZnIn_2S_4$ under visible light: a case study of rare earth modification, Int. J. Hydrogen Energy 39 (2014) 6335–6344.

[28] S. Shen, L. Zhao, X. Guan, L. Guo, Improving visible-light photocatalytic activity for hydrogen evolution over $ZnIn_2S_4$: a case study of alkaline-earth metal doping, J. Phys. Chem. Solid 73 (2012) 79–83.

[29] S. Shen, J. Chen, X. Wang, L. Zhao, L. Guo, Microwave-assisted hydrothermal synthesis of transition-metal doped $ZnIn_2S_4$ and its photocatalytic activity for hydrogen evolution under visible light, J. Power Sources 196 (2011) 10112–10119.

[30] R. Shi, H.F. Ye, F. Liang, Z. Wang, K. Li, Y. Weng, Z. Lin, W.F. Fu, C.M. Che, Y. Chen, Interstitial P-doped CdS with long-lived photogenerated electrons for photocatalytic water splitting without sacrificial agents, Adv. Mater. 30 (2018) 1705941.

[31] X. Yu, A. Shavel, X. An, Z. Luo, M. Ibanez, A. Cabot, Cu_2ZnSnS_4-Pt and Cu_2ZnSnS_4-Au heterostructured nanoparticles for photocatalytic water splitting and pollutant degradation, J. Am. Chem. Soc. 136 (2014) 9236–9239.

[32] Y.P. Yuan, L.W. Ruan, J. Barber, S.C.J. Loo, C. Xue, Hetero-nanostructured suspended photocatalysts for solar-to-fuel conversion, Energ. Environ. Sci. 7 (2014) 3934–3951.

[33] X. Zhang, Y.L. Chen, R.S. Liu, D.P. Tsai, Plasmonic photocatalysis, Rep. Prog. Phys. 76 (2013), 046401.

[34] H. Yan, J. Yang, G. Ma, G. Wu, X. Zong, Z. Lei, J. Shi, C. Li, Visible-light-driven hydrogen production with extremely high quantum efficiency on Pt–PdS/CdS photocatalyst, J. Catal. 266 (2009) 165–168.

[35] Y. Shemesh, J.E. Macdonald, G. Menagen, U. Banin, Synthesis and photocatalytic properties of a family of CdS-PdX hybrid nanoparticles, Angew. Chem. Int. Ed. 123 (2011) 1217–1221.

[36] F. Meng, J. Li, S.K. Cushing, M. Zhi, N. Wu, Solar hydrogen generation by nanoscale p–n junction of p-type molybdenum disulfide/n-type nitrogen-doped reduced graphene oxide, J. Am. Chem. Soc. 135 (2013) 10286–10289.

[37] L. Jia, D.H. Wang, Y.X. Huang, A.W. Xu, H.Q. Yu, Highly durable N-doped graphene/CdS nanocomposites with enhanced photocatalytic hydrogen evolution from water under visible light irradiation, J. Phys. Chem. C 115 (2011) 11466–11473.

[38] M. Zubair, I.-H. Svenum, M. Rønning, J. Yang, Facile synthesis approach for core-shell TiO_2–CdS nanoparticles for enhanced photocatalytic H_2 generation from water, Catal. Today 328 (2019) 15–20.

[39] Y. Qin, H. Li, J. Lu, F. Meng, C. Ma, Y. Yan, M. Meng, Nitrogen-doped hydrogenated TiO_2 modified with CdS nanorods with enhanced optical absorption, charge separation and photocatalytic hydrogen evolution, Chem. Eng. J. 384 (2020) 123275.

[40] L. Wang, H. Chen, L. Xiao, J. Huang, CuS/ZnS hexagonal plates with enhanced hydrogen evolution activity under visible light irradiation, Powder Technol. 288 (2016) 103–108.

430 Oxide free nanomaterials for energy storage and conversion applications

[41] B. Qiu, Q. Zhu, M. Du, L. Fan, M. Xing, J. Zhang, Efficient solar Light harvesting CdS/Co_9S_8 hollow cubes for Z-scheme photocatalytic water splitting, Angew. Chem. Int. Ed. 56 (2017) 2684–2688.

[42] X. Yao, X. Hu, W. Zhang, X. Gong, X. Wang, S.C. Pillai, D.D. Dionysiou, D. Wang, Mie resonance in hollow nanoshells of ternary TiO_2-Au-CdS and enhanced photocatalytic hydrogen evolution, Appl Catal B 276 (2020) 119153.

[43] W. Jiang, Y. Liu, R. Zong, Z. Li, W. Yao, Y. Zhu, Photocatalytic hydrogen generation on bifunctional ternary heterostructured $In_2S_3/MoS_2/CdS$ composite with high activity and stability under visible light irradiation, J. Mater. Chem. A 3 (2015) 18406–18412.

[44] M.H. Hsu, C.J. Chang, H.T. Weng, Effcient H_2 production using Ag_2S-coupled ZnO@ZnS core–shell nanorods decorated metal wire mesh as an immobilized hierarchical photocatalyst, ACS Sustain. Chem. Eng. 4 (2016) 1381–1391.

[45] K. Iwashina, A. Iwase, Y.H. Ng, R. Amal, A. Kudo, Z-schematic water splitting into H_2 and O_2 using metal sulfide as a hydrogen-evolving photocatalyst and reduced graphene oxide as a solid-state electron mediator, J. Am. Chem. Soc. 137 (2015) 604–607.

[46] Y. Liu, D. Huang, M. Cheng, Z. Liu, C. Lai, C. Zhang, C. Zhou, W. Xiong, L. Qin, B. Shao, Q. Liang, Metal sulfide/MOF-based composites as visible-light-driven photocatalysts for enhanced hydrogen production from water splitting, Coord. Chem. Rev. 409 (2020) 213220.

[47] W.Y. Lim, M. Hong, G.W. Ho, In situ photo-assisted deposition and photocatalysis of $ZnIn_2S_4$/transition metal chalcogenides for enhanced degradation and hydrogen evolution under visible light, Dalton Trans. 45 (2016) 552–560.

[48] S. Chen, G. Ma, Q. Wang, S. Sun, T. Hisatomi, T. Hisatomi, T. Higashi, Z. Wang, M. Nakabayashi, N. Shibata, Z. Pan, T. Hayashi, T. Minegishi, T. Takata, K. Domen, Metal selenide photocatalysts for visible-light driven Z-scheme pure water splitting, J. Mater. Chem. A 7 (2019) 7415–7422.

[49] H.L. Zhuang, R.G. Hennig, Computational search for single-layer transition-metal dichalcogenide photocatalysts, J. Phys. Chem. C 117 (2013) 20440–20445.

[50] R. da Silva, R. Barbosa, R.R. Mançano, N. Durães, R.B. Pontes, R.H. Miwa, A. Fazzio, J.E. Padilha, Metal chalcogenides janus monolayers for efficient hydrogen generation by photocatalytic water splitting, ACS Appl. Nano Mater. 2 (2019) 890–897.

[51] Y. Xu, Y. Huang, B. Zhang, Rational design of semiconductor-based photocatalysts for advanced photocatalytic hydrogen production: the case of cadmium chalcogenides, Inorg. Chem. Front. 3 (2016) 591–615.

[52] J. Xu, J. Wang, Z. Chen, X. Xia, S. Li, Z. Li, Boosting photocatalytic hydrogen generation of cadmium telluride colloidal quantum dots by nickel ion doping, J. Colloid Interface Sci. 549 (2019) 63–71.

[53] F. Wang, W. Wang, X. Wang, H. Wang, C. Tung, L. Wu, A highly efficient photocatalytic system for hydrogen production by a robust hydrogenase mimic in an aqueous solution, Angew. Chem. Int. Ed. 50 (2011) 3193–3197.

[54] C. Larquet, S. Carenco, Metal oxysulfides: from bulk compounds to nanomaterials, Front. Chem. 8 (2020) 179.

[55] M. Wang, M. Nakabayashi, T. Hisatomi, S. Sun, S. Akiyama, Z. Wang, Z. Pan, X. Xiao, T. Watanabe, T. Yamada, N. Shibata, T. Takata, K. Domen, Oxysulfide photocatalyst for visible-light-driven overall water splitting, Nat. Mater. 18 (2019) 827–832.

[56] N. Han, P. Liu, J. Jiang, L. Ai, Z. Shao, S. Liu, Recent advances in nanostructured metal nitrides for water splitting, J. Mater. Chem. A 41 (2018) 19912–19933.

[57] N. Arai, N. Saito, H. Nishiyama, Y. Inoue, K. Domen, K. Sato, Overall water splitting by RuO_2-dispersed divalent-ion-doped GaN photocatalysts with d10 electronic configuration, Chem. Lett. 35 (2006) 796–797.

Oxides free materials **Chapter | 16 431**

[58] K. Maeda, K. Teramura, N. Saito, Y. Inoue, K. Domen, Photocatalytic overall water splitting on gallium nitride powder, Bull. Chem. Soc. Jpn. 80 (2007) 1004–1010.

[59] K. Maeda, T. Takata, M. Hara, N. Saito, Y. Inoue, H. Kobayashi, K. Domen, GaN:ZnO solid solution as a photocatalyst for visible-light-driven overall water splitting, J. Am. Chem. Soc. 127 (2005) 8286–8287.

[60] K. Maeda, K. Teramura, T. Takata, M. Hara, N. Saito, K. Toda, Y. Inoue, H. Kobayashi, K. Domen, Overall water splitting on $(Ga_{1-x}Zn_x)(N_{1-x}O_x)$ solid solution photocatalyst: relationship between physical properties and photocatalytic activity, J. Phys. Chem. B 109 (2005) 20504–20510.

[61] K. Maeda, K. Teramura, N. Saito, Y. Inoue, K. Domen, Improvement of photocatalytic activity of $(Ga_{1-x}Zn_x)(N_{1-x}O_x)$ solid solution for overall water splitting by co-loading Cr and another transition metal, J. Catal. 243 (2006) 303–308.

[62] K. Maeda, H. Hashiguchi, H. Masuda, R. Abe, K. Domen, Photocatalytic activity of $(Ga_{1-x}Zn_x)(N_{1-x}O_x)$ for visible-light-driven H_2 and O_2 evolution in the presence of sacrificial reagents, J. Phys. Chem. C 112 (2008) 3447–3452.

[63] T. Hisatomi, K. Maeda, K. Takanabe, J. Kubota, K. Domen, Aspects of the water splitting mechanism on $(Ga_{1-x}Zn_x)(N_{1-x}O_x)$ photocatalyst modified with $Rh_{2-y}Cr_yO_3$ cocatalyst, J. Phys. Chem. C 113 (2009) 21458–21466.

[64] K. Maeda, A. Xiong, T. Yoshinaga, T. Ikeda, N. Sakamoto, T. Hisatomi, M. Takashima, D. Lu, M. Kanehara, T. Setoyama, Photocatalytic overall water splitting promoted by two different cocatalysts for hydrogen and oxygen evolution under visible light, Angew. Chem. Int. Ed. 112 (2010) 4190–4193.

[65] D. Wang, A. Pierre, M.G. Kibria, K. Cui, X. Han, K.H. Bevan, H. Guo, S. Paradis, A.-R. Hakima, Z. Mi, Wafer-level photocatalytic water splitting on GaN nanowire arrays grown by molecular beam epitaxy, Nano Lett. 11 (2011) 2353–2357.

[66] G. Hitoki, A. Ishikawa, T. Takata, J.N. Kondo, M. Hara, K. Domen, Ta_3N_5 as a novel visible light-driven photocatalyst ($\lambda < 600$ nm), Chem. Lett. 31 (2002) 736–737.

[67] M. Hara, G. Hitoki, T. Takata, J.N. Kondo, H. Kobayashi, K. Domen, TaON and Ta_3N_5 as new visible light driven photocatalysts, Catal. Today 78 (2003) 555–560.

[68] S.S.K. Ma, T. Hisatomi, K. Maeda, Y. Moriya, K. Domen, Enhanced water oxidation on Ta_3N_5 photocatalysts by modification with alkaline metal salts, J. Am. Chem. Soc. 134 (2012) 19993–19996.

[69] L. Yuliati, J.H. Yang, X. Wang, K. Maeda, T. Takata, M. Antonietti, K. Domen, Highly active tantalum(V) nitride nanoparticles prepared from a mesoporous carbon nitride template for photocatalytic hydrogen evolution under visible light irradiation, J. Mater. Chem. 20 (2010) 4295–4298.

[70] L. Tian, S. Min, F. Wang, Z. Zhang, Metallic vanadium nitride as a noble-metal-free cocatalyst efficiently catalyzes photocatalytic hydrogen production with CdS nanoparticles under visible light irradiation, J. Phys. Chem. C 123 (2019) 28640–28650.

[71] J. Wu, W. Walukiewicz, K. Yu, W. Shan, J. Ager Iii, E. Haller, H. Lu, W.J. Schaff, W. Metzger, S. Kurtz, Superior radiation resistance of $In_{1-x}Ga_xN$ alloys: full-solar-spectrum photovoltaic material system, J. Appl. Phys. 94 (2003) 6477–6482.

[72] J. Li, J. Lin, H. Jiang, Direct hydrogen gas generation by using InGaN epilayers as working electrodes, Appl. Phys. Lett. 93 (2008) 162107.

[73] M.G. Kibria, H.P. Nguyen, K. Cui, S. Zhao, D. Liu, H. Guo, M.L. Trudeau, S. Paradis, A.-R. Hakima, Z. Mi, One-step overall water splitting under visible light using multiband InGaN/GaN nanowire heterostructures, ACS Nano 7 (2013) 7886–7893.

[74] S. Liu, X. Meng, S. Adimi, H. Guo, W. Qi, J.P. Attfield, M. Yang, Efficient photocatalytic hydrogen evolution over carbon supported antiperovskite cobalt zinc nitride, Chem. Eng. J. 408 (2020) 127307.

432 Oxide free nanomaterials for energy storage and conversion applications

[75] Y. Moriya, T. Takata, K. Domen, Recent progress in the development of (oxy)nitride photocatalysts for water splitting under visible-light irradiation, Coord. Chem. Rev. 257 (2013) 1957–1969.

[76] M. Ahmed, G. Xinxin, A review of metal oxynitrides for photocatalysis, Inorg. Chem. Front. 3 (2016) 578–590.

[77] G. Hitoki, T. Takata, J.N. Kondo, M. Hara, H. Kobayashi, K. Domen, An oxynitride, TaON, as an efficient water oxidation photocatalyst under visible light irradiation ($\lambda < 500$ nm), Chem. Commun. (16) (2002) 1698–1699.

[78] H.N. Kim, Photocatalytic Water Splitting Reactions Based on Tantalum Oxynitrides, Literature Seminar, October, vol. 15, University of Illinois Urbana-Champaign, Urbana, IL, 2013.

[79] K. Maeda, K. Teramura, N. Saito, Y. Inoue, H. Kobayashi, K. Domen, Overall water splitting using (oxy)nitride photocatalysts, Pure Appl. Chem. 78 (2006) 2267–2276.

[80] Y. Lee, H. Terashima, Y. Shimodaira, K. Teramura, M. Hara, H. Kobayashi, K. Domen, M. Yashima, Zinc germanium oxynitride as a photocatalyst for overall water splitting under visible light, J. Phys. Chem. C 111 (2007) 1042–1048.

[81] Z. Zou, H. Arakawa, Direct water splitting into H_2 and O_2 under visible light irradiation with a new series of mixed oxide semiconductor photocatalysts, J. Photochem. Photobiol. A 158 (2003) 145–150.

[82] K. Kamata, K. Maeda, D. Lu, Y. Kako, K. Domen, Synthesis and photocatalytic activity of gallium–zinc–indium mixed oxynitride for hydrogen and oxygen evolution under visible light, Chem. Phys. Lett. 470 (2009) 90–94.

[83] B. Luo, R. Song, J. Geng, X. Liu, D. Jing, M. Wang, C. Cheng, Towards the prominent cocatalytic effect of ultra-small CoP particles anchored on g-C_3N_4 nanosheets for visible light driven photocatalytic H_2 production, Appl Catal B 256 (2019) 117819.

[84] Q. Sun, Z. Yu, R. Jiang, Y. Hou, L. Sun, L. Qian, F. Li, M. Li, Q. Ran, H. Zhang, CoP QD anchored carbon skeleton modified CdS nanorods as a co-catalyst for photocatalytic hydrogen production, Nanoscale 12 (2020) 19203–19212.

[85] Z. Sun, H. Zheng, J. Li, P. Du, Extra-ordinarily efficient photocatalytic hydrogen evolution in water using semiconductor nanorods integrated with crystalline Ni_2P cocatalysts, Energ. Environ. Sci. 8 (2015) 2668–2676.

[86] A. Indra, A. Acharjya, P.W. Menezes, C. Merschjann, D. Hollmann, M. Schwarze, M. Aktas, A. Friedrich, S. Lochbrunner, A. Thomas, M. Driess, Boosting visible-light-driven photocatalytic hydrogen evolution with an integrated nickel phosphide-carbon nitride system, Angew. Chem. Int. Ed. 56 (2017) 1653–1657.

[87] B. Qiu, Q. Zhu, M. Xing, J. Zhang, A robust and efficient catalyst of $Cd_xZn_{1-x}Se$ motivated by CoP for photocatalytic hydrogen evolution under sunlight irradiation, Chem. Commun. 53 (2017) 897–900.

[88] X. Yue, S. Yi, R. Wang, Z. Zhang, S. Qiu, Cobalt phosphide modified titanium oxide nanophotocatalysts with significantly enhanced photocatalytic hydrogen evolution from water splitting, Small 13 (2017) 1603301.

[89] X. Yue, S. Yi, R. Wang, Z. Zhang, S. Qiu, A novel and highly efficient earth-abundant Cu_3P with TiO_2 "P-N" heterojunction nanophotocatalyst for hydrogen evolution from water, Nanoscale 8 (2016) 17516–17523.

[90] S. Cao, Y. Chen, C. Wang, X. Lv, W.F. Fu, Spectacular photocatalytic hydrogen evolution using metal-phosphide/CdS hybrid catalysts under sunlight irradiation, Chem. Commun. 51 (2015) 8708–8711.

[91] J. Choi, D.A. Reddy, N.S. Han, S. Jeong, S. Hong, D.P. Kumar, J.K. Song, T. Kim, Modulation of charge carrier pathways in CdS nanospheres by integrating MoS_2 and Ni_2P for

Oxides free materials **Chapter | 16** **433**

improved migration and separation toward enhanced photocatalytic hydrogen evolution, Cat. Sci. Technol. 7 (2017) 641–649.

[92] X. Yan, Z. Jin, Y. Zhang, Y. Zhang, H. Yuan, Sustainable and efficient hydrogen evolution over a noble metal-free WP double modified $Zn_xCd_{1-x}S$ photocatalyst driven by visible-light, Dalton Trans. 48 (2019) 11122–11135.

[93] J. Tian, N. Cheng, Q. Liu, W. Xing, X. Sun, Cobalt phosphide nanowires: efficient nanostructures for fluorescence sensing of biomolecules and photocatalytic evolution of dihydrogen from water under visible light, Angew. Chem. Int. Ed. 54 (2015) 5493–5497.

[94] Y. Li, Z. Jin, H. Liu, H. Wang, Y. Zhang, G. Wang, Unique photocatalytic activities of transition metal phosphide for hydrogen evolution, J. Colloid Interface Sci. 541 (2019) 287–299.

[95] J. Dong, Y. Shi, C. Huang, Q. Wu, T. Zeng, W. Yao, A new and stable $Mo-Mo_2C$ modified g-C_3N photocatalyst for efficient visible light photocatalytic H_2 production, Appl. Catal. Environ. 243 (2019) 27–35.

[96] X. Ma, C. Ren, H. Li, X. Liu, X. Li, K. Han, W. Li, Y. Zhan, A. Khan, Z. Chang, C. Sun, H. Zhou, A novel noble-metal-free $Mo_2C-In_2S_3$ heterojunction photocatalyst with efficient charge separation for enhanced photocatalytic H_2 evolution under visible light, J. Colloid Interface Sci. 582 (2021) 488–495.

[97] Y.-X. Pan, T. Zhou, J. Han, J. Hong, Y. Wang, W. Zhang, R. Xu, CdS quantum dots and tungsten carbide supported on anatase–rutile composite TiO_2 for highly efficient visible-light-driven photocatalytic H_2 evolution from water, Cat. Sci. Technol. 6 (2016) 2206–2213.

[98] Y. Wang, W. Tu, J. Hong, W. Zhang, R. Xu, Molybdenum carbide microcrystals: efficient and stable catalyst for photocatalytic H_2 evolution from water in the presence of dye sensitizer, J. Mater. 2 (2016) 344–349.

[99] L. Cheng, X. Li, H. Zhang, Q. Xiang, Two-dimensional transition metal MXene-based photocatalysts for solar fuel generation, J. Phys. Chem. Lett. 10 (2019) 3488–3494.

[100] H. Wang, R. Peng, Z.D. Hood, M. Naguib, S.P. Adhikari, Z. Wu, Titania composites with 2D transition metal carbides as photocatalysts for hydrogen production under visible-light irradiation, ChemSusChem 9 (2016) 1490–1497.

[101] Y. Li, L. Ding, S. Yin, Z. Liang, Y. Xue, X. Wang, H. Cui, J. Tian, Photocatalytic H_2 evolution on TiO_2 assembled with Ti_3C_2 MXene and metallic $1T-WS_2$ as co-catalysts, Nano-Micro Lett. 12 (2020) 6.

[102] T. Su, Z.D. Hood, M. Naguib, L. Bai, S. Luo, C.M. Rouleau, I.N. Ivanov, H. Ji, Z. Qin, Z. Wu, Monolayer $Ti_3C_2T_x$ as an effective Co-catalyst for enhanced photocatalytic hydrogen production over TiO_2, ACS Appl. Energy Mater. 2 (2019) 4640–4651.

[103] Y. Li, L. Ding, Y. Guo, Z. Liang, H. Cui, J. Tian, Boosting the photocatalytic ability of g-C_3N_4 for hydrogen production by Ti_3C_2 MXene quantum dots, ACS Appl. Mater. Interfaces 11 (2019) 41440–41447.

[104] J. Ran, G. Gao, F.-T. Li, T.-Y. Ma, A. Du, S.-Z. Qiao, Ti_3C_2 MXene co-catalyst on metal sulfide photo-absorbers for enhanced visible-light photocatalytic hydrogen production, Nat. Commun. 8 (2017) 13907.

[105] R. Xiao, C. Zhao, Z. Zou, Z. Chen, L. Tian, H. Xu, H. Tang, Q. Liu, Z. Lin, X. Yang, In situ fabrication of 1D CdS nanorod/2D Ti_3C_2 MXene nanosheet Schottky heterojunction toward enhanced photocatalytic hydrogen evolution, Appl Catal B 268 (2019) 118382.

[106] L. Tie, S. Yang, C. Yu, H. Chen, Y. Liu, S. Dong, J. Sun, In situ decoration of ZnS nanoparticles with Ti_3C_2 MXene nanosheets for efficient photocatalytic hydrogen evolution, J. Colloid Interface Sci. 45 (2019) 63–70.

Chapter 17

Oxide-free materials for thermoelectric and piezoelectric applications

Jayaraman Theerthagiri[*], Seung Jun Lee[*], and Myong Yong Choi
Department of Chemistry and Research Institute of Natural Sciences, Gyeongsang National University, Jinju, South Korea

Chapter outline

1. Introduction   435
2. Oxide-free materials for thermoelectric applications   436
      2.1 Basic working function of a thermoelectric device   436
      2.2 Metal chalcogenides (MX; X = S, Se, Te)   437
      2.3 Thermoelectric metal carbides and nitrides   442
3. Oxide-free materials for piezoelectric applications   443
      3.1 Basic working function of a piezoelectric device   443
      3.2 Piezoelectric metal chalcogenides and nitrides   443
4. Conclusions   446
   Acknowledgments   446
   References   446

1. Introduction

With the fast advancement of industrialization, the scarcity, and weariness of energy sources, for example, utilization of fossil fuels like petroleum products, has become a serious issue that genuinely limits the improvement of society [1, 2]. Further, the burning of fossil fuels substantially increases CO_2 emissions and is observed to be a significant reason for global warming. Because of the constant growth of energy consumption, it is crucial to substitute fossil fuels with the progression of clean renewable energy resources [3, 4]. In this pursuit, renewable energy harvesting devices (thermoelectric and piezoelectric devices) have progressively turned into a focused topic in the current years among scientists and academicians due to being eco-friendly, having no CO_2 emission, and being maintenance-free [5, 6].

---

* These authors contributed equally to this work.

Oxide Free Nanomaterials for Energy Storage and Conversion Applications.
https://doi.org/10.1016/B978-0-12-823936-0.00006-1
Copyright © 2022 Elsevier Inc. All rights reserved.

436 Oxide free nanomaterials for energy storage and conversion applications

Thermoelectric devices are utilized to generate electrical energy from heat energy, which can be utilized to harness electrical energy, even from human body heat. When the difference in temperature, ΔT, at the sides of a thermoelectric generator (TEG) arises, it creates power as per the temperature difference. TEGs working is mainly based on the Seebeck effect, which demonstrates that the performance of the TEGs is mostly dependent on the temperature difference and the performance of the materials. The thermoelectric material performance utilized in TEGs is determined by the figure of merit (ZT) of the materials which is well-defined by the ratio of power factor ($S^2\sigma T$) and thermal conductivity (κ), where S is the Seebeck coefficient, σ denotes electrical conductivity, and T denotes absolute temperature. For high-efficiency TEGs, the materials should have high S, high σ, and low κ. TEGs have several advantages, such as silent operation, no moving parts, easy availability, and high reliability, which can be utilized for the powering of wireless sensors and small electronic devices [7, 8]. Among the various thermoelectric materials, Sb_2Te_3 and Bi_2Te_3 have been highly investigated as efficient thermoelectric materials [9, 10]. On the contrary, in piezoelectric devices, the electrical energy produced from kinetic energy changes because of the mechanical deformation of the material. Piezoelectric generators (PEGs) are used in actuators, electrochemical sensors, and measurements/monitoring systems [6, 11]. Thermoplastic polymers (such as polymethyl methacrylate [PMMA], polyvinylidene fluoride [PVDF], and polypropylene [PP]) and their nanocomposite materials have gained more attention in piezoelectric applications [12]. Among these, the polymer PVDF was utilized more due to its flexibility and advantageous piezoelectric properties [13].

In this chapter, the basics of working function and recent developments of non-oxide materials for thermo and piezoelectric applications are focused on. Special attention has been paid to the metal chalcogenides (sulfides, selenides, and tellurides), metal carbides, and nitrides. Finally, the future perspectives and challenges for the improvement of oxide-free thermo and piezoelectric materials are discussed.

2. Oxide-free materials for thermoelectric applications

2.1 Basic working function of a thermoelectric device

The production of electricity using heat energy is known as the thermoelectric effect. Based on Joules Law, a current (electricity)-carrying conductor generates heat directly proportional to the resistance of the conducting substrate and the square of the electricity passing over it. In the 1820s, Thomas J. Seebeck differently interpreted this law by taking two different metals where the junctions of the metals reach various temperatures. A voltage was developed between the junctions which were directly proportional to the difference in the heat. As a result of his investigation, the electricity produced because of the temperature difference at the junctions of two metals is called the Seebeck

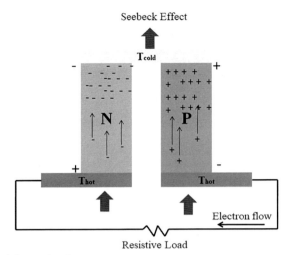

FIG. 1 Typical thermoelectric generators.

effect. The main applications of the Seebeck effect are the establishment of TEGs, which directly convert heat into electricity. In contrast, the Peltier effect that describes the thermoelectric cooler, which works in reverse of a TEG. The Peltier effect describes the heat absorption or dissipation at the junction of conducting materials. Based on the flow direction of the current, heat is either absorbed or dissipated by that point in the conducting material [5,14].

The main structure of TEGs consists of two primary junctions, a p-type (containing positive charge or holes) and an n-type (containing negative charge or electrons) semiconductor, which are connected by a metal strip that links them electrically in series (Fig. 1). The Seebeck effect arises because of the migration of charge carriers between the p-type and n-type semiconductors, where the holes are migrating to the n-type material from the p-type, and the electrons are migrating from the n-type to the p-type material. Charge carriers diffuse away from the hot side of the materials, leading to an accumulation of charge carriers at one end. Thus, the accumulation of charges produces a voltage that is proportional to the difference in temperature around the semiconductor materials [5].

Bi_2Te_3, Sb_2Te_3, and SiGe are some of the most commonly reported materials that are used in TEG applications. The subsequent sections of this chapter will discuss the advances in oxide-free materials for TEGs.

2.2 Metal chalcogenides (MX; X = S, Se, Te)

Transition metal chalcogenides are a series of materials comprised of transition metal and chalcogen elements (such as S, Se, Te), which show significant consideration in thermoelectric applications because of their high-power factor

438 Oxide free nanomaterials for energy storage and conversion applications

raised by their less covalent nature, low electronegativity, and reduced thermal conductivity. Furthermore, metal chalcogenides are easy to obtain in various structures, which benefits enhancing thermoelectric properties. Chen et al. [15] investigated the thermoelectric effect by doping Cu into β-In$_2$S$_3$ materials. A β-In$_{2-x}$Cu$_x$S$_3$ system with $x=0$–0.4 was synthesized by solid-state processing and pulsed current sintering. Doping Cu into β-In$_2$S$_3$ reduced the electrical resistivity with lattice thermal conductivity. The highest ZT of 0.51 was achieved for Cu $=0.05$ content at 700 K, and an average ZT of 0.31 was obtained in between 300 and 700 K, which was 1.3 times higher than that of pure β-In$_2$S$_3$ materials (0.3). Ge et al. [16] prepared Bi/Bi$_2$S$_3$ core-shell nanowire structured composite by the hydrothermal synthesis and showed the highest ZT value of 0.36 at 623 K after treating with hydrazine for 1 h. Treating with hydrazine enhanced the mobility ($\mu=11.5$ cm^2 V^{-1} S^{-1}) and carrier concentration ($n=20 \times 10^{19}$ cm^{-3}) of Bi/Bi$_2$S$_3$ samples. The untreated Bi/Bi$_2$S$_3$ materials showed $\mu=5.3$ cm^2 V^{-1} S^{-1}, and $n=1.9 \times 10^{19}$ cm^{-3}, which was higher than that of pristine Bi$_2$S$_3$ materials ($\mu=1.2$ cm^2 V^{-1} S^{-1}; $n=0.25 \times 10^{19}$ cm^{-3}). Furthermore, the increase of temperature improved the electrical conductivity (σ) of pure Bi$_2$S$_3$ from 5 S/m to 174 S/m, whereas initial σ values of 172 S/m and 3750 S/m are increased by 2 and 3 orders of magnitude for Bi/Bi$_2$S$_3$ treated with hydrazine for 1 h and 3 h, respectively. The resulting temperature dependence σ, Seebeck coefficient, and thermal power factor are shown in Fig. 2.

Ohta et al. [17] investigated the various stoichiometric and temperature effects of γ-Tb$_2$S$_{3-x}$ obtained by the reaction of Tb$_4$O$_7$ with CS$_2$ at 1323 K. Then, the obtained γ-Tb$_2$S$_{3-x}$ was heat-treated up to 1853 K and showed a decrease in S content. The γ-Tb$_2$S$_{3-x}$ displayed an n-type metallic nature in between the temperature of 300 and 1000 K due to the decrease of S content, along with resistivity and thermopower. However, the thermal conductivity was unchanged which might be because of phonon contribution, and the material achieved a maximum ZT of 0.17 at 873 K. Kumar et al. [18] investigated the effect of various substrate temperatures on Cu$_2$ZnSnS$_4$ (CZTS) films deposited using ultrasound wave-assisted CVD on the TEGs. The thermoelectric power factor for the CZTS prepared at 375°C showed a higher value of 7.1 µW K^{-2}, than that of lower temperatures at 275°C (0.4 µW K^{-2}) and 325°C (2.3 µW K^{-2}) which was because of an improved carrier concentration and higher mobility of charge carrier by high crystallinity and compactness of Cu$_{2-x}$S in CZTS films. Moghaddam et al. [8] developed Cu$_{5-x}$Zn$_x$FeS$_4$ and Cu$_{4.96}$Co$_{0.04}$Fe$_{1-x}$Zn$_x$S$_4$ systems via ball milling followed by hot pressing. In both crystalline systems, the Vickers microhardness of 167 Hv of pure bornite (Cu$_5$FeS$_4$) was increased to 250 Hv upon doping with Zn and Co. However, the thermal power factor of 0.27 mW m^{-1} K^{-2} decreased from pure bornite to Zn doping (0.25 mW m^{-1} K^{-2} for Cu$_{4.9}$Zn$_{0.1}$FeS$_4$), whereas increased power factor was observed by co-addition of Zn and Co (0.37 mW m^{-1} K^{-2} for Cu$_{4.96}$Co$_{0.04}$Zn$_{0.04}$Fe$_{0.96}$S$_4$) with a high ZT of 0.6 at 590 K. The doping strategy of Zn and crystalline structure of orthorhombic phase of bornite is shown in Fig. 3.

Oxide-free materials **Chapter | 17** 439

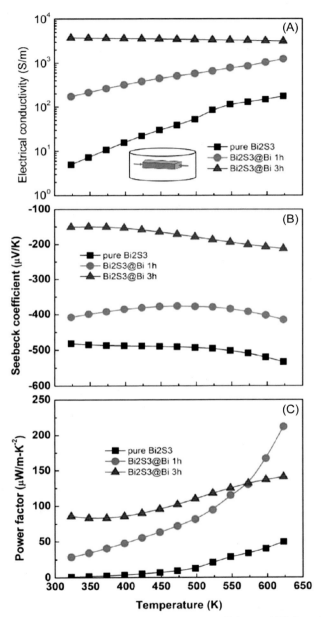

FIG. 2 The temperature dependence of (A) σ, (B) Seebeck coefficient, and (C) thermal power factor of pure Bi_2S_3, Bi/Bi_2S_3 composite treated with hydrazine for 1 h and 3 h [16]. Copyright (2020) American Chemical Society.

440 Oxide free nanomaterials for energy storage and conversion applications

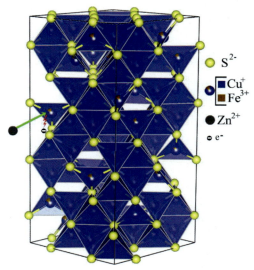

FIG. 3 The doping strategy of Zn and crystalline structure of orthorhombic phase of bornite [8]. Copyright (2020) Elsevier.

Metal selenide-based materials cover a broad range of thermoelectric applications due to their properties of various crystal structures, conduction types (ions or carriers), band gaps, and phase (polymorph or single-phase) configuration. The thermoelectric applications of some of the selected metal selenides for temperature are presented in Fig. 4 [19]. Patil et al. [20] fabricated Sb-incorporated Bi_2Se_3 thin films with a $Bi_{2-x}Sb_xSe_3$ system and used them as thermoelectric materials. The semiconducting nature of the fabricated thin films was examined using temperature-dependent electrical conductivity. Thermal conductivity of the $Bi_{2-x}Sb_xSe_3$ system was achieved a maximum of ZT was 0.036 at minimal Sb content of $X = 0.002$, and further increasing the Sb content decreased the thermoelectric behavior. The phononic influence on a semiconductor thermal conductivity is higher than the electronic influence. The phonon means free path is limited by the size of grains. The fine-grained crystalline structure holds high specific boundary scattering which leads to a reduction of thermal conductivity. Though more attention has been focused on reducing the size of grain for decreased thermal conductivity, it also subsequently decreases the electrical conductivity of the semiconductor material.

Among all materials, Bi_2Te_3 is mostly recognized as an effective thermoelectric material due to its potential to convert waste heat energy into electrical energy [9, 10]. In general, Bi behaves like metal, but in the chemical composition of Bi_2Te_3, Bi acts effectively as a semiconductor. The maximum ZT value obtained for the commercial Bi_2Te_3 material is around 1 which has been

Oxide-free materials **Chapter | 17** **441**

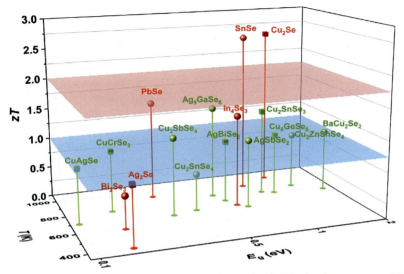

FIG. 4 Selected metal selenide materials with their thermoelectric ZT values for temperature [19]. Copyright (2020) Elsevier.

remained steadfast through the years. Recently, the physicochemical properties have been altered by doping or making composites. Sinduja et al. [21] investigated the effect of Cu ion implantation on Bi_2Te_3 toward the thermoelectric properties. The Seebeck coefficient test revealed that the conversion of n-type to p-type charge carriers upon Cu ion implantation. This behavior was because of the high density of defect complexes. Wu et al. [22] investigated why the mixed phase of bismuth telluride (Bi_2Te_3 and Bi_3Te_4) holds a higher charge carrier concentration 6.65×10^{20} cm^{-3}, low thermal conductivity (0.59 W/mK), and low resistivity (3.17×10^{-3} Ω cm). This behavior might be because of the mixed phases of telluride interfaces efficiently hindering phonon transport and causing low thermal conductivity. Jagadish et al. [23] developed multiwalled carbon nanotubes (MWCNTs) incorporated (wt% of 0.05–0.20) recycled carbon fiber/Bi_2Te_3 (MWCNT-RCF/Bi_2Te_3), and MWCNT-RCF/Bi_2S_3 composites and used them as TEG materials. The optimized content of MWCNT was 0.10 wt% for MWCNT-RCF/Bi_2Te_3 and 0.15 wt% for MWCNT-RCF/Bi_2S_3 which achieved thermoelectric properties of 1.044 and 0.849 $\mu W\ K^{-2m-1}$, respectively. Upon incorporation of MWCNTs, the difference in power factor among RCF/Bi_2Te_3 and RCF/Bi_2S_3 decreased from 52% to 19%. This is attributed to the reduced resistivity of Bi_2S_3 and increased thermoelectric properties. As a result of their investigation, it was proven that MWCNTs provide an effective electron transfer conductive pathway.

442 Oxide free nanomaterials for energy storage and conversion applications

2.3 Thermoelectric metal carbides and nitrides

Kato et al. [24] investigated the thermoelectric behavior of SiC-based semiconductors with polysilastylene (PSS) and Ag as a sintering compound and dopant, respectively. Ag is an excellent dopant to reduce the electrical resistance of SiC-based p-type thermoelectric semiconductors. Ag doping creates a high carrier concentration in the order of 10^3–10^4, which is higher than that of Al-doped SiC with conventional stimulant concentrations. An optimized content of Ag (2.0 wt%) and PSS (0.1 wt%) reached a maximum ZT of 4×10^{-4} at 700°C, which revealed that the SiC/Ag system can be applied for the high-temperature thermoelectric application. PSS helps to control the density of the sample, which is also a crucial factor in reducing the thermal conductivity and electrical resistivity. Fukuda et al. [25] developed a new quaternary layered carbide $(Zr_2 [Al_{3.56}Si_{0.44}]C_5)$, whose possess reaches the high power factor of 7.6×10^{-5} W m^{-1} K^{-2} than the corresponding triple-layer carbides $Zr_2Al_3C_4$ $(1.3 \times 10^{-5}$ W m^{-1} K$^{-2})$ and $Zr_3Al_3C_5$ $(5.2 \times 10^{-5}$ W m^{-1}K$^{-2})$. Thus, the enhanced power factor and electrical conductivity of quaternary layered carbide might be because of the development of preferred orientation using ceramic processing technique. Kitagawa et al. [26] fabricated SiC/Si$_3$N$_4$ using spark plasma sintering (SPS) process at 2000°C which showed n-type semiconducting behavior and led to increases in the carrier concentration with Si$_3$N$_4$ concentration. The addition of Si$_3$N$_4$ with SiC enhances the thermoelectric properties and the power factor gets maximum at 7% mass ratio of Si$_3$N$_4$ in SiC. As a result of their investigation, and increased Seebeck coefficient and σ of SiC/Si$_3$N$_4$ at high temperatures were beneficial for an efficient thermal power conversion application. Also, this statement was supported by other investigations, where a sintering technique was adopted for SiC-based materials to reduce the thermal conductivity of materials [24, 27]. However, the ZT value of SiC materials was low compared to other thermoelectric materials, which seemed necessary to establish structural and assembly controls of SiC-based materials [26]. Li et al. [28] prepared SiC/Bi$_2$Te$_3$ composite materials and investigated their thermoelectric behavior. As a result of SiC scattering in the Bi$_2$Te$_3$ matrix, the increased value of the Seebeck coefficient was obtained, and an optimal concentration of Bi$_2$Te$_3$ with 0.1 wt% SiC showed a higher power factor than pure Bi$_2$Te$_3$. The enhanced thermoelectric properties of SiC/Bi$_2$Te$_3$ using the scattering effect of SiC might be because of the reduced thermal conductivity and simultaneous increase in qualification number of Bi$_2$Te$_3$ by 18% with the addition of 0.1 wt%–0.24 vol% SiC. Furthermore, Yin et al. [29] investigated the thermoelectric property changes of nanopowder and nanowire structured SiC incorporated $Mg_{2.16}(Si_{0.3}Sn_{0.7})_{0.98}Sb_{0.02}$. The compressive strength and fracture hardness of the composite material with 0.8% SiC nanopowders and nanowires was improved by 50% and 30%, respectively. The thermoelectric properties were altered by tuning morphology and achieved a maximum ZT of 1.20 at 750 K.

MXenes is an interesting class of 2D materials containing intermediate metal carbides and nitrides that are currently under extensive study. The studies on theoretical calculations of MXenes thermoelectric properties are available, while the experimental reports are limited in the investigation level. Kim et al. [30] fabricated Mo-based MXenes (Mo_2CT_x, $Mo_2TiC_2T_x$, and $Mo_2Ti_2C_3T_x$) using the vacuum-assisted filtration method and investigated their thermoelectric properties. At high temperatures (800 K), $Mo_2TiC_2T_x$ showed a high electrical conductivity (1380 S/cm), n-type Seebeck coefficient ($-47.3\,\mu V\,K^{-1}$), and reached a maximum thermal power of $3.09 \times 10^{-4}\,W\,m^{-1}\,K^{-2}$ at 803 K. Duan et al. [31] prepared TiN incorporated $CoSb_{2.875}Te_{0.125}$ composites using SPS. The composite with 1.0 vol% of TiN reduced the lattice thermal conductivity and enhancing ZT (1.0 at 800 K) of the composite materials, which reached a 10% enhancement compared to the pure TiN sample. Thus, the enhanced thermoelectric properties of composite might be because of the improved fracture toughness and flexural strengths.

Furthermore, an improvement in thermoelectric performance of metal chalcogens and MXenes can be attained by doping, tailoring the surface of materials, making composites with other semiconductor-based nanomaterials, or polymeric compounds.

3. Oxide-free materials for piezoelectric applications

3.1 Basic working function of a piezoelectric device

The piezoelectric effect was discovered by Pierre Curie and Paul Jacques in 1880; however, the conversational effect was discovered by Gabriel Lippman in 1881 through the mathematical aspect of the theory. The use of the piezoelectric effect in manufacturing and industrial sensing applications started from the 1950s onwards. Piezoelectric products, either natural or man-made, can generate an electric charge when subjected to applied pressure and, conversely, generates a mechanical pressure in response to the applied electric field. The charges in the piezoelectric crystal are symmetrically set and uniform. The effects of the charges are properly canceled and there is no net charge on the crystal faces. When a force is applied to a piezoelectric object, it will generate a voltage across its opposite faces. Now the effects of the charges will not cancel each other and the net positive and negative charges will appear in the opposite faces of crystal [32]. A typical overview of the direct and indirect piezoelectric effect is shown in Fig. 5.

3.2 Piezoelectric metal chalcogenides and nitrides

The discovery of graphene in 2004 led to the discovery of numerous 2D materials. Due to their unique properties, 2D materials have developed a main research center over the past decade regarding their potential utilization in novel

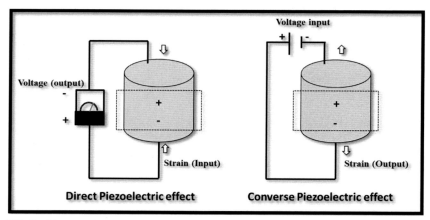

FIG. 5 Direct and indirect piezoelectric effect.

optoelectronic and electronic devices. Among the available 2D materials, transition metal dichalcogenides (TMDCs) have attracted considerable attention due to their tunable bandgap, ultra-thick and unique electronic, optical and mechanical properties, and simple manufacturing. Further, the performance of TMDCs can meet the growing demand for high-performance devices with metal dichalcogenide bandwidth, and excessive efforts have been paid to improve the physiochemical properties by doping [33–36], alloying [37], and incorporating heterostructures [38]. TMDC monolayers are atomically thin semiconductors of the MX_2 type, where M is a transition metal atom (such as Mo or W) and X is a chalcogen atom (such as S, Se, or Te). Here, one layer of M atoms is sandwiched between two layers of X atoms. Currently, 2D layer materials such as monolayer TMDCs, including MoS_2, MoS_2, WS_2, WSe_2, group IV monochalcogenides (GeSe and SnS), and group III–V binary compounds (AlSb, GaP, GaAs, InP, InAs, and InSb), have been experimentally stabilized or theoretically predicted to be piezoelectric [39]. Depending on the coordination and oxidation state of the metal atoms, TMDCs can be classified into metallic, semimetallic, or semiconductors. For example, WS_2 is a semiconductor, while $NbSe_2$ is a metal. A piezoelectric material should be an insulator or semiconductor with a sufficient bandgap to prevent leakage current [40].

In 2014, the first report of the piezoelectric effect with an odd number of atomic layers in MoS_2 was demonstrated by Wu et al. [41]. Structures with an even number of layers do not exhibit piezoelectric properties. As the number of layers increases, the piezoelectric effect decreases, and grain boundaries of the monolayer MoS_2 significantly enhanced its piezoelectric properties [42]. Kim et al. [43] used 2D semiconducting flexible piezoelectric MoS_2, which can efficiently harvest the mechanical energy using flexible piezoelectric N-doped graphene (NG). It was found that the output power derived from NG with

MoS$_2$ was 0.48% higher than that of pure NG with the same strain and strain velocity of 70 mm/s. Tan et al. [44] predicted that the multi-layered MoS$_2$ showed strong piezoelectricity compared to its monolayer. The maximum piezoelectric coefficient (e$_{11}$) of 0.457 c/m^2 was attained for the 5-layered materials which were 13% higher than that of monolayer MoS$_2$. When the number of layers exceeded 5, the piezoelectric coefficient decreased to 0.302 c/m^2. Han et al. [45] investigated the effect of MoS$_2$ shell size on the generation of piezoelectric power. In this study, they utilized Al$_2$O$_3$ deposited on polystyrene beads to avoid collapse of spherical structured MoS$_2$ shell. The piezoelectric MoS$_2$ shell generated a voltage of 1.2 V at 4.2 kPa. The power production was achieved by tapping minimum pressure from 0.3 kPa. The structural changes with a larger diameter MoS$_2$ shell are beneficial for the advancement of next-generation nano-piezoelectric power generators. The high diameter shell leads to high stress on the contact points. Sohn et al. [46] studied the effect of temperature on piezoelectric properties of the MoS$_2$ monolayer prepared using CVD by determining the transport behavior in the temperature range from 270 to 320 K. As a result, the piezoelectric effect was strengthened at low temperatures due to the low screening effect of the piezoelectric power produced in MoS$_2$. Interestingly, a family of TMDC layered piezoelectric materials was utilized for a wide range of potential applications in various fields. For example, MoSe$_2$ having a centrosymmetric structure with high piezoelectricity was used in toxic organic pollutant degradation [47] and fabrication of supercapacitors [48]. WS$_2$ was an efficient piezoelectric sensor [49], and WSe$_2$ was used in the driving of liquid crystal display without any other external power sources [50].

The 2D monolayer-based TMDC materials are piezoelectric, unlike their total parent crystals. Breaking the symmetry always leads to new phenomena, such as the Rashba effect and piezoelectric polarization, and makes them promising materials for various applications. This is because symmetry is an important factor influencing the properties of materials. Further, the asymmetric Janus 2D materials show new properties in piezoelectric polarization, which makes an excellent promise for their use in sensors, actuators, and other electro-mechanical devices [51, 52]. The outstanding piezoelectric performances of Janus 2D TMDCs can occur because of the properties of strong Rashba spin cleavage and second-harmonic generation response [51, 53–56]. To enhance the piezoelectric effect on group-III chalcogenide monolayers, Guo et al. [57] developed a series of Janus models including Ga$_2$SeTe, Ga$_2$STe, Ga$_2$SSe, In$_2$STe, In$_2$SSe, In$_2$SeTe, GaInSe$_2$, GaInS$_2$, and GaInTe$_2$. DFT calculations show that they have large piezoelectric coefficients of up to 8.47 pm/V, which was four-fold higher than corresponding perfect chalcogenide monolayers.

Nitride-based semiconductor materials and their heterostructures have been widely explored over the past three decades with various useful electronic and physical properties that led to applications in many technological devices. The

446 Oxide free nanomaterials for energy storage and conversion applications

advancement of GaN, AlN, and InN semiconductor materials toward an exploration of the effects of piezoelectric polarization and fabrication have laid a solid groundwork for the ecosystem of electronic and photonic devices, which is a growing field [58]. Among the others, AlN possesses an excellent piezoelectric property that is eagerly awaited the use in micro-electromechanical applications such as sensors, communication resonators, and energy harvesting [59–63].

4. Conclusions

The literary conclusions of this passage indicate that numerous possibilities in terms of sample structures and preparation methods lead to multiple paths to handle thermal and piezoelectric materials and to achieve performance improvements. New technologies have allowed researchers to develop highly flexible thermal and piezoelectric devices and electronics that have opened new doors in developing energy, biologically compatible, and durable harvests to generate the power needed for electronic devices. This chapter focuses on the basics of working function and recent developments of non-oxide materials for thermo and piezoelectric applications. Special attention has been paid to the metal chalcogenides (such as sulfides, selenides, and tellurides), metal phosphides, metal carbides, and nitrides. Finally, oxide-free materials produced by low-cost and large-scale methods offer a unique opportunity to combine material properties and use interface effects. This path, which relies on the technological advancement of preparation of compounds of oxide-free materials, is particularly promising for the exploitation of thermoelectric and piezoelectric properties. Although there has been a lot of development in both TEGs and PEGs independently, a hybrid system has not yet been adopted. This indicates that there is an opportunity for improvement in terms of the optimization of electronic parameters.

Acknowledgments

The authors Prof. M.Y. Choi and Dr. J. Theerthagiri acknowledge the financial support from the National Research Foundation of Korea (NRF), (2019H1D3A1A01071209). This work was supported by the Korea Basic Science Institute (KBSI) National Research Facilities & Equipment Center (NFEC) grant funded by the Korean government (Ministry of Education) (No. 2019R1A6C1010042, 2021R1A6C103A427). Dr. S.J. Lee acknowledges the financial support from the National Research Foundation of Korea (NRF), (2020R1I1A1A01065748).

References

[1] J. Theerthagiri, S.J. Lee, A.P. Murthy, J. Madhavan, M.Y. Choi, Fundamental aspects and recent advances in transition metal nitrides as electrocatalysts for hydrogen evolution reaction: a review, Curr. Opin. Solid State Mater. Sci. 24 (2020) 100805.

Oxide-free materials **Chapter | 17   447**

[2] J. Theerthagiri, J. Madhavan, S.J. Lee, M.Y. Choi, M. Ashokkumar, B.G. Pollet, Sonoelectrochemistry for energy and environmental applications, Ultrason. Sonochem. 63 (2020) 104960.

[3] J. Theerthagiri, A.P. Murthy, V. Elakkiya, S. Chandrasekaran, P. Nithyadharseni, Z. Khan, R. A. Senthil, R. Shanker, M. Raghavender, P. Kuppusami, J. Madhavan, M. Ashokkumar, Recent development on carbon based heterostructures for their applications in energy and environment: a review, J. Ind. Eng. Chem. 64 (2018) 16–59.

[4] J. Theerthagiri, R.A. Senthil, B. Senthilkumar, A.R. Polu, J. Madhavan, M. Ashokkumar, Recent advances in MoS2 nanostructured materials for energy and environmental applications—a review, J. Solid State Chem. 252 (2017) 43–71.

[5] S. Shittu, G. Li, Y.G. Akhlaghi, X. Ma, X. Zhao, E. Ayodele, Advancements in thermoelectric generators for enhanced hybrid photovoltaic system performance, Renew. Sustain. Energy Rev. 109 (2019) 24–54.

[6] L. Guo, Q. Lu, Potentials of piezoelectric and thermoelectric technologies for harvesting energy from pavements, Renew. Sustain. Energy Rev. 72 (2017) 761–773.

[7] K. Sunil, H.H. Singh, N., Khare, flexible hybrid piezoelectric-thermoelectric generator for harnessing electrical energy from mechanical and thermal energy, Energ. Conver. Manage. 198 (2019) 111783.

[8] A.O. Moghaddam, A. Shokuhfar, P. Guardia, Y. Zhang, A. Cabot, Substantial role of doping in the thermoelectric and hardness properties of nanostructured bornite, Cu_5FeS_4, J. Alloys Compd. 773 (2019) 1064–1074.

[9] H. Mamur, M.R.A. Bhuiyan, F. Korkmaz, M. Nil, A review on bismuth telluride (Bi_2Te_3) nanostructure for thermoelectric applications, Renew. Sustain. Energy Rev. 82 (2018) 4159–4169.

[10] K.K. Jung, J.S. Ko, Thermoelectric generator based on a bismuth-telluride alloy fabricated by addition of ethylene glycol, Curr. Appl. Phys. 14 (2014) 1788–1793.

[11] R. Hinchet, U. Khan, C. Falconi, S.W. Kim, Piezoelectric properties in two-dimensional materials: simulations and experiments, Mater. Today 21 (2018) 611–630.

[12] C. Wan, C.R. Bowen, Multiscale-structuring of polyvinylidene fluoride for energy harvesting: the impact of molecular-, micro- and macro-structure, J. Mater. Chem. A 5 (2017) 3091–3128.

[13] W. Xia, Z. Zhang, PVDF-based dielectric polymers and their applications in electronic materials, IET Nanodielectr. 1 (2018) 17–31.

[14] H. Meng, M. An, T. Luo, N. Yang, Thermoelectric applications of chalcogenides, in: Chalcogenide, Woodhead Publishing Series in Electronic and Optical Materials Thermoelectric Applications of Chalcogenides, 2020, pp. 31–56.

[15] Y.X. Chen, F. Li, W. Wang, Z. Zheng, J. Luo, P. Fan, T. Takeuchi, Optimization of thermoelectric properties achieved in Cu doped β-In_2S_3 bulks, J. Alloys Compd. 782 (2019) 641–647.

[16] Z.H. Ge, P. Qin, D.S. He, Z. Chong, D. Feng, Y.H. Ji, J. Feng, J. He, Highly enhanced thermoelectric properties of bi/Bi_2S_3 nanocomposites, ACS Appl. Mater. Interfaces 9 (2017) 4828–4834.

[17] M. Ohta, S. Hirai, T. Mori, Y. Yajima, T. Nishimura, K. Shimakage, Effect of non-stoichiometry on thermoelectric properties of γ-Tb_2S_{3-x}, J. Alloys Compd. 418 (2006) 209–212.

[18] K. Sunil, M.Z. Ansari, N., Khare, influence of compactness and formation of metallic secondary phase on the thermoelectric properties of Cu_2ZnSnS_4 thin films, Thin Solid Films 645 (2018) 300–304.

[19] T.R. Wei, C.F. Wu, F. Li, J.F. Li, Low-cost and environmentally benign selenides as promising thermoelectric materials, J. Mater. 4 (2018) 304–320.

[20] N.S. Patil, A.M. Sargar, S.R. Mane, P.N. Bhosale, Effect of Sb doping on thermoelectric properties of chemically deposited bismuth selenide films, Mater. Chem. Phys. 115 (2009) 47–51.

448 Oxide free nanomaterials for energy storage and conversion applications

[21] M. Sinduja, S. Amirthapandian, A. Masarrat, R. Krishnan, S.K. Srivastava, A. Kandasami, Investigations on morphology and thermoelectric transport properties of Cu^+ ion implanted bismuth telluride thin film, Thin Solid Films 697 (2020) 137834.

[22] Y.J. Wu, S.C. Hsu, Y.C. Lin, Y. Xu, T.H. Chuang, S.C. Chen, Study on thermoelectric property optimization of mixed-phase bismuth telluride thin films deposited by co-evaporation process, Surf. Coat. Technol. 394 (2020) 125694.

[23] P. Jagadish, M. Khalid, N. Amin, M.T. Hajibeigy, L.P. Li, A. Numan, N.M. Mubarak, R. Walvekar, A. Chan, Recycled carbon fibre/Bi2Te3 and Bi2S3 hybrid composite doped with MWCNTs for thermoelectric applications, Compos. Part B 175 (2019) 107085.

[24] K. Kato, K. Asai, Y. Okamoto, J. Morimoto, T. Miyakawa, Temperature and porosity dependence of the thermoelectric properties of SiC/Ag sintered materials, J. Mater. Res. 14 (5) (1999) 1752–1759.

[25] K. Fukuda, M. Hisamura, T. Iwata, N. Tera, K. Sato, Synthesis, crystal structure and thermoelectric properties of a new carbide Zr2 [Al3. 56Si0. 44] C5, J. Solid State Chem. 180 (6) (2007) 1809–1815.

[26] H. Kitagawa, N. Kado, Y. Noda, Preparation of N-type silicon carbide-based thermoelectric materials by spark plasma sintering, Mater. Trans. 43 (12) (2002) 3239–3241.

[27] S. Takeda, C.H. Pai, W.S. Seo, K. Koumoto, H. Yanagitda, Thermoelectric properties of porous β-SiC fabricated from rice hull ash, J. Cerma. Soc. Jpn. 101 (1175) (1993) 814–818.

[28] J.F. Li, J. Liu, Effect of nano-SiC dispersion on thermoelectric properties of Bi2Te3 polycrystals, Phys. Status Solidi A 203 (15) (2006) 3768–3773.

[29] K. Yin, X. Su, Y. Yan, H. Tang, M.G. Kanatzidis, C. Uher, X. Tang, Morphology modulation of SiC nano-additives for mechanical robust high thermoelectric performance Mg2Si1 − xSnx/ SiC nano-composites, Scr. Mater. 126 (2017) 1–5.

[30] H. Kim, B. Anasori, Y. Gogotsi, H.N. Alshareef, Thermoelectric properties of two-dimensional molybdenum-based MXenes, Chem. Mater. 29 (15) (2017) 6472–6479.

[31] B. Duan, P. Zhai, P. Wen, S. Zhang, L. Liu, Q. Zhang, Enhanced thermoelectric and mechanical properties of Te-substituted skutterudite via nano-TiN dispersion, Scr. Mater. 67 (4) (2012) 372–375.

[32] S. Mishra, L. Unnikrishnan, S.K. Nayak, S. Mohanty, Advances in piezoelectric polymer composites for energy harvesting applications: a systematic review, Macromol. Mater. Eng. 304 (2019) 1800463.

[33] N. Jena, S.D. Behere, A. De Sarkar, Strain-induced optimization of nanoelectromechanical energy harvesting and nanopiezotronic response in a MoS2 monolayer nanosheet, J. Phys. Chem. C 121 (17) (2017) 9181–9190.

[34] D. Dimple, N. Jena, A. Rawat, A. De Sarkar, Strain and PH facilitated artificial photosynthesis in monolayer MoS2 Nanosheet, J. Mater. Chem. A 5 (2017) 22265–22276.

[35] A.E. Maniadaki, G. Kopidakis, I.N. Remediakis, Strain engineering of electronic properties of transition metal Dichalcogenide monolayers, Solid State Commun. 227 (2016) 33–39.

[36] Y. Li, Y.-L. Li, C.M. Araujo, W. Luo, R. Ahuja, Single-layer MoS2 as an efficient photocatalyst, Cat. Sci. Technol. 3 (9) (2013) 2214–2220.

[37] T.L. Tan, M.F. Ng, G. Eda, Stable monolayer transition metal dichalcogenide ordered alloys with tunable electronic properties, J. Phys. Chem. C 120 (2016) 2501–2508.

[38] Z.Y. Zhang, M.S. Si, S.L. Peng, F. Zhang, Y.H. Wang, D.S. Xue, Bandgap engineering in van der Waals heterostructures of blue phosphorene and MoS2: a first principles calculation, J. Solid State Chem. 231 (2015) 64–69.

[39] K.A.N. Duerloo, M.T. Ong, E.J. Reed, Intrinsic piezoelectricity in two-dimensional materials, J. Phys. Chem. Lett. 3 (2012) 2871–2876.

Oxide-free materials **Chapter | 17** **449**

[40] M.M. Alyoruk, Y. Aierken, D. Cakır, F.M. Peeters, C. Sevik, Promising piezoelectric performance of single layer transition-metal dichalcogenides and dioxides, J. Phys. Chem. C 119 (2015) 23231–23237.

[41] W. Wu, L. Wang, Y. Li, F. Zhang, L. Lin, S. Niu, J. Hone, Piezoelectricity of single-atomic-layer MoS 2 for energy conversion and piezotronics, Nature 514 (2014) 470–474.

[42] M. Dai, W. Zheng, X. Zhang, S. Wang, J. Lin, K. Li, Y. Fu, Enhanced piezoelectric effect derived from grain boundary in MoS2 monolayers, Nano Lett. 20 (2019) 201–207.

[43] S.K. Kim, R. Bhatia, T.H. Kim, D. Seol, J.H. Kim, H. Kim, S.W. Kim, Directional dependent piezoelectric effect in CVD grown monolayer MoS2 for flexible piezoelectric nanogenerators, Nano Energy 22 (2016) 483–489.

[44] D. Tan, M. Willatzen, Z.L. Wang, Prediction of strong piezoelectricity in 3R-MoS2 multilayer structures, Nano Energy 56 (2019) 512–515.

[45] J.K. Han, S. Kim, S. Jang, Y.R. Lim, S.W. Kim, H. Chang, S. Myung, Tunable piezoelectric nanogenerators using flexoelectricity of well-ordered hollow 2D MoS2 shells arrays for energy harvesting, Nano Energy 61 (2019) 471–477.

[46] A. Sohn, S. Choi, S.A. Han, T.H. Kim, J.H. Kim, Y. Kim, S.W. Kim, Temperature-dependent piezotronic effect of MoS2 monolayer, Nano Energy 58 (2019) 811–816.

[47] M.H. Wu, J.T. Lee, Y.J. Chung, M. Srinivaas, J.M. Wu, Ultrahigh efficient degradation activity of single-and few-layered MoSe2 nanoflowers in dark by piezo-catalyst effect, Nano Energy 40 (2017) 369–375.

[48] P. Pazhamalai, K. Krishnamoorthy, V.K. Mariappan, S. Sahoo, S. Manoharan, S.J. Kim, A high efficacy self-charging MoSe2 solid-state supercapacitor using electrospun nanofibrous piezoelectric separator with Ionogel electrolyte, Adv. Mater. Interfaces 5 (2018) 1800055.

[49] J. Kim, E. Lee, T.K. An, Stable and high-performance piezoelectric sensor via CVD grown WS2, Nanotechnology 31 (2020) 445203.

[50] J.H. Lee, J.Y. Park, E.B. Cho, T.Y. Kim, S.A. Han, T.H. Kim, H. Ryu, Reliable piezoelectricity in bilayer WSe2 for piezoelectric nanogenerators, Adv. Mater. 29 (2017) 1606667.

[51] A.Y. Lu, H. Zhu, J. Xiao, C.P. Chuu, Y. Han, M.H. Chiu, Y. Wang, Janus monolayers of transition metal dichalcogenides, Nat. Nanotechnol. 12 (2017) 744–749.

[52] J. Zhang, S. Jia, I. Kholmanov, L. Dong, D. Er, W. Chen, J. Lou, Janus monolayer transition-metal dichalcogenides, ACS Nano 11 (2017) 8192–8198.

[53] R. Li, Y. Cheng, W. Huang, Recent progress of Janus 2D transition metal chalcogenides: from theory to experiments, Small 14 (2018) 1802091.

[54] C. Zhang, Y. Nie, S. Sanvito, A. Du, First-principles prediction of a room-temperature ferromagnetic Janus VSSe monolayer with piezoelectricity, ferroelasticity, and large valley polarization, Nano Lett. 19 (2019) 1366–1370.

[55] Y. Chen, J. Liu, J. Yu, Y. Guo, Q. Sun, Symmetry-breaking induced large piezoelectricity in Janus tellurene materials, Phys. Chem. Chem. Phys. 21 (2019) 1207–1216.

[56] J. Yang, A. Wang, S. Zhang, J. Liu, Z. Zhong, L. Chen, Coexistence of piezoelectricity and magnetism in two-dimensional vanadium dichalcogenides, Phys. Chem. Chem. Phys. 21 (2019) 132–136.

[57] Y. Guo, S. Zhou, Y. Bai, J. Zhao, Enhanced piezoelectric effect in Janus group-III chalcogenide monolayers, Appl. Phys. Lett. 110 (2017) 163102.

[58] D. Jena, R. Page, J. Casamento, P. Dang, J. Singhal, Z. Zhang, H.G. Xing, The new nitrides: layered, ferroelectric, magnetic, metallic and superconducting nitrides to boost the GaN photonics and electronics eco-system, Jpn. J. Appl. Phys. 58 (2019), SC0801.

[59] M.K. Kim, S.W. Yoon, Miniature piezoelectric sensor for in-situ temperature monitoring of silicon and silicon carbide power modules operating at high temperature, IEEE Trans. Ind. Appl. 54 (2017) 1614–1621.

450 Oxide free nanomaterials for energy storage and conversion applications

[60] M. Akiyama, T. Kamohara, K. Kano, A. Teshigahara, Y. Takeuchi, N. Kawahara, Enhancement of piezoelectric response in scandium aluminum nitride alloy thin films prepared by dual reactive cosputtering, Adv. Mater. 21 (2009) 593–596.

[61] R. Matloub, M. Hadad, A. Mazzalai, N. Chidambaram, G. Moulard, C.S. Sandu, P. Muralt, Piezoelectric Al1 − xScxN thin films: a semiconductor compatible solution for mechanical energy harvesting and sensors, Appl. Phys. Lett. 102 (2013) 152903.

[62] G.J. Lee, M.K. Lee, J.J. Park, D.Y. Hyeon, C.K. Jeong, K.I. Park, Piezoelectric energy harvesting from two-dimensional boron nitride nanoflakes, ACS Appl. Mater. Interfaces 11 (2019) 37920–37926.

[63] K. Tonisch, V. Cimalla, C. Foerster, H. Romanus, O. Ambacher, D. Dontsov, Piezoelectric properties of polycrystalline AlN thin films for MEMS application, Sensors Actuators A Phys. 132 (2006) 658–663.

Chapter 18

Future prospects of oxide-free materials for energy-related applications

Dhandapani Balaji, Kumar Premnath, and Madhavan Jagannathan
Solar Energy Lab, Department of Chemistry, Thiruvalluvar University, Vellore, India

Chapter outline

1. **Introduction**   451
2. **Transition metal-based oxide-free materials for hydrogen evolution reaction**   452
    2.1 Transition metal carbides   453

2.2 Transition metal nitrides   456
2.3 Transition metal phosphides   457
2.4 Transition metal sulfides   459
3. **Conclusions and future prospects**   462
References   463

1. Introduction

The application of traditional fuel and the subsequent environmental pollution are transferrable issues of the present century. The development of clean and renewable energy sources has become the main replacement for global energy demand and environmental pollution. Hydrogen is the superior renewable energy source that is pollution-free and non-toxic, making convenient alternator to another fuel source. The electrochemical water splitting is the most common method to generate hydrogen (H_2) gas which is dependent on efficient electrocatalyst in hydrogen evolution reaction (HER) [1–6]. However, a good electrocatalyst can influence the speed of catalytic reaction, possess unique stability, and requires low overpotential to derive the maximum current density towards HER. Although few noble-metal-based catalysts such as platinum, iridium, and ruthenium are still the more experimental option for HER, the high cost and scarcity of these electrocatalysts have lacked their further application. Hence, over the past years, several efforts have been made to synthesis efficient non-noble-metal electrocatalysts as a substitute for the noble-metal based electrocatalysts in HER [7–10].

Oxide Free Nanomaterials for Energy Storage and Conversion Applications.
https://doi.org/10.1016/B978-0-12-823936-0.00002-4
Copyright © 2022 Elsevier Inc. All rights reserved.

452 Oxide free nanomaterials for energy storage and conversion applications

In the recent decade, oxide-free materials such as transition metal sulfides, selenides, phosphides, carbides, and nitrides have received much attention due to their high abundance, low cost, environmental benignity, variable oxidation states, and excellent electrochemical activities [11–20]. Among these, transition metal carbides (TMCs) have been considered an extensive interest owing to their unique electronic structure, conductivity, durability, and band properties similar to Pt which make them suitable candidates for HER [21, 22]. Especially, few recent results clearly showed that W_2C is a considerably high active HER catalyst than WC as it contains higher electronic density states (DOS) in the fermi level and less negative free energy of hydrogen adsorption. The different TMCs are hopeful to be excellent electrocatalyst for HER but their corrosion and instability is considered as lacking in neutral and basic pH medium [23]. To solve this issue, heteroatom such as nitrogen is induced in the form of nitrides to resist corrosion and thereby improving the catalytic activity of metal carbides. Remarkably, Yang et al. fabricated Ni-Co-WC composite synthesized by pyrolysis process in presence of nickel and cobalt for excellent HER [24]. Chen et al. optimized an enhanced HER activity of the mutual loading of W_2C and WN on graphene supporters [25]. As per the previous reports, it has been clearly examined that the nitrogen loading makes local charge density and surface charge state due to the electron transfer of its 3p orbitals to vacant 3d orbitals and hence their activity is enhanced. In this context, metal nitrides are another attractive material for electrocatalytic HER as they resist corrosion in basic and neutral media. Hence, the combination of metal carbide with metal nitride could achieve more activity for HER with high durability [26, 27].

In this book chapter, we present a new class of transition metal precursors for assembling high-performance transition metal carbides, nitrides, phosphides, and sulfides for HER and examined facile strategies to synthesize materials with different structures to identify efficient electrocatalysts for HER. This book chapter explores three main discussion sections: in the first section, the basic HER electrochemical catalytic reactions are mostly catalyzed by transition metal carbides, nitrides, phosphides, and sulfides-based materials in water-splitting reactions. Secondly, the important techniques and methods used to improve the catalytic activity of the materials and the various parameters used for determining electrochemical performances will also be introduced. Finally, the future prospect of transition metal carbides, nitrides, phosphides, and sulfides will be discussed.

2. Transition metal-based oxide-free materials for hydrogen evolution reaction

The following section of these materials explores the different free oxide-free materials such as transition metal carbides, nitrides, phosphides, and sulfides in different pH environments. The reaction mechanism, electrochemical and structural properties are presented in detail.

2.1 Transition metal carbides

Transition metal carbides (TMC) have been widely attracted as an amazing candidate due to their high electrical conductivity and great electrochemical stability, which make them the most promising material for HER application. The crystal structure of all transition metal carbides has a similar structure and is located in the same size of carbon atom in the interstitial of the parent metals. Generally, most TMCs exhibits in three various forms viz., MC, M_2C, and M_3C. Among these, monocarbide like MC provides cubic structure while M_2C and M_3C have an arrangement of atoms to form hexagonal and orthorhombic structures.

Liao et al. [28] developed a highly stable nano-porous molybdenum carbide nanowires (np-Mo_2C NWs) for efficient hydrogen evolution reaction. The synthesis of np-Mo_2C NWs followed pyrolysis treatment with MoO_x and amine hybrid precursor which are having a sub-nanosized particle quasi homogenous environment that is suitable for forming a nano-crystallite-composed nanoporous structure. Therefore, the synthesized material provides a superior electrocatalytic activity due to its high surface area, nano-crystalline size and nano enrich porosity. All these parameters allowed the fast electron transfer kinetics within the electrocatalyst and also across the interface. Hence, the hydrogen evolution was observed at a low overpotential (70 mV) in a fixed current density in the case of the Mo_2C NWs catalyst. Emin et al. [29] fabricated metallic tungsten and tungsten carbide by taking hexacarbonyl tungsten source on graphite nanorod which was then heated at 900°C to form $W_2C@WC$. The obtained material was tested for the electrocatalytic activity in 0.5 M H_2SO_4 electrolytes. The HER activity was observed at an overpotential of 310 mV at 10 mA/cm^2 current density and the electrocatalyst showed good stability for about 1000 cycles. Chen et al. [30] investigated different transition metals (Fe, Co, Ni, Cu, Ag, and Pt) supported molybdenum carbide (Mo_2C) with simple easy precursors and carburization process. Among them, 2% Pt-loaded Mo_2C exhibited a higher activity due to the nano-surface structure in Mo_2C. The increase in catalyst loading amount resulted in the decrease in surface area and hence the HER activity was decreased. Consequently, 2% Pt doped Mo_2C revealed an efficient activity with a low-overpotential value of 79 mV and a Tafel slope of 55 mV/dec to attain 10 mA/cm^2 current density in 0.5 M H_2SO_4 electrolytes.

Li et al. [31] synthesized layered mesoporous structured molybdenum carbide with different dopant weights of N&C ($Mo_2C@CN$) under solid-state process for catalytic hydrogen evolution reaction in an acid environment. The structural properties clearly demonstrated that the formation of different mesopores ranging from 2 to 50 nm in the obtained $Mo_2C@CN$-0.6 material. The 0.6% N&C loaded material exhibited an efficient electrochemical behavior with potential values of 202 mV at 10 mA/cm^2 current density. The results suggested that the introduction of N and C in Mo_2C could be useful in modifying the HER kinetics.

454 Oxide free nanomaterials for energy storage and conversion applications

Li et al. [32] developed an easy carburization method to prepare a more synergistic copper-doped Mo$_2$C electrocatalyst. The added copper effectively induced the formation of the α-phase of Mo$_2$C and significantly raised the HER. The synergistic active area was obtained when the dopant ratio of Cu-Mo was 10:90 and at this composition, a superior HER activity than the commercial Pt/C was observed. Fu et al. [33] designed a heterogeneous nanomesh of V$_8$C$_7$ grown on high conductive graphene by one step epitaxial process and improved the catalytic activity for hydrogen evolution reaction in acidic and alkaline conditions. Further, a thermal annealing at a high temperature of 900°C favored the formation of the carbon defected V$_8$C$_7$ NMs/GR with more active sites for higher HER activity and the schematic representation is shown in Fig. 1. Due to the well-matched hybrid structure, an optimal overpotential of 52 mV in H$_2$SO$_4$ and 158 mV in KOH at a current density of 10 mA/cm^2 were observed.

The bimetallic transition metal carbide with exclusive properties provided a stable electrocatalytic hydrogen evolution reaction due to the coupling of heteroatom doped carbon [35]. Liu et al. synthesized Ni-Mo-based carbide nanowire on carbon cloth with (Ni$_3$Mo$_3$C@NPC NWs/CC) N, P-doped carbon matrix by electro-polymerization technique [36]. The pure bimetallic phase, unique nanowire structure, and self-supporting ability are found to be the reason for the good HER activity of these materials. That is, the Ni$_3$Mo$_3$C@NPC NWs/CC catalyst exhibited higher electrochemical activity in acid and alkaline electrolysis with a low overpotential of 161 and 215 mV needed to achieve 100 mA/cm^2 current density. Further, the catalyst material showed stable hydrogen evolution for 48 h. Similar HER study by using carbon-coated cobalt-tungsten carbide (Co$_6$W$_6$C) showed comparable activity in alkaline medium to that of commercial Pt/C catalyst. The optimal current density required a small overpotential of 73 mV which is close to noble Pt/C material. Further, the material showed excellent

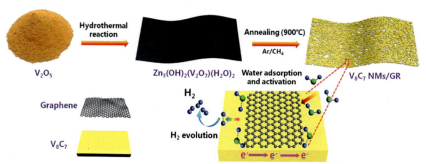

FIG. 1 Schematic illustration of the fabrication process for V$_8$C$_7$ NMs/GR and hydrogen generation on the surface of V$_8$C$_7$ NMs/GR [33]. *(Reprinted with permission from W. Fu, Y. Wang, H. Zhang, M. He, L. Fang, X. Yang, Z. Huang, J. Li, X. Gu, Y. Wang Epitaxial growth of graphene on V8C7 nanomeshs for highly efficient and stable hydrogen evolution reaction, J. Catal. 369 (2019) 47–53. Copyright 2021, Elsevier.)*

Future prospects of oxide-free materials **Chapter | 18** **455**

stability at a high current density of 30mV/cm² for 18h electrolysis under alkaline conditions. The reason for the significant activity is the cooperative effect of the nanosized cobalt-tungsten bimetallic composition. Meyer et al. [15] studied different transition metal electrocatalysts such as WC, Mo₂C, TaC, NbC prepared by a two-step oxidation-carburization method with various temperatures between 200°C and 400°C in the presence of molten KH₂PO₄ which is an electrolyte widely employed in a solid acid membrane electrolyze cells. The electrochemical activity was observed viz., in the order WC > Pt ≈ Mo₂C > NbC > TaC. In another study, Liu et al. [34] utilized the in-situ molten salt preparation method to fabricate tungsten carbide (WC coexisting with W₂C) nanocrystals anchored on carbon black (CB) and carbon nanotube (CNT). The composited carbon-based tungsten carbide (WC/C and WC/CNT) exhibited unique catalytic properties with an HER activity at an overpotential of 90mV and a Tafel slope of 69mV/dec at a standard current density of 10mA/cm². Further, WC/CNT showed excellent durability for the performed stability test 3000 cycles. The good dispersion and small particle size are reasons for the enhanced HER activity. Further, the increase in surface active sites can be clearly seen in the TEM image present in Fig. 2.

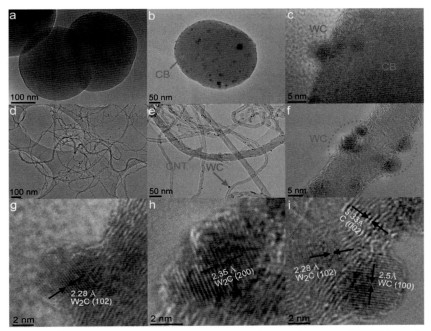

FIG. 2 (A) TEM image of as-received CB. (B, C) TEM images of as-prepared WC/CB sample. (D) TEM image of as-received CNTs. (E, F) TEM images of as-prepared WC/CNT sample. (G–I) HRTEM images of WC particles anchored on CB (G) and CNT (H, I) [34]. *(Reprinted with permission from C. Liu, Y. Wen, L. Lin, H. Zhang, X. Li, S. Zhang, Facile in-situ formation of high efficiency nanocarbon supported tungsten carbide nanocatalysts for hydrogen evolution reaction, Int. J. Hydrogen Energy 43 (2018) 1–9. Copyright 2021, Elsevier.)*

456 Oxide free nanomaterials for energy storage and conversion applications

2.2 Transition metal nitrides

Wang et al. [37] developed a facile strategy to design of $Ni_3N/Ni@C$ composite catalyst using urea-urea glass technique by controlling the calcination time at 450°C and then a small amount of Pt was fabricated on the prepared composite by a galvanic replacement method to improve the catalytic activity of $Ni_3N/Ni@C$ catalyst. It is noted that 0.45 wt% $Pt-Ni_3N/Ni@C$ exhibited an efficient activity with an optimal overpotential and Tafel slope value in the broad pH range of 0–14. The large surface area and high electrical conductivity are responsible for the superior electrocatalytic activity. Li et al. [38] proposed a novel preparation way to nano-porous Nb_2N catalyst from Nb_2O_5 which was obtained by simple anodization of Nb foil at room temperature with oxalic acid/HF electrolyte and observed an enhanced catalytic activity towards HER. As-prepared Nb_2N catalytic material exhibited a higher activity with a low overpotential of 92 mV/dec to meet a high current density of 326 mA/cm^2. The whole activity of the Nb_2N catalyst was found to be 4.2 times low when compared with pure Nb_2O_5.Wei et al. [39] fabricated a bimetallic MoVN thin film on a carbon sheet via a facile technique of co-sputtering and demonstrated its efficiency in alkaline and acid electrolyte conditions. The HER activity of MoVN film electrodes was observed to be higher than that of pure MoN and VN electrodes. η_{10} of 108 mV and a Tafel slope of 60 mV/dec was reported for the MoVN electrode material. Further, impressive cyclic durability of up to 3000 cycles at fixed potential was achieved. The availability of more active sites and synergistic effect was found to be the reason for the higher activity of the as-prepared MoVN electrocatalyst. Huang et al. [40] reported about Co nanodots embedded on nitrogen-doped CNT (N-CNT) grafted on hexagonal vanadium nitride (Co/N-CNT/VN). The co-assistance of Co embedded on N-CNT act as a conventional non-noble metal-based electrocatalyst for excellent HER in an alkaline environment. The hierarchical shape of Co/N-CNT/VN has revealed a high electrochemical behavior due to the doping of Co with the crystal size of 2 mm that created more active sites for hydrogen evolution. The HER activity of the Co/N-CNT/VN was noted at a low overpotential of 64 mV to attain 10 mA/cm^2 current density. The enhanced activity of Co/N-CNT/VN catalyst was due to the superior electron transport kinetics and more active edges on the catalyst. Gao et al. [41] studied the metallic Ni_3N nanosheets through a sintering method and investigated the tune-able electrochemical behavior to replace a noble metal of Pt. Herein, the Ni atom coupled with surrounded N-atom on the N-Ni surface played a predominant role to improve the activity of Ni_3N. The excellent catalytic activity was observed with a low overpotential (100 mV), Tafel slope (59 mA/dec) at a maximum current density of 100 mA/cm^2, due to the highly synergistic effect from Ni to N atom. Further, the electrode material was found to be stable for over 5000 cycles. Lai et al. [42] designed a strongly coupled nickel-cobalt nitrides/carbon complex nanocage (NiCoN/C) via a simple nitridation process and determined its catalytic activity towards HER in alkaline electrolyte. The NiCoN/C catalyst revealed a higher cathodic current of 10 mA at a low overpotential of 200 mV and the fabricated material maintained its stability for about 10 h of electrolysis. The obtained high catalytic behavior was due to high electronic conductivity between Co^{2+} to Co^0 to bonding of (Ni, Co)-N by strong coupling of the d-p-d orbital. Ojha et al. [43] discussed the binary composites of $FeNx/Mo_2N/CNT$ prepared by an in-situ process and utilized it as a non-precious metal-based electrocatalyst towards HER. The electrochemical activity of $FeNx/Mo_2N/CNT$ exceeded the catalytic properties with the combination of CNT and also showed better durability up to 48 h. The high synergistic effect was reasoned to the improved activity of this material. Ren et al. [44] developed a facile strategy to prepare an efficient electrocatalyst made of porous tungsten nitride nanowire loaded on carbon cloth (WN NW/CC) using N_2 plasma treatment and applied it in electrocatalytic HER application. WN NW/CC has acquired a low overpotential of 134 and 130 mV to achieve a maximum current density of $10 mA/cm^2$ in acidic and basic electrolytes respectively and the high synergistic effect with a favorable nanostructure is responsible for the enhanced HER activity. Yu et al. [45] synthesized mono and bimetallic transition metal nitrides and evaluated them to be potential electrocatalysts for hydrogen evolution reaction in a 1 M KOH solution. The monometallic CoN material was performed to explore an HER activity at a low onset potential of 95 and 212 mV to meet a current density of 10 and $100 mA/cm^2$ while overpotentials were found to be improved with Ni-doped CoN i.e., 48 and 149 mV for the same current density. The improved activity was mainly attributed to more active surface, the high synergistic effect from Co to Ni atom, and more charge-transfer efficiency with improved electrical conductivity. The tri-nickel based nitride catalyst has attracted much attention to perform electrochemical behavior towards HER [46]. The $Ni_3N@VN$-NF composite was prepared from NiV-LDH substrate under normal nitridation. When compared with the bare Ni_3N-NF and VN-NF electrocatalysts, the $Ni_3N@VN$-NF electrocatalyst showed an overpotential of 56 mV to achieve the maximum current density of $10 mA/cm^2$. Further, it exhibited excellent durability for over 50 h. In another study, $TiN@Ni_3N$ nanowire was designed and evaluated for HER by Zhang et al. [47]. The $TiN@Ni_3N$ nanowire implied a higher catalytic activity at a low onset potential of 15 mV and low Tafel slope of 42 mV/dec with excellent durability of less than 13% degradation even after 10 h of operation.

2.3 Transition metal phosphides

The transition metal phosphides (TMP) have been widely used for the hydrogen evolution reaction in a broad pH electrolyte solution. Phosphide-based materials have several advantages such as highly conductive, more active sites, and great stability.

458 Oxide free nanomaterials for energy storage and conversion applications

Highly earth-abundant Co-based electrocatalysts are involved in hydrogen energy research. The catalytic activity can be increased by the synthesis of nano-structured materials for hydrogen evolution reaction. Guo et al. [48] investigated the properties of CoP NS that exhibited high activity for HER. The CoP NS was synthesized by the following steps: At first, α-Co $(OH)_2$ was prepared by electrodeposition on Ni foam, then, it was heated with NaH_2PO_2 at 300°C under Ar atmosphere. The CoP is crystalline and it contains defects, which affords abundant active sites and facilitates the gas release which in turn enhances much high activity towards HER. For HER in 1.0 M KOH, the CoP NS achieved much higher (η of 86.6 mV at 10 mA/cm^2) HER activity. Moreover, Yan et al. [49] have observed much high stability and activity of CoP on carbon cloth (CC) electrocatalyst for HER. The CoPNS was prepared by the phosphidation process with NaH_2PO_2 at 300°C for 2 h under Ar atmosphere. The obtained CoPNS/CC appeared as nanosheet morphology with good crystalline nature, and that possibly enhanced its stability and activity for HER. In 1.0 M KOH, the electrocatalyst CoP NS/CC exhibited high durability with no measurable increase in overpotential even after 3000 cycles of CV. Further, this material also showed good HER activity in neutral and acidic conditions Saadi et al. [50] discussed an electrochemical cathodic deposition process to CoP as a film on a copper substrate in presence of a boric acid solution of Co^{2+} and $H_2PO_2^-$. The operando purification led to the formation of Co:P with a stoichiometric ratio of 1:1. The obtained surface of CoP thin film was densely packed with a micrometer in-size with the following characteristics (1) The as-deposited thin film showed the presence of quasi-spherical clusters on the surface. (2) The film underneath the particles appeared uniformly roughened. (3) Post-electrolysis film showed close-packed plateau-topped islands on the surface. (4) The tops of the mesa-like islands were flat and relatively smooth. The better electrochemical activity of CoP electrodeposits is inferred from an overvoltage of 85 mV at a current density of 10 mA/cm^2. Further, a significant oper and stability in acidic solution were characterized by an increase in η of 18 mV after 24 h of continuous operation. Ma et al. [51] fabricated CoP nanoparticle with reduced graphene oxide (RGO) of different compositions via deposition process denoted as CoP/RGO-0.18, CoP/RGO-0.36, CoP/RGO-0.54, CoP/RGO-0.72.it was dispersed Co_3O_4 nanoparticles are prepared and it was exclusively deposited on RGO sheets with the homogeneous distribution. The nanoparticles can be played as a "spacer," preventing the RGO sheets from aggregation and restacking. The average size of the prepared nanoparticle was 4.1 nm. Among different composition, the CoP/RGO-0.36 exhibited extraordinary electrochemical performance for HER. This composition exhibited a smaller Tafel slope than CoP, and the j_0 value was ca. 63.5 times higher than that of pure CoP catalyst. The enhanced catalytic activity of CoP/RGO catalysts can be attributed to the presence of RGO sheets, which increase the composite electrical conductivity and provide more active sites for HER. Xianwei et al. [52] developed CoP with N-doped carbon shell (CoP/Co$_2$P@NC-2) electrocatalyst by simple hydrothermal method. Further, particle-like $CoP/Co_2P@NC-2$ nanostructures are highly uniform with diameters of approximately 5–20 nm, the inner cobalt phosphide nanoparticles were surrounded by 6–9 layers of N-doped graphitized carbon shells with an interlayer spacing of 0.34 nm. The core-shell $CoP/Co_2P@NC-2$ showed an overpotential of 126 mV to achieve a current density of 10 mA/cm^2. When compared at different pH values, the catalyst exhibited high performance towards the HER in an alkaline medium. This study not only offered an efficient electrocatalyst for HER in all-pH but also opened a new avenue for the synthesis of transition metal phosphides encapsulated in nitrogen-doped carbon shells for energy storage and conversion systems. Jianshuai et al. [53] synthesized 3D hierarchical CoP nano-flowers as ternary nickel doped cobalt phosphide (Ni-Co-P) assembled by porous and ultrathin nano-sheets by a facile solvothermal reaction following a phosphidation procedure. The Ni-Co-P nanoflowers exhibited remarkable electrocatalytic HER performance, the low overpotentials of 83 and 92 mV at 10 mA/cm^2, small Tafel slopes of 46.6 and 49.6 mV/dec, in alkaline and acidic electrolytes respectively. This was attributed to the unique 3D hierarchical morphology and the modified electronic structure due to Ni incorporation. The superior activity and stability of novel Ni-Co-P nanoflowers have promising potential for application in the production of hydrogen fuel by water splitting. Xiao et al. [54] developed a simple and facile one-pot method of nickel phosphide nanowire array/Ni foam electrode (Ni-P NA/NF) by a direct phosphorization treatment of vapor solid growth mechanism. The morphology possessed the high-density arrays of vertically aligned nanowires. The obtained 3D Ni-P NA/NF nanowire showed better activity with a low cell voltage of 1.69 V to drive a current density of 10 mA/cm^2. The more active edge sites and three-dimensional structure are responsible for the higher activity for the water splitting reaction. Liu et al. [55] prepared robust NiCoP/CoP heterostructure material as an efficient electrocatalyst by hydrothermal and phosphidation process for hydrogen evolution reaction. It exhibited the nano-structured morphology with particle size (0.279 nm), and diameters of 40–60 nm anchored on the NF. The NiCoP-CoP/NF yielded a lower Tafel slope (91.3 mV/dec^1) and showed a larger double-layer capacitance (Cdl 120.9 mF/cm^2). The nano-size crystalline structure and synergistic effect of NiCoP-CoP/NF catalyst have played an important role in high HER and OER activities with good long term stability.

2.4 Transition metal sulfides

Transition metal sulfides (TMSs) are a great option to establish more activity and stability in HER. Recently, many research works are reported on sulfide-based materials showing promising electrocatalytic activities towards HER.

Dai et al. [56] have reported on Co-based MoS_2 electrocatalyst and synthesized it by a deposition-precipitation process with different transitional metal dopants such as Ni, Co, and Fe. The materials appeared as a sheet-like structure and showed better activity towards HER. Among the prepared catalysts, Co doped MoS_2 exhibited a higher electrochemical performance with a high exchange current density of $0.03\,mA/cm^2$, low onset potential of $90\,mV$, Tafel slope of $50\,mV/dec$, and noticed excellent stability for 10,000 cycles. Wang et al. [57] have prepared amorphous MoS_2 nanoflowers assembled by lamellar nanosheets by a facile hydrothermal method with an excellent morphological structure that is required for a good catalyst. The electrochemical measurements employing the as-prepared sulfide materials were tested towards the HER and OER performances. The MoS_2-220 (calcined at 220°C) showed the good HER activity with the most positive onset potential of $130\,mV$ and a current density of $13.8\,mA/cm^2$ at $300\,mV$, which is nearly 77 times larger than that of the commercial MoS_2 ($0.18\,mA/cm^2$). However, the amorphous MoS_2 sample exhibited a quite different catalytic activity with the Tafel values between 52 and $86\,mV/dec$. A smaller Tafel slope means a faster increase of the HER rate with increased potentials. To study the stability in an acidic environment, long-term potential cycling stability of MoS_2-220 was conducted by taking a potential scan from -0.4 to $0.1\,V$ for 1000 cycles with an accelerated scanning rate of $100\,mV/s$. Only a slight loss of HER activity was noted after 1000 cycles. Niyitanga et al. [58] showed an improved performance of carbon nanotube and molybdenum disulfide (CNT-MoS_2) composites for HER and a good conducting network as well as a large surface area were found to be responsible for the observed synergic effects between CNT and MoS_2. The electrochemical activity with CNT-MoS_2 composite was found at a low overpotential ($-0.14\,V$), much lower than MoS_2 itself ($-0.29\,V$), at a current density of $10\,mA/cm^2$. The CNT-MoS_2 composite showed a lower resistance ($24\,\Omega$) compared to the MoS_2 precursor ($1000\,\Omega$) at a frequency of $3000\,Hz$, demonstrated its improved conductivity. The CNT-MoS_2 composite also exhibited high stability and excellent durability.

Yin et al. [59] have synthesized Ni-MoS_2 catalyst using a simple hydrothermal method and MoS_2 submicron-flakes exhibited a typical 2D flat structure with an obvious thick stack, and the lateral size of $\sim500\,nm$. Ni-MoS_2 showed large uniform cauliflower-like nodules grains with clear facets and well-defined grain boundaries. Apart from the structure, the material exhibited a high catalyst activity towards HER i.e., an optimized overpotential at a high current density ($100\,mA/cm^2$ at $\eta=207\,mV$). Also, the stability of the material was good even after 1200 CV cycles. Dong et al. [60] discussed the structure-based electrocatalyst for HER. They synthesized NiS and Ni_3S_2 by an organic reflecting method. The NiS exhibited a nano-rod-like structure and Ni_3S_2 exhibited an octahedral structure. The average size of the NiS nanoparticles was $20\,nm$ wide and $50\,nm$ long, and that of the Ni_3S_2 nanoparticles was around $40\,nm$. In electrochemical activity, the NiS scored much higher activity compared to the Ni_3S_2. NiS showed a much lower overpotential at high current density. The Tafel slope of metal sulfide matched with Volmer step, the weak H-adsorption has been considered to be the main rate-determining step. Further, NiS showed good stability up to 5000 cycles. Hence, the nanorods like structured NiS have been proved as a

Future prospects of oxide-free materials **Chapter | 18** **461**

good electrocatalyst for HER. Arun et al. [61] described the growth of cobalt molybdenum sulfide of various compositions $Ni_{1-x}Mo_xS$ ($x = 0$, 0.04, 0.08, 0.16). The tightly packed nanocrystals are prepared by one-step process of simple electrodeposition method on fluorine-doped tin oxide (FTO) substrate as the highly active and low-cost HER electrocatalyst. The prepared electrocatalysts were characterized via various analytical techniques. The HER activity was evaluated through electrochemical methods such as CV and impedance analysis. The exhaustive electrochemical examinations show that the $Ni_{0.96}Mo_{0.04}S$ achieved the lowest potential value of 180 mV at 10 mA/cm^2 with the lowest Tafel slope value of 50 mV/dec towards HER. Furthermore, the active surface area of the as-deposited composite materials was also calculated by CV and found that catalyst possessed a much higher surface area than the bare catalyst. The energy and cost-efficient methods described in this study contributed to the contemporary synthetic designs and strategies for producing advanced electrocatalysts for HER. Zhang et al. [62] prepared the A-MoS$_x$ catalysts were prepared by the one-step arc-melting method. The catalyst mainly consisted of Mo_2S_3 and Mo_3S_4 and exhibited a superior activity in acid medium, with a low overpotential of 156 mV at $j = 10$ mA/cm^2, Tafel slope of 58 mV/dec, and a large exchange current density of 0.40 mA/cm^2. TEM images of the material confirmed the formation of a layer-like structure. The structure-based MoS$_x$ electrocatalyst was found to have much higher active sites to generate hydrogen from the water. Sun et al. [63] reported on a one-step general electrodeposition strategy followed to construct the TMs-based catalysts decorated with metal sulfide nanoparticles on Ni foam (M_xS_y/M/NF, M = Ni, Co, Fe) in the presence of thiourea. The as-prepared Ni_3S_2/Ni/NF, Co_xS_y/Co/NF, and Fe_xS_y/Fe/NF electrodes exhibited an enhanced catalytic activity towards HER due to the decoration of corresponding metal sulfide. The polarization curves recorded for the M_xS_y/M/NF materials showed the overpotential values of 45 mV, 89 mV, and 128 mV from 196 mV, 217 mV and 288 mV of their Ni/NF, Co/NF, and Fe/NF counterparts at 10 mA/cm^2 in 1.0 M KOH, respectively. The as-obtained Ni_3S_2/Ni/NF electrode obviously exhibited the 3D broccoli-like morphology, which is composed of the accumulation of numerous nanoparticles on nickel foam. The TEM characterization also conformed to the same morphological structures. Further, the HRTEM taken for the selected area showed two distinct lattice fringes. The set of 0.20 nm was consistent with the (111) plane of metallic Ni, and the other with a distance of 0.29 nm and an interplanar angle of 60 was assigned to the (110) plane of the hexagonal phase of Ni_3S_2. Finally, the electrochemical stability was recorded for the Ni_3S_2/Ni/NF. The chronopotentiometry results with a current density of 10 mA/cm^2 exhibited no evident degradation after 140,000 s, and no obvious change was observed in the polarization curves before and after the long-term stability tests. The results clearly illustrated the enhanced performance of Ni_3S_2/Ni/NF catalyst for HER and proved that sulfide-based materials can be suitable electrocatalysts for HER application due to their good morphological structure and crystalline nature.

3. Conclusions and future prospects

The enhancement of sustainable and eco-friendly energy systems requires catalysts that are stable, highly active, and earth-abundant. In the previous sections, different nanostructured metal carbides, nitrides, phosphides, and sulfides which act as an efficient electrocatalyst towards HER have been reported to decrease the usage of noble metal-based materials. Among them, doping of different transition metals (Fe, Co, Ni, Cu, Ag, and Pt) supported Mo_2C catalyst was found to possess significant HER activity due to the high active surface area. Further, the heteroatom doped transition metal carbide ($Ni_3Mo_3C@NPC$ NWs/CC) materials have shown even better HER activities than noble-metal-based electrode materials. Similarly, the carbon composited transition metal nitrides such as $Ni_3N@C$, NiCoN/C, and $FeNx/Mo_2N/CNT$ materials revealed an enhanced HER activity due to their high electrical conductivity. The carbon-based transition metal phosphide materials (NiCoP/CoP, CoP/RGO, and $CoP/Co_2P@NC$) are widely used as efficient electrocatalysts due to their low cost, easy fabrication, and high catalytic efficiency when compared to other non-noble metals. The low aggregation is achieved with doping of binary transition metals on carbon composites. The MoS_2 with different metals (Ni, Fe, Co) doped materials were used as the active electrocatalyst for the efficient HER process. The 2D and 3D MoS_2 materials have even provided a crucial HER activity owing to its low over-potential to drive maximum current density. These electrode materials are considered as more active materials that work at different pH's and derive greater HER efficiency with limited over potential range. These materials are established to be the most hopeful choices for future generation catalysts for energy systems. The main objective of these works is maximizing the level of active sites by doping of different transition metals and heteroatom on the metal surfaces, increasing the electrical conductivity of the catalyst matrix, and optimizing the available way for charge transports by constructing various nanostructures with maximum active surface areas. In this sequence, great improvements have been achieved by preparing different metal precursors for fabricating nanostructured metal carbides, nitrides, phosphides, and sulfides for energy conversions for their practical applications. Further, the synthesis of these transition metal-based materials from inexpensive compounds is required for large-scale application. Later, a novel synthetic strategy with desired metal dopant has to be developed to form the desired chemical nano-structure that will be more advantageous from the material design perspective. The catalytic behavior will be revealed more clearly by combining theoretical and experimental analysis and this will cover the way for preparing of metal-based catalysts with excellent activities and stabilities for energy conversion. We believe these pathways will enable the use of these transition metal and heteroatoms doped carbide, nitride, phosphide, and sulfide materials in real energy conversion systems.

References

[1] K. Fan, Y. Ji, H. Zou, J. Zhang, B. Zhu, H. Chen, Q. Daniel, Y. Luo, J. Yu, L. Sun, Hollow iron–vanadium composite spheres: a highly efficient iron-based water oxidation electrocatalyst without the need for nickel or cobalt, Angew. Chem. Int. Ed. 56 (2017) 3289–3293.

[2] D. Jang, S. Lee, S. Kim, K. Choi, S. Park, J. Oh, S. Park, Production of P, N Co-doped graphene-based materials by a solution process and their electrocatalytic performance for oxygen reduction reaction, ChemNanoMat 4 (2018) 118–123.

[3] J.R. McKone, S.C. Marinescu, B.S. Brunschwig, J.R. Winkler, H.B. Gray, Earth-abundant hydrogen evolution electrocatalysts, Chem. Sci. 5 (2014) 865–878.

[4] G.M. Guio, L.A. Stern, X. Hu, Nanostructured hydrotreating catalysts for electrochemical hydrogen evolution, Chem. Soc. Rev. 43 (2014) 6555–6569.

[5] M. Zeng, Y. Li, Recent advances in heterogeneous electrocatalysts for hydrogen evolution reaction, J. Mater. Chem. A 3 (2015) 14942–14962.

[6] P.C.K. Vesborg, B. Seger, I. Chorkendorff, Recent development in hydrogen evolution reaction catalysts and their practical implementation, J. Phys. Chem. Lett. 6 (2015) 951–957.

[7] T.R. Cook, D.K. Dogutan, S.Y. Reece, Y. Surendranath, T.S. Teets, D.G. Nocera, Solar energy supply and storage for the legacy and non-legacy worlds, Chem. Rev. 110 (2010) 6474–6502.

[8] M.G. Walter, E.L. Warren, J.R. McKone, S.W. Boettcher, Q. Mi, E.A. Santori, N.S. Lewis, Solar water splitting cells, Chem. Rev. 110 (2010) 6446–6473.

[9] M.R. Gao, Y.F. Xu, J. Jiang, S.H. Yu, Nanostructured metal chalcogenides: synthesis, modification, and applications in energy conversion and storage devices, Chem. Soc. Rev. 42 (2013) 2986–3017.

[10] P. Murthy, J. Madhavan, K. Murugan, Recent advances in hydrogen evolution reaction catalysts on carbon/carbon-based supports in acid media, J. Power Sources 398 (2018) 9–26.

[11] A. Wu, C. Tian, H. Yan, Y. Jiao, Q. Yan, G. Yang, Hierarchical MoS_2@MoP core-shell heterojunction electrocatalysts for efficient hydrogen evolution reaction over a broad pH range, Nanoscale 8 (2016) 11052–11059.

[12] D. Balaji, P. Arunachalam, K. Duraimurugan, J. Madhavan, J. Theerthagiri, A.M. Al-Mayouf, M.Y. Choi, Highly efficient $Ni_{0.5}Fe_{0.5}Se_2$/MWCNT electrocatalystfor hydrogen evolution reaction in acid media, Int. J. Hydrogen Energy 45 (2020) 7838–7847.

[13] Y.R. Liu, X. Li, G.Q. Han, B. Dong, W.H. Hu, X. Shang, Y.M. Chai, Y.Q. Liu, C.G. Liu, Template-assisted synthesis of highly dispersed MoS_2 nanosheets with enhanced activity for hydrogen evolution reaction, Int. J. Hydrogen Energy 42 (2017) 2054–2060.

[14] Y. Huang, C. Wang, H. Song, Y. Bao, X. Lei, Carbon-coated molybdenum carbide nanosheets derived from molybdenum disulphide for hydrogen evolution reaction, Int. J. Hydrogen Energy 43 (2018) 12610–12617.

[15] M. Simon, V.A. Nikiforov, M.I. Petrushina, K. Klaus, E. Christensen, J.J. Oluf, J.N. Bjerrum, Transition metal carbides (WC, Mo_2C, TaC, NbC) As potential electrocatalysts for the hydrogen evolution reaction (HER) at medium temperatures, Int. J. Hydrogen Energy 40 (2015) 2905–2911.

[16] X. Dong, H. Yan, Y. Jiao, D. Guo, A. Wu, G. Yang, X. Shi, C. Tian, H. Fu, 3D hierarchical V–Ni-based nitride hetero structure as a highly efficient pH-universal electrocatalyst for the hydrogen evolution reaction, J. Mater. Chem. A 7 (2019) 15823.

[17] W. Chen, T.J. Muckerman, E. Fujita, Recent developments in transition metal carbides and nitrides as hydrogen evolution electrocatalysts, Chem. Commun. 49 (2013) 8896–8909.

[18] Y. Men, P. Li, F. Yang, G. Cheng, S. Chen, W. Luo, Nitrogen-doped CoP as robust electrocatalyst for high-efficiency pH-universal hydrogen evolution reaction, Appl. Catal. Environ. 253 (2019) 21–27.

464 Oxide free nanomaterials for energy storage and conversion applications

[19] S. Yang, L. Chen, W. Wei, L.V. Xiaomeng, J. Xie, CoP nanoparticles encapsulated in three-dimensional N-doped porous carbon for efficient hydrogen evolution reaction in a broad pH range, Appl. Surf. Sci. 476 (2019) 749–756.

[20] H. Liang, C. Yang, S. Ji, N. Jiang, X. Yang, Cobalt-nickel phosphides@carbon spheres as highly efficient and stable electrocatalyst for hydrogen evolution reaction, Catal. Commun. 124 (2019) 1–5.

[21] W. Wang, C. Liu, D. Zhou, L. Yang, J. Zhou, D. Yang, In-situ synthesis of coupled molybdenum carbide and molybdenum nitride as electrocatalyst for hydrogen evolution reaction, J. Alloys Compd. 5 (2019) 230–239.

[22] Z. Chena, W. Gonga, S. Conga, Z. Wanga, G. Song, T. Pan, X. Tang, I. Chena, W. Lu, Z. Zhao, Eutectoid-structured WC/W_2C heterostructures: a new platform for long-term alkaline hydrogen evolution reaction at low overpotentials, Nano Energy 68 (2020), 104335.

[23] M.C. Weidman, D.V. Esposito, Y.C. Hsu, J.G. Chen, Comparison of electrochemical stability of transition metal carbides (WC, W_2C, Mo_2C) over a wide pH range, J. Power Sources 202 (2012) 11–17.

[24] Y. Yang, X. Zhu, B. Zhang, H. Yang, C. Liang, Electrocatalytic properties of porous Ni-Co-WC composite electrode toward hydrogen evolution reaction in acid medium, Int. J. Hydrogen Energy 44 (2019) 19771–19781.

[25] W.F. Chen, K. Sasaki, J.M. Schneider, C.H. Wang, J.T. Muckerman, E. Fujita, Nitride-stabilized tungsten carbide electrocatalysts for hydrogen evolution reaction, in: 66th Annual Meeting of International Society of Electrochemistry, 2015.

[26] H. Zhang, X. Chen, Z. Lin, L. Zhang, H. Cao, L. Yu, G. Zheng, Hybrid niobium and titanium nitride nanotube arrays implanted with nanosized amorphous rhenium–nickel: an advanced catalyst electrode for hydrogen evolution reactions, Int. J. Hydrogen Energy 45 (2020) 6461–6475.

[27] C. Wang, W. Qi, Y. Zhoub, W. Kuang, T. Azhag, T. Thomas, C. Jiang, S. Liu, M. Yangb, Ni-Mo ternary nitrides based one-dimensional hierarchical structures for efficient hydrogen evolution, Chem. Eng. J. 381 (2020) 122611.

[28] L. Liao, S. Wang, J. Xiao, X. Bian, Y. Zhang, M.D. Scanlon, X. Hu, Y. Tang, B. Liu, H.H. Giraultb, A nanoporous molybdenum carbide nanowire as an electrocatalyst for hydrogen evolution reaction, Energ. Environ. Sci. 7 (2014) 387.

[29] S. Emina, C. Altinkaya, A. Semerci, H. Okuyucu, A. Yildiz, P. Stefanov, Tungsten carbide electrocatalysts prepared from metallic tungsten nanoparticles for efficient hydrogen evolution, Appl. Catal. Environ. 236 (2018) 147–153.

[30] M. Chen, Y. Ma, Y. Zhou, C. Liu, Y. Qin, Y. Fang, G. Guan, X. Li, Z. Zhang, T. Wang, Influence of transition metal on the hydrogen evolution reaction over nano-molybdenum-carbide catalyst, Catalysts 8 (2018) 2942.

[31] Y. Li, Q. Huang, H. Wu, L. Cai, Y. Du, S. Liu, Z. Sheng, M. Wu, N-doped Mo_2C nano-block for efficient hydrogen evolution reaction, J. Solid State Electrochem. 23 (2019) 2043–2050.

[32] P. Li, D. Wu, C. Dai, X. Huang, C. Li, Z. Yin, S. Zhou, Z. Lv, D. Cheng, J. Zhu, J. Xu, X. Liu, Controlled synthesis of copper-doped molybdenum carbide catalyst with enhanced activity and stability for hydrogen evolution reaction, Catal. Lett. 149 (2019) 1368–1374.

[33] W. Fu, Y. Wang, H. Zhang, M. He, L. Fang, X. Yang, Z. Huang, J. Li, X. Gu, Y. Wang, Epitaxial growth of graphene on V_8C_7 nanomeshs for highly efficient and stable hydrogen evolution reaction, J. Catal. 369 (2019) 47–53.

[34] C. Liu, Y. Wen, L. Lin, H. Zhang, X. Li, S. Zhang, Facile in-situ formation of high efficiency nanocarbon supported tungsten carbide nanocatalysts for hydrogen evolution, Int. J. Hydrogen Energy 43 (2018) 1–9.

Future prospects of oxide-free materials **Chapter | 18 465**

[35] L. Guo, J. Wang, X. Teng, Y. Liu, X. He, Z. Chen, A novel bimetallic NiMo carbide nanowire array for efficient hydrogen evolution, ChemSusChem 11 (2018) 2717–2723.

[36] Y. Liu, G.D. Li, L. Yuan, L. Ge, H. Ding, D. Wang, X. Zou, Carbon-protected bimetallic carbide nanoparticles for highly efficient alkaline hydrogen evolution reaction, Nanoscale 7 (2015) 3130–3136.

[37] C. Wang, Y. Sun, E. Tian, D. Fu, M. Zhang, X. Zhao, W. Ye, Easy access to trace-loading of Pt on inert Ni_3N nanoparticles with significantly improved hydrogen evolution activity at entire pH values, Electrochim. Acta 320 (2019) 134597.

[38] Y. Li, J. Zhang, X. Qian, Y. Zhang, Y. Wang, R. Hu, C. Yao, J. Zhu, Nanoporous niobium nitride (Nb_2N) with enhanced electrocatalytic performance for hydrogen evolution, Appl. Surf. Sci. 427 (2018) 884–889.

[39] B. Wei, G. Tang, H. Liang, Z. Qi, D. Zhang, W. Hu, H. Shen, Z. Wang, Bimetallic vanadium-molybdenum nitrides using magnetron co-sputtering as alkaline hydrogen evolution catalyst, Electrochem. Commun. 93 (2018) 166–170.

[40] C. Huang, D. Wu, P. Qin, K. Ding, C. Pi, Q. Ruan, H. Song, B. Gao, H. Chen, P.K. Chu, Ultrafine Co nanodots embedded in N-doped carbon nanotubes grafted on hexagonal VN for highly efficient overall water splitting, Nano Energy 73 (2020) 104788.

[41] D. Gao, J. Zhang, T. Wang, W. Xiao, K. Tao, D. Xue, J. Ding, Metallic Ni_3N nanosheets with exposed active surface sites for efficient hydrogen evolution, RSC Adv. 5 (2015) 51961–51965.

[42] J. Lai, B. Huang, Y. Chao, X. Chen, S. Guo, Strongly coupled nickel–cobalt nitrides/carbon hybrid nanocages with Pt-like activity for hydrogen evolution catalysis, Adv. Mater. 31 (2018) 1805541.

[43] K. Ojha, S. Banerjee, A.K. Ganguli, Facile charge transport in $FeN_x/Mo_2N/CNT$ nanocomposites for efficient hydrogen evolution reactions, J. Chem. Sci. 129 (2017) 989–997.

[44] B. Ren, D. Li, Q. Jin, H. Cui, C. Wang, A self-supported porous WN nanowire array: an efficient 3D electrocatalyst for the hydrogen evolution reaction, J. Mater. Chem. A 5 (2017) 19072.

[45] L. Yu, S. Song, B. McElhenny, F. Ding, D. Luo, Y. Yu, S. Chen, Z. Ren, A universal synthesis strategy to make metal nitride electrocatalysts for hydrogen evolution reaction, J. Mater. Chem. A 7 (2019) 19728–19732.

[46] P. Zhou, D. Xing, Y. Liu, Z. Wang, P. Wang, Z. Zheng, X. Qin, X. Zhang, Y. Dai, B. Huang, Accelerated electrocatalytic hydrogen evolution of non-noble metal containing trinickel nitride by introduction of vanadium nitride, J. Mater. Chem. A 7 (2019) 5513–5521.

[47] Q. Zhang, Y. Wang, Y. Wang, A.M. Al-Enizi, A.A. Elzatahry, G. Zheng, Myriophyllum-like hierarchical $TiN@Ni_3N$ nanowire arrays for bifunctional water splitting catalyst, J. Mater. Chem. A 4 (2016) 5713–5718.

[48] P. Guo, Y.X. Wu, W.M. Lau, H. Liu, L.M. Liu, Porous CoP nanosheet arrays grown on nickel foam as an excellent and stable catalyst for hydrogen evolution reaction, Int. J. Hydrogen Energy 42 (44) (2017) 26995–27003.

[49] X.Y. Yan, S. Devaramani, J. Chen, D.L. Shan, D.D. Qin, Q. Ma, X.Q. Lu, Self-supported rectangular CoP nanosheet arrays grown on a carbon cloth as an efficient electrocatalyst for the hydrogen evolution reaction over a variety of pH values, New J. Chem. 41 (2017) 2436–2442.

[50] F.H. Saadi, A.I. Carim, E. Verlage, J.C. Hemminger, N.S. Lewis, M.P. Soriaga, CoP as an acid-stable active electrocatalyst for the hydrogen-evolution reaction: electrochemical synthesis, interfacial characterization and performance evaluation, J. Phys. Chem. C 118 (2014) 29294–29300.

[51] L. Ma, X. Shen, H. Zhou, G. Zhu, Z. Ji, K. Chen, CoP nanoparticles deposited on reduced graphene oxide sheets as an active electrocatalyst for the hydrogen evolution reaction, J. Mater. Chem. A 3 (2015) 5337–5343.

466 Oxide free nanomaterials for energy storage and conversion applications

[52] X. Lv, J.T. Ren, Y. Wang, Y.P. Liu, Z.Y. Yuan, Well-defined phase-controlled cobalt phosphide nanoparticles encapsulated in nitrogen-doped graphitized carbon shell with enhanced electrocatalytic activity for hydrogen evolution reaction at all-pH, ACS Sustain. Chem. Eng. 7 (2019) 8993–9001.

[53] J. Mu, J. Li, E.-C. Yang, X.J. Zhao, Three-Dimensional hierarchical nickel cobalt phosphide nanoflowers as an efficient electrocatalyst for the hydrogen evolution reaction under both acidic and alkaline conditions, ACS Appl. Energy Mater. 1 (2018) 3742–3751.

[54] J. Xiao, Q. Lv, Y. Zhang, Z. Zhang, S. Wang, One-step synthesis of nickel phosphide nanowire array supported on nickel foam with enhanced electrocatalytic water splitting performance, RSC Adv. 6 (2016) 107859–107864.

[55] D. Liu, M. Wang, Y. Chai, X. Wan, D. Cu, Robust NiCoP/CoP Heterostructures for highly efficient hydrogen evolution electrocatalysis in alkaline solution, ACS Catal. 9 (2019) 2618–2625.

[56] X. Dai, K. Du, Z. Li, M. Liu, Y. Ma, H. Sun, Y. Yang, Co-doped MoS_2 nanosheets with the dominant CoMoS phase coated on carbon as an excellent electrocatalyst for hydrogen evolution, ACS Appl. Mater. Interfaces 7 (2015) 27242–27253.

[57] D. Wang, Z. Pan, Z. Wu, Z. Wang, Z. Liu, Hydrothermal synthesis of MoS_2 nanoflowers as highly efficient hydrogen evolution reaction catalysts, J. Power Sources 264 (2014) 229–234.

[58] T. Niyitanga, P.E. Evans, T. Ekanayake, P.A. Dowben, H.K. Jeong, Carbon nanotubes-molybdenum disulfide composite for enhanced hydrogen evolution reaction, J. Electroanal. Chem. 845 (2019) 39–47.

[59] X. Yin, H. Dong, G. Sun, W. Yang, A. Song, Q. Du, G. Shao, Ni–MoS_2 composite coatings as efficient hydrogen evolution reaction catalysts in alkaline solution, Int. J. Hydrogen Energy 42 (2017) 11262–11269.

[60] D.Y. Chung, J.W. Han, D.H. Lim, J.H. Jo, S.J. Yoo, H. Lee, Y.E. Sung, Structure dependent active sites of Ni_xS_y as electrocatalysts for hydrogen evolution reaction, Nanoscale 7 (2015) 5157–5163.

[61] P. Arun, J. Theerthagiri, K. Premnath, J. Madhavan, K. Murugan, Single-step electrodeposited molybdenum incorporated nickel sulfide thin films from low-cost precursors as highly efficient hydrogen evolution electrocatalysts in acid medium, J. Phys. Chem. C 121 (2017) 1118–1126.

[62] L.F. Zhang, G. Ou, L. Gu, Z.J. Peng, L.N. Wang, H. Wu, A highly active molybdenum multisulfide electrocatalyst for the hydrogen evolution reaction, RSC Adv. 6 (2016) 107158–107162.

[63] Y. Sun, C. Huang, J. Shen, Y. Zhong, J. Ning, Y. Hu, One-step construction of a transition-metal surface decorated with metal sulfide nanoparticles: a high-efficiency electrocatalyst for hydrogen generation, J. Colloid Interface Sci. 558 (2019) 1–8. 2020.

Index

Note: Page numbers followed by *f* indicate figures and *t* indicate tables.

A

Aerogel
  carbon (CA), 78–79
  nanoalloy, 41–43
  noble metal, 35–36
  preparation, 40–41
Alkaline electrolysis cell (AEC), 331
Anion exchange method, 40–41
Annealing gas regulation, 32*f*
Aqueous gel polymer electrolyte (AGPEs), 119
Artificial photosynthesis, 410, 410*f*
Asymmetric supercapacitors
  electrode materials, 107
  vs. symmetric supercapacitor, 120–121
  transition metal chalcogenides, 96–102, 99*t*,
    100–101*f*
  transition metal nitrides (TMN), 102–107,
    104*t*, 106*f*

B

Borates, 332–333
Borides, 332

C

Carbazole based hole transporting materials,
  295–296, 296*f*
Carbon based composites
  transition metal phosphide (TMP), 191–193,
    192*f*
  transition metal selenide (TMSe), 187–189,
    190*f*
  transition metal sulphide (TMS), 186–187,
    188*f*
  transition metal telluride, 189–190
Carbon-based perovskite solar cells (PSC),
  299–301, 300*f*
Carbon-based supports
  advantages, 45
  noble metals, 45–46
  transition metal carbides (TMC), 46
  transition metal dichalcogenides (TMDC), 46
  transition metal nitrides (TMN), 46

Carbon electrode materials
  2D mesostructured carbon, 206–207
  carbon nanotubes (CNT), 206
  characteristics, 208, 209–210*t*
  charging-discharging plots, 207*f*, 208–211,
    211*f*
  energy efficiency (EE), 207
  minimal-architecture (MA), 207–208, 208*f*
  N_2-doped carbon (NOMC), 208–211
  redox reaction kinetics, 206
Carbon nanofibers (CNF), 182–185, 184*f*
Carbon nanotubes (CNT)
  carbon electrode materials, 206
  multi-walled (MWCNT), 79
  single-walled (SWCNT), 79
  sodium-ion batteries (SIBs), 182–186, 183*f*
  supercapacitor (SCs), 89
Carbon-supported Pt nanoparticle, 41
Charge carrier
  generation, 412
  separation, 412
Charge transfer engineering, 29–30
Co-doping heteroatoms. *See* Multiheteroatom
  doping
Composite based counter electrodes
  $NiCo_2S_4$/CNFs composites, 272–276, 277*f*
  Ni-MoSex@$CoSe_2$ CSNs, 279, 280*f*
  photocurrent density–voltage *(J–V)* curves,
    272–276, 275–276*f*
  photovoltaic parameters, 272–276, 273–274*t*
Composite photovoltaic technology.
    See Fourth-generation photovoltaic
    (PV) cells
Conventional stacked flexible
    supercapacitor, 121, 121*f*
CoP and CoX_2 nanoframes, 27–29
Copper chalcogenides, 294–295, 295*f*
Copper iodide (CuI), 293–294, 294*f*
Copper thiocyanate (CuSCN), 292–293
Counter electrodes (CE)
  composite based, 272–279, 273–274*t*,
    275–277*f*, 280*f*
  cyclic-voltammetry, 221–223, 222*f*

467

468 Index

Counter electrodes (CE) (*Continued*)
  development strategies, 220
  electrochemical impedance spectroscopy
    (EIS), 224–225, 225*f*
  function of, 260–261, 260*f*
  intensity-modulated photocurrent
    spectroscopy (IMPS), 226–227
  intensity-modulated photovoltage
    spectroscopy (IMVS), 226–227
  metal carbides, 235–249, 239*f*, 241–250*f*,
    250–254*t*
  metal nitrides, 229–235, 232*f*, 234*f*, 236–237*t*
  platinum (Pt), 259–260, 279
  selenides and tellurides, 267–271, 268*t*, 269*f*,
    271–272*f*
  sulfide based, 261–267, 262*t*, 263–264*f*,
    266–267*f*
  Tafel-polarization measurement, 223–224,
    224*f*
  transition metal compounds, 228–229,
    230–231*f*
Cyclic-voltammetry, 221–223, 222*f*

D

2D metal carbides. *See* MXenes
Direct methanol fuel cell, 44*f*
Dye-sensitized solar cell (DSSC)
  components of, 219–220
  composite based counter electrodes,
    272–279, 273–274*t*, 275–277*f*, 280*f*
  counter electrodes (CE), 220–227, 260–261,
    260*f*
  device architecture, 259–260
  metal carbides, 235–249, 239*f*, 241–250*f*,
    250–254*t*
  metal chalcogenide, 279
  metal nitrides, 229–235, 232*f*, 234*f*, 236–237*t*
  photovoltaic measurements, 227–228, 227*f*
  platinum (Pt) counter electrodes, 259–260, 279
  published articles, 220, 220–221*f*
  selenides and tellurides counter electrodes,
    267–271, 268*t*, 269*f*, 271–272*f*
  sulfide based counter electrodes, 261–267,
    262*t*, 263–264*f*, 266–267*f*
  third-generation solar cells, 67
  transition metal compounds (TMCs),
    228–229, 230–231*f*

E

Electrochemical battery
  disadvantages, 52
  limitations, 52–53

Li-S battery, 53–54
  oxidation-reduction (redox) reaction, 52
  redox process, 52
Electrochemical capacitors.
    See Supercapacitors (SCs)
Electrochemical double-layer capacitors
    (EDLCs)
  carbide-derived carbon, 78–79
  carbon aerogel (CA), 78–79
  carbon foams, 78–79
  carbon nanotubes (CNT), 79
  enacted carbon (AC), 79
  graphene, 79
Electrochemical impedance spectroscopy
    (EIS), 224–225, 225*f*
Electrochemical supercapacitors.
    See Supercapacitors (SCs)
Electrochemical water splitting.
    See Photoelectrochemical water
    splitting (PEC-WS)
Electron transport material (ETM)
  advantage, 289
  fullerenes, 289–290, 290*f*
  naphthalene-based, 290–291
  phenyl-C_{61}-butyric acid methyl ester
    (PCBM), 289–290, 291*f*
  tris (2,4,6-trimethyl-3-(pyridin-3-yl)phenyl)
    borane (3TPYMB), 290
Enacted carbon (AC), 79, 89–90
Exfoliant-assisted liquid phase exfoliation,
    44–45

F

First-generation photovoltaic (PV) cells, 309
First-generation solar cells
  monocrystalline silicon solar cells, 65
  polycrystalline silicon solar cells, 66
Flexible supercapacitor (SCs)
  advantages, 115–116
  applications, 116, 117*f*
  components of, 116
  design, 116, 117*f*
  device architectures, 120–122, 121*f*
  electrode materials, 122–137, 123*f*, 125*f*,
    126*t*, 129*f*, 130*t*, 133*f*, 135*f*, 136*t*, 138*t*
  hybrid-supercapacitors (HSCs), 116
  mechanical elements, 115–116
  solid-state electrolytes, 119–120
  substrate, 117–119
Fourth-generation photovoltaic (PV) cells, 310,
    311*f*
Fullerenes, 289–290, 290*f*

Index **469**

G

Graphene
electrochemical double-layer capacitors (EDLCs), 79
electrochemical properties, 154
graphene foam (GF), 153–154
material synthesis and cycling properties, 154, 155–156t
nanosized silicone, 153
Prussian blue (PB), 153–154
sodium-ion batteries (SIBs), 185–186, 185f
supercapacitors, 88–89

H

High entropy alloys (HEA)
definition, 13–14, 14f
features, 14–15
Hole transporting materials (HTMs)
carbazole based, 295–296, 296f
copper chalcogenides, 294–295, 295f
copper iodide (CuI), 293–294, 294f
copper thiocyanate (CuSCN), 292–293
lead sulfide (PbS), 298–299, 299f
PEDOT:PSS, 296–297, 297f
poly (3-hexylthiophene) (P3HT), 297–298, 298f
poly [bis(4-phenyl) (2,5,6-trimethylphenyl) amine] (PTAA), 292
Spiro-OMeTAD, 291–292, 293f
Hybrid capacitors (HCs)
asymmetric, 80
composite, 81
Hybrid-supercapacitors (HSCs), 57f, 59–60, 116
Hydrogen evolution reaction (HER)
catalytic activity, 341, 342–347t
challenges, 351–352
Co-Ni/MoS$_2$, 341–348
electrocatalysts, 11–12, 451
electrochemical water splitting, 451
ensemble effect, 378–379
metal phosphides (MPs), 378–379, 395
Ni$_2$P, 349
Ni-Co phosphides, 349
Ni-MoS$_2$, 348
transition metal carbides (TMC), 452–455
transition metal nitrides (TMN), 456–457
transition metal phosphides (TMP), 457–459
transition metal sulfides (TMS), 459–461
Hydrogen (H$_2$) production
pathways, 410, 411f
photocatalysts, 411
photosynthesis process, 409–410

I

Intensity-modulated photocurrent spectroscopy (IMPS), 226–227
Intensity-modulated photovoltage spectroscopy (IMVS), 226–227

L

Lead sulfide (PbS), 298–299, 299f
Li-S battery
disadvantages, 54
lithium-ion batteries (LIBs), 54
principles, 53–54, 53f
Lithium-ion batteries (LIBs)
components of, 150, 152
features, 149–150
graphene-based anodes, 153–154, 155–156t
history and technological evolution, 178, 179f
Li-S battery, 54
low cycling stability, 167
operating mechanism, 149–150, 151f, 152
performance of, 149–150, 151f
phosphide-based anodes, 162–166, 164f, 166t
selenide-based anodes, 154–158, 159t
sulfide-based anodes, 159–162, 163t
theoretical energy densities, 149–150, 151f
transition metal chalcogenides (TMCs), 150–152
types, 149–150

M

Metal alloys, 338–339
Metal carbides. *See* Transition metal carbides (TMC)
Metal chalcogenides. *See* Transition metal chalcogenides
Metal nitrides. *See* Transition metal nitrides (TMN)
Metal-organic frameworks (MOF), 34, 352
Metal phosphides. *See* Transition metal phosphides (TMP)
Metal pnictides
nitrides, 419–420, 420f
oxynitrides, 420–422, 421f
phosphides, 422–423
Metal sulfides. *See* Transition metal sulphides (TMS)
Methanol oxidation reaction (MOR), 43–44
Monocrystalline silicon solar cells, 65
Multiheteroatom doping
catalyst hybrid, 29–30
zeolitic imidazolate framework, 29–30, 30f

470 Index

Multijunction solar cells (MJSCs)
  atmospheric absorbing gases, 307–308, 308*f*
  electron-hole pairs, 308–309
  first-generation photovoltaic (PV) cells, 309
  fourth-generation photovoltaic (PV) cells, 310, 311*f*
  GaInP/InGaAs/Ge-based, 310–312, 312*f*
  III -V semiconductors, 312–320, 313–314*f*, 316–317*f*, 320*f*
  II-VI semiconductors, 321
  limitations, 310
  photovoltaic (PV) cells, 308–309
  power conversion efficiency (PCE), 309–310
  renewable energy technology, 307–308
  second-generation photovoltaic (PV) cells, 310
  solar energy, 307–308
  solar radiation spectrum, 307–308, 308*f*
  tandem solar cells, 310
  third-generation photovoltaic (PV) cells, 310
Multimetallic aerogels. *See* Aerogel
MXenes
  as cocatalysts, 425, 425*f*
  definition, 424
  properties, 427
  thermoelectric generator (TEG), 443

N

Nanoalloy aerogel, 41–43
Nanostructured nonoxide nanomaterials
  energy conversion skills, 2
  high entropy alloys (HEA), 13–15, 14*f*
  photocatalytic applications, 2
  solar energy, 1–2
  transition-metal borides (TMB), 6
  transition-metal carbides (TMC), 4–5
  transition metal nitrides (TMN), 5–6, 7*f*
  transition-metal phosphides (TMP), 3, 4*f*
  transition-metal sulphides (TMS), 6–15, 8*t*, 9*f*
Nanostructured oxide-free materials
  aerogels, 35–36
  anion exchange method, 40–41
  annealing gas regulation, 31–32, 32*f*
  carbon-based supports, 45–46
  carbon-supported Pt nanoparticle, 41
  CoP and CoX$_2$ nanoframes, 27–29
  direct methanol fuel cell, 43–44, 44*f*
  exfoliant-assisted liquid phase exfoliation, 44–45
  hydrothermal and electrochemical methods, 37–38

metal-organic frameworks (MOF), 34
multicomponent hybrid materials, 32–33, 33*f*
multiheteroatom doping, 29–30, 30*f*
nanoalloy aerogel, 41–43
one-dimensional (1D) multicomponent hybrid heterostructures, 36–37, 37*f*
one-dimensional (1D) nanoarchitecture, 30–31
polyaniline (PANI), 39–40
template synthesis method, 26–27, 28*f*
transition metal nitrides, 34–35
ultra-sonication, 38–39, 39*f*
Naphthalene-based electron transport material, 290–291

O

On-chip type flexible supercapacitor. *See* Planar type flexible supercapacitor
One-dimensional (1D) multicomponent hybrid heterostructures, 36–37, 37*f*
One-dimensional (1D) nanoarchitecture, 30–31
Organic gel polymer electrolytes (OGPEs), 119
Oxide free nanomaterials
  electron transport material (ETM), 288–291, 290–291*f*
  hole transporting materials (HTMs), 291–299, 293–299*f*
  transparent conductive oxide (TCO), 301–302
Oxygen evolution reaction (OER)
  catalytic activity, 342–347*t*
  challenges, 351–352
Oxygen reduction reaction (ORR)
  electrocatalyst, 11–12
  metal alloys, 338

P

Peltier effect, 436–437
Perovskite solar cells (PSC)
  carbon-based, 299–301, 300*f*
  charge transfer process, 288, 288*f*
  device configuration, 288, 288*f*
  electron transport material (ETM), 288–291, 290–291*f*
  hole transporting materials (HTMs), 291–299, 293–299*f*
  limitations, 288–289
  methylammonium lead iodide (MAPbI3), 287–288
  properties, 287–288
  third-generation solar cells, 68

Index **471**

transparent conductive oxide (TCO), 301–302
Phenyl-C_{61}-butyric acid methyl ester (PCBM), 289–290, 291*f*
Phosphide-based anodes
copper phosphide (Cu_3P), 165
germanium phosphide (GeP_3), 163–165, 164*f*
synthesis, morphological features and cycling performances, 165, 166*t*
Photocatalytic intermediate water splitting (PIWS), 410, 411*f*
Photocatalytic overall water splitting (POWS), 410, 411*f*
Photocatalytic partial water splitting (PPWS), 410, 411*f*
Photocatalytic water splitting (PWS)
artificial photosynthesis, 410, 410*f*
charge carrier generation, 412
charge carrier separation, 412
efficiency calculation, 413
energy levels, 412*f*
hydrogen (H_2) production, 409–410
light source, 413
metal carbides, 424–426
metal chalcogenides, 414–419
metal pnictides, 419–423
modifications, 411
photocatalytic intermediate water splitting (PIWS), 410, 411*f*
photocatalytic overall water splitting (POWS), 410, 411*f*
photocatalytic partial water splitting (PPWS), 410, 411*f*
reactor, 413
surface catalytic reaction, 412
Z-Scheme, 409–410
Photoelectrochemical water splitting (PEC-WS)
H_2 generation, 369–370, 371*f*
transition metal nitrides (TMN), 370–377, 372*f*
transition metal phosphides (TMP), 378–386
transition metal sulphides (TMS), 386–393
Multiheteroatom charge carrier
collection of, 64
photon absorption, 62–63, 63*f*
separation of, 63–64
simple solar cell model, 63, 64*f*
Photovoltaic (PV) technology
current-voltage *(I-V)* curve, 309, 309*f*
first-generation, 309
fourth-generation, 310, 311*f*

photoelectric effect, 308–309
power conversion efficiency (PCE), 308–309
power generation, 308
principle, 61–62
second-generation, 310
third-generation, 310
Piezoelectric generators (PEGs)
applications, 436
direct and indirect effect, 443, 444*f*
discovery of, 443
metal chalcogenides and nitrides, 443–446
Piezoelectric metal chalcogenides
Janus 2D materials, 445
MoS_2, 444–445
nitrides, 445–446
transition metal dichalcogenides (TMDCs), 443–444
Planar type flexible supercapacitor, 121, 121*f*
Platinum (Pt) counter electrodes, 259–260, 279
Poly (3-hexylthiophene) (P3HT), 297–298, 298*f*
Polyaniline (PANI), 39–40
Poly [bis(4-phenyl) (2,5,6-trimethylphenyl) amine] (PTAA), 292
Polycrystalline silicon solar cells, 66
Polymer-based/organic solar cells, 67
Power conversion efficiency (PCE)
carbon-based perovskite solar cells (PSC), 299–301, 300*f*
current-voltage *(I-V)* curve, 309
definition, 309
electron transport material (ETM), 290–291
Proton exchange membrane electrolysis cell (PEMEC), 331
Prussian blue analogs (PBA), 27–29
Pseudocapacitors (PCs)
conducting polymers (CPs), 79–80
disadvantage, 79–80
hybrid capacitors (HCs), 80–81
supercapacitors, 57*f*, 58–59, 59*f*
transition metal oxides (TMO), 80–81
Pulsed layer deposition (PLD), 10

R

Redox flow batteries (RFBs)
advantages, 201–203
electrolyte medium, 201–203
Renewable energy harvesting devices, 435

472 Index

S

Sandwich type flexible supercapacitor. *See* Conventional stacked flexible supercapacitor
Second-generation photovoltaic (PV) cells, 310
Second-generation solar cells
  amorphous Si (a-Si), 66
  cadmium telluride (CdTe), 66
  copper indium gallium diselenide (CIGS), 67
Seebeck effect, 436–437
Selenide-based anodes
  Bi_2Se_3, 154–156
  manganese selenide, 158
  micrograph images, $MoSe_2$, 156–158, 157*f*
  morphological feature and cycling performances, 158, 159*t*
  multiwalled carbon nanotubes (MWCNT), 156
Selenides and tellurides counter electrodes
  Co-Sn-S nanocages, 270, 271*f*
  cyclic voltammograms, 270, 272*f*
  $Ni_{0.85}Se$ spheres, 267–270, 269*f*
  photovoltaic parameters, 268*t*, 269*f*
Semiconductors
  II-VI, 321
  III-V, 312–320, 313–314*f*, 316–317*f*, 320*f*
Silicon-based solar cells. *See* First-generation photovoltaic (PV) cells
Silicon carbide (SiC), 103–105
Silicone cells. *See* First-generation solar cells
Sodium-ion batteries (SIBs)
  applications, 178–180, 181*f*
  carbon nanofibers (CNFs), 182–185, 184*f*
  carbon nanostructure, 180–181, 181*f*
  carbon nanotube (CNTs), 182–186, 183*f*
  graphene-based materials, 185–186, 185*f*
  graphite anode, 178
  history and technological evolution, 178, 179*f*
  vs. lithium-ion batteries (LIBs), 177–178
  publications in, 178, 180*f*
  standard hydrogen electrode (SHE), 178
  transition metal phosphides composites, 191–193, 192*f*
  transition metal selenide composites, 187–189, 190*f*
  transition metal sulfide composites, 186–187, 188*f*
  transition metal telluride composites, 189–190
Solar cells
  first-generation solar cells, 65–66

second-generation solar cells, 66–67
third-generation solar cells, 67–68
transition-metal sulfides, 13
Solar energy and solar cells
  charge carrier's generation, 62–63, 63–64*f*
  collection of, 64
  first-generation solar cells, 65–66
  photovoltaic device (PV), principle, 61–62
  second-generation solar cells, 66–67
  separation of, 63–64
  thin-film solar cells, 68–71
  third-generation solar cells, 67–68
Solid oxide electrolysis cell (SOEC), 331
Solid-state electrolytes
  aqueous gel polymer electrolyte (AGPEs), 119
  ionic liquids (ILs), 120
  organic gel polymer electrolytes (OGPEs), 119
  polyvinyl alcohol (PVA), 119
Spiro-OMeTAD, 291–292, 293*f*
Substrate, flexible supercapacitor
  paper-based, 118
  polymer, 118
  slurry-based electrodes, 118–119
  substrates, 117–118
Sulfide-based anodes
  electrochemical behaviors, 160–162, 161*f*
  manganese sulfide, 162
  morphological feature and cycling performances, 162, 163*t*
  tungsten sulfide, 162
Sulfide based counter electrodes
  CZTS, 265, 266*f*
  NiS_2 hierarchical microspheres, 261–265, 264*f*
  NiS nanoflower arrays, 261–265, 263*f*
  photocurrent density–voltage (*J–V*) curves, 265–267, 267*f*
  photovoltaic parameters, 261–265, 262*t*
Sulfide based supercapacitors
  Bi_2S_3, 86
  binary-based, 85
  cobalt sulfide (CoS), 83
  copper sulfide (CuS), 83, 84*f*
  iron sulfide (FeS2), 84, 85*f*
  La_2S_3, 86–87
  manganese cobalt sulfides (MCS), 86
  molybdenum disulfide (MoS_2), 84
  nickel sulfide (NiS), 83
  $NiCo_2S_4$, 85–86
  WS_2, 87

Supercapacitors (SCs)
  advantages, 54–55, 76
  vs. battery, 55, 55*t*
  carbon-based materials, 87
  characteristics, 56–57
  charging and discharging (CD), 76
  carbon nanotube (CNT), 89
  electrochemical double-layer capacitors
      (EDLCs), 57–58, 57–58*f*, 77–81
  electrolytes, 60–61
  enacted carbon (AC), 89–90
  Faradaic and non-Faradaic process, 80
  graphene, 88–89
  Helmholtz double layer formation, 76, 77*f*
  hybrid, 57*f*, 59–60
  material selection, 60
  metal-based chalcogenides, 82
  $NiMn_2O_4$/rGO/PANI nanocomposite
      materials, 87, 88*f*
  principle of, 55–56, 56*f*
  pseudocapacitors (PCs), 57*f*, 58–59, 59*f*, 79–80
  Ragone plot, 55–56, 56*f*
  structure, 77, 78*f*
  sulfide based, 82–87
  transition metal oxides (TMO), 81
  transition-metal sulfides, 12
Supercaps. *See* Supercapacitors (SCs)
Symmetric supercapacitors, 120–121

T
Tafel-polarization, 223–224, 224*f*
Tandem solar cells. *See* Multijunction solar
  cells (MJSCs)
Template synthesis method
  disadvantages, 26–27
  micro carbon spheres (CS), 26–27, 28*f*
Thermoelectric generator (TEG)
  advantages, 436
  Peltier effect, 436–437
  Seebeck effect, 436–437
  structure of, 437, 437*f*
  thermoelectric effect, 436–437
  thermoelectric metal carbides and nitrides,
      442–443
  thermoplastic polymers, 436
  transition metal chalcogenides, 437–441
Thermoelectric metal carbides, 442–443
Thermoelectric metal nitrides, 442–443
Thin-film solar cells
  challenges, 71
  n-type semiconductor, 68
  p-type semiconductor, 68–69

Third-generation photovoltaic (PV) cells, 310
Third-generation solar cells
  dye-sensitized solar cells (DSSC), 67
  perovskite solar cell (PSC), 68
  polymer-based/organic, 67
Titanium nitride (TiN), 127–128, 129*f*, 130*t*
Transition metal (TM)-based electrocatalysts
  alloys, 338–339
  borates, 332–333
  borides, 332
  carbides, 337–338
  chalcogenides, 333–334
  metal-organic frameworks (MOFs), 339–341
  nitrides, 336–337
  phosphates, 335–336
  phosphides, 334–335
Transition-metal borides (TMBs), 6
Transition metal carbides (TMC)
  adsorption results, 244–245, 245*f*
  bimetallic phase, 454–455
  bio-assisted porous carbon, 248–249, 248*f*
  carbides as counter electrodes, 249, 250–254*t*
  carburization method, 454
  carbon-based supports, 46
  carbon-supported composites, 240, 241*f*
  carburization method, 454
  Co_3O_4-WC-CN/rGO synthesis, 243–244,
      243*f*
  cobalt-titanium carbide nanoparticles (Co-
      TiC NPs), 240–242
  crystal structure, 453
  cyclic-voltammograms, 240, 242*f*, 244*f*,
      246–248, 248*f*
  disadvantages, 452
  electrocatalysts, 337–338, 452
  flexible supercapacitor (SCs), 122–126, 123*f*,
      125*f*, 126*t*
  Mo-based binary/ternary nanocomposites,
      249, 249–250*f*
  nano-porous molybdenum carbide nanowires
      (np-Mo2C NWs), 453
  one-step metal-catalyzed carbonization-
      nitridation strategy, 246–248, 247*f*
  one-step pyrolysis route, 246–248, 246*f*
  PCE variation plot, 240, 242*f*
  photocatalytic water splitting (PWS),
      424–426
  sintering temperature, 238
  structural morphology, 244–245, 245*f*
  surface active sites, 454–455, 455*f*
  Tafel polarization performance, 243–244,
      244*f*

474 Index

Transition metal carbides (TMC) (*Continued*)
  tungsten carbide (WC), 235–237
  vanadium carbide (VC), 242–243
  V_8C_7 NMs/GR, 454, 454*f*
Transition metal chalcogenides
  application, 97
  bandgap, 96–97
  Bi_2Te_3, 440–441
  crystalline structure of orthorhombic phase,
    bornite, 438, 440*f*
  Cu_2ZnSnS_4 (CZTS), 438
  doping strategy, Zn, 438, 440*f*
  electrocatalysts, 333–334
  flexible supercapacitor (SCs), 128–136, 133*f*,
    135*f*, 136*t*
  lithium-ion batteries (LIBs), 150–152
  metal selenide materials, 440, 441*f*
  multiwalled carbon nanotubes (MWCNTs),
    440–441
  oxysulfides, 418–419
  Seebeck coefficient, 437–438, 439*f*
  selenides, 417, 418*f*
  sulfides, 414–417, 416*f*
  tellurides, 418
  temperature dependence, 437–438, 439*f*
  thermal power factor, 437–438, 439*f*
  transition metal selenides (TMSe), 98–102,
    99*t*
  transition metal sulfides (TMS), 97–98,
    100–101*f*
Transition metal compounds, 228–229,
  230–231*f*
Transition metal dichalcogenides (TMDCs), 46,
  333–334, 443–444
Transition metal nitrides (TMN)
  applications, 370–371, 372*f*
  benefits of, 102
  binary, 374–377, 377*f*
  carbon-based supports, 46
  carbon fiber (*CF*), 229–231
  carbon nitride (CNx) thin films, 234–235
  catalytic activity and morphology, 234–235
  chemical synthesis method, 372–373
  2D nanostructured, 102–103, 104*t*
  electrocatalysts, 336–337
  electrochemical features, 102–103
  features, 370–371
  iron nitride (FeN), 233
  metallic layered double hydroxide (Me-
    LDH), 102–103
  molybdenum nitride (MoN), 233
  mono, 373, 374*t*, 375–376*f*

nanostructured oxide-free materials, 34–35
  Nb_2N catalyst, 456–457
  nickel-cobalt nitrides/carbon complex
    nanocage (NiCoN/C), 456–457
  nitrogen-doped CNT (N-CNT), 456–457
  photovoltaic parameters, 235, 236–237*t*
  physical synthesis method, 372–373
  synergistic effect, 456–457
  TiN@Ni_3N nanowire, 456–457
  titanium nitride-conductive carbon black
    (TiN-CCB), 231–233, 232*f*
  transition metal carbides (TMC), 103–105
  transition metal phosphides (TMP), 106*f*
  vanadium nitride (VN), 233–234
Transition metal oxides (TMO), 80–81
Transition metal phosphides (TMP)
  advantages, 457
  binary, 383–386, 385*f*
  Co-based electrocatalysts, 458–459
  electrocatalysts, 334–335
  electrodeposition method, 381
  engineering strategies, 378–379, 378*f*
  ensemble effect, 378–379
  flexible supercapacitor (SCs), 137, 138*t*
  gas-solid reaction method, 381
  high-energy mechanical milling route,
    381–383
  hydrogen evolution reaction (HER),
    378–379, 395
  hydrothermal assisted phosphidation process,
    381–383, 382*f*
  metalloid features, 105
  metal-organic precursor decomposition
    method, 379, 380*f*
  mono, 383
  Ni_2P NS/NF electrode, 105
  NiCoP nanoplates and graphene sheets,
    105–107, 106*f*
  pulsed laser deposition (PLD), 381–383
  sodium-ion batteries (SIBs), 191–193, 192*f*
  solid-state reaction method, 381
  solvothermal method, 380–381, 380*f*
  ternary nickel doped cobalt phosphide (Ni-
    Co-P), 458–459
  valence bands (VB), 395
Transition metal pnictides
  nitrides, 419–420, 420*f*
  oxynitrides, 420–422, 421*f*
  phosphides, 422–423
Transition metal selenides (TMSe)
  binary, 98–102
  $Co_{0.85}Se$ nanostructured composites, 98, 101*f*

cyclic voltammetric plots, 98, 100*f*
electrochemical features, 98
fabrication design, 167
sodium-ion batteries (SIBs), 187–189, 190*f*
Transition-metal sulphides (TMS)
  atomic arrangements, 6–8
  batteries, 12–13
  binary, 391–393, 394*f*
  bottom-up approach, 9*f*, 10–11
  carbon nanotube and molybdenum disulfide
    (CNT-MoS2), 459–460
  Co-based MoS_2 electrocatalyst, 459–460
  covalently bonded sulfides, 6–8
  electrocatalyst, ORR/HER, 11–12
  examples of, 8, 8*t*
  mono, 391, 392*f*
  Ni-MoS_2, 460–461
  photocatalyst, 12
  polarization curves, 460–461
  properties, 386
  sodium-ion batteries (SIBs), 186–187, 188*f*
  solar cells, 13
  supercapacitors, 12
  synthesis method, 386–390, 387–390*f*
  top-down approach, 8–9, 9*f*
Transition metal telluride composites, 189–190
Transparent conductive oxide (TCO), 301–302
Tris (2,4,6-trimethyl-3-(pyridin-3-yl)phenyl)
    borane (3TPYMB), 290

U

Ultracapacitors. *See* Supercapacitors (SCs)
Ultra-sonication, 39*f*

W

Water splitting reaction
  in acidic medium, 330
  alkaline electrolysis cell (AEC), 331
  in alkaline medium, 330
  challenges, 350–351
  hydrogen evolution reaction (HER), 351–352
  metal borates, 351
  metal-organic frameworks (MOFs), 352
  oxygen evolution reaction (OER), 351–352
  proton exchange membrane electrolysis cell
    (PEMEC), 331
  solid oxide electrolysis cell (SOEC), 331
  water-splitting electrolyzer, 330
Wire type flexible supercapacitor. *See* Yarn/
    fiber type flexible supercapacitor

Y

Yarn/fiber type flexible supercapacitor,
    121–122, 121*f*

Z

Zinc-bromine battery (ZBB)
  advantages, 201–203
  carbon electrode materials, 205–211,
    207–208*f*, 209–210*t*, 211*f*
  challenges, 211–212
  device configuration, 203–204, 204*f*
  EES, 201–203
  Nafion membrane, 205
  redox flow batteries (RFBs), 201–203
  supporting electrolytes, 205

Printed in the United States
by Baker & Taylor Publisher Services